T0189512

Lecture Notes in Computer Science 14333

Founding Editors

Gerhard Goos
Juris Hartmanis

The series Lecture Notes in Computer Science (LNCS), including its subseries Lecture Notes in Artificial Intelligence (LNAI) and Lecture Notes in Bioinformatics (LNBI), has established itself as a medium for the publication of new developments in computer science and information technology research, teaching, and education.

LNCS enjoys close cooperation with the computer science R & D community, the series counts many renowned academics among its volume editors and paper authors, and collaborates with prestigious societies. Its mission is to serve this international community by providing an invaluable service, mainly focused on the publication of conference and workshop proceedings and postproceedings. LNCS commenced publication in 1973.

Xiangyu Song · Ruyi Feng · Yunliang Chen ·
Jianxin Li · Geyong Min
Editors

Web and Big Data

7th International Joint Conference, APWeb-WAIM 2023
Wuhan, China, October 6–8, 2023
Proceedings, Part III

 Springer

Editors
Xiangyu Song 🆔
Peng Cheng Laboratory
Shenzhen, China

Yunliang Chen 🆔
China University of Geosciences
Wuhan, China

Geyong Min 🆔
University of Exeter
Exeter, UK

Ruyi Feng 🆔
China University of Geosciences
Wuhan, China

Jianxin Li 🆔
Deakin University
Burwood, VIC, Australia

ISSN 0302-9743   ISSN 1611-3349 (electronic)
Lecture Notes in Computer Science
ISBN 978-981-97-2386-7   ISBN 978-981-97-2387-4 (eBook)
https://doi.org/10.1007/978-981-97-2387-4

This Springer imprint is published by the registered company Springer Nature Singapore Pte Ltd.
The registered company address is: 152 Beach Road, #21-01/04 Gateway East, Singapore 189721, Singapore

Paper in this product is recyclable.

Preface

This volume (LNCS 14333) and its companion volumes (LNCS 14331, LNCS 14332, and LNCS 14334) contain the proceedings of the 7th Asia-Pacific Web (APWeb) and Web-Age Information Management (WAIM) Joint Conference on Web and Big Data, called APWeb-WAIM 2023. Researchers and practitioners from around the world came together at this leading international forum to share innovative ideas, original research findings, case study results, and experienced insights in the areas of the World Wide Web and big data, thus covering web technologies, database systems, information management, software engineering, knowledge graphs, recommend systems and big data.

The 7th APWeb-WAIM conference was held in Wuhan during 6–8 October 2023. As an Asia-Pacific flagship conference focusing on research, development, and applications in relation to Web information management, APWeb-WAIM builds on the successes of APWeb and WAIM. Previous APWeb conferences were held in Beijing (1998), Hong Kong (1999), Xi'an (2000), Changsha (2001), Xi'an (2003), Hangzhou (2004), Shanghai (2005), Harbin (2006), Huangshan (2007), Shenyang (2008), Suzhou (2009), Busan (2010), Beijing (2011), Kunming (2012), Sydney (2013), Changsha (2014), Guangzhou (2015), and Suzhou (2016); and WAIM was held in Shanghai (2000), Xi'an (2001), Beijing (2002), Chengdu (2003), Dalian (2004), Hangzhou (2005), Hong Kong (2006), Huangshan (2007), Zhangjiajie (2008), Suzhou (2009), Jiuzhaigou (2010), Wuhan (2011), Harbin (2012), Beidaihe (2013), Macau (2014), Qingdao (2015), and Nanchang (2016). The APWeb-WAIM conferences were held in Beijing (2017), Macau (2018), Chengdu (2019), Tianjin (2020), Guangzhou (2021), Nanjing (2022), and Wuhan (2023). With the ever-growing importance of appropriate methods in these data-rich times and the fast development of web-related technologies, APWeb-WAIM will become a flagship conference in this field.

The high-quality program documented in these proceedings would not have been possible without the authors who chose APWeb-WAIM for disseminating their findings. APWeb-WAIM 2023 received a total of 434 submissions and, after the double-blind review process (each paper received at least three review reports), the conference accepted 133 regular papers (including research and industry track) (acceptance rate 31.15%), and 6 demonstrations. The contributed papers address a wide range of topics, such as big data analytics, advanced database and web applications, data mining and applications, graph data and social networks, information extraction and retrieval, knowledge graphs, machine learning, recommender systems, security, privacy and trust, and spatial and multi-media data. The technical program also included keynotes by Jie Lu, Qing-Long Han, and Hai Jin. We are grateful to these distinguished scientists for their invaluable contributions to the conference program.

We would like to express our gratitude to all individuals, institutions, and sponsors that supported APWeb-WAIM2023. We are deeply thankful to the Program Committee members for lending their time and expertise to the conference. We also would like to acknowledge the support of the other members of the Organizing Committee. All of them helped to make APWeb-WAIM 2023 a success. We are grateful for the guidance of the honorary chair (Lizhe Wang), the steering committee representative (Yanchun Zhang) and the general co-chairs (Guoren Wang, Schahram Dustdar, Bruce Xuefeng Ling, and Hongyan Zhang) for their guidance and support. Thanks also go to the program committee chairs (Yunliang Chen, Jianxin Li, and Geyong Min), local co-chairs (Chengyu Hu, Tao Lu, and Jianga Shang), publicity co-chairs (Bohan Li, Chang Tang, and Xin Bi), proceedings co-chairs (David A. Yuen, Ruyi Feng, and Xiangyu Song), tutorial co-chairs (Ye Yuan and Rajiv Ranjan), CCF TCIS liaison (Xin Wang), CCF TCDB liaison (Yueguo Chen), Ph.D. consortium co-chairs (Pablo Casaseca, Xiaohui Huang, and Yanan Li), Web co-chairs (Wei Han, Huabing Zhou, and Wei Liu), and industry co-chairs (Jun Song, Wenjian Qin, and Tao Yu).

We hope you enjoyed the exciting program of APWeb-WAIM 2023 as documented in these proceedings.

October 2023

Yunliang Chen
Jianxin Li
Geyong Min
David A. Yuen
Ruyi Feng
Xiangyu Song

Organization

General Chairs

Guoren Wang   BIT, China
Schahram Dustdar   TU Wien, Austria
Bruce Xuefeng Ling   Stanford University, USA
Hongyan Zhang   China University of Geosciences, China

Program Committee Chairs

Yunliang Chen   China University of Geosciences, China
Jianxin Li   Deakin University, Australia
Geyong Min   University of Exeter, UK

Steering Committee Representative

Yanchun Zhang   Guangzhou University & Pengcheng Lab, China;
Victoria University, Australia

Local Co-chairs

Chengyu Hu   China University of Geosciences, China
Tao Lu   Wuhan Institute of Technology, China
Jianga Shang   China University of Geosciences, China

Publicity Co-chairs

Bohan Li   Nanjing University of Aeronautics and
Astronautics, China
Chang Tang   China University of Geosciences, China
Xin Bi   Northeastern University, China

Proceedings Co-chairs

David A. Yuen   Columbia University, USA
Ruyi Feng   China University of Geosciences, China
Xiangyu Song   Swinburne University of Technology, Australia

Tutorial Co-chairs

Ye Yuan   BIT, China
Rajiv Ranjan   Newcastle University, UK

CCF TCIS Liaison

Xin Wang   Tianjin University, China

CCF TCDB Liaison

Yueguo Chen   Renmin University of China, China

Ph.D. Consortium Co-chairs

Pablo Casaseca   University of Valladolid, Spain
Xiaohui Huang   China University of Geosciences, China
Yanan Li   Wuhan Institute of Technology, China

Web Co-chairs

Wei Han   China University of Geosciences, China
Huabing Zhou   Wuhan Institute of Technology, China
Wei Liu   Wuhan Institute of Technology, China

Industry Track Co-chairs

Jun Song	China University of Geosciences, China
Wenjian Qin	Shenzhen Institute of Advanced Technology CAS, China
Tao Yu	Tsinghua University, China

Program Committee Members

Alex Delis	University of Athens, Greece
Amr Ebaid	Google, USA
An Liu	Soochow University, China
Anko Fu	China University of Geosciences, China
Ao Long	China University of Geosciences, Wuhan, China
Aviv Segev	University of South Alabama, USA
Baoning Niu	Taiyuan University of Technology, China
Bin Zhao	Nanjing Normal University, China
Bo Tang	Southern University of Science and Technology, China
Bohan Li	Nanjing University of Aeronautics and Astronautics, China
Bolong Zheng	Huazhong University of Science and Technology, China
Cai Xu	Xidian University, China
Carson Leung	University of Manitoba, Canada
Chang Tang	China University of Geosciences, China
Chen Shaohao	China University of Geosciences, China
Cheqing Jin	East China Normal University, China
Chuanqi Tao	Nanjing University of Aeronautics and Astronautics, China
Dechang Pi	Nanjing University of Aeronautics and Astronautics, China
Dejun Teng	Shandong University, China
Derong Shen	Northeastern University, China
Dong Li	Liaoning University, China
Donghai Guan	Nanjing University of Aeronautics and Astronautics, China
Fang Wang	Hong Kong Polytechnic University, China
Feng Yaokai	Kyushu University, Japan
Giovanna Guerrini	University of Genoa, Italy
Guanfeng Liu	Macquarie University, Australia

Guoqiong Liao	Jiangxi University of Finance & Economics, China
Hailong Liu	Northwestern Polytechnical University, USA
Haipeng Dai	Nanjing University, China
Haiwei Pan	Harbin Engineering University, China
Haoran Xu	China University of Geosciences, Wuhan, China
Haozheng Ma	China University of Geosciences, Wuhan, China
Harry Kai-Ho Chan	University of Sheffield, UK
Hiroaki Ohshima	University of Hyogo, Japan
Hongzhi Wang	Harbin Institute of Technology, China
Hua Wang	Victoria University, Australia
Hui Li	Xidian University, China
Jiabao Li	China University of Geosciences, Wuhan, China
Jiajie Xu	Soochow University, China
Jiali Mao	East China Normal University, China
Jian Chen	South China University of Technology, China
Jian Yin	Sun Yat-sen University, China
Jianbin Qin	Shenzhen University, China
Jiannan Wang	Simon Fraser University, Canada
Jianqiu Xu	Nanjing University of Aeronautics and Astronautics, China
Jianxin Li	Deakin University, Australia
Jianzhong Qi	University of Melbourne, Australia
Jianzong Wang	Ping An Technology (Shenzhen) Co., Ltd., China
Jinguo You	Kunming University of Science and Technology, China
Jizhou Luo	Harbin Institute of Technology, China
Jun Gao	Peking University, China
Jun Wang	China University of Geosciences, Wuhan, China
Junhu Wang	Griffith University, Australia
K. Selçuk Candan	Arizona State University, USA
Krishna Reddy P.	IIIT Hyderabad, India
Ladjel Bellatreche	ISAE-ENSMA, France
Le Sun	Nanjing University of Information Science and Technology, China
Lei Duan	Sichuan University, China
Leong Hou U.	University of Macau, China
Li Jiajia	Shenyang Aerospace University, China
Liang Hong	Wuhan University, China
Lin Xiao	China University of Geosciences, Wuhan, China
Lin Yue	University of Newcastle, UK

Lisi Chen	University of Electronic Science and Technology of China, China
Lizhen Cui	Shandong University, China
Long Yuan	Nanjing University of Science and Technology, China
Lu Chen	Zhejiang University, China
Lu Qin	UTS, Australia
Luyi Bai	Northeastern University, China
Miaomiao Liu	Northeast Petroleum University, China
Min Jin	China University of Geosciences, Wuhan, China
Ming Zhong	Wuhan University, China
Mirco Nanni	CNR-ISTI Pisa, Italy
Mizuho Iwaihara	Waseda University, Japan
Nicolas Travers	Pôle Universitaire Léonard de Vinci, France
Peiquan Jin	University of Science and Technology of China, China
Peng Peng	Hunan University, China
Peng Wang	Fudan University, China
Philippe Fournier-Viger	Shenzhen University, China
Qiang Qu	SIAT, China
Qilong Han	Harbin Engineering University, China
Qing Xie	Wuhan University of Technology, China
Qiuyan Yan	China University of Mining and Technology, China
Qun Chen	Northwestern Polytechnical University, China
Rong-Hua Li	Beijing Institute of Technology, China
Rui Zhu	Shenyang Aerospace University, China
Runyu Fan	China University of Geosciences, China
Sanghyun Park	Yonsei University, South Korea
Sanjay Madria	Missouri University of Science & Technology, USA
Sara Comai	Politecnico di Milano, Italy
Shanshan Yao	Shanxi University, China
Shaofei Shen	University of Queensland, Australia
Shaoxu Song	Tsinghua University, China
Sheng Wang	China University of Geosciences, Wuhan, China
ShiJie Sun	Chang'an University, China
Shiyu Yang	Guangzhou University, China
Shuai Xu	Nanjing University of Aeronautics and Astronautics, China
Shuigeng Zhou	Fudan University, China
Tanzima Hashem	Bangladesh University of Engineering and Technology, Bangladesh

Tianrui Li	Southwest Jiaotong University, China
Tung Kieu	Aalborg University, Denmark
Vincent Oria	NJIT, USA
Wee Siong Ng	Institute for Infocomm Research, Singapore
Wei Chen	Hebei University of Environmental Engineering, China
Wei Han	China University of Geosciences, Wuhan, China
Wei Shen	Nankai University, China
Weiguo Zheng	Fudan University, China
Weiwei Sun	Fudan University, China
Wen Zhang	Wuhan University, China
Wolf-Tilo Balke	TU Braunschweig, Germany
Xiang Lian	Kent State University, USA
Xiang Zhao	National University of Defense Technology, China
Xiangfu Meng	Liaoning Technical University, China
Xiangguo Sun	Chinese University of Hong Kong, China
Xiangmin Zhou	RMIT University, Australia
Xiao Pan	Shijiazhuang Tiedao University, China
Xiao Zhang	Shandong University, China
Xiao Zheng	National University of Defense Technology, China
Xiaochun Yang	Northeastern University, China
Xiaofeng Ding	Huazhong University of Science and Technology, China
Xiaohan Zhang	China University of Geosciences, Wuhan, China
Xiaohui (Daniel) Tao	University of Southern Queensland, Australia
Xiaohui Huang	China University of Geosciences, Wuhan, China
Xiaowang Zhang	Tianjin University, China
Xie Xiaojun	Nanjing Agricultural University, China
Xin Bi	Northeastern University, China
Xin Cao	University of New South Wales, Australia
Xin Wang	Tianjin University, China
Xingquan Zhu	Florida Atlantic University, USA
Xinwei Jiang	China University of Geosciences, Wuhan, China
Xinya Lei	China University of Geosciences, Wuhan, China
Xinyu Zhang	China University of Geosciences, Wuhan, China
Xujian Zhao	Southwest University of Science and Technology, China
Xuyun Zhang	Macquarie University, Australia
Yajun Yang	Tianjin University, China
Yanfeng Zhang	Northeastern University, China

Yanghui Rao	Sun Yat-sen University, China
Yang-Sae Moon	Kangwon National University, South Korea
Yanhui Gu	Nanjing Normal University, China
Yanjun Zhang	University of Technology Sydney, Australia
Yaoshu Wang	Shenzhen University, China
Ye Yuan	China University of Geosciences, Wuhan, China
Yijie Wang	National University of Defense Technology, China
Yinghui Shao	China University of Geosciences, Wuhan, China
Yong Tang	South China Normal University, China
Yong Zhang	Tsinghua University, China
Yongpan Sheng	Southwest University, China
Yongqing Zhang	Chengdu University of Information Technology, China
Youwen Zhu	Nanjing University of Aeronautics and Astronautics, China
Yu Liu	Huazhong University of Science and Technology, China
Yuanbo Xu	Jilin University, China
Yue Lu	China University of Geosciences, Wuhan, China
Yuewei Wang	China University of Geosciences, Wuhan, China
Yunjun Gao	Zhejiang University, China
Yunliang Chen	China University of Geosciences, Wuhan, China
Yunpeng Chai	Renmin University of China, China
Yuwei Peng	Wuhan University, China
Yuxiang Zhang	Civil Aviation University of China, China
Zhaokang Wang	Nanjing University of Aeronautics and Astronautics, China
Zhaonian Zou	Harbin Institute of Technology, China
Zhenying He	Fudan University, China
Zhi Cai	Beijing University of Technology, China
Zhiwei Zhang	Beijing Institute of Technology, China
Zhixu Li	Soochow University, China
Ziqiang Yu	Yantai University, China
Zouhaier Brahmia	University of Sfax, Tunisia

Contents – Part III

Adaptive Graph Attention Hashing for Unsupervised Cross-Modal Retrieval via Multimodal Transformers

Yewen Li, Mingyuan Ge, Yucheng Ji, and Mingyong Li[✉]

School of Computer and Information Science, Chongqing Normal University, Chongqing 401331, China
limingyong@cqnu.edu.cn

Abstract. Unsupervised cross-modal hashing retrieval has been extensively studied due to its advantages in storage, retrieval efficiency, and label independence. However, there are still two obstacles to existing unsupervised methods: (1) Existing unsupervised methods suffer from inaccurate similarity as simple features do not describe fine-grained multimodal relationships. (2) Existing methods suffer from unbalanced multimodal learning due to the different coding capabilities of different modal networks. To address these obstacles, we devised an effective Adaptive Graph Attention Hashing (AGAH) for unsupervised cross-modal retrieval. Firstly, we use the multimodal transformer model CLIP to extract cross-modal fine-grained features and exploit multiple data similarities to mine similar information from different perspectives in multi-modal data and perform similarity enhancement. In addition, we present an adaptive graph attention hashing module to assist in generating hash codes, which uses an attention mechanism to learn relation-based similarity from image-text modality. It aggregates the essential neighborhood message of neighboring data nodes through the graph neural networks to generate more discriminative hash codes. Sufficient experiments on three benchmark datasets demonstrate that the proposed AGAH outperforms existing advanced unsupervised cross-modal hashing methods.

Keywords: cross-modal retrieval · deep hashing · unsupervised learning · graph neural networks · attention

1 Introduction

With the rapid development of social media, multimedia data in different forms such as text, images, audio, and video have exploded. People are no longer satisfied with unimodal data access and the need to retrieve data from multiple modalities is increasing. However, multimedia data is huge, heterogeneous and multi-dimensional, and the retrieval of multi-modal data requires a lot of time and storage space. Reducing storage space for multi-modal data and improving

© The Author(s), under exclusive license to Springer Nature Singapore Pte Ltd. 2024
X. Song et al. (Eds.): APWeb-WAIM 2023, LNCS 14333, pp. 1–15, 2024.
https://doi.org/10.1007/978-981-97-2387-4_1

retrieval performance are therefore essential. Of the many retrieval approaches, hash-based retrieval (HBR) [4,24] has drawn considerable attention for its low storage consumption and high retrieval efficiency.

The essential concept of cross-modal hash retrieval [3,26] is to learn hash transformations of different modalities using the information of sample pairs of different modalities, and then mapping the data of different modalities into Hamming binary space [13], while maintaining the similarity of data in the process of mapping (Original data with relatively similar semantics are encoded into a shared space where the distance between their hash codes is relatively similar). Cross-modal hashing can be divided into two types: supervised and unsupervised approaches. Supervised methods [9,15,27] use semantic tags to fill the heterogeneity gap and semantic disparity, resulting in often better retrieval precision. Jiang et al. first proposed a novel approach called Deep Cross-Modal Hashing (DCMH) [9], which integrates feature learning and hash code learning into an end-to-end learning framework. The unsupervised methods [1,13,19] eliminate the dependence on label information and only consider paired multimedia data. Due to the lack of jointly trained label information, unsupervised cross-modal hashing methods suffer from inaccurate training objectives and limited retrieval accuracy.

Recent years have seen great progress in deep learning-based unsupervised cross-modal hashing retrieval methods due to the strong feature extraction ability of deep neural networks [6,11,20]. the deep unsupervised cross-modal hashing methods [13,23,25] utilized deep neural networks to obtain feature representations of different modal data, thus achieving a large performance improvement. Liu et al. presented a joint similarity hashing (JDSH) [13] based on modal distributions, which adequately retained the semantic correlations between the data by designing a similarity weighting scheme.

While these unsupervised approaches obtain acceptable performance, the majority of them have problems with inaccurate similarity and modality imbalance, resulting in sub-optimal retrieval performance. Concretely, it is hard to fully weigh the complex data relevance with simple data features of different modalities. There is information loss as the original structure of the hash code is corrupted in the process of switching from real values to binarization. In addition, multimodal learning has imbalance problems due to modal disparity and data bias [21,22] and training efficiency remains limited. To solve these problems, we propose an efficient AGAH for large-scale unsupervised cross-modal retrieval, and the contributions are as follows:

1. We present an efficient unsupervised cross-modal hashing method that employs the Transformer-based multimodal model CLIP [5,16] to extract fine-grained cross-modal features. AGAH designs a cross-modal similarity enhancement module to integrate similarities across perspectives, resulting in a better guidance goal for hash code learning.
2. To mitigate the issue of multimodal learning imbalance, an adaptive graph attention module is devised as an auxiliary network to balance the learning ability of hash functions. The module employs an attention mechanism to

retain similar information and inhibit irrelevant information, and aggregates feature information of similar nodes through a graph convolutional neural network.

3. In addition, a semantic complementarity-based loss function is devised to minimize the information loss during hash code binarization. Sufficient experiments on three benchmark datasets show that the proposed approach surpasses other state-of-the-art cross-modal hashing methods.

2 Related Work

Cross-modal hashing approaches are categorized into two types: supervised methods for labeling multimodal data and unsupervised methods for pairing multimedia data. Unsupervised hashing methods are more valuable for research and promising for applications due to their label isolation. Deep neural networks [7,10] exhibit remarkable capabilities in encoding deep features of different modal data, and thus deep unsupervised cross-modal hashing retrieval draws an increasing amount of research.

One of the most iconic works is Aggregation-based Graph Convolutional Hashing (AGCH) [28], which consider that a single similarity metric is difficult to represent data relationships comprehensively, and develops an efficient aggregation strategy that uses multiple metrics to construct a more accurate learning affinity matrix. The high-order non-local hash (HNH) [29] considers the similarity relationships of multimodal data from both local and non-local perspectives, constructing a more comprehensive similarity matrix. DGCPN proposed by Yu et al. [25] retains the coherence of graph neighbors by combining information between the data and their neighbors, exploiting three different forms of data similarity and moderating the loss of retention of combined similarity. Deep adaptively-enhanced hashing (DAEH) [18] presents a policy with discriminative similarity guidance and adaptive augmented optimization, using information theory to discover weaker hash functions and augment them with additional teacher networks. Nonetheless, these unsupervised approaches are plagued by inaccurate similarity measurements, which results in restricted retrieval performance.

Inspired by multimodal transformers and related work [16], we exploit the multimodal transformer model CLIP to extract cross-modal features of images, which learns transferable vision models from natural language supervision. In addition, a multimodal similarity enhancement module is designed to fuse and enhance the similarity information of different modal data, which can effectively mitigate the problem of measuring inaccurate similarity of multimodal data.

3 Methodology

In this section, we first illustrate some notations that will be used in this paper. Subsequently, we describe our work in detail, including the notation and problem definition, framework overview, objective function.

3.1   Notation and Problem Definition

In order to achieve a greater understanding of the cross-modal retrieval task and the proposed approach, we first present some notational definitions used in this paper. Given a cross-modal dataset $O = \{I_i, T_i\}_{i=1}^{n}$, where I_i and T_i represent pairs of image-text, we divide the data into mini-batches of training samples $o = \{o_1, o_2, \cdots, o_j\}$. For each randomly sampled batch of training samples $\{o_k = [I_k, I_k]\}_{k=1}^{m}$, where m denotes the batch size, we use $f_I \in \mathbb{R}^{m \times d_I}$ and $f_T \in \mathbb{R}^{m \times d_T}$ for the image and text representations. Meanwhile, we denote the hash codes generated by the hash coding network as $B_I \in \{-1, +1\}^{m \times c}$ and $B_T \in \{-1, +1\}^{m \times c}$, the hash codes generated by the graph convolutional neural network as $B_I^g \in \{-1, +1\}^{m \times c}$ and $B_T^g \in \{-1, +1\}^{m \times c}$, where c represents the length of the hash code.

In the stage of constructing the similarity matrix, we first normalize f_I and f_T to $\widehat{f}_I$ and $\widehat{f}_T$, then we use cosine similarity to calculate the image and text modality similarity matrices $S_I = \cos(\widehat{f}_I, \widehat{f}_I)$ and $S_T = \cos(\widehat{f}_T, \widehat{f}_T)$, respectively, which in turn are used to describe the original image and the inherent similarity between textual data. Furthermore, we can consider the generated hash codes B_I and B_T as feature vectors that can only contain the vertices of a high-dimensional space. From this perspective, neighboring vertices correspond to similar hash codes; that is, the distance between two hash codes can be expressed in terms of their cosine angular distance, and the cosine distance of vectors $\overrightarrow{v_1}$ and $\overrightarrow{v_2}$ is defined as follows:

$$\cos(\overrightarrow{v_1}, \overrightarrow{v_2}) = \frac{(\overrightarrow{v_1})^T \overrightarrow{v_2}}{\|\overrightarrow{v_1}\|_2 \|\overrightarrow{v_2}\|_2} \in [-1, +1]^{m \times m}. \tag{1}$$

where $\|\cdot\|_2$ denotes the l_2-normalization of the vectors, and the cosine matrix of the samples reflects the cosine similarity relation between the hash codes. Cross-modal hashing improves retrieval speed and reduces storage consumption by projecting different modal data into a unified Hamming space. It is important to note that the original semantic similarity of the data is preserved in the data projection.

3.2   Framework Overview

As illustrated in Fig. 1, the AGAH framework is an end-to-end model, which contains four main modules: the deep feature extraction module, the multi-modal similarity enhancing module, the adaptive graph attention hashing module, and the hash code reconstructing module.

Deep Feature Extraction Module. The deep coding component consists of two main networks: the image coding network and the text coding network. The multimodal Transformer model, represented by CLIP, has proven to be more useful in terms of learning both text and image representations. In this paper, we adopt the CLIP encoder and Fully Connect (FC) as the backbone network, which can extract richer cross-modal semantic features. We denote the image

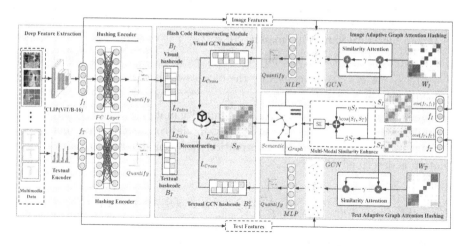

Fig. 1. An explanation of our proposed Adaptive Graph Attention Hash (AGAH) for unsupervised cross-modal retrieval, which is composed of four modules: deep feature extracting, multi-modal similarity enhancement, adaptive graph attention hashing, and hash code reconstructing module.

encoder as Enc_I, the text feature encoder as Enc_T, the symbols are expressed as follows:

$$f_I = Enc_I(I, \theta_I) \in \mathbb{R}^{m \times d_I}, F_T = Enc_T(T, \theta_T) \in \mathbb{R}^{m \times d_T} \qquad (2)$$

where I and T represent batches of image and text training samples. θ_I and θ_T represent the parameters of the visual and textual feature encoding network. Then we use MLP to learn the hash function, the formula is as follows:

$$H_I = FC_I(f_I, \theta_{HI}) \in [-1, +1]^{m \times c}, H_I = FC_T(f_T, \theta_{HT}) \in [-1, +1]^{m \times c}. \qquad (3)$$

Thus, we could embed rich semantic features of different modalities to provide a stronger description of the semantic similarity between the original data and further guide the learning of hash codes.

$$
\begin{aligned}
B_I &= \lim_{\alpha \to \infty} \tanh(\alpha H_I) \in [-1, +1]^{m \times c}, \\
B_T &= \lim_{\alpha \to \infty} \tanh(\alpha H_T) \in [-1, +1]^{m \times c}.
\end{aligned}
\qquad (4)
$$

where α denotes the number of iterations. As the number of iterations increases, the hyperbolic tangent function converges to a symbolic function: $\lim_{\alpha \to \infty} \tanh(\alpha x) = sign(x)$. The iterative approximate optimization strategy is used to mitigate information loss in the hash code binarization process.

Multi-modal Similarity Enhancing Module. As described in [13,17,19], building a similarity matrix using deep neural networks to capture the complementary and coexistence information of the original data is a superior method.

Specially, we use mini-batch visual features $f_I = \{f_i^I\}_{i=1}^m \in \mathbb{R}^{m \times d_v}$ to build the image modality similarity matrix $S_I = \{s_{ij}^I\}_{i,j=1}^m \in [-1, +1]^{m \times m}$, where $s_{ij}^I = cos(f_i^I, f_j^I)$. For the textual modality, we directly leverage the features $f_T = \{f_i^T\}_{i=1}^m \in \mathbb{R}^{m \times d_T}$ processed by the bag-of-words to create the text cosine similarity matrix $S_T = \{s_{ij}^T\}_{i,j=1}^m \in [-1, +1]^{m \times m}$, where $s_{ij}^T = cos(f_i^T, f_j^T)$. Subsequently, we use the visual modality similarity matrix S_I and textual modality similarity matrix S_T to construct a cross-modal cosine similarity matrix S_C that can preserve co-occurrence information between different instances. The equation is described as follows:

$$S_C = \{\frac{(s_{i*}^I)^T (s_{j*}^T)}{\|s_{i*}^I\|_2 \|s_{j*}^T\|_2}\}_{i,j=1}^m = \frac{(S_I)^T S_T}{\|S_I\|_2 \|S_T\|_2} = cos(S_I, S_T) \in [-1, +1]^{m \times m}. \quad (5)$$

where $(\cdot)^T$ indicates transposition of the matrix. In addition, we construct a semantic fusion matrix S_F that integrates information from different matrices, the formula is expressed as follows:

$$S_F = \eta S_I + \beta S_T + \varsigma S_C \in [-1, +1]^{m \times m},$$
$$s.t. \eta, \beta, \varsigma \geq 0, \eta + \beta + \varsigma = 1. \quad (6)$$

where η, β, and ς are balancing hyperparameters that trade off the degree of importance of similarity matrices between image and text modalities. Finally, we performed similarity enhancement on the fused affinity matrix S_F with the following formula:

$$E^+ = e^{\left(\frac{s_{ij} - s_{mean}}{s_{max} - s_{mean}}\right)}, E^- = e^{\left(-\frac{1}{2} \times \frac{s_{mean} - s_{ij}}{s_{mean} - s_{min}}\right)} \quad (7)$$

where $s_{max}, s_{min}, s_{mean}$ denote the maximum, minimum, and mean of the similarity matrix, respectively. The formula for the similarity matrix enhancement is as follows:

$$s_{ij} = \begin{cases} E^+ s_{ij}, & if \, s_{ij} > s_{mean}, \\ E^- s_{ij}, & if \, s_{ij} \leq s_{mean}. \end{cases} \quad (8)$$

After the similarity enhancement, the similarity enhancement matrix can be formed as $S_E = \{s_{ij}\}_{i,j=1}^m$. In comparison to former unsupervised approaches, this similarity enhancement provides an improved supervised signal for hash code learning by setting thresholds to bring similar data closer together and dissimilar data further apart.

Adaptive Graph Attention Hashing. The module employs an attention mechanism to learn the similarity matrix of adaptive modalities, the formula is as follows:

$$S_*^{att} = S_E + \gamma \widehat{W}_* S_E, * \in \{I, T\}. \quad (9)$$

where $\widehat{W}_I$ and $\widehat{W}_T$ represent the projection matrices of visual and textual modalities, γ is a trade-off parameter. and aggregates information between similar nodes through a GCN to generate more consistent hash codes. Subsequently, we pass the attention similarity matrix into a two-layer graph convolutional network that aggregates graph neighborhood correlations between similar nodes:

$$Z_*^{(1)} = \sigma_1 \left(\tilde{D}^{-1/2} S_*^{att} \tilde{D}^{-1/2} F_* \theta_{GI} \right),$$
$$Z_*^{(2)} = \left(\tilde{D}^{-1/2} S_*^{att} \tilde{D}^{-1/2} Z_*^{(1)} \theta_{GT} \right), * \in \{v, t\}. \tag{10}$$

where $D_{ii} = \sum_j s_{ij}$, θ_{GI} and θ_{GT} are the GCNs of parameters, σ_1 denotes the activation functions for the corresponding layers. $Z_*^{(i)}$ represents the output of the i-th layer of a visual and textual modality graph convolutional network. Hence, we could make use of the attention mechanism to learn the similarity between data. In the training process, the attention matrix is iteratively updated to maximize the similarity relationship between instances, and then the information from similar nodes is aggregated through a graph convolutional network to produce a more discriminative hash code. The hash code generated by graph convolution network is as follows:

$$B_v^g = \tanh\left(\alpha Z_v^{(2)}\right) \in [-1, +1]^{m \times c}, B_t^g = \tanh\left(\alpha Z_t^{(2)}\right) \in [-1, +1]^{m \times c}. \tag{11}$$

where α denotes the number of iterations, we used an iterative approximate optimization strategy to optimize the hash code. when $\lim_{\alpha \to \infty} \tanh(\alpha x) = sign(x)$. The problem of information loss and instability in the binarization process can be effectively mitigated by transforming the discrete problem into a series of continuous optimization problems.

Hash Codes Reconstructing Module. We construct the similarity matrices $S_I^B, S_T^B, S_C^B, S_G^{BI}, S_G^{BT}$ from the hash codes B_I, B_T, B_I^g, and B_T^g learned by the network, where $S_*^B = \cos(B_*, B_*), * \in \{I, T\}$, $S_G^{B*} = \cos(B_g^*, B_g^*), * \in \{I, T\}$, $S_G^{BC} = \cos(B_g^I, B_g^T)$. Finally, we construct the loss functions with them and the similarity enhancement matrix S_E. These loss function formulas are as follows:

$$L_{Intra} = \left\| \mu S_E - S_I^B \right\|_F^2 + \left\| \mu S_E - S_T^B \right\|_F^2,$$
$$L_{Cross} = \left\| \mu S_E - S_C^B \right\|_F^2 + \left\| \mu S_E - (S_C^B)^T \right\|_F^2 - \frac{1}{m} \sum (S_I^B \otimes S_T^B), \tag{12}$$
$$L_{Gcn} = \left\| \mu S_E - S_G^{BI} \right\|_F^2 + \left\| \mu S_E - S_G^{BC} \right\|_F^2.$$

where L_{Intra} and L_{Cross} denote the intra-modal loss and cross-modal loss, respectively. L_{Gcn} represents the graph convolution reconstructing loss. μ is a scale hyper-parameter that can regulate the quantization scope for the enhanced matrix, and the symbol $\otimes$ indicates the Hadamard matrix product.

3.3 Objective Function

The whole network's parameters are updated repeatedly by the backpropagation algorithm until the network narrows down and the reconstruction procedure of the hash code is completed. The total loss is calculated as follows:

$$\min_{B_I, B_T} L = \varepsilon L_{Intra} + L_{Cross} + \varphi L_{Gcn},$$

$$s.t. B_I, B_T \in [-1, +1]^{m \times c}. \tag{13}$$

where ε, φ are trade-off hyper-parameters. Minimizing the loss function described above leads to more consistent hash codes for similar data. AGAH generates high-quality hash codes by aiding the learning hash network through an adaptive graph attention network that efficiently captures the neighborhood structure and co-occurrence information of the original data.

4 Experiments

In this part, to evaluate the effectiveness of the proposed AGAH, comprehensive experiments were carried out on three multi-media benchmark datasets (MS COCO [12], MIRFLICKR-25K [8], and NUS-WIDE [2]). We summarize the statistics for the three datasets in Table 1. Firstly, we briefly introduce the datasets and evaluation metrics. Secondly, the proposed AGAH was compared with several advanced baseline methods, including DJSRH [19], JDSH [13], DSAH [23], HNH [29], DGCPN [25], DUCH [14], DAEH [18]. Finally, the proposed methods were empirically analyzed by parameter-sensitive analysis, ablation study.

Table 1. The Statistics of three Benchmark Datasets.

Datasets	MIRFLICKR-25K	NUS-WIDE	MS COCO
Database	25,000	186,577	123,287
Training	5,000	5,000	10,000
Testing	2,000	2,000	5,000
Labels	24	10	91

4.1 Evaluation Metrics

Cross-modal image-text retrieval focuses on two search tasks: "Text-query-Image $(T \rightarrow I)$" and "Image-query-Text $(I \rightarrow T)$". They use an instance of one modality as a query point to retrieve similar data from another modality in the database. In experiments, we employ two widely-used metrics for retrieval measurement: Mean Average Precision (MAP) and precision of top-N curve to measure the proposed model's retrieval performance compared with other methods. Precision and ranking information can be well reflected in the measurement methods. In particular, Given a query set $Q = [q_1, q_2, \cdots, q_M]$, the MAP is denoted as follows:

$$MAP = \frac{\sum_{i=1}^{M} AP(q_i)}{M}, AP = \frac{1}{L_q} \sum_{k=1}^{n} P(k) \Delta R(k). \tag{14}$$

where q_i indicates the query instances, and M indicates the total number of query instances. In addition, n represents the number of instances in the dataset, k denotes the number of instances returned during the search procedure and L_q indicates the number of data in the dataset associated with the query data, i.e. the total amount of data instances. $P(k)$ is the accuracy rate of the top k samples retrieved during the search procedure. $\Delta R(k)$ represents the recall value as the number of instances ranges from $k - 1$ to k. The average accuracy is defined as the average retrieval precision for a single query. top-N curve precision is also an important metric that represents precision over a different number of retrieval examples. top-N curve precision indicates the average precision of the top N-ranked retrieval results, which describes a model's generalization ability and comprehensive performance.

4.2 Implementation Details

In the experimental setting, the presented AGAH approach was achieved on a PyTorch platform and an NVIDIA RTX 3060 GPU with 32GB of RAM. Using the cross-validation method, the hyperparameters were determined as follows: $\{\eta = 0.5\}$, $\{\beta = 0.2, \varsigma = 0.3, \varphi = 1, \varepsilon = 0.1\}$ and $\mu = 1.4$ for all datasets, $\{\gamma = 0.45, \varphi = 0.15\}$ for MIRFLICKR-25K, $\{\gamma = 0.5, \varphi = 0.25\}$ for NUS-WIDE and MSCOCO. For the MAP evaluation, the number of ranked samples for MIRFLICKR-25K, NUS-WIDE and MSCOCO was set to 5000. For the optimization process of the network, we adopt the SGD and Adam optimizers with a learning rate of 0.01, weight decay of 5e-4, and momentum of 0.9. The batch size is set to 32 for three benchmark datasets at the training stage.

Table 2. The MAP@5000 results on cross-modal retrieval tasks (I→T indicates the image search text task and vice versa) and three datasets.

Task	Method	MIRFLICKR-25K				NUS-WIDE				MS COCO			
		16 bits	32 bits	64 bits	128bits	16 bits	32 bits	64 bits	128bits	16 bits	32 bits	64 bits	128bits
I→T	DJSRH [19]	0.6729	0.7015	0.7304	0.7443	0.5872	0.6715	0.7177	0.7437	0.7542	0.8156	0.8614	0.8614
	JDSH [13]	0.7254	0.7312	0.7524	0.7615	0.6781	0.7248	0.7434	0.7565	0.6905	0.7584	0.8884	0.8902
	DSAH [23]	0.6395	0.7663	0.7793	0.7898	0.7243	0.7530	0.7720	0.7780	0.8507	0.8813	0.9007	0.9005
	HNH [29]	0.7305	0.7449	0.7385	0.7211	0.6843	0.7215	0.7405	0.7374	0.8305	0.8552	0.8686	0.8502
	DGCPN [25]	0.7599	0.7815	0.7796	0.7880	0.7158	0.7456	0.7559	0.7538	0.8805	0.9020	0.9021	0.9063
	DUCH [14]	0.6670	0.6887	0.7064	0.7230	0.6866	0.7144	0.7282	0.7469	0.8472	0.8666	0.8767	0.8837
	DAEH [18]	0.7826	0.7940	0.8004	0.8047	0.7309	0.7542	0.7728	0.7794	0.8946	0.9029	0.9058	0.9104
	AGAH	**0.7902**	**0.8195**	**0.8248**	**0.8341**	**0.7565**	**0.7890**	**0.7950**	**0.8059**	**0.9166**	**0.9205**	**0.9290**	**0.9322**
T→I	DJSRH [19]	0.6756	0.6909	0.6985	0.7124	0.6010	0.6567	0.7076	0.7197	0.7593	0.8326	0.8621	0.8697
	JDSH [13]	0.6989	0.7192	0.7241	0.7354	0.6749	0.7155	0.7115	0.7181	0.7581	0.8296	0.8949	0.8952
	DSAH [23]	0.6462	0.7540	0.7593	0.7586	0.6688	0.7167	0.7484	0.7457	0.8546	0.8868	0.8904	0.8919
	HNH [29]	0.7234	0.7204	0.7060	0.7002	0.6711	0.6996	0.6962	0.6931	0.8398	0.8635	0.8669	0.8517
	DGCPN [25]	0.7273	0.7507	0.7571	0.7575	0.7023	0.7230	0.7426	0.7362	0.8807	0.8978	0.8991	0.9015
	DUCH [14]	0.6521	0.6684	0.6818	0.6972	0.6619	0.6943	0.7097	0.7130	0.8607	0.8855	0.8980	0.9032
	DAEH [18]	0.7607	0.7676	0.7743	0.7814	0.7132	0.7335	0.7485	0.7510	0.8882	0.8988	0.9007	0.9033
	AGAH	**0.7790**	**0.8018**	**0.8160**	**0.8272**	**0.7350**	**0.7472**	**0.7676**	**0.7697**	**0.9048**	**0.9057**	**0.9072**	**0.9172**

4.3   Comparison Results and Analysis

In experiments, we compare two cross-modal retrieval tasks: $I \to T$ and $T \to I$: using image query texts and vice versa. In this subsection, we compare the retrieval performance of all baselines and AGAH in terms of MAP and Top-N precision curves in the two retrieval tasks, respectively.

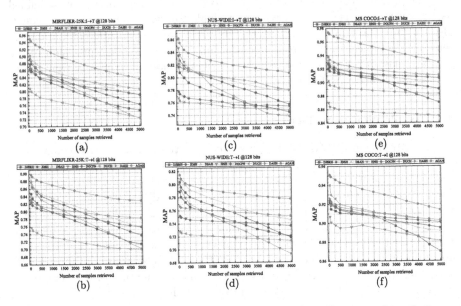

Fig. 2. The precision of top-N curves with 128 bits compared to several methods on three benchmark datasets.

MAP Comparison Results: Table 2 shows the MAP@5000 results for the proposed AGAH on three baseline datasets (MIRFlickr-25K, NUS-WIDE, and MS COCO) with hash code lengths from 16 to 128 bits compared to other state-of-the-art unsupervised cross-mode hashing methods. As can be seen from the data in Table 2, our proposed method outperforms all compared baselines. It is worth noting that the first four methods are traditional and the rest are deep neural network-based methods. The deep neural network-based methods achieve significant performance gains due to the powerful non-linear feature extraction capabilities of neural networks. Compared with some state-of-the-art unsupervised cross-mode hashing baselines, our method has about 1.5%-3% performance improvement, which confirms the superiority of the proposed AGAH. In addition, the improvement of our proposed method on the NUS-WIDE dataset is relatively small, about 0.6%-2.2%, because the NUS-WIDE dataset contains a small number of categories. The performance improvement of our method is more pronounced on MSCOCO with a large number of categories, which maintains good performance despite the low hash code length.

Top-N Precision Curves: Fig. 2 displays the top-N accuracy curves comparing the proposed method with all seven baseline methods on three multimedia datasets. top-N accuracy curves are plotted by varying the number of retrieved samples from 1 to 5000, and they reflect the fluctuation of the model in retrieval accuracy as the number of retrievals increases. As can be seen from the curves in Fig. 2, our method outperforms the baseline for all comparisons, which intuitively reflects the efficiency of our AGAH. It is worth noting that the top-N precision curve slowly decreases as the number of retrieved instances increases. A plausible explanation is that our proposed adaptive graph attention module can help learn hash codes and thus produce more high-quality hash codes. Finally, together with the MAP comparison results, the top-N accuracy curve can also illustrate that our proposed method mitigates the loss of accuracy in the binarization process, thus improving the retrieval performance and maintaining a high accuracy rate as the number of retrieved instances increases.

4.4 Ablation Study

In order to validate the validity and contribution of each module in our proposed approach, we designed five variants of the model to verify the impact of each module on the overall model. These variants of the model are detailed below:

1. AGAH-1: It shows that the variant model uses only image similarity.
2. AGAH-2: It shows that the variant uses only text similarity.
3. AGAH-3: It refers to the variant without the adaptive graph attention.
4. AGAH-4: AGAH-4 indicates that the variant does not employ an iterative approximate optimization strategy.
5. AGAH-5: It removes the attention mechanism from the model.

The MAP results for the different variants on the three multimedia datasets are shown in Table 3. Based on this, we could draw the following conclusions:

1. The analysis of Table 3 shows that each module plays an important role in the overall model. the performance degradation is most pronounced for AGAH-2 because language is human-refined information and the similarity matrix constructed from text is sparse. AGAH-1, however, uses only image features to construct the similarity matrix, but with less performance degradation. A possible explanation for this is that images contain richer and more fine-grained semantic information. the results of AGAH-1 and AGAH-2 can prove the effectiveness of our proposed multi-modal similarity augmentation module.
2. The adaptive graph attention hashing module also has an effect on the performance of the proposed AGAH. Concretely, the results from AGAH-3 and AGAH-5 show that both the graph convolutional neural network and the attention mechanism contribute to the performance improvement of the model by approximately 1.5%-2.5%.

Table 3. The MAP@5000 results on image-text retrieval tasks (I→T indicates the image search text task and vice versa) and datasets.

Task	Method	Configuration	MIRFLICKR-25K		NUS-WIDE		MS COCO	
			32bits	128bits	32bits	128bits	32bits	128bits
I→T	AGAH-1	$S = S_v$	0.7660	0.8155	0.7415	0.7954	0.9030	0.9266
	AGAH-2	$S = S_t$	0.7578	0.7975	0.7495	0.7722	0.8643	0.9036
	AGAH-3	$-(\varphi = 0.15)$	0.7965	0.8107	0.7776	0.7998	0.8981	0.9137
	AGAH-4	$-(\alpha=1)$	0.7960	0.8123	0.7711	0.7910	0.9036	0.9121
	AGAH-5	$-(\gamma = 0.45)$	0.7969	0.8225	0.7738	0.7875	0.9045	0.9152
	AGAH	**FULL**	**0.8195**	**0.8341**	**0.7890**	**0.8059**	**0.9205**	**0.9322**
T→I	AGAH-1	$S = S_v$	0.7603	0.8053	0.7339	0.7553	0.8754	0.9098
	AGAH-2	$S = S_t$	0.7561	0.7787	0.7411	0.7551	0.8600	0.8867
	AGAH-3	$-(\varphi = 0.15)$	0.7954	0.8041	0.7451	0.7654	0.8537	0.8876
	AGAH-4	$-(\alpha=1)$	0.7830	0.8131	0.7531	0.7647	0.8852	0.9053
	AGAH-5	$-(\gamma = 0.45)$	0.7757	0.8000	0.7352	0.7550	0.8867	0.9042
	AGAH	**FULL**	**0.8018**	**0.8272**	**0.7472**	**0.7697**	**0.9057**	**0.9172**

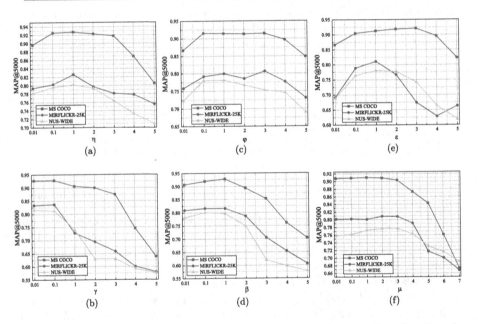

Fig. 3. The parameters sensitivity analysis for $\eta, \varphi, \varepsilon, \gamma, \beta, \mu$ on three benchmark datasets at 128 bits.

4.5   Parameter Sensitivity Analysis

Several hyperparameters that may affect the results of the proposed method were analyzed, and the results are displayed in Fig. 3. The analysis was performed by the controlled variable method, i.e. one parameter was changed in the experimental setup and the values of the other parameters were fixed. The effect of image and text similarity on the performance of the model was moderated by η and β respectively. It was observed that η and β remained relatively stable in the range of 0.01 to 2, while a significant decrease occurred when they were greater than 2. γ, a trade-off parameter for balancing attention weights, remains stable around $[0.1, 1]$, with a significant drop when $\gamma > 1$. Therefore, an appropriate adjustment of similarity can yield satisfactory results. analysis of the results in Fig. 3 shows that our method is not sensitive to the choice of ε and φ in the range $[0.1, 2]$ and that ε and φ weigh the contribution of in-model loss and graph convolution loss, and appropriate adjustment can lead to the best performance of the model. μ is a scale hyperparameter that adjusts the quantization range of the matrix. In conclusion, sensible parameter tuning can maintain the advanced retrieval performance of the model, and the proposed method is robust to hyperparameters within a suitable range.

4.6 Convergence Testing

In this subsection, we examine the convergence and training efficiency of the proposed AGAH on three baseline datasets. Fig. 4(a) shows the final loss function convergence curve at 16-bit hash code length, and Fig. 4(b) shows the variation curve of MAP with an increasing number of iterations. From the results of Fig. 4,

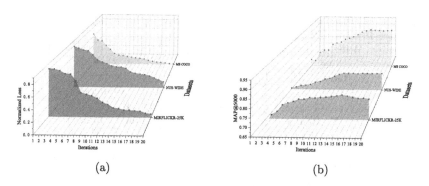

(a)   (b)

Fig. 4. The loss function convergence curves and MAP change curves of the proposed AGAH on three widely-used multimedia datasets at 16-bit code length.

the following can be concluded: Firstly, as the number of optimization iterations increases, the loss function diminishes gradually, and the results show that the optimization process can improve the encoding ability of the hash function. In addition, the loss function can converge to the optimal result after tens of iterations, indicating that our method reduces the consumption of training time and improves training efficiency. Finally, the results show that the proposed network converges to the optimal point in a few tens of iterations, validating the suitability of our proposed network for unsupervised hash retrieval tasks.

5   Conclusion

In this paper, we propose an effective and novel adaptive graph attention hashing method named AGAH for unsupervised cross-modal hashing retrieval tasks. Specifically, we apply a combination of the multimodal model CLIP and graph neural networks to unsupervised image-to-text hashing. We have designed a multimodal similarity enhancement module to enhance the similarity of the data, which helps to improve the accuracy of retrieval. In addition, we design a semantic complementarity-based loss function to reduce the information loss during hash code binarisation. Finally, a well-designed graph adaptive attention module assists in balancing the learning of the hash network and alleviates the problem of unbalanced multimodal learning. Extensive experiments on three benchmark datasets show that the proposed method outperforms several representative state-of-the-art approaches.

References

1. Cheng, M., Jing, L., Ng, M.K.: Robust unsupervised cross-modal hashing for multimedia retrieval. ACM Trans. Inform. Syst. (TOIS) **38**(3), 1–25 (2020)
2. Chua, T.S., Tang, J., Hong, R., Li, H., Luo, Z., Zheng, Y.: Nus-wide: a real-world web image database from national university of singapore. In: Proceedings of the ACM International Conference on Image and Video Retrieval, pp. 1–9 (2009)
3. Duan, Y., Chen, N., Zhang, P., Kumar, N., Chang, L., Wen, W.: Ms2gah: multi-label semantic supervised graph attention hashing for robust cross-modal retrieval. Pattern Recogn. **128**, 108676 (2022)
4. Gong, Q., Wang, L., Lai, H., Pan, Y., Yin, J.: Vit2hash: unsupervised information-preserving hashing. arXiv preprint arXiv:2201.05541 (2022)
5. Guzhov, A., Raue, F., Hees, J., Dengel, A.: Audioclip: extending clip to image, text and audio. In: ICASSP 2022-2022 IEEE International Conference on Acoustics, Speech and Signal Processing (ICASSP), pp. 976–980. IEEE (2022)
6. He, K., Zhang, X., Ren, S., Sun, J.: Deep residual learning for image recognition. In: Proceedings of the IEEE Conference on Computer Vision and Pattern Recognition, pp. 770–778 (2016)
7. Huang, G., Liu, Z., Van Der Maaten, L., Weinberger, K.Q.: Densely connected convolutional networks. In: Proceedings of the IEEE Conference on Computer Vision and Pattern Recognition, pp. 4700–4708 (2017)
8. Huiskes, M.J., Lew, M.S.: The mir flickr retrieval evaluation. In: Proceedings of the 1st ACM International Conference on Multimedia Information Retrieval, pp. 39–43 (2008)
9. Jiang, Q.Y., Li, W.J.: Deep cross-modal hashing. In: Proceedings of the IEEE Conference on Computer Vision And Pattern Recognition, pp. 3232–3240 (2017)
10. Khan, S., Naseer, M., Hayat, M., Zamir, S.W., Khan, F.S., Shah, M.: Transformers in vision: A survey. ACM Comput. Surv. (CSUR) **54**(10s), 1–41 (2022)
11. Krizhevsky, A., Sutskever, I., Hinton, G.E.: Imagenet classification with deep convolutional neural networks. Commun. ACM **60**(6), 84–90 (2017)
12. Lin, T.-Y., et al.: Microsoft COCO: common objects in context. In: Fleet, D., Pajdla, T., Schiele, B., Tuytelaars, T. (eds.) ECCV 2014. LNCS, vol. 8693, pp. 740–755. Springer, Cham (2014). https://doi.org/10.1007/978-3-319-10602-1_48

13. Liu, S., Qian, S., Guan, Y., Zhan, J., Ying, L.: Joint-modal distribution-based similarity hashing for large-scale unsupervised deep cross-modal retrieval. In: Proceedings of the 43rd International ACM SIGIR Conference on Research and Development in Information Retrieval, pp. 1379–1388 (2020)

14. Mikriukov, G., Ravanbakhsh, M., Demir, B.: Deep unsupervised contrastive hashing for large-scale cross-modal text-image retrieval in remote sensing. arXiv preprint arXiv:2201.08125 (2022)

15. Qu, L., Liu, M., Wu, J., Gao, Z., Nie, L.: Dynamic modality interaction modeling for image-text retrieval. In: Proceedings of the 44th International ACM SIGIR Conference on Research and Development in Information Retrieval, pp. 1104–1113 (2021)

16. Radford, A., et al.: Learning transferable visual models from natural language supervision. In: International Conference on Machine Learning, pp. 8748–8763. PMLR (2021)

17. Shen, X., Zhang, H., Li, L., Liu, L.: Attention-guided semantic hashing for unsupervised cross-modal retrieval. In: 2021 IEEE International Conference on Multimedia and Expo (ICME), pp. 1–6. IEEE (2021)

18. Shi, Y., et al.: Deep adaptively-enhanced hashing with discriminative similarity guidance for unsupervised cross-modal retrieval. IEEE Trans. Circ. Syst. Video Technol. (2022)

19. Su, S., Zhong, Z., Zhang, C.: Deep joint-semantics reconstructing hashing for large-scale unsupervised cross-modal retrieval. In: Proceedings of the IEEE/CVF International Conference on Computer Vision, pp. 3027–3035 (2019)

20. Vaswani, A., et al.: Attention is all you need. Adv. Neural Inform. Process. Syst. **30** (2017)

21. Wang, X., et al.: Modality-balanced embedding for video retrieval. In: Proceedings of the 45th International ACM SIGIR Conference on Research and Development in Information Retrieval, pp. 2578–2582 (2022)

22. Wu, N., Jastrzebski, S., Cho, K., Geras, K.J.: Characterizing and overcoming the greedy nature of learning in multi-modal deep neural networks. In: International Conference on Machine Learning, pp. 24043–24055. PMLR (2022)

23. Yang, D., Wu, D., Zhang, W., Zhang, H., Li, B., Wang, W.: Deep semantic-alignment hashing for unsupervised cross-modal retrieval. In: Proceedings of the 2020 International Conference on Multimedia Retrieval, pp. 44–52 (2020)

24. Yang, W., Wang, L., Cheng, S.: Deep parameter-free attention hashing for image retrieval. Sci. Rep. **12**(1), 1–20 (2022)

25. Yu, J., Zhou, H., Zhan, Y., Tao, D.: Deep graph-neighbor coherence preserving network for unsupervised cross-modal hashing. In: Proceedings of the AAAI Conference on Artificial Intelligence, vol. 35, pp. 4626–4634 (2021)

26. Zhan, Y.W., Luo, X., Wang, Y., Xu, X.S.: Supervised hierarchical deep hashing for cross-modal retrieval. In: Proceedings of the 28th ACM International Conference on Multimedia, pp. 3386–3394 (2020)

27. Zhang, D., Wu, X.J., Xu, T., Kittler, J.: Watch: two-stage discrete cross-media hashing. IEEE Trans. Knowl. Data Eng. (2022)

28. Zhang, P.F., Li, Y., Huang, Z., Xu, X.S.: Aggregation-based graph convolutional hashing for unsupervised cross-modal retrieval. IEEE Trans. Multimedia **24**, 466–479 (2021)

29. Zhang, P.F., Luo, Y., Huang, Z., Xu, X.S., Song, J.: High-order nonlocal hashing for unsupervised cross-modal retrieval. World Wide Web **24**(2), 563–583 (2021)

Answering Property Path Queries over Federated RDF Systems

Ningchao Ge[1], Peng Peng[2], Jibing Wu[1(✉)], Lihua Liu[1], Haiwen Chen[1], and Tengyun Wang[1]

[1] Laboratory for Big Data and Decision, National University of Defense Technology, Changsha, China
{geningchaoge,wujibing,liulihua,chenhaiwen13,wangtengyun18}@nudt.edu.cn
[2] Hunan University, Changsha, China
hnu16pp@hnu.edu.cn

Abstract. Property path query can get graph nodes meeting complex conditions with more concise expressions for some SPARQL basic graph patterns, which is an important new query type in SPARQL 1.1. However, property path query is only applied in centralized RDF system, not in a federated RDF system. In this paper, a MinDFA-based property path query method over federated RDF system is proposed, called FPPQO (Federated Property Path Query Optimization). FPPQO first decomposes a federated property path query into multiple subqueries. Then, a Thompson-based MinDFA algorithm is used to construct MinDFA corresponding to the property path expression of each subquery. Finally, the query execution strategy base on B-DFS is used to search for MinDFA matching. During the matching process, the possible circular matching problem is eliminated by adopting the alternating buffer marking mechanism. Experimental results on datasets of different sizes show that the proposed scheme is effective and scalable.

1 Introduction

Resource Description Framework (RDF) [13] is a framework proposed by World Wide Web Consortium (W3C) for describing resources on the Web. RDF describes information as a triple pattern in the form of <subject, predicate, object> or <subject, attribute, value>, which can be easily read and understood by computers. With RDF became a standard model for data interchange on the Web, it was widely used to construct RDF datasets in various fields [6,32,38]. SPARQL Protocol and RDF Query Language (SPARQL 1.0) [30] were acknowledged by W3C to retrieve and manipulate data stored in RDF format. SPARQL 1.1 [12] was proposed to support and meet the query requirements in more complex scenarios. Property path query is one of the most important updates of SPARQL 1.1. A property path is a possible route through a graph between two graph nodes. The ends of the path may be RDF terms or variables. Variables can not be used as part of the path itself, only the ends. The form of property path query is shown in Fig. 1. Among that, $R = foaf : knows+$ is the property

X. Song et al. (Eds.): APWeb-WAIM 2023, LNCS 14333, pp. 16–31, 2024.
https://doi.org/10.1007/978-981-97-2387-4_2

path express of the property path query Q, which consists of predicates and regular operators. Property path query allows for more concise expressions for some SPARQL basic graph patterns and it also has the ability to match connectivity of two resources by an arbitrary length path. Because of the convenience and importance of this query type, researchers have done a lot of related works [14, 22, 24, 36].

```
PREFIX foaf: <http://xmlns.com/foaf/0.1/>
PREFIX :    <http://example/>
SELECT ?person
{
  :x foaf:knows+ ?person
}
```

Fig. 1. A property path query example Q in SPARQL 1.1: finding all the people :x connects to via the foaf:knows relationship.

Federated RDF systems [16, 28, 35] were put forward to integrate and provide transparent access over many RDF datasets. Federated RDF system is a distributed RDF system, which consists of a control node and a series of RDF sources. These RDF sources are distributed independently on different servers, and provide SPARQL query interfaces of RDF datasets. Because SPARQL is a query language designed for centralized RDF system, it cannot be directly applied to federated RDF system. Therefore, it is an urgent task to realize federated SPARQL query in a federated RDF system. However, the existing research mainly realized basic query [18, 33, 35], top-k query [16] and multi-query [27] in SPARQL 1.0, while the complex queries in SPARQL 1.1 has not been paid attention.

To this end, in this paper, we propose a MinDFA-based property path query method over federated RDF system. The property path expression in property path query is constructed as MinDFA based on Thompson algorithm, and the property path query problem can be transformed into MinDFA-based search matching problem. In the matching process of MinDFA, the optimization algorithm based on B-DFS is adopted to improve the matching efficiency.

In this paper, our key contributions are as follow:

- We propose a fast MinDFA construction algorithm based on Thompson to transform property path expression into MinDFA. Besides, unnecessary repeated construction can be avoided by adopting isomorphic multiplexing technology.
- We propose a MinDFA search matching algorithm based on B-DFS, which can quickly match vertex pairs that meet the rules over federated RDF system.
- We have done a lot of experiments on datasets of different sizes, and the results show that the proposed scheme is effective and scalable.

2   Related Work

Federated SPARQL Query Optimization. The federated SPARQL query execution process consists of three steps: data sources localization, query decomposition and subqueries execution. The main differences of the related query optimization methods lie in the specific implementation methods of the three steps. B. Quilitz et al. [31] use a kind of metadata called "service description" to assist the quick localization of data sources to improve the query efficiency of federated SPARQL. O. Görlitz et al. [18] utilize Verbal of Linked Datasets (VOID) as auxiliary metadata of data sources localization. Different from the above text description method, A. Harth et al. [20] and F. Prasser et al. [29] adopt a data structure called QTree as metadata. The QTree inherits the structure of RTree [10], and the data source information is stored in its leaf nodes. The efficient traversal of the tree structure is used to realize the quick location of the data source. M. Saleem et al. [33] caches the mappings of properties to subjects and objects of each data source into a set, and takes the set as metadata. In another achievement, M. Saleem et al. [34] uses a hash table based on data distribution as metadata. On the basis of M. Saleem et al. [33], E. C. Ozkan et al. [26] created a special index to reduce the space overhead caused by the large number mapping of properties to subjects and objects. G. Montoya et al. [23] and N. Ge et al. [15,16] store metadata by designing a cost model, and generate the optimal query plan based on the cost model to minimize the overall cost. Especially, A. Schwarte et al. [35] does not need to use metadata, but adopts ASK query to locate data source. Furthermore, J. M. Arenas [5] and C. B. Aranda et al. [4] use Federation extension in SPARQL 1.1 to realize data source location. C. B. Aranda et al. [3,9] analyzed the feasibility of using the new SERVICE keyword in SPARQL 1.1 to locate data sources. Some papers [17,19,25,37] improve the overall query efficiency by optimizing the query decomposition and subquery rewriting.

Property Path Query Optimization. Extensive researches use views [11] or rare labels [22] to answer property path queries on graphs. M Nolé et al. [24] realize the evaluation of attribute path query by using parallel framework based on Brzozowski's derivatives [8] and Antimirov's partial derivatives [2]. W. Xin et al. [39] first devise a dynamic programming approach to evaluate local and partial answers. Then, an automata-based algorithm is proposed to assemble the partial answers into the final results. S. Wadhwa et al. [36] design a random-walk based sampling algorithm, which is backed by theoretical guarantees on its expected quality. Y. Xin et al. [40] propose an automata-based distributed algorithm under the provenance-aware semantics using the Pregel graph parallel computing framework.

All the above methods will produce a large number of intermediate results, which will bring huge space overhead. In addition, these methods are effective for centralized RDF system, but not on federated RDF system. So, we propose a MinDFA-based property path query optimization method over federated RDF system in this paper. To the best of our knowledge, it is the first work to implement and optimize property path query over federated RDF system.

3   Problem Formulation

Before introducing the proposed method formally, we need to give some definitions of related terms, concepts and problems in order to understand the relevant contents.

Definition 1. (RDF Term and RDF Dataset). *The set of RDF Terms,* $\mathcal{RDF}_T$*, is* $\mathcal{I} \cup \mathcal{RDF}_{\mathcal{L}} \cup \mathcal{RDF}_{\mathcal{B}}$*, where* $\mathcal{I}$ *is the set of all IRIs,* $\mathcal{RDF}_{\mathcal{L}}$ *is the set of all RDF Literals and* $\mathcal{RDF}_{\mathcal{B}}$ *is the set of all blank nodes. An RDF dataset is a set:* $\{G, (<u_1>, G_1), (<u_2>, G_2), ..., (<u_n>, G_n)\}$*, where G and each G_i are graphs, each* $<u_i>$ *is an IRI and is distinct. Among that, G is called the default graph.* $(<u_i>, G_i)$ *are called named graphs.*

Definition 2. (Triple Pattern and Basic Graph Pattern). *A triple pattern is a member of the set:* $TP = (\mathcal{RDF}_T \cup \mathcal{V}) \times (\mathcal{I} \cup \mathcal{V}) \times (\mathcal{RDF}_T \cup \mathcal{V})$*, where $\mathcal{V}$ is the set of query variables. And a basic graph pattern is a set of triple patterns:* $BGP = \{TP_1, TP_2, ..., TP_n\}$*.*

Definition 3. (Federated RDF System). *A federated RDF system can be defined as:* $F = \{S, f, L\}$*, where (1)* $S = \{S_1, S_2, ..., S_n\}$*, it represents the collection of all RDF datasets in the federated system; (2)* $f : S \rightarrow 2^G$ *is a mapping that associates each source with a graph of RDF graph G; (3)* $L : u_i \rightarrow S$ *is a mapping of IRI u_i to its RDF datasets location $L(u) \in S$. $L(u)$ is also called the host source of u_i.*

Definition 4. (Property Path and Property Path Pattern). *A Property Path is a sequence of triples, t_i in sequence ST, with $n = length(ST) - 1$, such that, for $i = 0$ to n, the object of t_i is the same term as the subject of t_{i+1}. Among that, t_0 is called the start of the path, and t_n is called the end of the path. A property path is a path in graph G if each t_i is a triple of G. Let R be the set of all property path expressions, a property path pattern is a member of the set:* $PPPs = (\mathcal{RDF}_T \cup \mathcal{V}) \times R \times (\mathcal{RDF}_T \cup \mathcal{V})$*.*

Definition 5. (Match of Federated Property Path SPARQL Query (FPPSQ)). *A federated property path SPARQL query can be expressed as:* $FPPSQ = <Q, V, R, F>$*, where R and V is the property path expression and query variable of query Q, respectively. F is the federated RDF system of federated property path query Q. The match of Q over the federated RDF system F is the set of vertex pairs $<v_i, v_j>$, where the v_i and v_j are entities in F and the path from v_i to v_j conforms to the rule of attribute path expression R.*

Problem: The problem to be studied in this paper is defined as follows: given a federated RDF system $F = \{S, f, L\}$ and a federated property path SPARQL query $FPPSQ = <Q, V, R, F>$, find out the match of $FPPSQ$ accurately and efficiently.

4   The Proposed Method

In this section, we will introduce the property path query optimization method over federated RDF system in detail. The overall framework of the scheme is shown in the Fig. 2, which mainly includes three modules: (1) query decomposition and source localization, (2) Thompson-based MinDFA construction and (3) query execution strategy based on B-DFS.

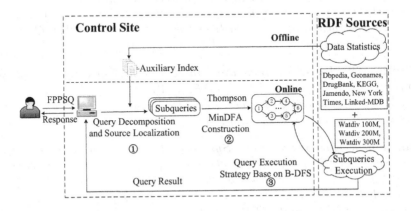

Fig. 2. The Overall Framework of Property Path Query over Federated RDF System.

4.1   Query Decomposition and Source Localization

For an $FPPSQ$ submitted by a user, it needs to be decomposed into subqueries that can be executed on a single RDF source. In order to improve the efficiency of query decomposition and source localization, we design an auxiliary index to guide the query decomposition. The auxiliary index is a hash map: $PS = \{< P_1, \{S_{P1}\} >, < P_2, \{S_{P2}\} >, ..., < P_n, \{S_{Pn}\} >\}$, which is statistically obtained and maintained in the offline phase. Its keys $\{P_1, P_2, ..., P_n\}$ are constant predicates of each RDF source in a federated RDF system, and the value $\{S_{Pi}\}$ of key P_i is the RDF sources identifiers set where the predicate P_i appears. The auxiliary index generation algorithm is shown in Algorithm 1, and its time complexity is $O(|E|)$, where $|E|$ is the total number of edges of each RDF sources.

Algorithm 1: Auxiliary Index Generation

Input: A federated RDF system $F = \{S, f, L\}$;

Output: Auxiliary index PS;

1 Initialize an empty hash map $PS = \{\}$;

2 **for** $i = 1$ to $|S|$ **do**

3   **for** $j = 1$ to $|S_i|$ **do**

4         $D_{Pij} = PS.get(P_{ij})$;

5   **if** $D_{Pij} == null$ **then**

6   Initialize an hash set $D_{Pij} = \{identifier(S_i)\}$;

7   **else**

8            $D_{Pij}.add(identifier(S_i))$;

9         $PS.put(P_{ij}, D_{Pij})$;

10 Return PS;

For each triple pattern TP of a federated property path SPARQL query, (1) its data sources are mapped to all RDF sources of federated RDF system $F = \{S, f, L\}$: $D(TP) = S$, if its predicate is a variable; (2) its data sources are mapped to the RDF sources identifiers set where the predicate P appears: $D(TP) = PS.get(P)$, if its predicate is a IRI constant P; (3) its data sources are mapped to the union of the RDF sources identifiers sets where each IRI constant P_i of property path expression R appear: $D(TP) = PS.get(P_1) \cup PS.get(P_2) \cup ... \cup PS.get(P_n)$, if its predicate is a property path expression R. In addition, we can use ASK query to prune data sources $D(TP)$ if the subject or object of the triple pattern is constant.

Firstly, we will get the set of source localization: $\{< TP_1, D(TP_1) >, < TP_2, D(TP_2) >, ..., < TP_n, D(TP_n) >\}$, for a federated property path SPARQL query with n triple patterns. Then, we group the set according to the value of $D(TP_i)$. If the count of $D(TP_i)$ and $D(TP_j)$ is one respectively and their value is same, the TP_i and TP_j can be assigned to a group. Finally, the triple patterns $\{TP_1, TP_2, ..., TP_n\}$ in each group combine to form a subquery, and the source of the subquery is $D(TP_1)$. The query decomposition and source localization algorithm is shown in Algorithm 2, and its time complexity is $O(n)$, where n is the number of triple patterns of the federated property path SPARQL query.

Algorithm 2: Query Decomposition and Source Localization

Input: A federated property path SPARQL query $PPSQ = <Q, V, R, F>$, auxiliary index PS;

Output: Subqueries set $Subs$;

1 Initialize an empty hash map $Map = \{\}$;
2 Get all triple patterns of Q to set TP;
3 **for** $i = 1$ *to* $|TP|$ **do**
4      $D(TP_i) = getSource(TP_i)$;
5      $Map.put(TP_i, D(TP_i))$;
6 Initialize an empty hash map $Map\_Reverse = \{\}$;
7 **for** *Entry entry : Map.enrtySet()* **do**
8      $Group = Map\_Reverse.get(entry.getValue())$;
9   **if** $Group == null$ **then**
10   Initialize an empty set $Group = \{\}$;
11     $Group.add(entry.getKey())$;
12     $Map\_Reverse.put(entry.getValue(), Group)$;
13 Initialize an empty set $Subs = \{\}$;
14 **for** *Entry entry : Map_Reverse.enrtySet()* **do**
15   Initialize a SPARQL query sub;
16     $sub.setTriplePattern(entry.getValue())$;
17     $sub.setSource(entry.getKey())$;
18     $Subs.add(sub)$;
19 Return $Subs$;

4.2 Thompson-Based MinDFA Construction

Inspired by the similarity between property path expression and regular expression, we propose a property path query method based on MinDFA search. This scheme needs to transform the property path expression into the corresponding MinDFA. In this subsection, the Thompson-based MinDFA construction method will be described in detail. There are two steps to construct MinDFA: transforming regular expression into NFA and transforming NFA into DFA and minimizing it.

(1) **Transforming regular expression into NFA.** Because property path syntax is similar to regular syntax, we can easily transform a property path expression into a regular expression. In addition to transforming the symbol, we need to maintain an letter-to-IRI mapping, which facilitates recovery of SPARQL statements during regular matching in Sect. 4.3. Figure 3 show an example of transforming property path expression into regular expression. On the left is the expression conversion result, and on the right is the letter-to-IRI mapping.

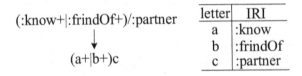

letter	IRI
a	:know
b	:frindOf
c	:partner

Fig. 3. An example of transforming property path expression into regular expression.

Then, we employ Thompson algorithm to convert regular expressions into NFA. For the example expression in the Fig. 3, its NFA constructed by Thompson algorithm is shown in the left of Fig. 4.

(2) Transforming NFA into DFA and minimizing it. For a NFA constructed by Thompson algorithm, we adopt the subset construction method base on ε closure to transform NFA into DFA, then the MinDFA will be obtained by eliminating invalid state and merging equivalent state as shown in the right of Fig. 4.

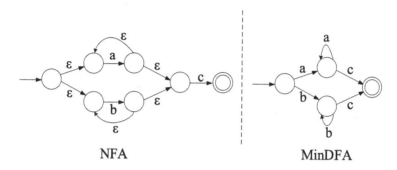

NFA   MinDFA

Fig. 4. The NFA and MinDFA of regular expression in the Fig. 3.

Different from ordinary MinDFA, according to the characteristics of property path query, we define some storage elements for each status node T_i in MinDFA, including: whether it is the end status $isEnd_i$, current status query variable V_i, current query variable result set R_i, the mapping of status transition letters to next status g, and the data sources of each state transition query $D(T_i)$. These elements of each node will be used during the matching process of automata in Sect. 4.3. Besides, unnecessary repeated construction can be avoided by adopting isomorphic multiplexing technology. For two federated property path SPARQL query with same regular expressions transforming from respective property path expressions, they can share a MinDFA, and only need to maintain different letter-to-IRI mappings.

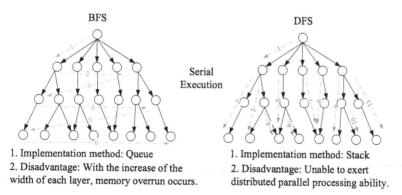

BFS   DFS

Serial Execution

1. Implementation method: Queue
2. Disadvantage: With the increase of the width of each layer, memory overrun occurs.

1. Implementation method: Stack
2. Disadvantage: Unable to exert distributed parallel processing ability.

Fig. 5. The matching process based on BFS and DFS.

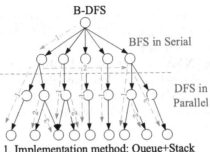

1. Implementation method: Queue+Stack
2. Advantage: Increase parallel processing ability.

Fig. 6. The matching process based on B-DFS.

4.3   Query Execution Strategy Base on B-DFS

Generally, automata matching methods are divided into two categories: automata matching based on depth-first search (DFS) and automata matching based on breadth-first search (BFS). These two kinds of methods have achieved good performance in query optimization of property path over centralized RDF system. Figure 5 gives the matching process based on BFS and DFS. BFS is usually implemented by queues. The disadvantage of this method is that with the increase of the width of each layer, the memory will exceed the limit. DFS is usually implemented by stack. There is a common problem between this method and the above matching method based on BFS: the serial execution strategy can not exert the parallel processing ability in distributed environment.

Therefore, we propose a MinDFA matching method based on B-DFS over federated RDF system, which realizes fast matching by adopting parallel processing mode, as shown in the Fig. 6. This method first adopts BFS matching strategy (not limited to the first layer in the Fig. 6) until the elements in the queue reach a certain scale. Then, DFS matching strategy is implemented on the elements in the queue in parallel processing mode.

Algorithm 3: A MinDFA Matching Algorithm based on B-DFS

Input: A federated RDF system: $F = \{S, f, L\}$, a MinDFA: $M = \{T, \Sigma, g\}$, the letter-to-IRI mapping: Map;

Output: MinDFA matching result R;

1 Initialize an empty set R;

2 Initialize an empty stack ST;

3 //Deal with initial state T_0 base on BFS

4 **for** $i=0$ to $|\Sigma_{T0}|$ **do**

5       $R_0 = R_0 \cup nextQuery(T_0, \ Map.get(\Sigma_{T0}\_i), \ g(T_0, \Sigma_{T0}\_i))$;

6       $ST.add(g(T_0, \Sigma_{T0}\_i))$;

7 //Deal with remain state base on DFS

8 **for** $i=0$ to $|R_0|$ **do**

9   //parallel processing

10   **while** $|ST| > 0$ **do**

11          $Current\_T = ST.pop()$;

12   **for** $i=0$ to $|\Sigma_{Current\_T}|$ **do**

13   **if** $g(Current\_T, \Sigma_{Current\_T}\_i) == Current\_T$ **then**

14                $R_{Current\_T} = R_{Current\_T} \cup selfQuery(Current\_T,$

15                   $Map.get(\Sigma_{Current\_T}\_i), \ Current\_T)$;

16   **else**

17                $R_{Current\_T} = R_{Current\_T} \cup nextQuery(Current\_T,$

18                   $Map.get(\Sigma_{Current\_T}\_i), \ g(Current\_T, \Sigma_{Current\_T}\_i))$;

19                $ST.put(g(T_0, \Sigma_{Current\_T}\_i))$;

20   **if** $Current\_T.isEnd == true$ **then**

21             $R.add(< R_0\_i, \ R_{Current\_T} >)$;

22 Return R;

The MinDFA matching process based on B-DFS is shown in Algorithm 3. We first get the result R_0 of initial state T_0 base on BFS, and put its next state into a stack ST (Line 4–6). Then, DFS-based method is adopted to match subsequent states for each element in R_0 in parallel (Line 8-21). Among them, there are two important functions: $selfQuery()$ (Line 14) and $nextQuery()$ (Line 17). The $selfQuery()$ is used to obtain the query result for the cyclic state transition (the next state is the current state). The $nextQuery()$ is used to obtain query results with different transition statu from current statu. Finally, if the end state is encountered during the matching process, its query results are merged into the final result set R (Line 20-21).

5 Evaluation

This section will evaluate the performance of the proposed scheme FPPQO, including: experimental environment, the effectiveness evaluation and the scalability evaluation.

5.1   Experimental Environment

In order to sufficiently evaluate the effectiveness, efficiency and scalability of the proposed scheme, we build a federated RDF system and implement the proposed scheme on this system. The federated RDF system consists of six cloud servers, one of which serves as a control site and the rest as RDF data sites. At each RDF data site, an RDF dataset management framework is deployed, named Sesame [7], which can conveniently manage RDF datasets and provide the unified SPARQL access interface. Based on the Sesame data management framework, we deploy the **WatDiv** [1] datasets on five data sites.

WatDiv: WatDiv is a synthetic dataset proposed by Waterloo University of Canada, which consists of two components: the data generator and the query (and template) generator. WatDiv can generate datasets of different sizes according to the needs of users. In this paper, we generate WatDiv datasets of four sizes: WatDiv 100T (one hundred thousand triple patterns), WatDiv 10M (ten million triple patterns), WatDiv 50M and WatDiv 100M. Then, METIS [21] data partitioning scheme is used to divide each size of dataset into five subdatasets to form federated WaDiv datasets.

Because existing federated RDF systems, such as SPLENDID [18], HiBIS-CuS [33] and FedX [35], do not support federated property path query, we cannot compare the performance with existing methods. Therefore, in the following, we mainly evaluate the performance of the optimization strategy proposed in this paper, which mainly includes three aspects: evaluating the performance comparison of five property path query symbols under the same scale dataset, evaluating the performance and resource consumption of different matching strategies of MinDFA, and evaluating the performance robustness of five attribute path query symbols under different scale datasets. Therefore, we set up five groups of property path queries for five typical property path query symbols.

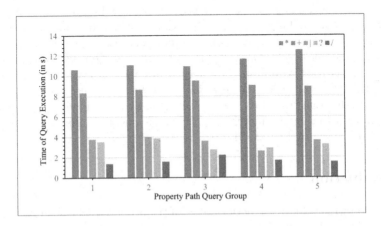

Fig. 7. The performance comparison of five property path query symbols under the same scale dataset.

5.2   Performance Comparison of Five Property Path Query Symbols

In order to evaluate the query performance of different property path symbols, we tested the execution time of five groups of property path queries of five typical property path query symbols on the data set of WatDiv 100M, and the evaluation results are shown in Fig. 7. The experimental results show that the performance of five typical property path query symbols is quite different under the same size data set. Among them, ZeroOrMore path query "*" and OneOrMore path query "+" have the longest average query response time and the worst query performance. Because there are self-circulating queries in these two types of property path queries, and the path length cannot be estimated. Secondly, the Alternative path query "|" and ZeroOrOne path query "?" have better query performance. Sequence path query "/' has the best performance. Because, among the five property path query symbols, Sequence path query "/" is closer to SPARQL basic query, and SPARQL basic query has the highest query efficiency in an RDF system.

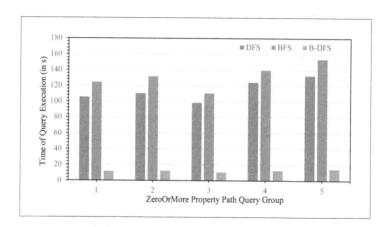

Fig. 8. The performance of different matching strategies of MinDFA.

5.3   Performance and Resource Consumption of Different Matching Strategies of MinDFA

In Sect. 4.3, we propose a B-DFS-based MinDFA optimized matching execution strategy. In order to evaluate the effectiveness of this optimization strategy, we have simultaneously implemented the property path query processing scheme of MinDFA matching strategy based on BFS and the property path query processing scheme of MinDFA matching strategy based on DFS. The following will compare the performance of these three MinDFA matching strategies from two aspects. Here, we choose ZeroOrMore path query "*", which has the highest complexity of property path query.

(1) Firstly, based on the data set WatDiv 100M of the same scale, we run five groups of ZeroOrMore path queries "*" on the systems based on DFS, BFS

and B-DFS respectively, and get their query response time, as shown in Fig. 8. Experimental results show that due to the parallelism of MinDFA matching strategy based on B-DFS, its query performance has been significantly improved.

(2) In Sect. 4.3, we mentioned that BFS execution strategy will occupy a lot of memory because it needs a large cache space. Here, we run five groups of ZeroOrMore path queries "*" on three MinDFA matching strategy systems, count their memory consumption and take the average value. The results are shown in Table 1. The experimental results verify our analysis in the solution: among the three MinDFA matching strategies, the memory consumption of MinDFA matching strategy based on BFS is the largest, and that of MinDFA matching strategy based on DFS is the smallest. MinDFA matching strategy based on B-DFS takes into account the characteristics of BFS and DFS at the same time, so its memory consumption lies between them. Compared with the MinDFA matching strategy based on DFS, the memory cost of our proposed solution is increased by 20%, but the final query response time is reduced by nearly 10 times.

Table 1. The resource consumption of different matching strategies of MinDFA.

Matching Strategy of MinDFA	Average Memory Consumption (M)
DFS	53.8
BFS	124.3
B-DFS	64.6

5.4   Performance Robustness of Five Property Path Query Symbols

In order to evaluate the performance change trend of five property path symbol queries under different scale data sets, we run five groups of property path queries of five typical property path query symbols on the federated distributed RDF systems of WatDiv 100M, WatDiv 200M and WatDiv 300M respectively. Finally, the average query execution time of five groups of queries of each property path query symbol is taken separately. The results are shown in Fig. 9. Here, we define an indicator: Scale Growth Rate (SGR), which represents the ratio of SPARQL property path query execution time growth multiple S(QET) to dataset size growth multiple S(D).

$$SGR = \frac{S(QET)}{S(D)} \tag{1}$$

The experimental results show that with the increase of the dataset size of the federated distributed RDF system, the average query execution time of five property path symbol queries also increases. Among them, the scale growth rate of ZeroOrMore path query "*" and OneOrMore path query "+" is greater than 1, which indicates that these two types of property path queries increase exponentially with the increase of dataset size. Because there are a large number of self-circulating queries in these two types of attribute path queries, the larger the dataset, the exponential increase in the complexity of self-circulating structure. However, the scale growth rate of the other three types of property path queries is less than 1, which indicates that they increase linearly with the increase of dataset size.

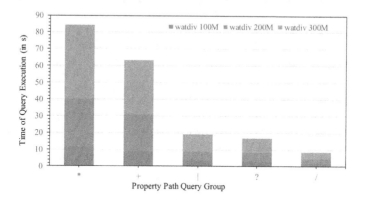

Fig. 9. The robustness of different property path symbols.

6   Conclusion

In this paper, we realized and optimized the property path query over federated RDF system. The proposed scheme mainly realizes efficient property path matching based on MinDFA and B-DFS. For a property path SPARQL query, it is first decomposed into several subqueries. Then, the corresponding MinDFA is constructed for each subquery containing the property path expression. Finally, a query execution method based on B-DFS was employed to get the final result. Eventually, a large number of experimental results show that the proposed method has good performance and scalability.

Acknowledgment. This work was supported by the Science and Technology Major Projects of Changsha City (No. kh2103003), Hunan Provincial Natural Science Foundation of China under grant (2022JJ30165).

References

1. Aluç, G., Hartig, O., Özsu, M.T., Daudjee, K.: Diversified stress testing of RDF data management systems. In: ISWC, pp. 197–212 (2014)
2. Antimirov, V.: Partial derivatives of regular expressions and finite automaton constructions. TCS (1996)
3. Aranda, C.B., Arenas, M., Corcho, S.: Semantics and Optimization of the SPARQL 1.1 Federation Extension. Springer-Verlag (2011)

4. Aranda, C.B., Arenas, M., Corcho, Ó., Polleres, A.: Federating queries in SPARQL 1.1: syntax, semantics and evaluation. J. Web Semant. **18**(1), 1–17 (2013)
5. Arenas, J.M.: Federation and navigation in SPARQL 1.1. reasoning web. Semantic Technol. Adv. Query Answering (2013)
6. Auer, S., Bizer, C., Kobilarov, G., Lehmann, J., Cyganiak, R., Ives, Z.: DBpedia: a nucleus for a web of open data. In: Aberer, K., et al. (eds.) ASWC/ISWC -2007. LNCS, vol. 4825, pp. 722–735. Springer, Heidelberg (2007). https://doi.org/10.1007/978-3-540-76298-0_52
7. Broekstra, J., Kampman, A., van Harmelen, F.: Sesame: a generic architecture for storing and querying RDF and RDF schema. In: ISWC, pp. 54–68 (2002)
8. Brzozowski, Janusz, A.: Derivatives of regular expressions. J. ACM **11**(4), 481–494 (1964)
9. Buil-Aranda, C., Polleres, A., Umbrich, J.: Strategies for executing federated queries in SPARQL1.1. In: Mika, P., et al. (eds.) ISWC 2014. LNCS, vol. 8797, pp. 390–405. Springer, Cham (2014). https://doi.org/10.1007/978-3-319-11915-1_25
10. Cai, M., Revesz, P.: Parametric R-tree: an index structure for moving objects. In: Proceedings of the COMAD (2000)
11. Calvanese, D., Giacomo, G.D., Lenzerini, M., Vardi, M.Y.: Answering regular path queries using views. In: ICDE (2002)
12. Consortium, WWW., et al: Sparql 1.1 overview (2013)
13. Dan, B., Guha, R.V.: Resource Description Framework. RDF) Schema Specification, Proposed Recommendation (2000)
14. Dassow, J., Manea, F., Mercaş, R.: Regular languages of partial words. Inf. Sci. **268**, 290–304 (2014)
15. Ge, N., Qin, Z., Peng, P., Li, M., Zou, L., Li, K.: A cost-driven top-k queries optimization approach on federated rdf systems. IEEE TBD **9**(2), 665–676 (2023). https://doi.org/10.1109/TBDATA.2022.3156090
16. ge, N., Qin, Z., Peng, P., Zou, L.: FedTopK: top-K queries optimization over federated rdf systems. In: DASFAA, pp. 595–599 (2021)
17. Goasdoue, F., Kaoudi, Z., Manolescu, I., Quiane-Ruiz, J.A., Zampetakis, S.: CliqueSquare: flat plans for massively parallel RDF queries. In: ICDE (2015)
18. Görlitz, O., Staab, S.: SPLENDID: SPARQL endpoint federation exploiting VOID descriptions. In: COLD (2011)
19. Hammoud, M., Rabbou, D.A., Nouri, R., Beheshti, S.M.R., Sakr, S.: DREAM: distributed RDF engine with adaptive query planner and minimal communication. VLDB **8**(6), 654–665 (2015)
20. Harth, A., Hose, K., Karnstedt, M., Polleres, A., Sattler, K., Umbrich, J.: Data summaries for on-demand queries over linked data. In: WWW, pp. 411–420 (2010)
21. Karypis, G., Kumar, V.: Multilevel graph partitioning schemes. In: ICPP, pp. 113–122 (1995)
22. Koschmieder, A., Leser, U.: Regular path queries on large graphs. In: Ailamaki, A., Bowers, S. (eds.) SSDBM 2012. LNCS, vol. 7338, pp. 177–194. Springer, Heidelberg (2012). https://doi.org/10.1007/978-3-642-31235-9_12
23. Montoya, G., Skaf-Molli, H., Hose, K.: The Odyssey approach for optimizing federated SPARQL queries. In: ISWC, pp. 471–489 (2017)
24. Nolé, M., Sartiani, C.: Regular path queries on massive graphs. In: SSDBM, pp. 1–12 (2016)
25. Nomikos, C., Gergatsoulis, M., Kalogeros, E., Damigos, M.: A Map-Reduce algorithm for querying linked data based on query decomposition into stars. In: EDBT (2014)

26. Ozkan, E.C., Saleem, M., Dogdu, E., Ngomo, A.N.: UPSP: unique predicate-based source selection for SPARQL endpoint federation. In: PROFILES@ESWC (2016)
27. Peng, P., Ge, Q., Zou, L., Özsu, M.T., Xu, Z., Zhao, D.: Optimizing multi-query evaluation in federated RDF systems. TKDE (2019)
28. Peng, P., Zou, L., Özsu, M.T., Zhao, D.: Multi-query optimization in federated RDF systems. In: DASFAA, pp. 745–765 (2018)
29. Prasser, F., Kemper, A., Kuhn, K.A.: Efficient distributed query processing for autonomous RDF databases. In: EDBT, pp. 372–383 (2012)
30. PrudHommeaux, E.: SPARQL query language for RDF (2008). http://www.w3.org/TR/rdf-sparql-query/
31. Quilitz, B., Leser, U.: Querying distributed RDF data sources with SPARQL. In: ESWC, pp. 524–538 (2008)
32. Rebele, T., Suchanek, F., Hoffart, J., Biega, J., Kuzey, E., Weikum, G.: YAGO: a multilingual knowledge base from wikipedia, wordnet, and geonames. In: Groth, P., et al. (eds.) ISWC 2016. LNCS, vol. 9982, pp. 177–185. Springer, Cham (2016). https://doi.org/10.1007/978-3-319-46547-0_19
33. Saleem, M., Ngomo, A.N.: HiBISCuS: hypergraph-based source selection for SPARQL endpoint federation. In: ESWC, pp. 176–191 (2014)
34. Saleem, M., et al.: TopFed: TCGA tailored federated query processing and linking to LOD. J. Biomed. Semant. **5**, 47 (2014)
35. Schwarte, A., Haase, P., Hose, K., Schenkel, R., Schmidt, M.: FedX: optimization techniques for federated query processing on linked data. In: ISWC, pp. 601–616 (2011)
36. Wadhwa, S., Prasad, A., Ranu, S., Bagchi, A., Bedathur, S.: Efficiently answering regular simple path queries on large labeled networks. In: SIGMOD, pp. 1463–1480 (2019)
37. Wang, D., Zou, L., Zhao, D.: Top-k queries on RDF graphs. Inf. Sci. **316**, 201–217 (2015)
38. Wishart, D.S., et al.: Drugbank: a comprehensive resource for in silico drug discovery and exploration. Nucleic Acids Res. **34**, D668–D672 (2006)
39. Xin, W., Wang, J., Zhang, X.: Efficient distributed regular path queries on RDF graphs using partial evaluation. In: The 25th ACM International (2016)
40. Xin, Y., Xin, W., Di, J., Wang, S.: Distributed efficient provenance-aware regular path queries on large RDF graphs. In: DASFAA (2018)

Distributed Knowledge Graph Query Acceleration Algorithm

Peifan Shi[1], Youhuan Li[1(✉)], Wenjie Li[2], and Xinhuan Chen[3]

[1] College of Computer Science and Electronic Engineering, Hunan University,
Changsha, China
{spf,liyouhuan}@hnu.edu.cn
[2] Peking University Chongqing Research Institute of Big Data, Chongqing, China
liwenjiehn@pku.edu.cn
[3] Tencent Inc., Shenzhen, China
chenxinhuancxh@gmail.com

Abstract. As the era of big data continues to evolve, the scale of knowledge data that needs to be processed in reality is enormous, and the single-machine model is incapable of handling queries on large-scale knowledge graph data. Therefore, distributed clusters are necessary to improve processing capability. The core of the existing approaches is all by splitting the large-scale graph data into multiple copies, distributing each copy to different machines for processing, and finally merging the results. However, these approaches suffer from two problems: (i) the result of knowledge graph merging is huge, far exceeding the final result itself, resulting in a lot of data transfer overhead during the distributed merging phase; (ii) the parallelism of algorithms is limited to the physical level of machine parallelism in task partitioning and lacks computational logic parallelism, such as the merging phase, which does not achieve good parallelism. To address these issues, we propose a distributed framework for offline index construction and online SPARQL query processing framework to achieve parallel accelerated processing. Our approach can more efficiently filter candidate solutions that do not match the result, reducing the size of the results to be merged and leading to a reduction in computational and communication costs. Additionally, we also introduce additional parallelism in the mutual merging phase to improve computational efficiency and system throughput.

Keywords: Knowledge Graph · Distributed Query · Query Acceleration

1 Introduction

With the rapid emergence and complexity of data on the Internet, a model that can more flexibly represent large amounts of complex data has been developed—the Resource Description Framework (RDF) data model. The RDF was proposed by the World Wide Web Consortium (W3C) as a unified standard for representing resources on the Web and the relationships between them through a collection of triples. The triple serves as the fundamental unit of knowledge representation in the knowledge graph, with each piece of knowledge forming a node in the graph. This graphical approach facilitates the management of knowledge and enables data modeling, storage, and query mining. Subsequently, W3C defined the SPARQL query language for RDF datasets, allowing for more efficient retrieval and manipulation of data stored in RDF format.

In recent years, various query systems for RDF have been developed, including gStore [23], SW-Store [1], RDF-3X [17], and others. However, as the amount of RDF data increases, many existing SPARQL query systems are standalone algorithms that face challenges in terms of low time efficiency and poor spatial scalability when dealing with larger-scale RDF data. Some distributed algorithms may experience data redundancy during the intermediate merging stage, which can lead to inefficient querying.

To address these challenges and enable efficient processing of SPARQL queries on massive RDF data, this paper presents a distributed knowledge graph query acceleration algorithm. The algorithm consists of two main modules: the offline module and the online module. The offline module involves the construction of indexes, including the encoding of vertex base code and vertex-edge information, and the hierarchical clustering of global encoding sets based on MapReduce computation in Hadoop to form the index of encoding BitSet-Tree structure. The online module involves the parallel filtering of queries through distributed indexes to obtain candidate solutions. Then, a distributed parallel merge algorithm is used to process the query results in a highly parallel manner and return the query solution. Overall, the proposed algorithm provides an efficient and scalable solution for processing SPARQL queries on large-scale RDF datasets.

Our main contributions can be summarized as follows:

- We propose an innovative query method to achieve distributed query acceleration. This method designs a BitSet-Tree structure using vertex labels and neighbor information, along with master and slave node BFS parallel join queries. By efficiently filtering out candidate solutions that do not match the results, this method reduces the size of the results to be merged, resulting in lower computational and communication costs.
- The offline module of this algorithm adopts a bottom-up distributed hierarchical clustering method when building the indexed BitSet-Tree. This approach avoids structural interdependencies such as splitting when constructing BitSet-Tree from the top-down. Thus, it enables the construction of BitSet-Tree on larger RDF datasets with good spatial scalability.
- The online module of this algorithm is designed to perform SPARQL queries based on the parallel algorithm of MPI. The access mode of BFS allows the slave nodes to be fully joined in parallel while the master nodes are globally joined. This ensures high query efficiency despite communication delay overhead during the join process.

2   Related Works

In recent years, significant research has been conducted in both academia and industry to address the SPARQL query problem of massive RDF data. This has led to the development of numerous RDF database systems, which can be broadly classified into centralized and distributed systems [21]. Centralized systems, such as RDF-3X [17], SW-Store [1], and gStore [23] store and process RDF data on a single node. While these systems do not incur communication overhead when querying, the query efficiency is typically limited by the computational power and memory capacity of a single machine. Distributed systems, on the other hand, manage RDF data across multiple nodes and usually use a set of clusters for querying, thereby meeting the needs of complex SPARQL queries with the help of technologies such as Hadoop and Spark. Unlike centralized systems, distributed RDF systems are characterized by a larger aggregated memory capacity and higher processing power. However, cross-node queries may cause significant intermediate data shuffling when answering complex SPARQL queries, which can degrade query performance. In this paper, we focus on SPARQL queries based on RDF data under distributed systems and briefly introduce the relevant RDF query algorithms under distributed systems.

Early distributed RDF data query processing systems such as YARS2 [10], Clustered Jena [18], and Virtuoso [5] systems use a master/slave model architecture and support simple SPARQL join query operations. To improve the scalability of the system, Zoi Kaoudi et al. [12,13] from the University of Athens, Greece, proposed an RDF data query processing system based on a P2P network architecture in which the RDF data is stored in a distributed hash table (DHT). However, frequent data transfers between compute nodes during SPARQL queries can cause significant network communication overhead. Moreover, the above system has low parallelism in query processing and cannot fully utilize all computational resources. For this reason, Sairam Gurajada et al. of the Planck Institute for Informatics, Germany, proposed the TriAD [9] system to improve the query processing performance. Similar to the RDF-3X [17] system, TriAD builds B+ tree indexes for all permutations of the triple <s,p,o> and proposes the Summary Graph technique to build a global index. During query processing, the TriAD system utilizes multi-threading and parallel computation processing techniques based on the Asynchronous MPI protocol for query processing to increase the parallelism of distributed join operations.

Further, the emergence of inexpensive hardware devices has significantly influenced the evolution of distributed data processing architectures. For instance, the Google File System [6], a distributed file system developed by Jeffrey Dean and Sanjay Ghemawat of Google Inc., and the MapReduce [4] distributed programming model for massive data processing in large-scale clusters have become widely used, along with their open-source versions HDFS and Hadoop. Many systems have been designed and developed using Hadoop and HDFS for storing RDF data and processing SPARQL queries, to leverage the high scalability and availability of the MapReduce system framework. The offline module of the algorithm proposed in this paper is based on this.

In addition to file storage, many systems use Key-Value systems to store and process RDF data, such as CumulusRDF [14], H2RDF [19], and AMADA [3]. Key-Value systems can store and process RDF data at a triple granularity level. Among all the Key-Value based RDF query processing systems, Trinity.RDF system [22] designed based on the distributed Key-Value graph processing system Trinity [20] has better performance. It uses the MPI protocol for parallel communication during query execution. While several other graph data processing systems can handle RDF, including PowerGraph [7] and Pregel [16], they are not specifically optimized for SPARQL query processing and therefore may not offer optimal query performance.

3   Offline Module for Distributed Construction of Indexes

The offline module of our proposed algorithm mainly primarily consists of two main parts. The first part involves MapReduce-based data preprocessing, which is further divided into two sub-parts: graph ID mapping and generation of vertex base code. The second part involves encoding-oriented distributed hierarchical clustering construction, which mainly includes two clustering methods: Canopy clustering and KMeans clustering.

3.1   MapReduce-Based Data Pre-processing

Graph ID Mapping. The traditional approach to ID assignment is a stand-alone implementation where IDs are assigned to entities and predicates separately in a serial manner, and duplicate checking is required for each tuple assignment. In contrast, the distributed allocation method used in this paper leverages Hadoop's powerful sorting capability in MapReduce to first de-duplicate and then assign IDs to the de-duplicated results, which are then stored in HBase. Therefore, it can significantly improve index space and building efficiency.

The MapReduce Job for ID assignment consists of two phases: the Map phase and the Reduce phase. Taking the subject entity as an example, the input RDF dataset is read, and the MapReduce Job is started. The system splits the input subject data file into multiple pieces and assigns them to numerous map tasks. Each call to the map function will yield intermediate results in the form of 0 or more <subID, subject> pairs. The shuffle phase reads the intermediate results from the map phase and sorts them to merge <subID, subject> pairs with the same subID into a list of subIDs and corresponding subjects, i.e., <subID, subject list>. In the reduce phase, multiple <subID, subject list> data are obtained for each reduce task. A single reduce task will call the reduce function for each <subID, subject list> data, and the reduce function will output 0 or more <subID, subject> pairs as the result. Finally, all the outputs of each reduce task, i.e., entities, edges, and corresponding IDs, are saved in HBase.

Generation of Vertex Base Code. The edges in RDF triples are associated with predicates and their weights are mapped to fixed-length bitsets using predicate encoding. The base encoding of the edges is obtained using the BKDR string hash function. As a pair of entities can be part of multiple RDF triples with different predicates, the bitwise OR operation is used to combine the base codes of multiple edges from subject to object to obtain the final base codes of the edges from subject to object. Similarly, the bitsets of the base codes of all the adjacent edges of an entity and the neighboring strings are combined using the bitwise OR operation to obtain the base codes of the entity.

Since the computation of base codes for triples is independent, the MapReduce computational framework is suitable for building the base codes of corresponding entities. The RDF file is split and the base codes for each triple are encoded. Then, the base codes of the same entity are aggregated to the Reduce node for the bitwise OR operation to obtain the final base code of the entity.

3.2   Coding-Oriented Construction of Distributed Hierarchical Clustering

BitSet-Tree. Compared with the current top-down construction algorithm of building BitSet-Tree by classification, we propose a bottom-up distributed algorithm of building BitSet-Tree by clustering. The current BitSet-Tree construction algorithm is difficult to build the BitSet-Tree in a distributed manner due to the organizational structure, and thus the BitSet-Tree construction in the single machine case will face spatial challenges when the data volume is relatively large. Our proposed distributed BitSet-Tree construction algorithm determines the initial cluster center value by Canopy clustering of the base code, and then uses the cluster center of each cluster as the parent node of the current cluster by K-means clustering to build the BitSet-Tree bottom-up. The core of the algorithm lies in implementing Canopy clustering and K-means clustering of base codes through MapReduce to cluster more similar base codes together.

The algorithm utilizes entities as vertices and determines the distance between vertices based on the Hamming distance between the corresponding base codes. K values are defined using the size of the vertex set N and the maximum number of child nodes M of each node of the BitSet-Tree being constructed, $K = [N/M]$. The initial K points are selected using Canopy clustering.

If the final BitSet-Tree has L layers, with the bottom layer being layer 0 and the top layer being layer $L-1$, $(L-1)$ K-means algorithms are executed. The K_i centers obtained from the i_{th} clustering will be all the vertices of the layer i, which are also the input vertices of the $(i+1)_{th}$ clustering. When building from layer i to layer $i+1$, K_{i+1} vertices are randomly selected from the vertices of layer i, and then the final K_{i+1} centers are obtained by multiple iterations as all the vertices of layer $i+1$. In each iteration, each map task matches the input vertices with the closest center and outputs the center as the key and the current input vertex as the value. This aggregates the points corresponding to the same center into the same Reduce, forming a clustering cluster, and finds the new center in the clustering cluster. The new center has a new base code, which is the result of all the base codes of the current cluster obtained by a bitwise OR operation. While the size of each cluster is not easy to control, each node of the BitSet-Tree has a specific size limit for query efficiency reasons. Therefore, if a cluster exceeds a certain size, it can be partitioned by hierarchical clustering. This approach effectively controls the number of clusters and the maximum size of individual clusters, solving the issue of poorly controlled K-means cluster size. Finally, the results of the hierarchical clustering are stored in HBase in a distributed manner.

Edge Tables. The edge tables play a crucial role in filtering the index part of the algorithm and the join in the query part, where the read efficiency of Hbase is also critical. Each read of HBase caches a whole row of data in the data table. Based on this, the edge tables structure of the indexing part is designed as a series of collections of rows <entity, OutEdgeList, InEdgeList>, where the two edge lists represent the outgoing and incoming edge lists of the corresponding entity, and the corresponding edge base codes are stored.

The construction of the edge tables only requires a MapReduce Job. For the outgoing edges, only the subID of the six-tuple is used as the key, and the corresponding objID and the base code of the edge obtained by encoding the current six-tuple are used as the pair <objID, edge base code> as the value. In other words, all the outgoing edges of the entity are gathered into the same Reduce node, and the same is done for incoming edges. If two nodes at level 0 in the BitSet-Tree have edges, the corresponding weights are also added between the parents of the two nodes. The specific implementation is that, in the Map task, in addition to the outgoing and incoming edges of the current entity, the outgoing and incoming edges of the corresponding parent node of the current entity are also output, and all output pairs are marked with the corresponding levels. This ensures that in the Reduce task, the outgoing and incoming edges of all the entities in each level of the BitSet-Tree are obtained. In the Reduce phase, the same edges of the outgoing and incoming edge sets associated with the entity are subjected to bitwise OR operation and loaded into HBase according to the HBase storage structure of the above edge tables described above.

4 Online Module for Distributed Parallel Processing of SPARQL Queries

The online module consists of a master node and several slave nodes, as shown in Fig. 1a. The master node receives the user input queries and provides feedback to the user once the query processing result is obtained. Upon receiving the query graph from the user, the master node sends it to each slave node. The slave nodes retrieve the query from their loaded BitSet-Tree to obtain the candidate set. They then join the candidate set by accessing the edges of the query graph in BFS order to obtain the result of a one-step join. The result of the one-step join is sent to the master node, which performs a global join on the results and finally returns the result to the user.

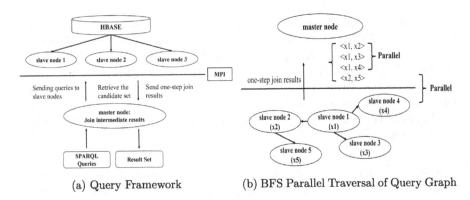

(a) Query Framework   (b) BFS Parallel Traversal of Query Graph

Fig. 1. Online SPARQL Query

4.1 Splitting and Loading of BitSet-Tree

At the beginning of query processing, each slave node is allocated a percentage of the BitSet-Tree on a specific allocation policy. For example, if there are 10 subtrees in the BitSet-Tree and 5 slave nodes in the cluster, each slave node will be assigned 2 subtrees for MPI parallel processing. This helps in retrieving candidate sets quickly and in parallel during query processing. The algorithm used in this paper distributes the task equally by splitting the tree branches, ensuring load balancing and good scalability.

Due to the need for space expansion, each slave node only loads a certain percentage of the BitSet-Tree. Therefore, the BitSet-Tree needs to be linearly split. The splitting approach used in this paper splits all nodes of a layer of the BitSet-Tree. The resulting subset, along with all its subtrees rooted in it, is included as part of the BitSet-Tree split, and all subtrees are loaded into memory after the split. To ensure full utilization of computational resources, the number of nodes in the split layer should be more than the number of slave nodes. The split layer should also be the highest of all layers that satisfy this condition, which improves the pruning effect and retrieval filtering effect. In addition to the subtree of the BitSet-Tree, preload data is also loaded with the one-step neighbors associated with all nodes in the loaded subtree within the current slave node, which is used to pre-filter the candidate set after the retrieval and before the Shuffle. After splitting the BitSet-Tree, each slave node loads the edges associated with the nodes in the subtree of the BitSet-Tree it split.

4.2 Candidate Solution Acquisition

After the master node receives the SPARQL queries from the user, it organizes and encodes them into a graph, which is the same as the encoding method used for offline index construction. Subsequently, the encoded query graph is distributed to each slave node through the master node. The slave nodes then perform retrieval and filtering operations in parallel. Specifically, starting from the top layer of the graph, each layer retrieves solely the child nodes of all nodes that match the base code of a candidate variable in the previous layer. The nodes matching the base code of the corresponding candidate variable in these child nodes are then employed as the current candidate nodes to achieve vertical filtering. This process enables efficient pruning of irrelevant data, allowing for more effective and faster query processing.

One-Step Neighborhood Filtering. During each layer of retrieval, the candidate nodes can be filtered by their one-step neighbors. Concretely, all nodes in the neighbor set of a candidate node that do not match the base codes of the candidate variables corresponding to the query graph can be filtered out. For example, if x_1 and x_2 are two candidate variables with edges in the query graph, each having its own corresponding base code after the graph encoding, a candidate node node1 of x_1 can be filtered out if none of its neighbor's base codes cover the base code of x_2, as node1 cannot match the final result set.

Literal Filtration. When the bottom level is retrieved, the algorithm obtains the candidate set of each candidate variable, which has been filtered by the one-step neighbors. After retrieving to the bottom level, the algorithm performs Literal filtering, further filtering the candidate sets. In addition to edges between variables, the query graph has Literal edges, which are edges where one end is a string. In such cases, the algorithm checks for the existence of the corresponding Literal edge for each candidate vertex to satisfy the Literal edge constraint while satisfying the edge constraint of the candidate variables. If the Literal corresponds to an entity, the algorithm queries the ID of the corresponding entity and looks for the existence of the corresponding one-step neighbor in the candidate vertex set that is exactly that entity. If the Literal corresponds to a string, the algorithm encodes the Literal edge into the base code of an edge and looks for the existence of the edge covering that base code in the candidate vertex edge set. If none of the corresponding edges exist in the candidate vertex set, the candidate vertex is filtered out. This process is called Literal filtering after retrieval and is performed in parallel by each slave node. One-step neighbor filtering satisfies the constraints among the candidate variables, while Literal filtering satisfies the requirements of the Literal edges of the candidate variables in the query graph. The overall base code matching filtering is fully filtered by the base code index for the final join.

4.3   Shuffle

Each slave node in the system after retrieval filtering may contain multiple candidate sets of candidate variables, and the candidate sets of each candidate variable may be distributed across multiple slave nodes. This state can result in significant communication overhead during the join phase, as the endpoints of the same edge may be distributed in multiple slave nodes. Consequently, processing a certain edge in the join phase may require multiple sources to send candidate sets to multiple targets, leading to a large communication delay and inefficiency. To address this problem, the algorithm performs a Shuffle process after retrieval and filtering. This involves aggregating the candidate sets of the same candidate variables to the same slave node and aggregating the candidate sets of different candidate variables to different slave nodes. In this way, for a pair of connected candidate variables in the join phase, only one candidate set needs to be transmitted, and the candidate vertices are transmitted at most once in the Shuffle process. This is more efficient than the case where multiple sources pass multiple targets. The Shuffle process is mainly divided into two parts. In the first part, all the slave nodes inform the master node of the candidate sets of candidate variables they contain. The master node then collects the global candidate set distribution and feeds it to all slave nodes using MPI functions such as MPI_Gather and MPI_Bcast, which are optimized for efficiency. In the second part, each node sends or receives the candidate sets according to its own candidate sets and completes the Shuffle.

4.4   Merge Splicing of Candidate Vertices

The joining of candidate sets in this algorithm mainly is achieved through two parallel processing dimensions. First, communication between different machines in the cluster is joined in parallel through the access structure of BFS. Second, one-step join is performed at the slave nodes, with the master node receiving the one-step join results from the slave nodes and performing global merging and splicing to obtain the final results, which largely improves the time efficiency of the algorithm.

After obtaining the filtered candidate set and Shuffle, the algorithm joins the candidate sets of all candidate variables through the query graph to get the final result set. This step is the core of the distributed query algorithm, and the results obtained through the filtering of Literal and the join of graph structure are the ones that satisfy the query graph. The BFS access structure is chosen because it allows all the outgoing edges of the same vertex can be processed in parallel. The current access vertex sends its own candidate vertices to all the neighbors. These vertices are then performed one-step join in parallel, where the candidate sets of two variables are paired, checking whether there is an edge and whether the edge base code can cover the edge between two vertices in the query graph. Since the base code mapping of the predicate is one-to-one, the edge coverage guarantees the existence of the corresponding edge in the query graph. The candidates that do not match are filtered out.

Under the MPI framework, all nodes run the same code and execute BFS. However, only vertices associated with the currently accessed edges are sent, received, and joined. The global join after the one-step join is handled by the master node, which receives all the one-step join results and keeps the intermediate results. This approach reduces the communication overhead of transmitting intermediate join results multiple times between slave nodes and allows the master node to perform the global join while the slave node performs the one-step join in BFS. As a result, it enables parallel execution and improves time efficiency. Finally, the master node feeds back the final query result to the user.

In the BFS process, when a slave node accesses an edge of a query graph, if the candidate sets corresponding to the outgoing and incoming vertices of the current edge are not local, the slave node does not communicate with other nodes. The execution time of this code is almost negligible, so it can be assumed that each slave node will run fast enough to access the process an edge related to itself. Therefore, in addition to processing all edges of the same access vertex in parallel, edges without associated vertices in the BFS are also processed in parallel, leading to sufficient parallelism for the overall query. As shown in Fig. 1b, the master node integrates the results of the one-step join while the slave node performs the one-step join. The three outgoing edges of slave node 1 (x_1) are processed in parallel in the one-step join between the slave nodes, and the slave node 5 (x_5) also processes the edges related to it, such as (x_2, x_5), with almost negligible time cost. Thus, the parallelism of the algorithm is high enough.

5   Experiments

5.1   Experimental Settings

All experiments were conducted on a DELL-R740-E38S server with two CPUs, 10 cores and 40 threads. Based on this, we deployed an OpenStack cloud platform and set up a distributed cluster of 10 cloud hosts. The configuration of the cloud host is shown in Table 1, which is mainly divided into a master node and 9 slave nodes. The hardware resource configuration of cloud host is shown in Table 2.

Table 1. OpenStack Cloud Host Configuration

Instance Type	Quantity	Hadoop Role	VCPUs	RAM Memory	Disk
master	1	NameNode	4	16 GB	80G
slave	9	DataNode	2	16 GB	80G

We evaluated the proposed query acceleration algorithm using RDF datasets, including standard synthetic datasets and real datasets (Table 3). The standard synthetic datasets include WatDiv[2] and LUBM [8], which allow users to define their own datasets and generate datasets of different sizes. In this experiment, we used three different scales of WatDiv datasets and LUBM datasets (i.e. WatDiv10, WatDiv50, WatDiv100, and LUBM10, LUBM50, LUBM100). The real datasets include DBpedia [15] and YAGO [11], which are extracted from websites such as Wikipedia. The DBpedia dataset used in the experiment mainly came from DBpedia 3.6, and two different sizes of YAGO datasets (i.e. YAGO1, YAGO2) were used. For the single dataset experiment, we used six SPARQL query statements for online querying. Some SPARQL statements are from the dataset's official website's SPARQL test queries. The same type of dataset with different sizes used the same SPARQL statements. However, the SPARQL statements for YAGO1 and YAGO2 differed slightly.

Table 2. Hardware Resources

Resource Name	Configuration Version
OS	CentOS Linux 7
JAVA	1.8.0
OpenStack	Train
Hadoop	3.2.2
ZooKeeper	3.6.3
Hbase	2.4.5
Thrift	0.15.0
GCC/G++	8.3.0
MPICH	3.2.1

Table 3. Statistics of Datasets

Dataset	Entities	Triples	Properties
WatDiv10	101046	1098818	86
WatDiv50	498760	5492375	86
WatDiv100	996986	11008025	86
LUBM10	207631	1272953	18
LUBM50	1083833	6656560	18
LUBM100	2181789	13409395	18
DBpedia	2329624	13795664	1100
YAGO1	2987544	12304136	91
YAGO2	13406742	82219046	96

5.2   Results and Discussion

The experiment consisted of two main modules: offline index construction and online SPARQL query processing. For the offline module, all 10 machines were used, while for the online module, we compared the results of running on a single machine, 3 machines, 6 machines, and 9 machines. The efficiency of online query processing was evaluated based on the loading efficiency of BitSet-Tree and Edge Tables, as well as the computation efficiency.

Offline Index Construction Performance. The offline module involves vertex ID mapping, BitSet-Tree and Edge Tables construction. We report the distributed experimental results for each dataset in Table 4, including space occupation and time consumption. The space and time complexities are dependent on the characteristics and size of the graph. Generally, as the size of the RDF graph increases, the space and time requirements also increase. Even so, using distributed clusters is more efficient than using single machines.

Table 4. Experimental Results of Offline Index Construction

Dataset	Space Occupancy	Time Consumption
WatDiv10	361.82 M	2046 s
WatDiv50	1.83G	5312 s
WatDiv100	3.47G	9187 s
LUBM10	245.91 M	2977 s
LUBM50	2.12G	6163 s
LUBM100	4.87G	10361 s
DBpedia	3.99G	33582 s
YAGO1	3.37G	23621 s
YAGO2	19.53G	92528 s

Online SPARQL Query Performance. The online module comprises the loading of BitSet-Tree and Edge Tables, candidate vertex filtering, shuffling, and candidate vertex join. We show the online query results in Figs. 2, 3, 4, 5 and 6, which mainly include computational efficiency, loading efficiency of BitSet-Tree and Edge Tables, and the relative speedup ratios. It should be noted that the online query experiment for the YAGO2 dataset on a single machine (Figs. 5b, 5c, and 6) was not feasible due to insufficient memory. From the figures, we can draw the following conclusions:

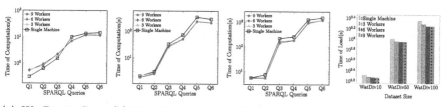

(a) WatDiv10 Com- (b) WatDiv50 Com- (c) WatDiv100 Com- (d) WatDiv Loading
putation Efficiency   putation Efficiency   putation Efficiency   Efficiency

Fig. 2. Comparison of WatDiv Query Efficiency

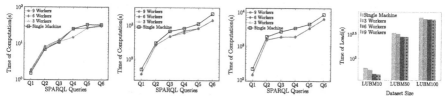

(a) LUBM10 Com- (b) LUBM50 Com- (c) LUBM100 Com- (d) LUBM Loading
putation Efficiency   putation Efficiency   putation Efficiency   Efficiency

Fig. 3. Comparison of LUBM Query Efficiency

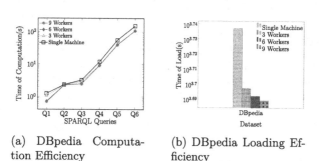

(a) DBpedia Computa- (b) DBpedia Loading Ef-
tion Efficiency   ficiency

Fig. 4. Comparison of DBpedia Query Efficiency

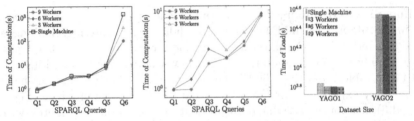

(a) YAGO1 Computation (b) YAGO2 Computation (c) YAGO Loading Effi-
Efficiency   Efficiency   ciency

Fig. 5. Comparison of YAGO Query Efficiency. Experimental data for the YAGO2 dataset on a single machine is unavailable due to memory limitations.

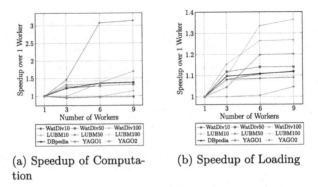

(a) Speedup of Computa-   (b) Speedup of Loading
tion

Fig. 6. Relative computation and loading speedup ratios for datasets with varying numbers of workers. The computation speedup is the average across six SPARQL queries. The YAGO2 dataset could not be executed on a single machine due to memory limitations, so its speedup ratio was calculated with three workers.

1) In the horizontal comparison of a single dataset, for the load section, the acceleration ratio of the distributed load is generally greater than 1, indicating a more significant acceleration effect compared to single machine load. For the calculation section, the performance varies with different SPARQL statements. Typically, distributed acceleration is more significant for SPARQL statements with a large number of query result sets and long single machine execution time. However, for SPARQL statements with a small number of query result sets and short single machine execution time, the distributed acceleration may slow down the speed due to the time cost of MPI message passing between multiple machines.

2) In the vertical comparison of a single dataset, for the same type of dataset using the same SPARQL statement, the distributed acceleration effect becomes more obvious with the increase of dataset size, and the proportion of SPARQL statements with acceleration ratio exceeding 1 also increases.

6   Conclusions

The proposed distributed knowledge graph query acceleration algorithm has good spatial scalability in the index construction part of the offline module, including the generation of encoding for vertex-edge information and the construction of the BitSet-Tree structure. Additionally, the use of HBase as the management platform for all offline data ensures linear spatial scalability for data management. The online query part loads a certain percentage of BitSet-Tree and related edge data from each node, and the memory also scales linearly at this stage. Consequently, the distributed knowledge graph query acceleration algorithm proposed in this paper demonstrates good spatial scalability.

Regarding query efficiency, the results of SPARQL query statements demonstrate that the proposed algorithm has high time efficiency. Moreover, the time efficiency of the distributed algorithm is more apparent with the greater complexity of the query graph. This is due to the high parallelism achieved in each major stage of the query for a complex query graph, resulting in significant time reduction and improved efficiency.

Future follow-up work will focus on the following aspects:

1) Running on larger datasets. The current computational resources still allow the algorithm to run on larger RDF datasets. The larger the dataset and the more complex the query, the more advantages of the distributed algorithm in terms of time and space will be highlighted.
2) Optimize the partitioning strategy of BitSet-Tree. The current BitSet-Tree is only randomly and equally partitioned by the number of subtrees. In the future, we will consider optimizing this part by assigning similar subtrees to the same slave nodes, which will reduce the dispersion of the candidate set of candidate variables, and thus reduce the communication overhead during Shuffle.

Acknowledgements. This work was supported by NSFC (No. 62102142), Hunan Provincial Natural Science Foundation of China (No. 2022JJ40093) and Science and Technology Major Projects of Changsha City (No. kh2103003).

References

1. Abadi, D.J., Marcus, A., Madden, S.R., Hollenbach, K.: SW-store: a vertically partitioned DBMS for semantic web data management. VLDB J. **18**(2), 385–406 (2009)
2. Aluç, G., Hartig, O., Özsu, M.T., Daudjee, K.: Diversified stress testing of RDF data management systems. In: Mika, P., et al. (eds.) ISWC 2014. LNCS, vol. 8796, pp. 197–212. Springer, Cham (2014). https://doi.org/10.1007/978-3-319-11964-9_13
3. Bugiotti, F., Goasdoué, F., Kaoudi, Z., Manolescu, I.: RDF data management in the amazon cloud. In: Proceedings of the 2012 Joint EDBT/ICDT Workshops, pp. 61–72 (2012)

4. Dean, J., Ghemawat, S.: MapReduce: simplified data processing on large clusters. Commun. ACM **51**(1), 107–113 (2008)
5. Erling, O., Mikhailov, I.: RDF support in the virtuoso DBMS. In: Pellegrini, T., Auer, S., Tochtermann, K., Schaffert, S. (eds.) Networked Knowledge - Networked Media. Studies in Computational Intelligence, vol. 221, pp. 7–24. Springer, Heidelberg (2009). https://doi.org/10.1007/978-3-642-02184-8_2
6. Ghemawat, S., Gobioff, H., Leung, S.T.: The Google file system. In: Proceedings of the Nineteenth ACM Symposium on Operating Systems Principles, pp. 29–43 (2003)
7. Gonzalez, J.E., Low, Y., Gu, H., Bickson, D., Guestrin, C.: {PowerGraph}: distributed {graph-parallel} computation on natural graphs. In: 10th USENIX Symposium on Operating Systems Design and Implementation (OSDI 12), pp. 17–30 (2012)
8. Guo, Y., Pan, Z., Heflin, J.: LUBM: a benchmark for owl knowledge base systems. J. Web Semant. **3**(2–3), 158–182 (2005)
9. Gurajada, S., Seufert, S., Miliaraki, I., Theobald, M.: Triad: a distributed shared-nothing RDF engine based on asynchronous message passing. In: Proceedings of the 2014 ACM SIGMOD International Conference on Management of Data, pp. 289–300 (2014)
10. Harth, A., Umbrich, J., Hogan, A., Decker, S.: YARS2: a federated repository for querying graph structured data from the web. In: Aberer, K., et al. (eds.) ASWC/ISWC -2007. LNCS, vol. 4825, pp. 211–224. Springer, Heidelberg (2007). https://doi.org/10.1007/978-3-540-76298-0_16
11. Hoffart, J., Suchanek, F.M., Berberich, K., Weikum, G.: Yago2: a spatially and temporally enhanced knowledge base from Wikipedia. Artif. Intell. **194**, 28–61 (2013)
12. Kaoudi, Z., Koubarakis, M., Kyzirakos, K., Miliaraki, I., Magiridou, M., Papadakis-Pesaresi, A.: Atlas: storing, updating and querying RDF (S) data on top of DHTs. J. Web Semant. **8**(4), 271–277 (2010)
13. Kaoudi, Z., Kyzirakos, K., Koubarakis, M.: SPARQL query optimization on top of DHTs. In: Patel-Schneider, P.F., et al. (eds.) ISWC 2010. LNCS, vol. 6496, pp. 418–435. Springer, Heidelberg (2010). https://doi.org/10.1007/978-3-642-17746-0_27
14. Ladwig, G., Harth, A.: CumulusRDF: linked data management on nested key-value stores. In: The 7th International Workshop on Scalable Semantic Web Knowledge Base Systems (SSWS 2011), vol. 30 (2011)
15. Lehmann, J., et al.: DBpedia-a large-scale, multilingual knowledge base extracted from Wikipedia. Semant. Web **6**(2), 167–195 (2015)
16. Malewicz, G., et al.: Pregel: a system for large-scale graph processing. In: Proceedings of the 2010 ACM SIGMOD International Conference on Management of Data, pp. 135–146 (2010)
17. Neumann, T., Weikum, G.: RDF-3x: a RISC-style engine for RDF. Proc. VLDB Endow. **1**(1), 647–659 (2008)
18. Owens, A., Seaborne, A., Gibbins, N., et al.: Clustered TDB: a clustered triple store for Jena (2008)
19. Papailiou, N., Konstantinou, I., Tsoumakos, D., Koziris, N.: H2RDF: adaptive query processing on .RDF data in the cloud. In: Proceedings of the 21st International Conference on World Wide Web, pp. 397–400 (2012)
20. Shao, B., Wang, H., Li, Y.: Trinity: a distributed graph engine on a memory cloud. In: Proceedings of the 2013 ACM SIGMOD International Conference on Management of Data, pp. 505–516 (2013)

21. Wylot, M., Hauswirth, M., Cudré-Mauroux, P., Sakr, S.: RDF data storage and query processing schemes: a survey. ACM Comput. Surv. (CSUR) **51**(4), 1–36 (2018)
22. Zeng, K., Yang, J., Wang, H., Shao, B., Wang, Z.: A distributed graph engine for web scale RDF data. Proc. VLDB Endow. **6**(4), 265–276 (2013)
23. Zou, L., Mo, J., Chen, L., Özsu, M.T., Zhao, D.: gStore: answering SPARQL queries via subgraph matching. Proc. VLDB Endow. **4**(8), 482–493 (2011)

Truth Discovery of Source Dependency Perception in Dynamic Scenarios

Xiu Fang, Chenling Shen, Guohao Sun[✉], Hao Chen, and Yating Tang

Donghua University, Shanghai 201600, China
{xiu.fang,ghsun}@dhu.edu.cn, {2212626,2222842,2212646}@mail.dhu.edu.cn

Abstract. In the era of big data, obtaining large amounts of data from different sources has become increasingly easy. However, conflicts may arise among the information provided by these sources. Therefore, various truth discovery methods have been proposed to solve this problem. In practical applications, information may be generated in chronological order, such as daily or hourly updates on weather conditions in a particular location. As a result, the truth of an object and the reliability of sources may dynamically change over time. Besides, there may be dependencies among data sources and the dependencies are stable in the short term. However, existing truth discovery methods for dynamic scenarios ignore the continuity of source dependencies in the short term. To address this issue, we study the source dependency detection and the problem of data sparsity caused by removing dependent sources in dynamic scenarios, and propose an incremental model based on source dependency detection, namely SDPTD, which can dynamically update object truth values and source weights and detect source dependencies when new data arrive. Experiments on two real-world datasets and synthetic datasets demonstrate the effectiveness and efficiency of our proposed method.

Keywords: Truth Discovery · Source Dependency · Dynamic Data

1 Introduction

In the era of big data, people can access vast amounts of data from various sources. However, due to the uncertainty of the data and the unreliability of the sources, conflicts among the data may arise, which can lead to misunderstandings and even harm [13,14]. Therefore, truth discovery methods have been developed to help determine the truthfulness and credibility of information [7]. The simplest method is majority voting, which assumes equal reliability of all sources, ignoring the fact that different sources have varying reliability. This means that when the majority of sources are of low quality, the truth may not be identified. Hence, various truth discovery methods have been proposed that take into account factors such as the reliability of sources, dependencies between sources, and difficulty of objects. Although these methods differ in many ways, they are all based on the principle that a reliable source usually provides trustworthy information, and information supported by multiple credible sources is more likely to be true.

© The Author(s), under exclusive license to Springer Nature Singapore Pte Ltd. 2024
X. Song et al. (Eds.): APWeb-WAIM 2023, LNCS 14333, pp. 48–63, 2024.
https://doi.org/10.1007/978-981-97-2387-4_4

Most existing methods are designed for static data [14,17]. With the further information explosion, we witness more and more information generated in chronological order. For example: the weather conditions in a certain area may be updated hourly or daily. In dynamic scenarios, the truth value of objects and the reliability of data sources are no longer fixed, but change over time. In recent years, some advanced methods have been proposed to capture these changes [6,8,15,16]. However, these methods either assume that the data sources are independent of each other, or assume that the source dependencies are independent at different timestamps.

In reality, it is not uncommon to observe sources replicating or crawling data from other sources to enhance information richness or even maliciously mislead audiences. Ignoring dependencies among data sources may result in malicious copiers being assigned with high reliability or to identify false values supported by dependent sources as true. Furthermore, there may be stability in the dependencies among data sources across consecutive time periods, i.e., data sources with dependencies at the previous moment also have a higher possibility of having dependencies at the current moment. Considering this effect, we can eliminate the outliers that appear at the current time and also solve the problem of inaccurate capture of source dependencies caused by the sparse problem.

In this paper, we propose an incremental model for source dependency perception to address the problem of source dependency detection in dynamic scenarios. The model considers the stability of source dependencies in the short term, and we detect source dependencies in the current moment based on the Bayesian model and the source dependencies in the previous moment, and remove the claims provided by the dependent sources. To resolve the potential data sparsity caused by independent source selection, we utilize the continuity of object truth values to fill in the missing claims. Specifically, we utilize the values of adjacent timestamps as virtual sources to support our claims.

The contributions of this paper are summarized as follows:

(i) We considered the short-term stability of source dependencies. Based on the Bayesian model, we used the source dependencies of the previous timestamp as a reference to calculate the source dependencies of the current timestamp.
(ii) We utilize the identified truths from the adjacent timestamps to deal with the missing claims caused by removing dependent sources.
(iii) We evaluate the performance of our method on both real-world and synthetic datasets, and the results demonstrate its superiority in terms of both accuracy and efficiency.

2 Problem Setting

2.1 Notation Definition

Assuming there are T timestamps, at time t ($t \in T$), a group of sources represented by S provide claims on a group of objects represented by O. We denote the claim provided by source s for object o at time t as $c_{o,t}^s$. So the set of all claims provided by source s at time t is denoted by $C_t^s = \{c_{o,t}^s\}_{o \in O}$. $|C_t^s|$ represents the number of claims provided by source s at time t. a_t^s is the cumulative quantity provided by the source up to time t, $a_t^s = \sum_{\tau=1:t} |C_\tau^s|$, the initial value is denoted by a_0^s. In this paper, we assume that the claims provided by the source are numeric data that can reflect continuous changes in the timestamp.

After collecting all the information at time t, we need to aggregate it to obtain a trustworthy result. Let $\hat{c}_{o,t}^*$ represent the inferred truth of object o at time t, and $\hat{C}_t^*$ represent the set of all inferred truths of objects at time t, $\hat{C}_t^* = \{\hat{c}_{o,t}^*\}_{o \in O}$. In addition, let $c_{o,t}^*$ represent the ground truth of object o at time t.

In the inference process, it is also necessary to estimate the reliability of the source and the dependency relationships between sources. We denote the reliability of source s as r_s, and denote all source reliability as R. Due to the correlation between source reliability and the error made by the source, we introduce $\epsilon_{o,t}^s$ to measure the difference between the claim provided by source s on object o at time t and the truth, i.e., $\epsilon_{o,t}^s = |c_{o,t}^s - c_{o,t}^*|$. Thus, we can use ϵ_t^s to represent the sum of errors of all objects provided by source s at time t, $\epsilon_t^s = \sum_{o \in O} \epsilon_{o,t}^s$, and e_t^s represents the cumulative error generated by source s up to time t, $e_t^s = \sum_{\tau=1:t} \epsilon_\tau^s$, the initial value is e_0^s.

For objects that have only one true value, there are usually multiple erroneous values in the collected data. It is common for two sources to provide the same correct value, however, if two sources often provide the same erroneous value, it is very likely that there is a dependency relationship between them. We use $s_i \sim s_j$ to denote that source s_i and s_j are dependent. If s_i copies from s_j, denoted by $s_i \rightarrow s_j$, s_i is regarded as a dependent source, and s_j is regarded as an independent source.

The dependency relationship between two sources s_i and s_j at time t is quantified by the dependency probability $p_{i,j}^t$ at time t, where $p_{i,j}^t$ ranges from 0 to 1. The closer the value of $p_{i,j}^t$ is to 1, the more likely there is a dependency between the two sources. If $p_{i,j}^t = 0$, source s_i and s_j are independent of each other. We use P^t to represent the set of dependency probabilities of all sources at time t, where $P^t = \{p_{i,j}^t | i, j \in S, i \neq j, p_{i,j}^t \geq 0\}$.

2.2 Task Description

During a period that has T timestamps, streaming data arrive in chronological order. For each time point $t \in T$, given a set of objects of interest (i.e., O), conflicting information is collected from a set of correlated sources (i.e., S). Our task is to find a trustworthy value $\hat{c}_{o,t}^*$ for each object o, making it as close as possible to its ground truth $c_{o,t}^*$ while capturing source dependencies.

3 Preliminary

In this section, we first introduce the Bayesian model for source dependency detection in static scenarios. We will generalize the problem to dynamic scenarios in Sect. 4. Then, we provide preliminary knowledge regarding optimization-based truth discovery on dynamic data.

3.1 Source Dependency Detection Based on Bayesian Model

We use the Bayes model to calculate the dependency probability $p_{i,j}$ between two sources (s_i and s_j) to determine the dependency relationship between them. To eliminate dependent sources, the direction of the dependency needs to be determined. Generally, sources with lower reliability are more likely to copy information from sources with higher reliability.

Taking sources s_i and s_j as an example, we classify the objects for which both sources provide claims into three categories: (i) objects with the same correct claim provided by s_i and s_j, denoted as O_c; (ii) objects with the same erroneous claim provided by s_i and s_j, denoted as O_e; and (iii) objects with different claims provided by s_i and s_j, denoted as O_d. In addition, ψ denotes the observed outcomes among the three sets.

In the case of two independent sources, denoted by $s_i \perp s_j$, the source reliability r_s represents the probability that the independent source provides the correct claim. Considering that there is only one true value for the object, the probability that sources s_i and s_j provide the same correct claim is

$$P_c = P(o \in O_c | s_i \perp s_j) = r_i \cdot r_j. \tag{1}$$

Assuming that the distribution of erroneous claims follows a uniform distribution, n_e represents the number of erroneous claims for an object. Therefore, for independent sources, each erroneous claim has an equal probability of occurrence, i.e., $\frac{1-r_j}{n_e}$. The probability that source s_i and s_j provide the same erroneous claim is given by:

$$P_e = P(o \in O_e | s_i \perp s_j) = \frac{(1 - r_i) \cdot (1 - r_j)}{n_e}. \tag{2}$$

Therefore, the probability that source s_i and s_j provide different claims is:

$$P_d = P(o \in O_d | s_i \perp s_j) = 1 - P_c - P_e. \tag{3}$$

According to the assumption of independence, the observed conditional probability is:

$$P(\psi | s_i \perp s_j) = \prod_{o \in O_c} P_c \cdot \prod_{o \in O_e} P_e \cdot \prod_{o \in O_d} P_d. \tag{4}$$

Consider the case where there is a dependency between two sources, taking $s_i \rightarrow s_j$ as an example. Let c represent the probability that s_i copies information from s_j. When sources s_i and s_j provide the same claim, there are two possibilities: the first is that source s_i copies the claim from source s_j, in this case, the probability of source s_i providing the correct claim is r_j, and the probability of providing an erroneous claim is $(1 - r_j)$. The second is that source s_i independently provides the claim, in this case, the probability of providing the correct claim and the erroneous claim is the same as when the two sources are independent. Therefore, we have

$$P(o \in O_c | s_i \rightarrow s_j) = c \cdot r_j + (1 - c) \cdot P_c, \tag{5}$$

$$P(o \in O_e | s_i \rightarrow s_j) = c \cdot (1 - r_j) + (1 - c) \cdot P_e. \tag{6}$$

Finally, the probability that two sources provide different claims is:

$$P(o \in O_d | s_i \to s_j) = (1 - c) \cdot P_d. \tag{7}$$

Therefore, when sources are dependent, the observed conditional probability is:

$$P(\psi | s_i \to s_j) = \prod_{o \in O_c} [c \cdot r_j + (1 - c) \cdot P_c] \cdot$$
$$\prod_{o \in O_e} [c \cdot (1 - r_j) + (1 - c) \cdot P_e] \cdot \prod_{o \in O_d} [(1 - c) \cdot P_d]. \tag{8}$$

According to Bayes Model, we have

$$p_{i,j} = [1 + \frac{1 - \eta}{\eta} \cdot \prod_{o \in O_c} \frac{P_c}{c \cdot r_j + (1-c) \cdot P_c} \cdot \prod_{o \in O_e} \frac{P_e}{c \cdot (1-r_j) + (1-c) \cdot P_e} \cdot \prod_{o \in O_d} \frac{1}{1-c}]^{-1}. \tag{9}$$

where $\eta = P(s_i \to s_j)$ is the prior probability of dependency between source s_i and s_j.

3.2 Truth Discovery Framework Based on Optimization Model

Based on the principle of truth discovery, an optimization model can be used to infer the reliability of sources and the true values of objects. Therefore, we establish the following optimization model:

$$\min_{R, C^*} \sum_{t \in T} l_t$$
$$l_t = \mu \sum_{s \in S} r_s \sum_{o \in O} (c_{o,t}^s - \hat{c}_{o,t}^*)^2 \tag{10}$$
$$0 \le r_s \le 1$$

where μ is used as a balancing parameter in the loss function, and l_t measures the weighted distance between the claim provided by a source and the estimated truth.

In this optimization problem, there are two sets of variables: source reliability R and estimated truth C^*. We can use the coordinate descent method to solve this optimization problem, by fixing the value of one set of variables and solving for the other set of variables. When the source reliability R is fixed, we can infer the true value $\hat{c}_{o,t}^*$ of the object:

$$\hat{c}_{o,t}^* = \frac{\sum_{s \in S} r_s \cdot c_{o,t}^s}{\sum_{s \in S} r_s} \tag{11}$$

It can be seen from Eq. (11) that it is the result of weighted aggregation, where the claims provided by the sources with high reliability have a greater impact on the final result, which is in line with the basic principle of truth discovery.

However, when calculating the source reliability, it is necessary to revisit all data from previous timestamps, which undoubtedly increases the time cost and reduces efficiency. In order to improve efficiency, [8] has demonstrated the equivalence between the solution based on the optimization model and the maximum a posteriori estimation, and proposed the following method for calculating the source weights at time t:

$$r_s = \frac{2\alpha - 2 + \sum_{t \in T} |C_t^s|}{2\beta + \mu \sum_{t \in T} \sum_{o \in O} (\epsilon_{o,t}^s)^2} \tag{12}$$

where α and β are two parameters introduced in the maximum a posteriori estimation, μ is the weighting parameter in the loss function. It can be observed from this equation that the reliability of a source is inversely proportional to its error, meaning that the closer the claims provided by a source are to the estimated truth, the more reliable the source is.

The number of claims and the error of the source in Eq. (12) can be converted to the cumulative quantity of claims and cumulative error of the source. Let $a_0^s = 2\alpha - 2$ and $e_0^s = 2\beta$, so Eq. (12) can be transformed into the following form:

$$r_s = \frac{a_{t-1}^s + |C_t^s|}{e_{t-1}^s + \mu \sum_{o \in O} (\epsilon_{o,t}^s)^2} \tag{13}$$

According to Eq. (13), instead of revisiting data from previous timestamps, we only need to calculate the number of claims provided by the source and the error at time t.

4 Methodology

4.1 Source Dependency Detection in Dynamic Scenarios

In dynamic scenarios, the dependencies among sources may change. Dependent sources may keep copying information from the same sources, or different sources, or even gradually stop copying and become independent sources. On the other hand, independent sources may change their mind and start to collect information from other sources. So we need to calculate the dependency probability between sources at every timestamp. According to the method of source dependency detection based on the Bayesian model in static scenario introduced in Sect. 3.1, at time t, we can calculate P_c^t, P_e^t, and P_d^t according to Eqs. (1)–(3):

$$P_c^t = P(o \in O_c^t | s_i \perp s_j) = r_i \cdot r_j. \tag{14}$$

$$P_e^t = P(o \in O_e^t | s_i \perp s_j) = \frac{(1 - r_i) \cdot (1 - r_j)}{n}. \tag{15}$$

$$P_d^t = P(o \in O_d^t | s_i \perp s_j) = 1 - P_c^t - P_e^t. \tag{16}$$

According to Eq. (9), the dependency probabilities can be calculated based on the observations at time t:

$$\hat{p}_{i,j}^t = [1 + \frac{1-\eta}{\eta} \cdot \prod_{o \in O_c^t} \frac{P_c^t}{c \cdot r_j + (1-c) \cdot P_c^t} \cdot$$
$$\prod_{o \in O_e^t} \frac{P_e^t}{c \cdot (1-r_j) + (1-c) \cdot P_e^t} \cdot \prod_{o \in O_d^t} \frac{1}{1-c}]^{-1}. \tag{17}$$

However, since the distribution of data is uneven at different timestamps, the amount of data collected at different moments may be different. It is noted that in dynamic scenarios, source dependencies may exhibit large variations over the long term, but usually, they show stability in the short term, i.e., sources that have dependencies at time $(t-1)$ are more likely to have dependencies at time t. Therefore, based on the short-term stability, we introduce a smoothing factor γ that incorporates the dependency probability at time $(t-1)$ into the calculation of the dependency probability at time t.

$$p_{i,j}^t = \gamma \cdot p_{i,j}^{t-1} + (1-\gamma) \cdot \hat{p}_{i,j}^t. \tag{18}$$

After calculating the dependency probabilities at time t, if $p_{i,j}^t > \theta$, we remove the dependent sources. This may cause sparsity in the dataset. In Sect. 4.2, we propose a new dynamic incremental model framework to address this sparsity issue.

4.2   Dynamic Incremental Model Framework

To address the issue of data sparsity caused by removing dependent sources, which typically provide a wealth of information, we leverage the relative stability of time series data in practical applications. Specifically, we use the data from the two timestamps immediately preceding and following the current timestamp t as reference values when estimating the true value. Although we have the true value estimated for timestamp $(t-1)$, the value for timestamp $(t+1)$ is unknown, so we introduce a predicted value $\hat{c}_{o,t}^{prep}$. This approach helps maintain the effective sample size of the dataset.

In dynamic scenarios, data changes are continuous, meaning that the changes between the previous and current timestamps are relatively stable. As a result, it can be assumed that the value at the previous time point will have a certain impact on the value at the current time point. To estimate the predicted value at the current time point, we use an exponential smoothing model, as follows:

$$\hat{c}_{o,t+1}^{prep} = \rho \hat{c}_{o,t}^* + (1-\rho)\hat{c}_{o,t}^{prep} \tag{19}$$

where $\hat{c}_{o,t}^*$ is the true value estimated for object o at the previous timestamp, $\hat{c}_{o,t}^{prep}$ is the predicted value for object o at the previous timestamp, and ρ is the smoothing coefficient, which takes values between 0 and 1. The larger the value of ρ, the greater the weight of the new data, and the smaller the weight of the predicted value.

To solve the data sparsity problem, we introduce the estimated value at time $(t-1)$ and the predicted value at time $(t+1)$ as the values provided by two virtual sources into the model of Eq. (10) to obtain the new optimization model as follows:

$$\min_{R,C^*} \sum_{t \in T} l_t$$

$$l_t = \mu \sum_{s \in S} r_s \sum_{o \in O} (c_{o,t}^s - \hat{c}_{o,t}^*)^2 + \mu \lambda_1 \sum_{o \in O} (\hat{c}_{o,t-1}^* - \hat{c}_{o,t}^*)$$

$$+ \mu \lambda_2 \sum_{o \in O} (\hat{c}_{o,t+1}^{prep} - \hat{c}_{o,t}^*) \tag{20}$$

$$0 \leq r_s \leq 1$$

Substituting Eq. (19) into the loss function, the third term in l_t can be transformed as: $\mu \lambda_2 \sum_{o \in O} (\hat{c}_{o,t+1}^{prep} - \hat{c}_{o,t}^*) = \mu \lambda_2 \sum_{o \in O} (1-\rho)(\hat{c}_{o,t}^{prep} - \hat{c}_{o,t}^*) = \mu \lambda_2 (1-\rho) \sum_{o \in O} (\hat{c}_{o,t}^{prep} - \hat{c}_{o,t}^*)$.

Therefore, the loss function l_t can be expressed as:

$$l_t = \mu \sum_{s \in S} r_s \sum_{o \in O} (c_{o,t}^s - \hat{c}_{o,t}^*)^2 +$$

$$\mu \lambda_1 \sum_{o \in O} (\hat{c}_{o,t-1}^* - \hat{c}_{o,t}^*) + \tag{21}$$

$$\mu \lambda_2 (1-\rho) \sum_{o \in O} (\hat{c}_{o,t}^{prep} - \hat{c}_{o,t}^*)$$

where λ_1 and λ_2 can be regarded as the reliabilities of two virtual sources, the addition of these two terms do not affect the estimation of the source reliabilities, which is still calculated according to Eq. (13). However, they do have an impact on the estimation of the ground truth of the target. Therefore, we have:

$$\hat{c}_{o,t}^* = \frac{\sum_{s \in S} r_s c_{o,t}^s + \lambda_1 \hat{c}_{o,t-1}^* + \lambda_2 (1-\rho) \hat{c}_{o,t}^{prep}}{\sum_{s \in S} r_s + \lambda_1 + \lambda_2 (1-\rho)} \tag{22}$$

4.3   Truth Discovery with Source Dependency Perception

In this part, we propose the framework of truth discovery with source dependency perception(SDPTD) in dynamic scenarios. Algorithm 1 presents the pseudocode of our proposed method.

In Algorithm 1, we initialize the estimated and predicted values to the mean value at time $t = 1$. We set the cumulative number of claims and errors to $a_0^s = 2\alpha - 2$ and $e_0^s = 2\beta$. The source reliability is initialized to $1/|S|$, the dependency probability is set to 0.5. Lines 4–7 estimate the predicted and true values for each object at time t. Line 10 calculates the dependency probability between two sources. In lines 11–17, the dependent and independent sources are judged based on the dependency probability and source reliability, and the dependent

Algorithm 1. SDPTD

Input: C_t, θ, α, β, ρ, γ, λ_1, λ_2

Output: C_t^*

1: initialize $\hat{c}_{o,0}^{prep} = \hat{c}_{o,0}^* = \frac{1}{|S_o|} \sum_{s \in S_o} c_{o,1}^s$, $a_0^s = 2\alpha - 2$, $e_0^s = 2\beta$, $r_s = 1/|S|$, $P^0 = 0.5$;

2: **for** each time $t \in T$ **do**

3:     **while** the convergence condition is not satisfied **do**

4:         **for** each object $o \in O$ **do**

5:             calculate the predicted value $\hat{c}_{o,t}^{prep}$ according to Eq. (19);

6:             calculate the truth $\hat{c}_{o,t}^*$ according to Eq. (22);

7:         **end for**

8:         **for** each source $s_i \in S$ **do**

9:             **for** each source $s_j \in S$, and $i \neq j$ **do**

10:                 calculate the dependency probability $p_{i,j}^t$ according to Eq. (18);

11:                 **if** $p_{i,j}^t > \theta$ **then**

12:                     **if** $r_i < r_j$ **then**

13:                         remove s_i from S;

14:                     **else**

15:                         remove s_j from S;

16:                     **end if**

17:                 **end if**

18:             **end for**

19:         **end for**

20:         **for** each source $s \in S$ **do**

21:             calculate the cumulative error e_t^s and the cumulative number of claims a_t^s;

22:             the source reliability r_s according to Eq. (13);

23:         **end for**

24:     **end while**

25: **end for**

---

sources are removed. After removing the dependent sources, the cumulative error and the cumulative number of claims for the independent sources are calculated in lines 20–23, and the source reliability is estimated from them.

We use an incremental model, which conforms to the general principles of truth discovery in the process of estimating truth and source reliability. The method does not require access to data on past timestamps and scans the data only once on the current timestamp, which improves time efficiency and is applicable to dynamic scenarios, and we also demonstrate in our experiments that the source reliability eventually converges to the true source reliability.

5   Experiment

5.1   Experimental Setup

Datasets. In this part, we used two real-world datasets, namely the weather dataset and the stock dataset, and generated synthetic datasets in our experiments. Here we describe these datasets in detail.

- **Weather Dataset**: We used descriptions of weather conditions for 29 cities from 16 websites, containing the attributes of temperature, humidity, and visibility, for a period of one week (from January 28 to February 4, 2010), with a total of 56,643 records.
- **Stock Dataset**: The stock dataset contains a total of 925,583 records for 1,000 stocks provided by 48 sources between July 11 and July 29, 2011, and we selected a continuous type of attribute data, i.e., open prices.
- **Synthetic Dataset**: We generated different synthetic datasets based on different proportions of dependent sources. We simulated 20 sources providing claims for 100 objects with continuous values for a total of 40,000 records at 20 timestamps.

Baseline. We selected several classic truth discovery methods, including static truth discovery methods and dynamic truth discovery methods.

- **TruthFinder** [17] is the first method proposed to use iterative updates of source weights and truths, and calculates the probability of a claim being correct by computing the reliability of the sources.
- **CRH** [5] is a truth discovery method that handles heterogeneous data types, and also uses an optimization model to estimate truth and source reliabilities.
- **Investment** [11] calculates source weights based on a nonlinear function that aggregates the weighted sum of the trustworthiness of each claim.
- **3-Estimates** [3] introduces a set of parameters to capture the difficulty of each claim, in order to better distinguish trustworthy from untrustworthy claims.
- **DynaTD+ALL** [8] is an incremental model that introduces smoothing and decay factors to improve the estimate of the true values in dynamic scenarios.

We also computed the mean and median values for each object, which aggregated the data irrespective of the source reliability. As DynaTD is presently the most widely used method in dynamic scenarios, we selected only DynaTD+ALL as the baseline for comparison. We did not employ Dong's method [2] for comparison as it was proposed earlier and has not been extensively used in dynamic scenarios. Additionally, we devised a truth discovery method, SDPTD_withoutDep, that neglects source dependencies to establish the significance of source dependency detection in dynamic scenarios.

Metrics. For numerical data, we chose RMSE and MAE to evaluate the accuracy of the estimated values. RMSE calculates the root mean square value between the estimated values and the true values, and MAE calculates the mean of the absolute differences between the estimated values and the true values.

The smaller the values of RMSE and MAE, the more accurate the estimated values. In addition, we also compared the running time of different algorithms.

5.2 The Results on Real-World Datasets

Results Analysis: We report the experimental results of SDPTD and baseline methods on real-world datasets in Table 1. SDPTD outperforms the baseline methods in terms of RMSE and MAE because the method takes into account the uneven distribution of data over timestamps in dynamic scenarios and is also able to consider the dependencies between sources, thus improving the accuracy of source dependency detection and truth estimation. Among the baseline methods, Mean and Median perform the worst because they do not consider the reliability of sources. Investment and 3-Estimates are designed for categorical data and thus perform poorly on numerical datasets. TruthFinder and CRH process data in batches and ignore the temporal order of the data, resulting in suboptimal performance. DynaTD+ALL considers smoothing and decay factors in dynamic scenarios but neglects the source dependencies, making it less effective than our proposed method. We also demonstrate that methods that do not consider source dependencies perform worse than SDPTD.

In terms of time efficiency, DynaTD+ALL and our proposed method have similar run times for Mean and Median. However, our proposed method takes more time because it calculates the dynamic changes in source dependencies. Other truth discovery algorithms require iterations to reach convergence, which makes them very time-consuming.

Table 1. Comparison of different algorithms on real datasets: the best performance values are in bold.

Method	Weather Dataset			Stock Dataset		
	RMSE	MAE	Time	RMSE	MAE	Time
Mean	47.038	1.6221	1.4	0.582	0.2824	3.5
Median	41.3357	1.6774	**1.3**	0.512	0.2628	**2.9**
TruthFinder	1.4475	0.6774	2.0	0.019	**0.0024**	9.6
CRH	1.4371	0.6742	2.8	0.013	0.0033	9.8
Investment	2.1278	1.0051	9.9	0.011	0.0029	21.8
3-Estimate	2.7806	2.0524	9.8	0.026	0.0124	21.1
DynaTD+ALL	2.326	1.1191	1.7	0.012	0.0029	3.8
SDPTD	**1.3357**	**0.6733**	1.9	**0.010**	0.0026	4.1
SDPTD_withoutDep	1.3403	0.6782	1.7	0.014	0.0031	3.7

The Influence of Parameters λ_1 and λ_2: Considering the two virtual sources in our algorithm, the parameters λ_1 and λ_2 represent the reliability of these sources, which affect the truth inference. To study the impact of each parameter on the results, we fixed one parameter and varied the other. With λ_2 set to 0.4, the experimental results on two datasets were obtained by changing λ_1 from 0.4 to 0.8, as shown in Fig. 1. In the Weather dataset, the best results were obtained when λ_1 was around 0.6, while in the Stock dataset, the smallest error was achieved when λ_1 was around 0.75. The results from both datasets suggest that the values of the last timestamp can influence the current estimation.

Similarly, with λ_1 set to 0.6, Fig. 2 shows the experimental results on two datasets as λ_2 varies from 0.4 to 0.8. In both datasets, the best experimental results are obtained when λ_2 is around 0.45. This may be because there is some error in predicting the value of the next timestamp, and the reliability of the virtual source providing the predicted values is not very high.

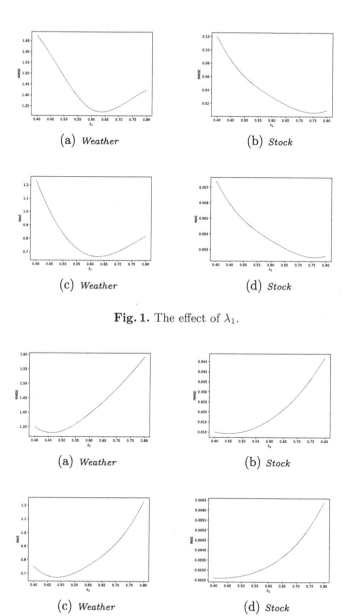

(a) *Weather*   (b) *Stock*

(c) *Weather*   (d) *Stock*

Fig. 1. The effect of λ_1.

(a) *Weather*   (b) *Stock*

(c) *Weather*   (d) *Stock*

Fig. 2. The effect of λ_2.

5.3   The Results on Synthetic Datasets

Datasets Generation: We simulated the claims of 20 sources with different reliability at 20 timestamps for the cases of 0, 10%, 20% and 30% of dependent sources, and generated 2000 claims for each source at each timestamp according to the true source reliability, so that the errors follow a normal distribution $N(0, 1/r_s^2)$.

Table 2. Comparison of different algorithms on synthetic datasets: the best performance values are in bold.

Method	0			10%			20%			30%		
	RMSE	MAE	Time	RMSE	MAE	Time	RMSE	MAE	Time	RMSE	MAE	Time
Mean	0.0971	0.0773	0.7	0.0780	0.0619	0.7	0.1030	0.0826	**0.7**	0.1047	0.0826	**0.5**
Median	0.0883	0.0691	0.7	0.0796	0.0687	**0.6**	0.0623	0.0605	**0.7**	0.0674	0.0757	0.7
TruthFinder	1.0461	0.9478	1.4	0.0647	0.0513	1.3	0.1519	0.1211	1.4	0.2156	0.1711	1.4
CRH	0.0517	0.0409	1.3	0.0649	0.0512	1.3	0.1519	0.1211	1.4	0.2157	0.1711	1.3
Investment	1.0461	0.9478	3.9	0.9127	0.7776	3.8	1.0799	0.9665	4.0	1.0333	0.8999	3.9
3-Estimate	1.0460	0.9477	3.8	0.6698	0.5207	3.8	0.7906	0.6054	3.9	0.3449	0.2443	3.7
DynaTD+ALL	0.1801	0.0328	**0.6**	0.1836	0.0508	**0.6**	0.1881	0.0543	0.8	0.1826	0.0586	0.7
SDPTD	**0.0282**	**0.0188**	0.7	**0.0495**	**0.0379**	0.7	**0.0621**	**0.0404**	0.8	**0.0671**	**0.0455**	0.7
SDPTD_withoutDep	0.0282	0.0189	0.7	0.0504	0.0383	0.7	0.0623	0.0406	**0.7**	0.0673	0.0457	0.6

Results Analysis: Table 2 presents the experimental results on the synthetic datasets. The data indicate that our proposed method performs well on datasets with varying proportions of dependent sources. This may be because our proposed method considers the problem of uneven data distribution in dynamic scenarios and is able to better identify the dependency sources. As the number of dependent sources increases, the performance of the proposed method decreases, as dependent sources replicate a significant amount of incorrect information, leading to substantial errors in the aggregated results.

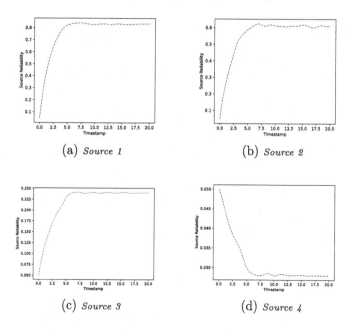

Fig. 3. The source reliability covergence.

Source Reliability Convergence Study: We chose four sources with different reliabilities from a dataset containing 30% dependent sources to investigate the convergence of source reliability. Figure 3 illustrates the variations in the reliability of the four sources at different timestamps. It is observed that the reliability of the sources tends to stabilize and converge to the true reliability from the 5th timestamp, demonstrating the efficacy of our algorithm.

6   Related Work

Early truth discovery algorithms mainly focus on static scenarios without considering the temporal dimension. Yin et al. [17] formally defined the problem of truth discovery and proposed the TruthFinder algorithm, which uses Bayesian analysis for iterative computation of source reliability and corresponding truth values. Since then, various truth discovery algorithms [9,13] have been continuously proposed for different scenarios, considering various influencing factors such as source reliability, sources relationships, object difficulty, and the number of true values. Traditional truth discovery methods perform well on static data. When dealing with data streams, it is time-consuming to conduct the truth discovery process at each timestamp. To improve time efficiency, several methods are designed for dynamic data. Li et al. [8] proposed an incremental truth discovery framework that dynamically updates true values of objects and source weights as new data arrives, introducing smooth and decay factors to reflect the time relationship. To balance accuracy and efficiency, Li et al. [6] designed an ASRA method that introduces adjustable parameters. When data arrives in the form of streams, Yao et al. [16] proposed an optimization framework that includes both truth discovery and time series analysis to improve accuracy. Yang et al. [15] incorporate object correlation into the truth discovery model, proposing a probabilistic graphical model. However, these methods ignore the stability of source dependencies in the short term, leading to errors in the judgment of dependent sources and inaccurate estimates of source reliability. The proposed SDPTD aims to identify the truth about objects that change over time while being aware of source dependencies that are affected by short-term stability.

In truth discovery, source dependencies are not a new research point of view. In static scenarios, Dong et al. [1] proposed in 2009 to use the Bayesian analysis to determine source dependencies assuming a single true value for a data object. Zhang et al. [18] proposed an unsupervised probability model that considers source correlation as a prior and applies to both numerical and categorical data. Wang et al. [12] analyzed the characteristics of multi-truth discovery and proposed an integrated Bayesian method MBM to solve the multi-truth problem by considering the different meanings of mutual exclusion between values and the interrelationship between sources. Jiang et al. [4] estimated task true values based on worker dependency and accuracy in the crowdsourcing field. Ma et al. [10] proposed IEMTD, an algorithm that infers the reliability of each individual, the dependence of observation results, and the truthfulness of events in social perception. In dynamic scenarios, Dong et al. [2] studied replication relationships and truth search with known update history, and built HMM models to determine source replication relationships. Existing studies of source dependencies ignore their stability in the short term and may lead to inaccurate detection of source dependencies.

7   Conclusion

We propose an incremental truth discovery framework called SDPTD, which considers the stability of source dependencies in the short term. To solve the data sparsity problem arising from the removal of dependent sources, we aggregate estimates from the previous timestamp and predictions from the next timestamp to fill in the missing claims in the incremental model. Experimental results on both real-world and synthetic datasets demonstrate that our proposed method outperforms the state-of-the-art truth discovery methods in terms of accuracy while maintaining time efficiency.

Acknowledgements. This work was supported by Fundamental Research Funds for the Central Universities (No. 23D111204, 22D111210), Shanghai Science and Technology Commission (No. 22YF1401100), and National Science Fund for Young Scholars (No. 62202095).

References

1. Dong, X.L., Berti-Equille, L., Srivastava, D.: Integrating conflicting data: the role of source dependence. VLDB **2**(1), 550–561 (2009)

2. Dong, X.L., Berti-Equille, L., Srivastava, D.: Truth discovery and copying detection in a dynamic world. VLDB **2**(1), 562–573 (2009)
3. Galland, A., Abiteboul, S., Marian, A., Senellart, P.: Corroborating information from disagreeing views. In: WSDM, pp. 131–140 (2010)
4. Jiang, L., Niu, X., Xu, J., Yang, D., Xu, L.: Incentivizing the workers for truth discovery in crowdsourcing with copiers. In: ICDCS, pp. 1286–1295. IEEE (2019)
5. Li, Q., Li, Y., Gao, J., Zhao, B., Fan, W., Han, J.: Resolving conflicts in heterogeneous data by truth discovery and source reliability estimation. In: SIGMOD, pp. 1187–1198 (2014)
6. Li, T., Gu, Y., Zhou, X., Ma, Q., Yu, G.: An effective and efficient truth discovery framework over data streams. In: EDBT, pp. 180–191. Springer (2017)
7. Li, Y., et al.: A survey on truth discovery. In: SIGKDD, vol. 17, no. 2, pp. 1–16 (2016)
8. Li, Y., et al.: On the discovery of evolving truth. In: SIGKDD, pp. 675–684 (2015)
9. Li, Y., Sun, H., Wang, W.H.: Towards fair truth discovery from biased crowd-sourced answers. In: SIGKDD, pp. 599–607 (2020)
10. Ma, L., Tay, W.P., Xiao, G.: Iterative expectation maximization for reliable social sensing with information flows. Inf. Sci. **501**, 621–634 (2019)
11. Pasternack, J., Roth, D.: Knowing what to believe (when you already know something). In: Coling 2010, pp. 877–885 (2010)
12. Wang, X., Sheng, Q.Z., Fang, X.S., Yao, L., Xu, X., Li, X.: An integrated Bayesian approach for effective multi-truth discovery. In: CIKM, pp. 493–502 (2015)
13. Wang, Y., Wang, K., Miao, C.: Truth discovery against strategic Sybil attack in crowdsourcing. In: SIGKDD, pp. 95–104 (2020)
14. Yang, J., Tay, W.P.: An unsupervised Bayesian neural network for truth discovery in social networks. TKDE **34**(11), 5182–5195 (2021)
15. Yang, Y., Bai, Q., Liu, Q.: A probabilistic model for truth discovery with object correlations. KBS **165**, 360–373 (2019)
16. Yao, L., et al.: Online truth discovery on time series data. In: SDM, pp. 162–170. SIAM (2018)
17. Yin, X., Han, J., Yu, P.S.: Truth discovery with multiple conflicting information providers on the web. In: SIGKDD, pp. 1048–1052 (2007)
18. Zhang, H., et al.: Influence-aware truth discovery. In: CIKM, pp. 851–860 (2016)

Truth Discovery Against Disguised Attack Mechanism in Crowdsourcing

Xiu Fang, Yating Tang, Guohao Sun$^{(\boxtimes)}$, Chenling Shen, and Hao Chen

Donghua University, Shanghai 201600, China
{xiu.fang,ghsun}@dhu.edu.cn, {2212646,2212626,2222842}@mail.dhu.edu.cn

Abstract. Crowdsourcing is an effective paradigm for recruiting online workers to perform intelligent tasks that are difficult for computers to complete. More and more attacks bring challenges to crowdsourcing systems. Although the truth discovery method can defend against common attacks to a certain extent, the real scene is much more complex. Malicious workers can not only improve their reliability by agreeing with normal workers on tasks that are unlikely to be overturned, but also gather together to launch more effective attacks on tasks that are easily overturned. This disguised attack is smarter and harder to defend. To solve this problem, we propose a new defense framework TD-DA (**T**ruth **D**iscovery against **D**isguised **A**ttack) composed of truth discovery and task allocation. In the truth discovery phase, we quantify the aggressiveness and reliability of workers on the golden task based on the sigmoid function. In the task allocation phase, the Weighted Arithmetic Mean (WAM) is used to estimate the allocation probability of golden tasks to avoid the shortage of golden tasks. Extensive experiments on real-world datasets and synthetic datasets demonstrate that our method is effective against disguised attacks.

Keywords: Crowdsourcing · Truth Discovery · Task Allocation · Disguised Attack Defense

1 Introduction

Crowdsourcing is a technique that utilizes Internet platforms to bring people together and solve problems that cannot be solved by computers alone. According to the intuition that multiple heads are better than one, the platforms always employ more than one worker to answer each task, and aggregate all the answers to obtain the crowd wisdom and avoid biases. Due to the openness of these platforms and varying quality of workers, conflicts are widely observed in the answers. The process which aims at resolving the conflicts is referred to as truth discovery [15,16], which has been widely studied in data mining area. Early research assumes that workers are truthful and independent, rendering these methods ineffective when dealing with malicious workers. Malicious workers can intentionally submit incorrect answers, leading to the manipulation of the final aggregated result, especially when they collude with each other. Therefore, several defense mechanisms have been developed to prevent such attacks.

© The Author(s), under exclusive license to Springer Nature Singapore Pte Ltd. 2024
X. Song et al. (Eds.): APWeb-WAIM 2023, LNCS 14333, pp. 64–79, 2024.
https://doi.org/10.1007/978-981-97-2387-4_5

The copy detection approach [8] considers the dependencies among workers and can effectively defend against blind copy attacks, where all malicious workers behave uniformly. [18] proposes a cluster-based Sybil defense framework (SADU), which identifies Sybil workers by analyzing their comparable behaviors without prior knowledge of the attack. These defense measures can enhance the resistance of crowdsourcing systems against simple attacks.

With the evolution of attack methods, disguised attacks [11] has emerged as a prevalent means of attack, with attackers becoming more sophisticated. When the chance of a successful attack is low, attackers may collaborate with regular workers to increase their weight and blend in with the crowd. However, when the probability of a successful attack is high, attackers would unite all malicious workers to vote for the second-highest answer, maximizing their success rates. In such scenarios, the existing defense measures are no longer effective in detecting attackers, substantially undermining the accuracy of answer aggregation.

We propose a new defense framework, TD-DA (**T**ruth **D**iscovery against **D**isguise **A**ttack), to tackle disguised attacks. TD-DA consists of two parts: truth discovery and task assignment. In the truth discovery phase, we analyze worker behaviors on golden tasks using a sigmoid function to quantify their aggressiveness and reliability. We then classify workers as malicious, reliable, or undecided, based on the known proportion of malicious workers. In the task assignment phase, we assign tasks to workers with different identities and use the Weighted Arithmetic Mean (WAM) to estimate the assignment probability of golden tasks, mitigating the problem of golden task shortage. TD-DA aims to enhance the accuracy and reliability of worker task assignments, making crowdsourcing systems more robust and secure.

In summary, the main contributions of this paper are as follows:

(i) We propose a novel defense framework called TD-DA to detect and suppress disguised attacks in crowdsourcing platforms. This framework uses batch mode truth discovery and task assignment. To the best of our knowledge, our approach is the first to address the issue of disguised attacks in crowdsourcing platforms.

(ii) We propose a behavior-based truth discovery algorithm that analyses workers' behaviors on golden tasks, using a sigmoid function to quantify their aggressiveness and reliability. We also propose a differentiated task assignment method based on Weighted Arithmetic Mean (WAM), which assigns different tasks according to the status of the workers to avoid a shortage of golden tasks.

(iii) Through a comparison with baseline methods on two real-world datasets and one synthetic dataset, we demonstrate the superior accuracy and efficiency of our method.

2   Related Work

Data Attack: Many crowdsourcing systems use truth discovery to detect and tolerate malicious worker behavior, but as attackers use more sophisticated methods, its effectiveness is challenged. One common attack method is Sybil attack

[2], where attackers create multiple fake identities (known as Sybil nodes) to impersonate multiple workers and influence the aggregation results to increase their profits. Availability attacks [14] involves attackers sending large amounts of meaningless requests to occupy system resources, reducing system performance and availability, affecting worker efficiency and normal system operation. Target attacks [4] involves attackers deliberately providing incorrect annotations to influence the aggregation results and data quality of the system, disrupting normal system operation and affecting crowdsourcing task results. Data attacks [5] involves attackers providing false or inaccurate data to interfere with the normal operation of the system and affect the aggregation results, such as providing incorrect annotation results or maliciously modifying data.

Defense Mechanism: To resist a variety of data attacks, effectively assess the quality of workers, and improve the accuracy of label aggregation, people also propose several defense methods. Copy detection [19] explores the interdependence between source dependencies and truth values to defend against blind copy-type attacks. The Dawid-Skene model [6] jointly estimates the true label of the project and the reliability of each worker, which can resist naive attackers to a certain extent. [18] is the first work to solve Sybil attacks in a general crowdsourcing environment. By adopting a clustering-based method called SADU. However, disguised attacks [11] have emerged as a new threat where malicious workers coordinate with normal workers to disguise themselves and gain higher weights. Existing defense mechanisms rely on ground truth discovery [9], copy detection [19], and clustering methods [18], but we are the first to propose a defense framework against disguised attacks.

3   Preliminary

3.1   Disguised Attack Mechanism

In a crowdsourcing system that includes a truth discovery module, malicious workers can strategically disguise their attacks to maximize their effectiveness and avoid being detected by the system with lower weights. In a disguised attack system, the attacker organizes a group of malicious workers to carry out the attack. Through the disguised attack mechanism, an attacker can access all normal workers' responses to each task without being detected. The attacker knows the number of malicious workers in advance and keeps in touch with them. To perform the attack, the attacker sorts the options in reverse order according to the number of normal workers who chose each option. The attacker then calculates the sum of the number of malicious workers and the number of votes for the second-ranked option, and compares it to the number of votes for the first-ranked option. If the former is larger, the attacker instructs the malicious workers to jointly choose the second-ranked option to launch the attack. Otherwise, malicious workers pretend to be normal workers to gain higher weights. Specifically, the distribution of option votes of malicious workers will be the same as that of normal workers, the malicious workers pose as normal workers to obtain higher weights.

Example: Given 110 workers answer a task with four options. Initially, the distribution of option votes of normal workers is (A,B,C,D) = (10,0,49,41). If we apply majority voting, C should be estimated as the truth. Now, an attacker who controls 10 malicious workers intend to attack the platform. If the attacker ask all the malicious workers to choose option D, the distribution of answers would become (A,B,C,D) = (10,0,49,51). As a result, the attack would be successful with the estimated truth being turned into D. However, if the initial answer distribution is (A,B,C,D) = (10,0,80,10), the results will not be changed even if all malicious workers choose option D. In this case, the attacker would rather ask the malicious workers to disguise their intentions. Therefore, the answer distribution of malicious workers would be the same as that of normal workers, i.e., (A,B,C,D) = (1,0,8,1). Thus, the final answer distribution is (11,0,88,11), and option C remains the most frequently chosen option.

3.2   Problem Formulation

In this paper, we consider the single-option problem. The set of workers on crowdsourcing platforms is denoted by W, where $W' \subset W$ represents the set of malicious workers, and $|W|$ and $|W'|$ respectively represent the total number of workers and the number of malicious workers. Therefore, we can define the overall proportion of malicious workers as $\frac{|W'|}{|W|}$. The set of tasks is denoted by T, where $\widetilde{T} \subset T$ represents the set of golden tasks with known ground truth, and $|T|$ and $|\widetilde{T}|$ respectively represent the total number of tasks and the number of golden tasks.

Each task has L options, only one of which is the correct answer. We rank the options in a descending order according to the number of workers who select them, with l^i denotes the i-th option and $|l^i|$ denotes the number of workers who choose l^i. Each task is assigned to K workers, and each worker is required to provide a label from L options and return it to the crowdsourcing platform. We denote the answers provided by worker $w(w \in W)$ on task $t(t \in T)$ as a vector x_t^w of length L. The elements of the vector correspond to each option of the task, and conform to the order of the options. Each element takes a value of 0 or 1, 1 means that the worker has selected this option, and 0 means that the worker has not selected this option. Since each worker can only choose one option for each task, if the worker chooses the i-th option, the i-th element in the vector will be 1 and all other elements will be set to 0. The set of answers collected from all $|T|$ tasks of all $|W|$ workers is denoted by $X = \{x_t^w\}_{w,t=1}^{|W|,|T|}$.

We detect malicious workers by examining their responses to golden tasks, where $\widehat{x}_t$ denotes the ground truth of golden tasks. Worker aggressiveness and reliability are denoted by a_w and r_w, respectively. Initially, all workers are in the "undetermined" state, and both a_w and r_w are set to 0. The reliability threshold is represented by γ, and a worker is classified as "reliable" when their $r_w > \gamma$. Similarly, the aggressiveness threshold is represented by ζ, and a worker is classified as "malicious" when their $a_w > \zeta$.

By performing the truth discovery process, the weight q_w of each worker and the aggregate truth x_t^* of each task are calculated. $X^* = \{x_t^*\}_{t=1}^{|T|}$ is used to represent the set of aggregate true values of all tasks, $Q = \{q_w\}_{w=1}^{|Q|}$ is used to represent the set of weights of all workers. $T_w \subseteq T$ is the set of tasks answered by the worker w, and $W_t \subseteq W$ is the set of workers answering the task t.

It is important to note that the worker reliability r_w and worker weight q_w serve different purposes: the reliability score reflects the trust gained by worker through answering golden tasks, while the worker weight estimates the accuracy of the worker based on the aggregated labels.

3.3   Truth Discovery

The truth discovery algorithm assigns higher weights to workers who consistently provide accurate answers, and selects answers from these high-weight workers as more likely to be true. This paper concentrates on the truth discovery problem for categorical data and implements the CRH method [10] to tackle it. The following are the specific steps:

Optimization Objective Function: CRH uses an optimization model to estimate the aggregation value and worker weights. The objective is to minimize the weighted deviation between observed results and the true values while ensuring the rationality of worker weights. To transform worker weights into positive values, CRH adopts an exponential mechanism. The sum of the exponents of all worker weights is required to be equal to 1:

$$\min_{X^*,Q} f(X^*, Q) = \sum_{w \in W} q_w \sum_{t \in T_w} d(x_t^w, x_t^*)$$
$$\text{s.t.} \sum_{w \in W} \exp(-q_w) = 1, \tag{1}$$

where $d(\cdot)$ is a loss function used to measure the distance between the answer x_t^w and the estimated truth x_t^*. It is defined as follows:

$$d(x_t^w, x_t^*) = (x_t^w - x_t^*)^T (x_t^w - x_t^*) = \sum_{l=1}^{L} \left(x_t^{w(l)} - x_t^{*(l)}\right)^2, \tag{2}$$

where $x_t^{w(l)}$ and $x_t^{*(l)}$ represent the l-th element in the corresponding label vector x_t^w and vector x_t^*, respectively. To solve the optimization problem, we conduct the following two steps iteratively until meeting the convergence condition.

Step1. Truth inference: In this step, the worker's weight q_w is fixed, and the aggregate truth value x_t^* of task t is updated according to:

$$x_t^* = \frac{\sum_{w \in W_t} q_w \cdot x_t^w}{\sum_{w \in W_t} q_w}. \tag{3}$$

Step2. Worker weight estimation: In this step, the aggregated truth value of the task x_t^* is fixed, and the worker's weight q_w is updated according to:

$$q_w = \log\left(\frac{\sum_{w \in W} \sum_{t \in T_w} d(x_t^w, x_t^*)}{\sum_{t \in T_w} d(x_t^w, x_t^*)}\right). \tag{4}$$

It can be seen from Eq. (4) that the smaller the distance d, the closer the worker's answer is to the true value, and the greater the weight of the worker.

4 Methodology

4.1 Behavior-Based Truth Discovery

Determine Whether the Task is Attacked: At the very beginning when the crowdsourcing system first launches, worker identities are initially marked as "undetermined". When workers send requests to the system, each of them would be randomly assigned to a golden task. These golden tasks are divided into two categories: correctly answered and incorrectly answered. Depending on the category, worker behavior is classified as either a successful attack or disguised action. For correctly answered golden tasks, we acknowledge the contribution of the malicious worker whether they choose to attack or perform a disguise because the result is correct. For incorrectly answered golden tasks, we considered the attack valid only if the malicious worker chose to attack. Therefore, our goal is to identify those golden tasks that were answered incorrectly and attacked successfully.

Given a task if the difference between the number of respondents for the highest and second-highest options after an attack is greater than the number of malicious workers in the task, i.e., $|l^1| - |l^2| > \frac{|W'|}{|W|} \times K$, the task is considered unattacked. Otherwise, the task is judged to be attacked. Still use the previous example, in the first case, the answer distribution was (A, B, C, D) = (10, 0, 49, 51), $|l^1| = 51$, $|l^2| = 49$, $|W'|=10$, $|W|=110$, $K=110$, so we have $|l^1| - |l^2| < \frac{|W'|}{|W|} \times K$, this task would be judged to be attacked. In the second case, the answer distribution was (A, B, C, D) = (11, 0, 88, 11), $|l^1| - |l^2| = 88 - 11 = 77$, $\frac{|W'|}{|W|} \times K = \frac{10}{110} \times 110 = 10$, 77>10, so this task will be unattacked.

Additionally, we considered that the proportion of malicious workers on each task follows the overall proportion of malicious workers, which maximizes the efficiency of disguised attacks. Experimental results on two real-world datasets in Sect. 5.2 show that the proportion of malicious workers on each task follows a normal distribution relative to the overall proportion of malicious workers.

Worker Aggressiveness Estimation: Capturing the malicious behavior of worker w on golden tasks who are both answered incorrectly and judged to have successfully attacked, defined as:

$$a_w = \frac{2}{1+e^{-\sum_{t \in T_w}(1-\mathbb{I}(x_t^w, \hat{x}_t)) \cdot b_t}} - 1. \tag{5}$$

This formula uses the sigmoid function to convert the worker's aggressiveness into a value between 0 and 1. where $\hat{x}_t$ is the ground truth of the golden task t, and $\mathbb{I}(\cdot)$ is the indicator function that takes the value 1 when the worker answers the golden task correctly, and 0 otherwise. The variable b_i is computed using the following formula to determine whether the golden task is attacked:

$$b_t = \begin{cases} 1 & |l_t^1| - |l_t^2| \le \frac{|W'|}{|W|} \times K \\ 0 & |l_t^1| - |l_t^2| > \frac{|W'|}{|W|} \times K, \end{cases} \tag{6}$$

Only when worker w gives an incorrect label on golden task t and the golden task is attacked, $(1 - \mathbb{I}(x_t^w, \widehat{x}_t)) \cdot b_t = 1$, which is considered evidence that the worker has conducted a disguised attack. The evidence is accumulated by summing up $(1 - \mathbb{I}(x_t^w, \widehat{x}_t)) \cdot b_t$ for all golden tasks assigned to the worker. The more evidence there is, the more suspicious the worker is, and the output value of the sigmoid function is closer to 1.

Worker Reliability Estimation: Based on the number and accuracy of golden tasks answered by worker w, we measure our level of trust in them, defined as:

$$r_w = \left(\frac{2}{1 + e^{-|\widetilde{T_w}|}} - 1 \right) \cdot p_w, \tag{7}$$

where the first term of the equation considers the number of golden tasks answered by worker w, and p_w represents their accuracy on golden tasks, which is calculated by:

$$p_w = \begin{cases} \frac{1}{L} & |\widetilde{T_w}| = 0 \\ \frac{\sum_{t \in \widetilde{T_w}} \mathbb{I}(x_t^w, \widehat{x}_t)}{|\widetilde{T_w}|} & otherwise, \end{cases} \tag{8}$$

The sigmoid function is used to constrain the value range of r_w between 0 and 1. r_w will only be relatively high if worker w answers multiple golden tasks with high accuracy.

Behavior-Based Truth Discovery(BTD): Integrates worker aggressiveness and reliability estimation into the truth discovery process.

Truth Inference: Revise Eq. (3) and aggregate truth values x_t^* for each task:

$$x_t^* = \frac{\sum_{w \in W_t} (1 - a_w) \cdot q_w \cdot x_t^w}{\sum_{w \in W_t} (1 - a_w) \cdot q_w}. \tag{9}$$

This formula regulates the worker weights by introducing their aggressiveness. Each worker has a $1 - a_w$ probability of being an independent worker. q_w is initialized with p_w, If a worker has never answered a golden task, their weight will be a random guess with a probability of $1/L$. If a worker has answered a golden task, their weight will be determined by a standard aggregation formula on the golden tasks. Specifically, Eq. (3) is a special case of Eq. (9).

Worker weight estimation: Optimize Eq. (4) and update weight q_w for each worker:

$$q_w = \log \left(\frac{\sum_{w \in W} \sum_{t \in T_w} c_t \cdot d(x_t^w, x_t^*)}{\sum_{t \in T_w} d(x_t^w, x_t^*)} \right), \tag{10}$$

where c_t is the average reliability of the worker set W_t who answered task t, computed by

$$c_t = \frac{\sum_{w \in W_t} r_w}{|W_t|}. \tag{11}$$

This formula measures the resistance of each task by using the average reliability of workers who participated in task t as c_t. When there are no malicious workers and all workers have a reliability score of 1, Eq. (10) simplifies to Eq. (4).

The pseudo-code of the proposed BTD method is shown in Algorithm 4.1.

Algorithm 1. Behavior-based Truth Discovery Algorithm (BTD)

Input: X,R,A,P,C
Output: X^*
1: Initialize each worker weight q_w to p_w;
2: **while** X^* is not converged **do**
3:    **for** each task $t \in T$ **do**
4:       Aggregate the truth value of task x_t^* using Eq.(9)
5:    **end for**
6:    **for** each worker $w \in W$ **do**
7:       Update the weight of worker q_w using Eq.(10)
8:    **end for**
9: **end while**
10: **return** X^*,Q

The termination criterion is set such that the algorithm stops if the change in aggregated labels between the latest two iterations is less than a predefined threshold, or if the maximum number of iterations is reached. The input includes the worker label set X, the worker's aggressiveness set A, the reliability set R, the accuracy set on golden tasks P, and the resistance set C against attacks on tasks. The output is the aggregated true value set X^*. When the convergence condition is not satisfied, the process of inferring true values and estimating worker weights iterates. The convergence condition is set to stop the algorithm if the change in aggregated labels between the two most recent iterations is less than a predefined threshold or if the maximum number of iterations has been reached.

4.2   Task Assignment Based on WAM

We evaluate the reliability and aggressiveness level of "undetermined" workers by utilizing golden tasks. However, this approach is subject to two main challenges. Firstly, during the initial phase, a large number of pending workers may result in a shortage of golden tasks. Secondly, frequent allocation of golden tasks may increase the likelihood of workers discovering the nature of these tasks, thus jeopardizing the integrity of the process. To overcome these challenges, we introduce a new task assignment method that's based on the Weighted Arithmetic Mean (WAM).

Given a worker w, we apply the weighted arithmetic mean method, where the worker aggressiveness a_w and reliability r_w serve as weights, to conduct task assignment. Specifically, the probability of assigning a golden task to worker w is calculated as follows:

$$g(\alpha) = \alpha \cdot (1 - r_w) + (1 - \alpha) \cdot a_w. \tag{12}$$

The coefficient α is a weight factor between 0 and 1, which ensures that any worker with relatively high aggressiveness and low reliability has a higher probability of being assigned to a golden task. By doing so, the golden task could help us to accurately classify the worker. When the crowdsourcing platform launches, a_w and r_w of worker w are initialized to 0. In this case, α serves as the probability of assigning a golden task to a new worker.

Algorithm 4.2 presents our task assignment based on WAM.

Algorithm 2. Differentiated Task Assignment Algorithm(DTA)

Input: w:A requesting worker
Output: X
1: **if** w is in the "undetermined" workers **then**
2:     The probability $g(\alpha)$ of assigning a golden task in $\widetilde{T}$ computed by Eq.(12);
3:     **if** w is not assigned any golden task **then**
4:         Assign an uncompleted normal task in T;
5:     **end if**
6: **else**
7:     Assign an uncompleted normal task in T;
8: **end if**

When a new worker is to be assigned a task, the DTA is entered. We first determine their identity state. Initially, all workers are "undetermined", and the probability of assigning a golden task to the worker is computed using Eq. (12) in Steps 1–2. If no golden task is assigned to the worker, an uncompleted normal task is assigned in Steps 3–4. If $r_w > \gamma$, the worker is considered "reliable" and is only assigned uncompleted normal tasks without being tested with a golden task again in Steps 6–7. If $a_w > \zeta$, the worker is considered "malicious" and no task is assigned to them.

4.3   TD-DA Framework

Algorithm 3. TD-DA Framework

Input: X: label set of workers
Output: X^*: aggregated truth of tasks
1: **while** not all tasks in T are completed **do**
2:      **while** batch condition B is not met **do**
3:          Worker $w_j \in W$ requests or labels a task.
4:          **if** w_j sends a request **then**
5:              Assign w_j a task by $DTA(w_j)$;
6:          **else if** w_j labels a golden task in $\widetilde{T}$ **then**
7:              Update parameters s_j, r_j, p_j(Eq.5-7) and parameter c_i(Eq.11);
8:              **if** s_j passes the aggressiveness threshold ζ **then**
9:                  Remove the label of w_j in T;
10:             **end if**
11:         **else**
12:             Add the label to X;
13:         **end if**
14:     **end while**
15:     Update X^* by BTD;
16: **end while**
17: **return** X^*;

The TD-DA framework is presented by Algorithm 4.3, which runs BTD and DTA in batches. TD-DA responds differently to each action of worker w. When w requests, a task will be assigned using DTA(w) in Steps 4–5. If w labels a golden task, the worker and task parameters will be updated in Steps 6–7, and the label of w on normal tasks in T will be removed once a_w passes the aggressiveness threshold in Steps 8–9. If w labels a normal task, the worker label matrix X will be updated. When the batch condition is met, the aggregated true value set X^* will be updated using BTD.

5   Experiments

5.1   Experiment Setting

Datasets: We used two real-world crowdsourcing datasets from AMT, named DOG and NLP, and a synthetic dataset named SYN.

DOG Dataset: In that dataset, the task was to recognize a breed for a given dog. The dataset contained 807 tasks and 109 workers. Each task had four options, and each task was assigned to 10 staff members to answer. The dataset contained a total of 8,070 records.

NLP Dataset: In that dataset, the task was to determine whether the sentiment of the tweet was positive or negative in a crowdsourcing survey. It contained

1,000 tweets (tasks), each answered by 20 workers. In total, 85 workers provided 20,000 records.

SYN Dataset: Both real-world datasets consist of a set of tuples (w,t,x_t^w), representing that the worker w provides the answer label x_t^w on the task t. Therefore, we also created a synthetic dataset SYN by generating (w,t,x_t^w), and set the probability of workers answering tasks correctly as 0.8. We varied several parameters in our experiments, including the number of tasks being set from 1,000–6,000 (default 3,000), the number of workers being set from 100–600 (default 300), the number of workers assigned to each task being set from 10–50 (default 10), and the number of options of each task being set from 2–10 (default 4).

In addition, we used the method in [11] to inject different proportions of malicious workers into three datasets and generated multiple attack datasets.

Baseline Methods: The current defense mechanisms primarily relied on truth discovery methods, and there was no dedicated defense mechanism against disguised attacks. As a result, we selected state-of-the-art truth discovery methods that were representative for comparison purposes.

Majority Voting: For each task, it outputted the option provided by the most workers as the estimated truth without iteration.

CRH [10]: The method optimized the weights of workers and estimated the true value of the problem through iterations, with the aim of minimizing the weighted deviation between the observed results from workers and the true value.

TruthFinder [17]: The authors proposed the source consistency hypothesis and implication and utilized Bayesian analysis to iteratively estimate source reliability and identify truth values.

2-Estimates [3]: The method discussed the single truth assumption, which assumed that each object had only one true value, and employed complementary voting.

Sums [7]: This method was employed in hyperlink network environments to extract information and filter search topics to identify the "authoritative" information source on a given topic.

Average-Log [12]: The approach involved creating a framework that enabled the consideration of new and important factors in trust decisions.

TD-DEP [1]: TD-DEP was an ACCU model used for copy detection, specifically in the context of strategic Sybil attacks. Sybil workers that consistently shared labels were identified as copying from the same source.

Guess-LCA [13]: The Guess-LCA method represented source reliability with a set of potential parameters, providing more information to end users and offering some resistance against common attackers.

Evaluation Metrics: We evaluated the effectiveness of our disguised attack defense method by comparing its accuracy with the baseline methods. Since the primary objective of crowdsourcing was to obtain accurate results, the algorithm's estimated output should be as close to the ground truth as possible to be considered effective.

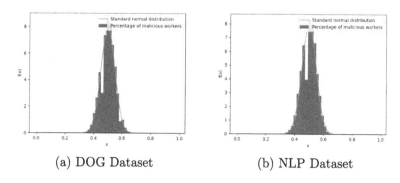

(a) DOG Dataset   (b) NLP Dataset

Fig. 1. The proportion of malicious workers follows normal distribution.

5.2   Verification of the Proportion of Malicious Workers

In this paper, we proposed that the proportion of malicious workers on each task followed a normal distribution, with a mean equal to the overall proportion of malicious workers. To verify this hypothesis, we assumed a 50% overall proportion of malicious workers and used the central limit theorem to accumulate errors in proportions of malicious workers across all tasks from two real-world datasets (DOG and NLP).

Our experimental results, illustrated in Fig. 1, were compared with the probability density function of the standard normal distribution, indicating a strong correlation between the histogram and the probability density function. Consequently, we could use the mean and standard deviation computed here to describe the overall proportion of malicious workers and the extent to which the proportion deviated from the overall proportion on each task.

5.3   Experiment on Real-World Datasets

Figure 2 displayed the accuracy of our approach on both the DOG and NLP datasets, under different parameter conditions. $(B, \alpha, \zeta, \gamma)$.

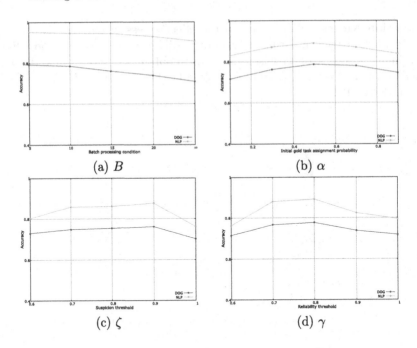

Fig. 2. Accuracy under different parameter conditions.

Batch Processing Condition B: In Fig. 2(a), a smaller batch processing condition led to slightly higher aggregation accuracy but incurred higher computational cost. Conversely, increasing the batch processing condition increased computational cost. An infinitely large batch processing condition with only one batch slightly decreased aggregation accuracy. We used 100 as the default batch processing condition for our experiments.

Initial Golden Task Assignment Probability α: In Fig. 2(b), a maximum uncertainty was achieved with an initial golden task assignment probability of 0.5, resulting in the highest aggregation accuracy when considering both the aggressiveness score and reliability score simultaneously.

Aggressiveness Threshold ζ and Reliability Threshold γ: Figure 2(c) and (d) demonstrated that the aggregation accuracy decreased when the aggressiveness score threshold was too high or the reliability score threshold was too low. Overly high aggressiveness score thresholds mislabeled normal workers as "malicious," while overly low reliability score thresholds mislabeled malicious workers as "reliable," both affecting the aggregation result. Our experiments showed that the best aggregation accuracy is achieved when the aggressiveness threshold is 0.9 and the reliability threshold is 0.8.

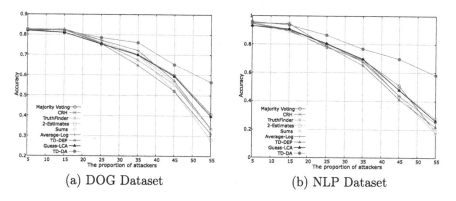

(a) DOG Dataset   (b) NLP Dataset

Fig. 3. Accuracy under different proportion of malicious workers.

Figure 3 presents a comparison of the TD-DA method and eight baseline methods on the DOG and NLP datasets at various proportions of malicious workers. As the proportion of malicious workers increased, the overall accuracy showed a decreasing trend. The accuracy of all baseline methods was slightly higher than the accuracy after the task was attacked, indicating their effectiveness in resisting masquerade attacks. In contrast, TD-DA consistently achieved the best accuracy, and its defense efficacy improved with an increase in the proportion of attackers. This was because our method was the only one that could judge and restrict the behavior of malicious workers.

5.4   Experiment on Synthetic Datasets

We studied the impact of task number(N), worker number(M), number of workers assigned to each task(K), and label size(L) on the performance of our TD-DA algorithm in the synthetic dataset. The proportion of malicious workers was adjusted during the experiments. We randomly selected a varying proportion of workers as malicious workers and ran the algorithm 50 times to calculate the average performance.

Task Number N: Figure 4(a) showed that increasing the number of tasks improved the detection of malicious workers, as it provided more data for aggregation and detection, thereby enhancing the accuracy of the algorithm.

Worker Number M: Figure 4(b) demonstrated that a higher number of workers enhanced defense effectiveness. With more workers, more data could be collected, making it harder for malicious workers to manipulate task outcomes. Moreover, increasing the number of workers could also enhance the diversity and stability of task results, further strengthening the defense against attacks.

Number of Workers Assigned to Each Task K: Figure 4(c) indicated that increasing the number of workers assigned to each task made capturing the suspicious scores of malicious workers easier and reduced their influence. This

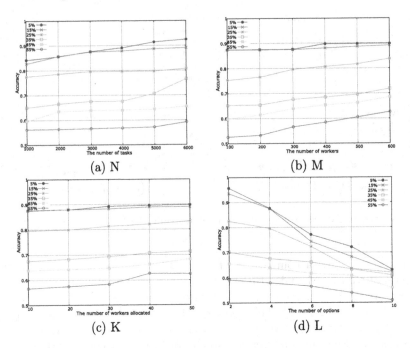

Fig. 4. Accuracy under different proportion of malicious workers.

was because a larger K brought the distribution of malicious workers on golden tasks closer to the overall distribution. Additionally, involving more workers in a task provided more data for analysis, which improved the accuracy of TD-DA.

Label Size L: Figure 4(d) illustrates that a larger label size increased sensitivity to attacks and affected the detection of malicious workers. On the other hand, a smaller label size mitigated the impact of malicious workers but also resulted in some normal workers being mistakenly labeled as "malicious." The default label size was set to 4, and experiments were conducted based on this.

6   Conclusion

This paper proposes a new defense framework called TD-DA to counter disguised attacks in crowdsourcing systems. TD-DA consists of two phases, i.e., truth discovery and task allocation. The proposed allocation algorithm DTA can improve the accuracy of task allocation, and the truth discovery BTD can quantify the reliability of workers and detect malicious behaviors. Experimental results show its superiority to existing methods. Future research can investigate TD-DA's effectiveness in more complex crowdsourcing scenarios involving numerical tasks, address the issue of golden tasks, and explore its combination with other defense mechanisms. Overall, TD-DA provides a promising solution for defending against disguised attacks and opens up new avenues for research in this area.

Acknowledgement. This work was supported by Fundamental Research Funds for the Central Universities (No. 23D111204, 22D111210), Shanghai Science and Technology Commission (No. 22YF1401100), and National Science Fund for Young Scholars (No. 62202095).

References

1. Dong, X.L., Berti-Equille, L., Srivastava, D.: Integrating conflicting data: the role of source dependence. VLDB **2**(1), 550–561 (2009)
2. Druschel, P., Kaashoek, F., Rowstron, A.: Peer-to-Peer Systems: First International Workshop, IPTPS 2002, Cambridge, MA, USA, 7–8 March 2002, Revised Papers, vol. 2429. Springer, Heidelberg (2003). https://doi.org/10.1007/3-540-45748-8
3. Galland, A., Abiteboul, S., Marian, A., Senellart, P.: Corroborating information from disagreeing views. In: WSDM, pp. 131–140 (2010)
4. Hooi, B., Song, H.A., Beutel, A., Shah, N., Shin, K., Faloutsos, C.: Fraudar: bounding graph fraud in the face of camouflage. In: SIGKDD, pp. 895–904 (2016)
5. Huberman, B.A., Romero, D.M., Wu, F.: Crowdsourcing, attention and productivity. JIS **35**(6), 758–765 (2009)
6. Kaghazgaran, P., Caverlee, J., Squicciarini, A.: Combating crowdsourced review manipulators: a neighborhood-based approach. In: WSDM, pp. 306–314 (2018)
7. Kleinberg, J.M.: Authoritative sources in a hyperlinked environment. JACM **46**(5), 604–632 (1999)
8. Li, X., Dong, X.L., Lyons, K.B., Meng, W., Srivastava, D.: Scaling up copy detection. In: ICDE, pp. 89–100. IEEE (2015)
9. Li, Y., et al.: A survey on truth discovery. SIGKDD **17**(2), 1–16 (2016)
10. Li, Y., et al.: Conflicts to harmony: a framework for resolving conflicts in heterogeneous data by truth discovery. TKDE **28**(8), 1986–1999 (2016)
11. Miao, C., Li, Q., Su, L., Huai, M., Jiang, W., Gao, J.: Attack under disguise: an intelligent data poisoning attack mechanism in crowdsourcing. In: Proceedings of the 2018 World Wide Web Conference, pp. 13–22 (2018)
12. Pasternack, J., Roth, D.: Making better informed trust decisions with generalized fact-finding. In: IJCAI (2011)
13. Pasternack, J., Roth, D.: Latent credibility analysis. In: Proceedings of the 22nd international conference on World Wide Web, pp. 1009–1020 (2013)
14. Spatscheck, O., Peterson, L.L.: Defending against denial of service attacks in scout. In: OSDI, vol. 99, pp. 59–72 (1999)
15. Xiao, H., Wang, S.: A joint maximum likelihood estimation framework for truth discovery: a unified perspective. IEEE Trans. Knowl. Data Eng. (2022)
16. Yang, J., Tay, W.P.: An unsupervised Bayesian neural network for truth discovery in social networks. IEEE Trans. Knowl. Data Eng. (2021)
17. Yin, X., Han, J., Yu, P.S.: Truth discovery with multiple conflicting information providers on the web. In: SIGKDD, pp. 1048–1052 (2007)
18. Yuan, D., Li, G., Li, Q., Zheng, Y.: Sybil defense in crowdsourcing platforms. In: CIKM, pp. 1529–1538 (2017)
19. Zhang, H., et al.: Influence-aware truth discovery. In: CIKM, pp. 851–860 (2016)

Continuous Group Nearest Group Search over Streaming Data

Rui Zhu$^{(\boxtimes)}$, Chunhong Li, Anzhen Zhang, Chuanyu Zong, and Xiufeng Xia

Shenyang Aerospace University, Shenyang, China
{zhurui,azzhang,zongcy,xiaxiufeng}@sau.edu.cn

Abstract. Group nearest group query(GNG for short) is an important variant of NN search. Let $\mathcal{D}$ be the $d-$multi-dimensional object set, $GQ\langle k, Q \rangle$ be a GNG with Q containing a set of d-multi-dimensional query points. The target of GNG is to select k object points O_Q from $\mathcal{D}$ such that the total distance between these query points and their NNs in O_Q is minimal. In this paper, we study GNG in a very dynamical data environment, i.e., continuous GNG query(CGNG for short) over sliding window, which has many applications. To the best of our knowledge, it is the first time to study the problem of CGNG over sliding window.

In this paper, we propose a novel framework named KMPT(short for K-Means Partition Tree-based framework) for supporting CGNG. The key behind KMPT is to partition query points into a group of k subsets, generate a group of k virtual points based on objects in these subsets, and reduce the CGNG problem to continuous NN search over data stream. In order to efficiently support continuous NN search, we first partition objects in the window into a group of sub-windows based on their arrived order. We then form a group of quad-tree based indexes to maintain objects' position information in each partition, form an R-tree based index to evaluate which objects have a chance to become query result objects in the near future, and finally achieve to goal of using a small number of objects to support query processing. The comprehensive experiments on both real and synthetic data sets demonstrate the superiority in both efficiency and quality.

Keywords: Streaming Data · Continuous Group Nearest Group Query · Dominance · Partition

1 Introduction

Nearest neighbor search(NN search for short) is a fundamental problem in the domain of spatio-temporal database [12]. It also has many variants. Among all types of variants, group nearest group query(GNG for short) is an important one [4,15]. Let $\mathcal{D}$ be the $d-$multi-dimensional object set, $GQ\langle k, Q \rangle$ be a query with Q being the d-multi-dimensional query point set. The target of GNG is to select k object points from $\mathcal{D}$, i.e., denoted as O_Q, such that the total distance between these query points and their NNs in O_Q is minimal. Here, NN refers to the smallest Euclidean distance sum between object and query points.

X. Song et al. (Eds.): APWeb-WAIM 2023, LNCS 14333, pp. 80–95, 2024.
https://doi.org/10.1007/978-981-97-2387-4_6

Take an example in Fig. 1(a). There are 12 objects and 4 query points. Since the parameter $k = 2$, we should return 2 objects to the system. Under the case we select $\{o_1, o_2\}$, the distance sum between $\{o_1, o_2\} \in \mathcal{D}$ to their corresponding NN query points equals to $(1 + 3) + (2 + 3) = 9$, which is minimal. As a result, the GNG $GQ\langle 2, \{q_1, q_2, q_3, q_4\}\rangle$ returns $\{o_1, o_2\}$ to the system.

In this paper, we study the problem of continuous GNG(CGNG for short) over sliding window. Formally, a sliding window W, denoted by the tuple $\langle n, s \rangle$, contains a set of n objects in the window [8]. Whenever the window slides, s objects arrive in the window, and another set of s objects expire from the window. A CGNG, denoted as the tuple $\langle n, s, Q, k \rangle$, monitors the window W, which returns k objects with the smallest distance sum to the system whenever the window slides. In other words, whenever the window slides, the query returns k objects $\{o_1, o_2, \cdots, o_k\}$, denoted as O_Q, from the query window such that the total distance sum from query objects to the corresponding NN object points is minimized.

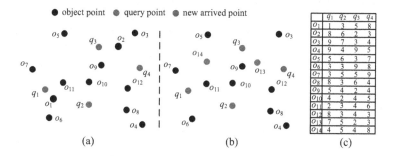

Fig. 1. The running example of CGNG with $k = 2, s = 2$

Back to the example in Fig. 1(a). A continuous GNG $\langle 12, 2, Q, 2 \rangle$ retrieves two objective points, i.e., $\{o_1, o_2\}$ with the smallest distance sum being 9. When the window slides, as shown in Fig. 1(b), objects o_1 and o_2 expire from the window, o_{13} and o_{14} flow into the window. At that moment, the query results are $\{o_{11}, o_{13}\}$ with the distance sum being $(2+3)+(2+3)=10$.

This type of query has many important applications [2,14]. In a take-out delivery system, when the platform tries to find riders for a group of merchants, it can find suitable riders from the rider set by issuing a GNG. The locations of their submissions could be regarded as query points. In addition, "online-riders", denoted as $r\langle b, d, loc \rangle$, have to be awaited for being assigned. Here, b refers to the moment r begins to await for assignment, and d refers to the deadline that r awaits for assignment. Accordingly, riders could be regarded as streaming data in the window. Via using CGNG, the system could assign suitable riders for delivery.

The state-of-the-art efforts mainly focus on GNG over static data, they cannot efficiently work under streaming data environment. As discussed in Sect. 2, these efforts do not consider how to efficiently maintain newly arrived/expired objects so as to support query processing. In addition, this problem is NP-hard. Although a few approximate algorithms are proposed, the running cost of these efforts is still high, and cannot return high-quality results in real time. Above all, an approximate algorithm that can effectively work under data stream in the premise of returning high-quality results is desired.

In this paper, we propose a novel framework named KMPT to support CGNG over sliding window. We first partition query points into a group of partitions, and then form a group of virtual points based on the partitioning result. In this way, we monitor these virtual points' NN, and use monitoring results to support query processing. In other words, we reduce the costly CGNG problem to the problem of continuous NN search over sliding window. Obviously, the running cost could be reduced a lot. Above all, the contributions of this paper are as follows.

- (i) We propose a novel query named continuous group nearest group over sliding window. To the best of our knowledge, this is the first time to study the problem of CGNG over sliding window.
- (ii) We partition objects in the window into a group sub-windows based on objects' arrived order. In each sub-window, we use a quad-tree based index with bounded height to organize streaming data. Via using the partition, we only need to incrementally maintain quad-trees corresponding to the first and last partition. Accordingly, we could use as low as running costs to maintain streaming data.
- (iii) We further propose a group of algorithms to reduce the cost of monitoring virtual points' NN. For one thing, we can use the partition result to evaluate the earliest moment each object has chance to become a virtual point's NN, finally achieve the goal of using a small number of objects to support query processing. In addition, the partition result further reduces the NN searches to the problem of range queries in many cases with the search radius being very small values.

The rest of this paper is as follows. Section 2 reviews the related work and proposes the problem definition. Section 3 explains the framework KMPT. Section 4 evaluates the performance of KMPT. Section 5 concludes this paper.

2   Preliminary

In this section, we first review some important existing results about various types of queries over d-dimensional objects including k-nearest neighbor query and the group nearest neighbor queries. Then, we introduce the problem definition.

2.1   Related Works

k-**Nearest Neighbor Query** is a fundamental problem in databases, data mining, and information retrieval research [1, 10]. Let $\mathcal{D}$ be the set of d-dimensional objects, and q be a kNN query. q searches on $\mathcal{D}$, returns k objects with minimal distances to q to the system.

Among all efforts, Lee et al. [6] propose the algorithm $PKNN_{gird}$. Its core idea is to find the k-nearest neighbors of each object in advance. It uses a unique property to support query processing, that is, two objects may share many neighbors if they are spatially close to each other. In this way, when a kNN search is submitted, it can first find NN of q, i.e., which is o. Then, the algorithm can find other query results based on o and its KNN. Miao et al. [9] propose a novel index named LαB. It uses the grid data structure to partition objects and can support KNN queries over incomplete data. Then, the algorithm named LP is proposed. It uses two pruning parameters, i.e., α and distance value, to eliminate the objects that do not satisfy the query conditions, which greatly improves the query efficiency.

Group Nearest Group Query was proposed by Papadias et al. [11]. Their idea is to find the minimum bounding rectangle of Q to prune the search space. Here, Q refers to the one point in space with a small value of distance sum to all query points. Xu at el. [13] proposed an algorithm based on *hill climbing scheme*, called ADM, which is proposed to support continuous group nearest group query. As it is an NP-hard problem, ADM only searches results in candidate sets that could make positive contributions by local distance constraints. Moreover, ADM utilizes the initial stored information to avoid computing from scratch at every update timestamp. Deng et al. [3] proposed the algorithm named SHR, which uses the hierarchical blocks of data points at a high level. Then, it finds an intermediate solution and refines the guided search direction at a low level so as to prune irrelevant subsets. It performed well in query efficiency and quality.

2.2   Problem Definition

In this section, we first introduce the concept of sliding windows. Generally, A sliding window, denoted by the tuple $\langle n, s \rangle$, refers to a set of objects contained in the window with n being the number of objects contained in the window, and s being the number of objects that flow into(or expire from) the window when the window slides.

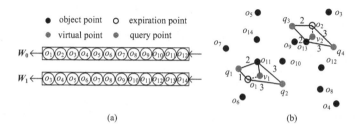

(a)   (b)

Fig. 2. The Running Example of Problem Definition under $k = 2, s = 2$

Definition 1: Continuous Group Nearest Group Query. Let $q\langle \mathcal{Q}, k \rangle$ be the continuous group Nearest Group query(CGNG for short). It monitors the window and returns k objects that the distance sum among query points in $\mathcal{Q}$ and these k objects are minimum. Here, the distance sum between query points and these k objects is calculated based on Eq. 1, $\mathcal{Q}\{q_1, q_2, \cdots, q_t\}$ refers to the query point set with the set size being t, and $dist(p_i, q_j)$ refers to Euclidean distance between the points p_i and q_j.

$$d = \sum_{i=1}^{k} \sum_{q_j \in Q} dist(p_i, q_j) \tag{1}$$

Take the example in Fig. 2. Let q be a CGNG with k being 2 and $\mathcal{Q}$ being the query point set that contains 4 query points $\{q_1, q_2, q_3, q_4\}$. We can calculate the distance among all these 4 query points and objects in the window according to definition 1. Because the distance sum among $\{q_1, q_2, q_3, q_4\}$ to o_1 and o_2 is minimal in the window W_0, i.e., equaling $1 + 3 + 2 + 3 = 9$, we return $\{o_1, o_2\}$ to the system. When the window slides in W_1, objects o_1 and o_2 expire from the window, o_{13} and o_{14} flow into the window. At that moment, the query result objects are updated to $\{o_{11}, o_{13}\}$, i.e., equaling $2+3+2+3=10$.

3   The Framework KMPT

In this section, we propose a novel framework named KMPT(short for K-Means-based Partiton Tree) for supporting CGNG over streaming data. In the following, we first explain the framework overview.

3.1   The Basic Idea

As stated in Sect. 2.2, we understand that the problem CGNG is NP-hard. In order to guarantee the real-time requirement of the system, the target of our framework is using, a low, running cost to return high-quality approximate results to the system whenever the window slides. The intuition behind the algorithm is to partition query points into a group of subsets based on the distance relationship among them [7]. Based on the partition result, we generate a group of virtual points. In this way, we monitor these virtual points' NNs and use these virtual points' NNs to support CGNG.

Formally, we use the well known algorithm named K-MEANS [5] to partition the query point set into k subsets $\{Q_1, Q_2, \cdots, Q_k\}$. After partitioning, query points in each subset are close to each other. Then, we access each subset Q_i, calculate the center of objects in each Q_i, use these center points as virtual points, i.e., $\{v_1, v_2, \cdots, v_k\}$. After forming these virtual points, we monitor these virtual points' NNs and use these objects to support query processing.

Backing to the example in Fig. 2. Let the parameter k be 2, and the query points be $\{q_1, q_2, q_3, q_4\}$. Via using the K-Means algorithm, query points q_1 and q_2 are grouped into one subset, and query points q_3 and q_4 are grouped into another subset. The nearest neighbours of v_1 and v_2(denoted as $\mathsf{NN}(v_1)$ and $\mathsf{NN}(v_2)$) under the window W_0 are o_1 and o_2, we accordingly return $\{o_1, o_2\}$ to the system. When the window slides from W_0 to W_1, objects $\{o_1, o_2\}$ expire from the window, and objects $\{o_{13}, o_{14}\}$ flow into the window. Query results are updated to $\{o_{11}, o_{13}\}$. In other words, $\mathsf{NN}(v_1)$ and $\mathsf{NN}(v_2)$ are updated to o_{11} and o_{13} respectively.

Discussion. Our main idea is monitoring virtual points' NNs, using these objects to support query processing. The benefit is that it is not necessary to consider the distance relationship between objects and query points. In this way, the running cost could be reduced a lot. In other words, we reduce the costly problem CGNG to a group of continuous NN searches over sliding window. In the following, we will explain how to efficiently maintain these virtual points' NNs over sliding windows.

3.2   The Initialization Algorithm

In this section, we will explain the initialization algorithm. It mainly contains 2 steps, including Partition-based index Construction, and Candidate NN Set Initialization.

Partition-Based Index Construction. It first partitions the whole window into m sub-windows with equal side lengths. Formally, let $P(W, m)$ be a partition that partitions the window into a group of sub-windows $\{P_0, P_1, P_2, \cdots, P_{m-1}\}$, such that: (i) given two partitions P_i and P_{i+1}, they have the same side-length, i.e., $|P_i| = |P_{i+1}|$ with $|P_i|$ being the side-length of P_i; and (ii) given any two objects $o \in P_i$ and $o' \in P_{i+1}$, $T(o)$ is always smaller than $T(o')$ with $T(o)$ being the arrived order of o.

After partition, we form a group of quad-tree $\{T_0, T_1, P_2, \cdots, T_{m-1}\}$ based on objects in $\{P_0, P_1, P_2, \cdots, P_{m-1}\}$ respectively. Note, as our proposed algorithm, is an approximate algorithm, in order to guarantee the height of each quad-tree is not too high, the height of each quad-tree should be bounded by $2 \log |P_i|$. In other words, when we form the quad-tree T_i, we do not further divide leaf nodes with depth no smaller than $2 \log |P_i|$. Alternatively, we retain at most k objects with the largest arrived orders in such leaf nodes respectively. Take an example in Fig. 3. The current window W_0 consists of 12 objects, which are divided into 3 sub-windows $\{P_0, P_1, P_2\}$. We then form 3 quad-trees $\{T_0, T_1, T_2\}$ based on objects in P_0, P_1 and P_2 respectively.

Candidate NN Set Initialization. After the $\{T_0, T_1, T_2, \cdots, T_{m-1}\}$ training, we are going to form a group of object sets $\{C(v_1), C(v_2), \cdots, C(v_k)\}$ for virtual points $\mathcal{V}\{v_1, v_2, \cdots, v_k\}$ respectively. Here, objects in $C(v_i)$ are regarded as candidate nearest neighbors of v_i. In other words, we select a group of objects for v_i from the window to form $C(v_i)$. Whenever the window slides, we should make sure the nearest neighbor of v_i must be contained in $C(v_i)$.

In the following, we are going to explain how to form $C(v_i)$ for the virtual point v_i. We first search on T_{m-1} for finding m nearest neighbors of v_i. Let the query result set be R_{m-1}. We then use *dominance relationship* to delete

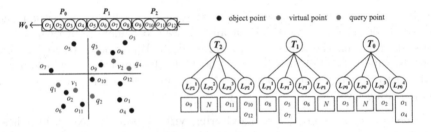

Fig. 3. Position and Arrived Order based Partition when Initialization.

Algorithm 1: The Baseline Algorithm

Input: The Query Point Set $\mathcal{Q}$, The query Window W, k

Output: Result Object Set O_Q

1   $\{Q_1, Q_2, \cdots, Q_k\}$ ←KMeans($\mathcal{Q}, k$);

2   $\mathcal{V}\{v_1, v_2, \cdots, v_k\}$ ←getCenter($\{Q_1, Q_2, \cdots, Q_k\}$);

3   $\mathcal{P}$ ←partition(W);

4   $\{T_0, T_1, \cdots, T_{m-1}\}$ ←initQuadTree(W);

5   **for** i *from* $m-1$ *to* 0 **do**

6   T_i ←formQuadTree(P_i);

7      $\{nn_1^i, nn_2^i, \cdots, nn_k^i\}$ ←getmNN($T_i, \{v_1, v_2, \cdots, v_k\}$);

8   reform(NN);

9   $O_Q \leftarrow \{nn_1^1, nn_2^1, \cdots, nn_k^1\}$;

---

meaningless objects in R_{m-1}, which helps us further reduce the space cost. After deleting meaningless objects in R_{m-1}, we insert the reminders into the set $C(v_i)$. As stated in Definition 1, given two objects o and o' in R_{m-1}, if o' is dominated by o under v_i, o' cannot become the neighbor of v_i before it expires from the window so we can remove it. As the algorithm is simple, we skip the details for saving space. For simplicity, in the following, if an object o is not dominated by any object under v_i, we call o as a meaningful object under v_i. Otherwise, we call o a meaningless object under v_i.

Definition 1. Dominate. *Let o and o' be two objects in the window. If $T(o) > T(o')$ and $D(o, v_i) < D(o', v_i)$, we say o dominates o' under the virtual point v_i.*

After accessing T_{m-1}, we go on scanning objects in T_{m-2}. We submit a range-constraint-based NN search on T_{m-2}, find m'(bounded by m) nearest neighbours of v_i. Here, the center of the searching range is v_i, and the searching radius is $D(v_i, SP^m)$ with $D(v_i, SP^m)$ being the distance between v_i and the m-th nearest neighbor of v_i in the scanned partitions, i.e., when we access P_{m-2}, the scanned partition is P_{m-1}. When we process P_{m-3}, the scanned partition is $P_{m-1} \cup P_{m-2}$, and etc. In other words, if more than m objects are contained in the search range, the corresponding query result set R_{m-2} contains m nearest neighbors of v_i. Otherwise, R_{m-2} includes all objects within the search range.

After the search, we insert meaningful objects in R_{m-2} into $C(v_i)$, further using *dominance relationship* to delete meaningless objects. From then on, we repeat the above operations to process other objects in the window. In particular, when searching on the quad-tree T_0 corresponding to P_0, we insert all meaningful elements contained in the searching region into $C(v_i)$.

As is shown in Fig. 4. Take the example of forming $C(v_1)$ for the virtual point v_1 as the example. We first search on T_2 for finding 3 nearest neighbours of $v_1(\{o_{11}, o_{10}, o_9\})$, and then we insert meaningful objects into $C(v_1)$, i.e., o_{11}. Next, we submit a range-constraint-based NN search on T_1 with center being v_1, and searching radius being $D(v_1, P_2^3) = D(v_1, o_9) = 6$. Query results are o_6 and o_7. As o_7 is dominated by o_{11}, we only insert o_6 into $C(v_1)$. When we access T_0, we submit a range query with searching radius being $D(v_1, (P_1 \cup P_2)^3) = D(v_1, o_{10}) = 3$, and insert the query result object o_2 into $C(v_1)$. Finally, we form $C(v_1)$ based on based in $\{o_{11}, o_6, o_2\}$. Similarly, $C(v_2)$ is $\{o_9, o_{12}, o_8\}$.

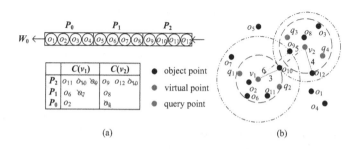

Fig. 4. The Candidate NN Set Forming using the Partition

After forming $C(v_i)$ for each $v_i \in V$, in order to support incremental maintenance under data stream, we next form a group of cubes $\{c_1, c_2, \cdots, c_k\}$ with center $\{v_1, v_2, \cdots, v_k\}$ and diagonal length $\{2C_m(1, m-1, v_1), 2C_m(1, m-1, v_2), \cdots, 2C_m(1, m-1, v_k)\}$. Here, $C_m(1, m-1, v_i)$ refers to the distance between v_i and its nearest neighbor among all objects contained in the partitions ranging from P_1 to P_{m-1}. Lastly, we form another Index R-Tree $\mathcal{I}_q$ to maintain these cubes. The function of the index $\mathcal{I}_q$ will be explained in Sect. 3.3.

We want to highlight two points. Firstly, for each partition $P_i(i \neq 0)$, we only select at most m objects as candidate NNs. For the other objects in P_i, before P_i become the first partition, they cannot become query results. In this way, we only need to maintain the meaningful objects in P_0 and at most m nearest neighbors of v_i in the other partitions to support query processing before P_0 turns to empty. Secondly, it is significant to find a suitable m. If m is large, the running cost of maintaining $C(v_i)$ is high. By contrast, when objects in P_0 are all expired from the window, if some objects in $P_0 - C(v_i)$ are still meaningful under v_i, we may have to re-scan T_1 for finding these objects, which may lead highly running cost.

Theorem 1. *Let D be a set of real values that are randomly partitioned into m subsets $\{D_1, D_2, \cdots, D_m\}$. Each subset contains m integers, i.e., $|D| = m \cdot m$. d_1^m refers to the m-th largest value contained in D_1, $d_{D-D_1}^1$ refers to the largest value among all elements in $D - D_1$. The probability of d_1^m being larger than $d_{D-D_1}^1$ is $m!/A_{mm}^{m+1}$.*

Proof. We can reduce this problem by calculating the probability of the first m real values with the largest values among all elements in D are all contained in D_1. After calculating, the probability is $m!/A_{mm}^{m+1}$.

We can find a suitable m based on Eq. 2. Based on the 3-sigma rule, when P_1 becomes the first partition, if the probability of re-searching could be avoided under any virtual point is larger than 0.95, this event almost cannot happen in most cases in real applications. Therefore, we can find the suitable m based on Eq. 2.

$$(\frac{m!}{A_{mm}^{m+1}})^k > 0.95 \tag{2}$$

3.3 The Incremental Maintenance Algorithm

After the window slides, newly arrived objects flow into P_m. Expired objects are deleted from P_0. When the number of objects in P_m achieves to $\frac{N}{m}$, the partition P_m turns to full. The first partition P_0 turns empty. Besides, we should incrementally maintain the candidate set $\mathcal{C}$. In the following, we will explain the algorithm's details. It is based on the following observation.

Observation 1. *Let $C_m(1, m-1, v_i)$ be the distance between v_i and its nearest neighbor among all objects contained in the partitions ranging from P_1 to P_{m-1}, o be an object in the window. If $D(o, v_i)$ is no smaller than $C_m(1, m-1, v_i)$, o cannot become the nearest neighbour of v_i before objects in P_0 expire from the window. Otherwise, o must be contained in the cube c_i corresponding to v_i.*

Our algorithm mainly contains 3 steps, including *range query-based evaluation*, *batch-based candidate set maintenance*, and *group-based re-searching*.

Step-I: Range Query-based Evaluation. It is used for evaluating which newly arrived objects have a chance to become nearest neighbors of these virtual points before objects in P_0 expire from the window(See lines 1–6). Let o be a newly arrived object, and I_c be the R-Tree used for maintaining cubes corresponding to these virtual points. We search on I_c for finding cubes containing o. For each cube c_i, we first insert o into the corresponding candidate NN set $C(v_i)$ as it may become a query result object before objects in P_0 expire from the window. Next, we delete elements in $C(v_i)$ that are dominated by o. Lastly, we should shrink the diagonal length of c_i to $2D(v_i, o)$.

Step-II: Batch-based Candidate Set Maintenance. After the partition P_m turns to full, the algorithm enters into the second step(Lines 7–15). We first form the quad-tree T_m based on the objects in P_m. Next, for each virtual point v_i, we search on T_m for finding its m nearest neighbors. We then insert them into $C(v_i)$ and delete meaningless objects contained in $C(v_i)$. Note, we have inserted some objects in P_m into $C(v_i)$ before step-II is invoked. Therefore, if more than m objects in P_m have been inserted into $C(v_i)$, this step could be skipped.

Step-III: The Re-Searching. After Step-II, we should evaluate whether existing meaningful objects are contained in $C(v_i) - P_1$ for each virtual point v_i(Lines 16-21). We can achieve this goal via the following observation, that is, let $D(v_i, P_1^m)$ be the distance between v_i and its m-th nearest neighbor in P_1. If it is dominated by at least one object in $P - P_1$, all objects in $P_1 - C(v_i)$ are meaningless under v_i. Otherwise, we should research meaningful objects in P_1 based on the virtual point v_i. At that moment, we submit a range query on T_1, find all objects contained in the searching region, and finally insert them into $C(v_i)$. Here, the center of the searching range is v_i, and the searching radius is $D(v_i, P - P_1)$, with $D(v_i, P - P_1)$ being the distance between v_i and the nearest neighbor of v_i in the partitions ranging from P_2 to P_m.

Algorithm 2: The IM-Algorithm

Input: The R-Tree $\mathcal{I}_c$, the set $C(v_i)$, the object o
Output: Updating Result Object Set $C(v_i)$

1   $c_i \leftarrow I_c(o)$;
2   $count_i \leftarrow 0$;
3   $C(v_i) \leftarrow o$;
4   $count_i$++;
5   Delete the object that is dominated by o in $C(v_i)$;
6   $DL(c_i) \leftarrow 2D(v_i, o)$;
7   **if** $|P_m| = \frac{N}{m}$ **then**
8      $T_m \leftarrow initQuadTree(P_m)$;
9   **for** i *from* 1 *to* k **do**
10         $(nn_i^1, nn_i^2, \cdots, nn_i^m) \leftarrow getsmNNs(T_m, v_i)$;
11   **if** $count_i = m$ **then**
12   break;
13   **else**
14           $C(v_i) \leftarrow$ insert $\{nn_i^1, nn_i^2, \cdots, nn_i^{m-count_i}\}$;
15   Delete the object that is dominated by the newly inserted object in $C(v_i)$;
16         $D(v_i, P_1^m) \leftarrow$ the distance between v_i and m_{th} in P_1 ;
17         $D(v_i, P - P_1) \leftarrow$ the distance between v_i and o_i in $P - P_1$;
18   **if** $D(v_i, P_1^m) \geq D(v_i, P - P_1)$ *and* $T(v_i, P_1^m) \leq T(v_i, P - P_1)$ **then**
19   continue;
20   **else**
21           $C(v_i) \leftarrow RangeQuery(T_i, v_i)$;

22 **return** $C(v_i)$;

The Cost Analysis. When processing a new object, the height of each T_i is bounded by $2\log|P_i|$(constant). The insertion cost into T_i is $\mathcal{O}(1)$. Finding dominated objects for the new object is $\mathcal{O}(1)$ as each object is deleted once. With a bounded partition amount of m, the overall cost of processing a new object is $\mathcal{O}(n)$. Processing an expired object costs $\mathcal{O}(1)$. Thus, the overall cost for processing one object is $\mathcal{O}(n)$.

4 The Experiment

In this section, we conduct extensive experiments to demonstrate the efficiency of the KMPT. In the following, we first explain the settings of our experiments and then report our findings.

4.1 Experiment Settings

Data Set. In total, three datasets are used in our experiments, one real dataset namely DIDI, and two synthetic datasets namely SYN-R and SYN-U respectively. DIDI contains 1 G location information from ShangHai/Beijing over two years, with the original size being 30 GB. Each record contains three attributes (i.e., taxi Id, location longitude, location latitude). In the synthetic dataset SYN-R, objects obey normal distribution. In the synthetic dataset SYN-U, objects obey uniform distribution.

Parameters. In our experiment, we evaluate the performance of different algorithms under four parameters, including N, Q, s, and k. Here, N refers to the window size, Q refers to the number of query points, s refers to the flow rate of the sliding window, and k refers to the number of groups of query points. The parameter settings are listed in Table 1 with the default values bolded.

Table 1. Parameter Settings

Parameter	value
N	200KB, 400KB, **600KB**, 800KB,1MB
Q	5, **10**, 15, 20, 25 ($\times$ 100)
s	2%, 4%, **6%**, 8%, 10% ($\times$ N)
k	10, 15, **20**, 25, 30 ($\times$ 10)

Performance Metrics. In our experiment, we evaluate algorithm differences via the following metrics: initialization time, update time, and accuracy rate as the main performance indicators. Here, the initialization time refers to the time spent on building the index and querying results, the update time refers to the average time of updating the index and querying results, and the accuracy rate refs to $1 - \frac{|acc\_D - app\_D|}{acc\_D}$. Here, the acc_D refers to the accuracy distance, and the app_D refers to the approximate distance.

Competitors. In addition to our algorithm, we also implement EHC and SHR approaches [3] for answering the CGNG query as competitors. Compared with the original EHC algorithm, We assume that, in each time unit, all points are moving and returning results to the system. In other words, the original algorithm was implemented over static data, we apply it under data stream. EHC updates the results of all windows whenever the window slides.

4.2 Performance Comparison

Initialization Running Time Comparison. We compare the performance of KMPT with its competitors. Initially, we evaluate the running time of the initial procession using different algorithms and N values, with default parameters. In our experiments from Fig. 5(a)–(c), EHC and SHR require significant initialization time as they scan all points in each sub-window. In contrast, KMPT has the shortest initialization time. This is because KMPT divides the sliding window into sub-windows to calculate non-dominated objects, reducing the number of objects points to be scanned. KMPT outperforms other algorithms in terms of runtime due to shorter partitioning time and fewer computations required.

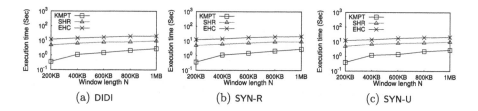

Fig. 5. Initialization time comparison under different N

Updating Running Time Comparison. In our second set of experiments in Fig6(a)–(c), we compare the performance of different algorithms at varying N values. KMPT consistently outperforms the others, as previously observed. As N increases, EHC's running time significantly increases because it needs to access all data points when the window slides. While SHR performs better than EHC due to pruning methods, it struggles to provide an accurate threshold for virtual center points, resulting in accessing many data points during window slides. In contrast, KMPT has a much lower running time than EHC and SHR as it calculates fewer objects within the window. Larger data sets naturally require more processing time, making the running time of these two algorithms relatively higher compared to other data sets.

In our third set of experiments in Fig. 7(a)–(c), we assess the performance of different algorithms with varying Q values, using default parameters. KMPT

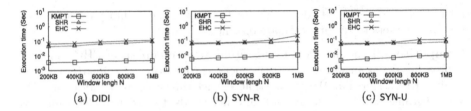

Fig. 6. Updating time comparison under different N

consistently outperforms the others. Similar to previous observations, KMPT efficiently scans fewer points. Despite the increasing number of query points, they are grouped based on their location. In contrast, EHC's running time rapidly increases and is highly sensitive to the distribution of query points. For instance, under SYN-R dataset, EHC's running time is significantly higher than KMPT. This is because the data distribution in SYN-R is uneven, whereas KMPT is less sensitive to data distribution, resulting in a slower increase in running time and a significantly smaller value compared to the other two algorithms.

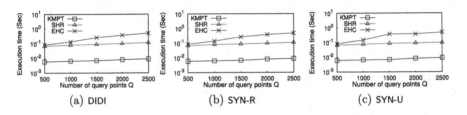

Fig. 7. Updating time comparison under different Q

In our fourth set of experiments in Fig. 8(a)–(c), we assess the performance of different algorithms with varying s values, using default parameters. KMPT continues to outperform the others. We also observe that as s increases, the differences in running time among the three algorithms become smaller. This is because a larger s requires our algorithm to maintain more points, making it less distinct from the other algorithms. However, since s is typically not high in most applications, KMPT remains the most efficient in the majority of cases. When the candidate object set, which is not dominated in the window, is small enough, KMPT runs faster.

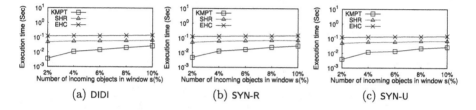

Fig. 8. Updating time comparison under different s

In our fifth set of experiments, we evaluated different algorithms under various k values. KMPT consistently outperformed EHC and SHR across all three datasets in Fig. 9(a)–(c). For instance, KMPT's running time was only 1.7% of EHC and 5.7% of SHR. This significant improvement is due to KMPT considering spatiotemporal correlation among query points and dividing them into k groups. By finding only one object from each group, KMPT efficiently reduces computations, speeds up queries, and provides a partitioned area with tight bounds.

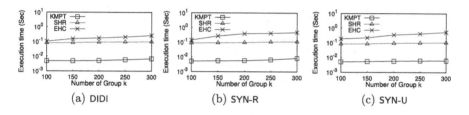

Fig. 9. Updating time comparison under different k

Accuracy Rate Comparison. In the accuracy evaluation, we examined different parameters k and s. Table 2 shows that the accuracy rate of KMPT improves with higher values of s. This is because a larger k means more objects are maintained, resulting in a larger candidate set, which improves accuracy. Similarly, increasing the value of s also enhances accuracy. A larger s means more precise points are included, but it comes at the cost of higher computational overhead in maintaining the candidate set.

Table 2. Accuracy Rate Comparison

Dataset	Accuracy Rate Analysis(k)					Accuracy Rate Analysis (s)				
	100	150	200	250	300	500	1000	1500	2000	2500
DIDI	0.913	0.921	0.934	0.946	0.951	0.921	0.928	0.934	0.938	0.947
SYN-R	0.836	0.897	0.924	0.936	0.937	0.842	0.899	0.926	0.938	0.941
SYN-U	0.851	0.899	0.936	0.938	0.943	0.838	0.908	0.912	0.922	0.936

5 Conclusion

In this paper, we propose a novel framework named KMPT for supporting CGNG over sliding window. It reduces the costly CGNG to the problem of continuous NN search. In order to support continuous NN search, we propose a group of partition-based algorithms, which help us with a small number of objects to support query processing. We have conducted extensive experiments to evaluate the performance of KMPT on several datasets under different distributions. The results demonstrate the superior performance of KMPT.

Acknowledgements. This paper is partly supported by the National Key Research and Development Program of China (2020YFB1707901), the National Natural Science Foundation of Liao Ning (2022-MS-303, 2022-MS-302, and 2022-BS-218), the National Natural Science Foundation of China (62102271, 62072088, Nos. U22A2025, 62072088, 62232007, 61991404), and Ten Thousand Talent Program (No. ZX20200035).

References

1. de Berg, S., Staals, F.: Dynamic data structures for k-nearest neighbor queries. Comput. Geom. **111**, 101976 (2023)
2. Cui, N., et al.: Towards multi-user, secure, and verifiable $k$nn query in cloud database. IEEE Trans. Knowl. Data Eng. (2023)
3. Deng, K., Sadiq, S., Zhou, X., Xu, H., Fung, G.P.C., Lu, Y.: On group nearest group query processing. IEEE Trans. Knowl. Data Eng. **24**(2), 295–308 (2010)
4. Deng, K., Xu, H., Sadiq, S., Lu, Y., Fung, G.P.C., Shen, H.T.: Processing group nearest group query. In: 2009 IEEE 25th International Conference on Data Engineering, pp. 1144–1147. IEEE (2009)
5. Krishna, K., Murty, M.N.: Genetic k-means algorithm. IEEE Trans. Syst. Man Cybern. Part B (Cybernetics) **29**(3), 433–439 (1999)
6. Lee, J.M.: Fast k-nearest neighbor searching in static objects. Wirel. Pers. Commun. **93**(1), 147–160 (2017)
7. Li, T., Chen, L., Jensen, C.S., Pedersen, T.B., Gao, Y., Hu, J.: Evolutionary clustering of moving objects. In: 2022 IEEE 38th International Conference on Data Engineering (ICDE), pp. 2399–2411. IEEE (2022)
8. Li, T., Gu, Y., Zhou, X., Ma, Q., Yu, G.: An effective and efficient truth discovery framework over data streams. In: Proceedings of the 20th International Conference on Extending Database Technology, pp. 180–191. Springer (2017)

9. Miao, X., Gao, Y., Chen, G., Zheng, B., Cui, H.: Processing incomplete k nearest neighbor search. IEEE Trans. Fuzzy Syst. **24**(6), 1349–1363 (2016)
10. Nandy, S.C., Das, S., Goswami, P.P.: An efficient k nearest neighbors searching algorithm for a query line. Theor. Comput. Sci. **299**(1–3), 273–288 (2003)
11. Papadias, D., Shen, Q., Tao, Y., Mouratidis, K.: Group nearest neighbor queries. In: Proceedings. 20th International Conference on Data Engineering, pp. 301–312. IEEE (2004)
12. Papadias, D., Tao, Y., Mouratidis, K., Hui, C.K.: Aggregate nearest neighbor queries in spatial databases. ACM Trans. Database Syst. (TODS) **30**(2), 529–576 (2005)
13. Xu, H., Lu, Y., Li, Z.: Continuous group nearest group query on moving objects. In: 2010 Second International Workshop on Education Technology and Computer Science, vol. 1, pp. 350–353. IEEE (2010)
14. Yang, K., Tian, C., Xian, H., Tian, W., Zhang, Y.: Query on the cloud: improved privacy-preserving k-nearest neighbor classification over the outsourced database. In: World Wide Web, pp. 1–28 (2022)
15. Yiying, J., Liping, Z., Feihu, J., Xiaohong, H.: Groups nearest neighbor query of mixed data in spatial database. J. Front. Comput. Sci. Technol. **16**(2), 348 (2022)

Approximate Continuous Skyline Queries over Memory Limitation-Based Streaming Data

Yunzhe An$^{(\boxtimes)}$, Zhu Zhen, Rui Zhu, Tao Qiu, and Xiufeng Xia

Shenyang Aerospace University, Shenyang, China
{anyunzhe,zhurui,qiutao,xiaxiufeng}@sau.edu.cn

Abstract. Continuous skyline query over sliding window is an important problem over streaming data. The query returns all skyline objects to the system whenever the window slides. Existing efforts include exact-based algorithms and approximate-based algorithms. Their key idea is to find objects that cannot become skyline objects before they expire from the window, delete them, and use reminders to support query processing. However, the space cost of all existing efforts is high, and cannot work under memory limitation-based streaming data, i.e., a general environment in real applications.

In this paper, we define a novel query named $\rho-$approximate continuous skyline query(ρ-ACSQ), which returns error-bounded answers to the system. Here, ρ is a threshold, which can bind the error ratio between approximate and exact results. In order to support ρ-ACSQ, we propose a novel framework named $\rho-$SEAK(short for $\rho-$<u>S</u>elf-adaptive <u>E</u>rror-based <u>A</u>pproximate <u>S</u>kyline). It can self-adaptively adjust ρ based on the distribution of streaming data, and achieve the goal of supporting ρ-ACSQ over memory limitation-based streaming data. Theoretical analysis indicates that even in the worst case, both the running cost and space cost of $\rho-$SEAK are all unrelated with data scale.

Keywords: Streaming Data · Continuous Skyline query · Dominance · Memory Limitation

1 Introduction

Continuous skyline query over data stream is an important problem in the domain of streaming data management, which has been studied over 18 years [1–3]. Formally,let $o[1, \cdots, d]$ and $o'[1, \cdots, d]$ be two $d-$dimensional objects in the object set $\mathcal{D}$ in $[0, 1]^d$ space. We say o dominates o' if $\forall\ u \in [1, d]$, $o'[u] \leq o[u]$, and $\exists\ v \in [1, d], o'[v] < o[v]$. Here, $o[u]$ refers to the coordinate of o in the dimension u. In particular, if an object $o \in \mathcal{D}$ is not dominated by any object in $\mathcal{D}$, o is regarded as a *skyline object*. All skyline objects in $\mathcal{D}$ form the skyline set $\mathcal{SK}$.

A continuous skyline query q, expressed by the tuple $q\langle n, s \rangle$, monitors the window W, which returns skyline objects whenever the window slides. The query window can be time-based or count-based [4,5]. For simplicity, in this paper, we

X. Song et al. (Eds.): APWeb-WAIM 2023, LNCS 14333, pp. 96–110, 2024.
https://doi.org/10.1007/978-981-97-2387-4_7

only focus on count-based window. However, our techniques also can be applied to answer skyline queries over time-based sliding window. Under this setting, n refers to the number of objects contained in the window, s refers to the number of objects arriving in/expiring from the window whenever the window slides. A natural property of s is it partitions objects in the window into a group of sub-windows $\{s_0, s_1, \cdots, s_{m-1}\}$ with m being n/s, s_i being the subset of objects in the window with arrived order i. In other words, a continuous skyline query q monitors a window containing n objects. Whenever the window slides, s objects arrive in the window, and another s objects expire from the window, q retrieves skyline objects $\{sk_1, sk_2, \cdots, sk_t\}$ from the query window to the system.

Based on whether *exact* query results are required, the state-of-the-art efforts can be divided into *exact*-based [1–5] and *approximate*-based algorithms [5–8]. Their common idea is monitoring meaningful objects in the window, incrementally maintaining them, and using them to support query processing. Here, given one object o, if o is not dominated by any other object with arrived order larger than it, it has a chance to become a query result object, and we call it as a meaningful object. Otherwise, it cannot become a query result object before it expires from the window, and could be deleted. However, as discussed in Sect. 2, the space cost of all exact algorithms is high, i.e., $\mathcal{O}(n)$ in the worst cases. Some approximate algorithms are proposed for reducing space cost. They are allowed to return deviation-bounded results to the system via introducing a user-defined "error threshold". Compared with exact algorithms, they can use fewer candidates for supporting an approximate skyline search. However, such a threshold is usually set in advance, which cannot find a suitable threshold based on streaming data distribution.

Note, in real applications, the memory resource of streaming data [9,13] management system is usually limited. Using linear space cost for supporting a signal query is usually unacceptable, especially under memory limitation-based streaming data, i.e., a general environment in real applications. Consider the forest fire monitoring system. The sensor measures the temperature, humidity, and wind speed at various locations, collects a vast quantity of data, monitors the most likely fire site, and returns it to the user in real-time. Because the memory size of each sensor is very limited, it is impossible to maintain all monitored information in each sensor, leading that useful data(meaningful objects) may be incorrectly discarded. It leads to failure to return high-quality query results to the system.

In this paper, we propose a novel framework named $\rho-$SEAK(short for $\rho-$Self-adaptive Error-based Approximate Skyline) to support approximate continuous *skyline* search(ρ-ACSQ for short) over memory limitation-based streaming data with ρ being an error threshold ranging from 0 to 1. Let T_M be the maximal number of objects the system can maintain. We use a quad-tree T to maintain objects in the query window. When the window slides, we insert meaningful objects into T, and delete meaningless objects from T. When the number of objects in the quad-tree achieves T_M, we calculate ρ, and further delete a group of objects based on ρ. Let the exact skyline set $\mathcal{SK}$ be $\{sk_1, sk_2, \cdots, sk_s\}$, approximate skyline set be $\mathcal{MK}\{mk_1, mk_2, \cdots, mk_r\}$. $\forall sk \in \mathcal{SK}$, there exists at least one element $mk \in \mathcal{MK}$ satisfying $\forall i \in [1, d], mk[i] + \rho \geq sk[i]$.

Table 1. Frequent Notations

$T(o)$	the arrived order of o
$\mathcal{SK}$	the exact skyline objects set
$\mathcal{MK}$	the approximate skyline objects set
$|T|$	the number of objects maintained by T
T_M	the maximal number of objects the system can load
$|O_W|$	all objects in the window
C_ρ	the set of objects in the $\rho-\mathsf{CSS}$
$o.p$	the prediction moment of o
c_M	the maximal number of objects contained in leaf nodes
c_{med}	the median of cubes' side-length
$|e.c|$	the side-length of e

Challenges. In order to make $\rho-\mathsf{SEAK}$ effectively work, we should overcome the above challenges. Firstly, as the distribution of streaming data is timely changed, it is difficult to find a suitable ρ. Secondly, even though ρ could be found, it is difficult to form a small object subset to support an approximate skyline search. To deal with the above challenges, the contributions of this paper are as follows.

- (i)We propose a novel query named ρ-approximate continuous skyline query with ρ being a threshold. It is used for bounding the error ratio between exact and approximate results;
- (ii) We propose a cube-based algorithm to find a reasonable ρ. The key behind the algorithm is using a quad-tree to partition objects into a group of cubes. Based on the partition result, we can find a suitable ρ based on the side-length distribution of these cubes. Then we propose a partition-based algorithm for further enhancing the query results quality.
- (iii) We propose a group of algorithms to support incremental maintenance. They fully use the parameter ρ for keeping the space cost as low as possible. In addition, we can guarantee the running cost of processing each object is bounded by $\mathcal{O}(d)$.

The rest of this paper is as follows. Section 2 reviews the related work and proposes the problem definition. Section 3 explains the framework $\rho-\mathsf{SEAK}$. Section 4 evaluates the performance of $\rho-\mathsf{SEAK}$. Section 5 concludes this paper.

2 Preliminary

2.1 Related Works

Many efforts have studied the problem of skyline queries over data stream. We only review existing works that are relevant to ours in the following, including i) traditional skyline query algorithms; ii) exact skyline query algorithms over data stream; and iii) approximate skyline query algorithms.

Traditional Skyline Query Algorithms. Among them, BNL [10] is the first one that can support skyline queries. The algorithm named DC [10] uses the key of *divide-and-conquer* to support skyline queries. Another algorithm named NN [11] supports skyline queries via nearest neighbor search. BBS [12] calculates skyline objects via using the index R-tree. Compared with NN, BBS only needs to access a small number of objects.

Skyline Query Over Data Stream. Lin et al. [1] propose an interval tree-based algorithm. Compared with maintaining all objects in the window, it uses *domination relationship* and *arrived order relationship* to find meaningless objects, delete them, and use the reminders(called meaningful objects) to support query processing. In the worst cases, the meaningful object amount may be linear to the window size. Tao et al. [5] propose two algorithms named Lazy and Eager to support skyline queries. Sarkas et al. [3] use geometric arrangements to maintain categorical skylines over data stream.

Approximate Skyline Query Algorithms. Bai et al. [6] propose a new definition named ρ-dominance. Specially, given two $d-$dimensional objects o and o', $\forall i$, if $\frac{o[i]}{o'[i]} \leq \rho$, and $\exists j, \frac{o[j]}{o'[j]} < \rho$, o ρ-dominates o'. They then propose a framework named BBDS to support query processing. Liang et al. [7] propose an algorithm named EMCFTA for pruning skyline objects within an acceptable range of difference. The approximate threshold ρ is defined as the difference in any dimension between the values of two objects. As a result, approximate skyline query results are obtained with acceptable quality loss. Last but not least, Xiao et al. [8] propose a sampling-based approximate skyline algorithm.

Discussion.BNL, DC, NN and BBS are all based on static data. They cannot work under data stream. Exact algorithms for continuous skyline queries are sensitive to data distribution, dimension, and scale. In the worst-case scenario, the space cost of these algorithms is linear to the window length; additionally, they cannot work in a memory limitation-based streaming data environment. Approximate algorithms about skyline queries are also based on static data. Furthermore, because the parameter ρ is "pre-defined", it cannot be changed in response to the distribution of streaming data. Thus, finding an algorithm that can work efficiently over memory limitation-based streaming data is critical.

2.2   Problem Definition

A continuous skyline query q, expressed by the tuple $q\langle n, s\rangle$, monitors the objects in the window W. Whenever the window W slides, the query q returns all skyline objects in the window to the system. Take an example in Fig. 1. o_8 is not dominated by any object in the window. It is regarded as a skyline object. Similarly, another 5 skyline objects are $\{o_1, o_4, o_6, o_7, o_9\}$. Accordingly, $\{o_1, o_4, o_6, o_7, o_8, o_9\}$ form the skyline object set. When the window slides to W_1, objects in s_0 expires from the current window, objects in s_4, i.e., $\{o_{13}, o_{14}, o_{15}\}$ flow into the window. Skyline objects in W_1 are updated to $\{o_4, o_6, o_7, o_8, o_9\}$.

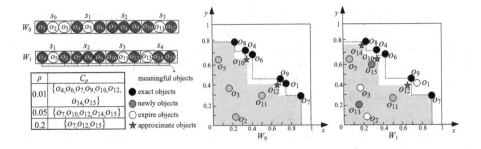

Fig. 1. Running Example of Problem Definition($s = 3, \rho = 0.05$).

Note, in real applications, algorithms have to incrementally maintain all meaningful objects in the window. The number of meaningful objects in the window is sensitive to the data distribution, dimension, and scale, which may be so large that the system is unable to load all of them. Therefore, it is significant to study the problem of approximate continuous skyline query($\rho-$ACSQ for short) over memory limitation-based streaming data.

Definition 1. ρ**-ACSQ.** *A ρ-ACSQ query q, denoted as $\langle \rho, n, s \rangle$, monitors objects in the window. Whenever the window slides, q returns a set of approximate skyline objects $\mathcal{MK} = \{mk_1, mk_2, \cdots, mk_r\}$ to the system. Compared with exact skyline set $\mathcal{SK} = \{sk_1, sk_2, \cdots, sk_s\}$ of the window, $\forall sk \in \mathcal{SK}$, there exists at least one element $mk \in \mathcal{MK}$ satisfying $\forall i \in [1, d], mk[i] + \rho \geq sk[i]$.*

Back to the example in Fig. 1. Under the window W_0, the skyline set SK_0 is $\{o_1, o_4, o_6, o_7, o_8, o_9\}$. For the object $o_{10}(0.36, 0.66)$, compared with $o_6(0.41, 0.67)$, as $0.36 + 0.05 = 0.41 \geq 0.41, 0.66 + 0.05 = 0.71 > 0.67$. o_{10} is regarded as a $\rho-$approximate skyline object. Similarity, o_{12} is also a $\rho-$approximate skyline object. Based on the definition 1, $\{o_7, o_8, o_{10}, o_{12}\}$ form a $\rho-$approximate skyline object set. When the window slides to W_1, the object set $\{o_7, o_{10}, o_{12}, o_{14}\}$ could be regarded as a $\rho-$approximate skyline object set under W_1. Note, the $\rho-$approximate skyline object set is not unique. Under the window W_1, we also could use $\{o_6, o_7, o_{12}, o_{14}\}$ as a $\rho-$approximate skyline object set.

3   The Self-adaptive-based Framework $\rho-$SEAK

This section proposes a novel framework named $\rho-$SEAK(short for $\rho-$Self-adaptive Error-based Approximate Skyline) to support ρ-ACSQ over memory limitation-based streaming data.

3.1   The ρ-CSS Definition

Definition 2. ρ**-dominate.** *Given any two objects o and o' in the window W, if they satisfy the following two conditions, we say o ρ-dominates o'.*

- (i) $T(o)$ is no smaller than $T(o')$ with $T(o)$ being the arrived order of o;
- (ii) $\forall j \in [1, d], o[j] + \rho \geq o'[j]$.

In order to support ρ−ACSQ, we further propose the concept of ρ−CSS(short for ρ-Candidate Skyline Set). Our target is to use, as few as, objects in the window to support query processing.

Definition 3. ρ-CSS. *The ρ−CSS(short for ρ−Candidate Skyline Set), i.e.,denoted as C_ρ, is a subset of objects in the window such that: $\forall o' \in O_W − C_\rho$, at least one object in C_ρ ρ−dominates o'.*

Back to the example in Fig. 1. In W_0, the object $o_9(0.69, 0.46)$ is ρ-dominated by the object $o_{12}(0.68, 0.42)$ as (i) $T(o_9) < T(o_{12})$, and (ii) $0.68 + 0.05 = 0.73 > 0.69$ and $0.42 + 0.05 = 0.47 > 0.46$. Moreover, $o_1(0.72, 0.41)$ is also ρ-dominated by o_{12}. Similarly, o_{10} ρ-dominates o_4 and o_6. Accordingly, we could use the set $\{o_7, o_8, o_{10}, o_{12}\}$ as a ρ-CSS. When the window slides to W_1, the ρ-CSS is updated to $\{o_7, o_{10}, o_{12}, o_{14}, o_{15}\}$. Similar with the running example discussed in Sect. 2.2, the ρ-CSS under each window is also not unique. Therefore, the key of ρ−SEAK is to develop an algorithm that can form ρ-CSS via as few as objects.

Discussion. It is significant to find a reasonable parameter ρ. Figure 1 shows the ρ-CSS based on different ρ in W_1 when $C_M = 5$. When ρ=0.01, $|C_{0.01}| = 8$ with $C_{0.01}$ being $\{o_4, o_6, o_7, o_9, o_{10}, o_{12}, o_{14}, o_{15}\}$, i.e., larger than 5. When $\rho = 0.05$, $|C_{0.05}| = 5$ with $C_{0.05}$ being $\{o_7, o_{10}, o_{12}, o_{14}, o_{15}\}$, where we can make sure that we can load all elements in $C_{0.05}$ to the memory, as well as return high quality query results to the system. When ρ=0.2, $|C_{0.2}| = 3$ with $C_{0.2}$ being $\{o_7, o_{12}, o_{15}\}$. Obviously, $|C_{0.2}|$ decreases significantly. However, the larger the ρ, the smaller the space cost but the lower the quality of query results.

3.2   The Initialization Algorithm

In this section, we mainly discuss how to form ρ−CSS C_ρ based on objects in the query window. The algorithm contains 3 phases: Exact, ρ−Selection, and ρ−Approximate. Phase-I uses exact algorithm to find candidate skyline objects in the window when we start to form ρ−CSS.

Phase-I: Exact. Let the query window W_0 be partitioned into $\{s_0, s_1, \cdots, s_{m-1}\}$ based on the parameter s. We first reversely scan objects in s_{m-1}, execute the algorithm BBS to find local skyline objects in s_{m-1}(Line 3). The others could be deleted. Next, we form a quad-tree T to maintain local skyline objects in s_{m-1}. Here, in this paper, we assume that each leaf node in T contains *one and only one* object. In addition, an object $o \in s_{m-1}$ is called as a local skyline object if it is not dominated by anyone contained in s_{m-1}.

After processing objects in s_{m-1}, the algorithm reversely scans s_{m-2}, calculates local skyline set in s_{m-2}. After calculation, we try to insert local skyline objects in s_{m-2} into the quad-tree T. During the insertion, for each local skyline object $o \in s_{m-2}$, the above operations should be done based on the domination

---

Algorithm 1: The Initialization Algorithm

Input: Stream Data S, the Maximal Memory M

Output: ρ-CSS C_ρ

1 $\mathcal{P} \leftarrow$ formPartition(S);
2 **for** i *from* m *to* 1 **do**
3      $SK_i \leftarrow$ BBS(s_i);
4   **for** j *from* 1 *to* $|SK_i|$ **do**
5   **if** SK_{ij} *is not* $\rho-dominated$ *by any object in* T **then**
6              $T \leftarrow$ insert(SK_{ij}, ρ);
7          $O_{Dom} \leftarrow$ findDom(T);
8   **for** j *from* 1 *to* $|O_{Dom}|$ **do**
9   **if** $O_{Domj}.t < T(SK_{ij})$ **then**
10                 $O_{Domj}.t \leftarrow T(SK_{ij})$;
11   **if** $|T| = M$ **then**
12             $\rho \leftarrow$ getMedain$(T) \cdot 2$;
13             $T \leftarrow$ merge(T, ρ);
14             $T \leftarrow$ delete(T, ρ);

15 **return** C_ρ;

---

relationship among objects. Here, given an object o, $o.p$ refers to the prediction moment of o. It records the earliest moment when o has a chance to become a skyline object. For example, let o be dominated by o' with o' being the object with the largest arrived order among all elements that can dominate o. At the moment o' expires from the window, o has a chance to become a skyline object. Thus, $o.p$ is set to the moment o' expires from the window.

- (i) If o is dominated by at least one object $o' \in T$, it implies o cannot become a skyline object before o expire from the window. Therefore, we delete o directly for saving space. Otherwise, we insert it into T(Line 5–6);
- (ii) If existing an object $o'' \in T$ dominated by o, it means o'' cannot become a skyline object before o expires from the window. Accordingly, we set $o''.p$ to $T(o)$. As o'' also has a chance to become a skyline object, we regard it as a candidate skyline object, still retain it in T(Line 7–10). Lastly, we insert o into the quad-tree T.

We repeat the above operations to handle other objects in $\{s_{m-3}, s_{m-4}, \cdots\}$, delete non-skyline objects, and insert candidate skyline objects (or skyline objects) into the quad-tree T. In particular, when $|T|$ achieves to T_M, we have to delete some objects in T from the memory via finding a reasonable error threshold ρ. Here, T_M refers to the maximal number of objects the system can load, and $|T|$ refers to the number of objects maintained by T. At that moment, the algorithm enters into Phase-II.

Take an example in Fig. 2. The algorithm first accesses objects in s_3, uses the algorithm BBS to find local skyline objects in s_3, i.e., $\{o_{10}, o_{12}\}$. We then

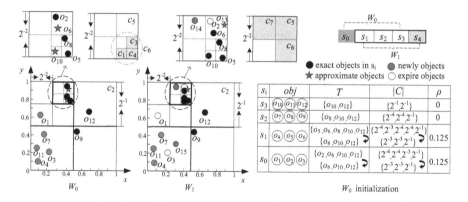

Fig. 2. The Running Example of the Initialization($s = 3, N = 12, C_M = 5$).

form a quad-tree T based on $\{o_{10}, o_{12}\}$. Next, we find local skyline objects in s_2, i.e., $\{o_8, o_9\}$. For o_9, since it is dominated by o_{12}, we delete it. For o_8, as o_8 dominates o_{10}, we set $o_{10}.p$ to $T(o_8)$, which means o_{10} may has a chance to become a skyline object until o_8 expired from the window. After processing objects in s_1, we find that $|T|$ achieves to 5, and we should find a suitable ρ.

Phase-II: ρ−Selection. The parameter ρ is computed based on Theorem 1. It implies we can find a reasonable parameter ρ based on cubes corresponding to leaf nodes in T.

Specially, let $\{e_1, e_2, \cdots, e_M\}$ be leaf nodes of the quad-tree T, $\{c_1, c_2, \cdots, c_M\}$ be cubes that are used for bounding objects contained in the leaf nodes of the quad-tree T. We first scan cubes in $\{c_1, c_2, \cdots, c_M\}$, and then calculate the *median*, i.e.,denoted as c_{med}, of these cubes' side-length, use $c_{med} \cdot 2$ as the parameter ρ(Line 11). In Fig. 2, after processing objects in s_1, let the side-length of cubes corresponding to these leaf nodes be $\{2^{-4}, 2^{-4}, 2^{-4}, 2^{-3}, 2^{-1}\}$, the median value is 2^{-4}. We set ρ to $2^{-4} \cdot 2$, which is 2^{-3}.

Phase-III: ρ−Approximate. From then on, the algorithm enters into the third phase. We access the quad-tree T. For each node e with $|e.c|$ being c_{med}, if it has child nodes, we access its child nodes, find the object o with the largest arrived order, and finally delete the other objects(Line 13–14). In other words, we regard e as a leaf node, and only remain o in e. As is stated in Theorem 1, o must ρ−dominate other objects contained in e's child nodes. Here, $e.c$ refers to the cube that is used as the boundary for bounding objects contained in e, and $|e.c|$ refers to the side-length of $e.c$.

Furthermore, as is stated in Observation 1, for each non-leaf node e with $|e.c|$ being c_{med}, it has at least two children. In other words, $e.c$ contains at least 2 objects. Accordingly, we can remove at least $\frac{1}{2} \cdot \frac{T_M}{2}$ objects from the memory. In Fig. 2, we delete objects o_5 and o_8, and delete leaf nodes e_4 and e_3, only maintain e_6 with $|e_6.c| = 2^{-3}$.

Theorem 1. *Let o and o' be two objects contained in the cube c with $|c|$ being ρ. If $T(o) \geq T(o')$, o must $\rho-$dominate o'.*

Proof. As o and o' are contained in the same cube with side-length ρ, $\forall i \in [1, d]$, $|o[i] - o'[i]| \leq \rho$ is guaranteed. In other words, $o[i] + \rho$ must be no smaller than $o'[i]$ for each $i \in [1, d]$. In addition, as $T(o) \geq T(o')$, o must $\rho-$dominate o'. The proof is complete.

Observation 1 *Let $\{c_1, c_2, \cdots, c_M\}$ be cubes that are used for bounding objects contained in these leaf nodes. Our algorithm makes sure that at least $\frac{T_M}{4}$ objects could be deleted.*

After deleting meaningless objects in T, we go on processing the other objects in the window. Let s_i be the set of objects we should process. The algorithm BBS is applied again for finding local skyline objects in s_i. Next, we insert these local skyline objects into T. The difference is if an object o is inserted into a leaf node e with $|e.c| = c_{med}$, we delete o as it must be $\rho-$dominated by the one contained in e. When $|T|$ achieves to T_M again, we repeat the $\rho-$Selection to update ρ, and delete the corresponding objects based on the new ρ. After all objects in the window are processed, the algorithm is terminated. In Fig. 2, when we try to insert o_2 into T, as it is contained in $e_5.c$, $|e_5.c| = 0.125$ and e_5 is not empty, we delete o_2 directly.

3.3   The Incremental Maintenance Algorithm

In this section, we first discuss how to process newly arrived/expired objects. Next, we explain the parameter ρ selection under data stream.

Object Insertion and Deletion. When the window slides from W_0 to W_1, objects in s_0 expires from W, and objects in s_m flow into the window. We first execute the algorithm BBS to find local skylines in s_m. We next insert each local skyline object $o \in s_m$ into a leaf node e of the quad-tree T. After insertion, we search on T for finding elements $\rho-$dominated by o via Observation 2, and deleting this kind of objects. Lastly, we remove expired objects from T, and report approximate skyline objects to the system. Here, $e.p$ and $e'.p$ refers to the left-down coordinate of $e.c$ and $e'.c$ respectively.

Observation 2 *Let o and o' be two objects contained in the leaf nodes $e, e' \in T$ respectively. o' must be $\rho-$dominated by o under the following 3 cases:*

- *(i) If $e = e'$ and $|e.c| = |e'.c| = \rho$, o' must be $\rho-$dominated by o;*
- *(ii) If o dominates o', o' must be $\rho-$dominated by o;*
- *(iii) If $|e.c| = |e'.c| = \rho$ and $e.p[i] \geq e'.p[i] (i \in [1, d])$, o $\rho-$dominates o'.*

The intuition behind case (i) of Observation 2 is explained in Theorem 1. For case (ii) of Observation 2, as the arrived order of o must be larger than that of o', and o dominates o', o' must be $\rho-$dominated by o. For case (iii) of Observation 2, if $e.p[i] > e'.p[i] (i \in [1, d])$, o must dominate objects contained in

---

Algorithm 2: The Incremental Maintenance Algorithm

Input: A set of cube side lengths C, Quad Tree, the Maximal Memory M, s_m
Output: ρ-Skyline Set Sk_ρ

1   $LSK_m \leftarrow$ BBS$(s_m), T \leftarrow T - O_{exp}$;
2   **for** *each object* $o \in LSK_m$ **do**
3   Leaf node $e \leftarrow$ findLeafNode(T, o);
4      $o.p \leftarrow$ prediction(T, o);
5   **if** $e \neq \emptyset$ **then**
6         $e \leftarrow \emptyset$;
7      $e \leftarrow e \cup o$;
8      $O_{dom} \leftarrow$ findObjDom(T, o);
9      $e_{dom} \leftarrow$ findNodeDom(T, e);
10     $T \leftarrow T - O_{dom} - e_{dom}$;
11   **if** $|T| \leq M$ **then**
12   Phase-II and Phase-III of the Algorithm 1;
13   **if** $|T| = \frac{M}{4}$ **then**
14     $\rho \leftarrow \frac{\rho}{2}$;
15   **return** MK;

$e'.p$. If $e.p[i] = e'.p[i]$, $o[i] + \rho$ must be larger than $o'[i]$. Thus, objects contained in $e'.p$ must be ρ−dominated by o.

Take an example in Fig. 2. After the window slides from W_0 to W_1, objects in s_0 expired from the window, and objects in s_4, i.e., $\{o_{13}, o_{14}, o_{15}\}$ flow into the query window. We first find local skyline objects in s_4, i.e., $\{o_{13}, o_{14}\}$. Then, we insert o_{13} into $e_5 \in T$. As $|e_5.c| = 2^{-3}$ and e_5 contains o_6, o_6 is ρ−dominated by o_{13}. In addition, as o_{13} dominates o_{10}, o_{10} also must be ρ−dominated by o_{13}. Lastly, as $e_5.p[i] \geq e_7.p[i](i \in [1, d])$ is satisfied, o_{14} is ρ−dominated by o_{13}. We can delete the corresponding objects $\{o_6, o_{10}, o_{14}\}$ directly.

The ρ Selection Algorithm. We should dynamically adjust ρ so as to guarantee all elements in C_ρ are loaded to the memory, as well as return high-quality query results to the system. The intuition behind the algorithm is if the number of objects in T is large, we should enlarge ρ so as to keep $|T|$ no larger than T_M. Otherwise, it implies the current ρ may be so large that most objects are deleted, where we should reduce ρ so as to enhance the query result quality.

The ρ selection algorithm is simple. Specially, we monitor the number of objects in T. When $|T|$ achieves to T_M, we re-execute phase II-III of the initialization algorithm to find a new ρ and delete more objects. If $|T|$ reduces to $|C| = \frac{M}{4}$, we update ρ be $\frac{\rho}{2}$.

3.4   The Partition-Based Optimization Algorithm

This algorithm is proposed based on an useful observation. That is, the larger the $o.p$ of an object o, the less importance it is. In many case, o may becomes a meaningless object after some objects flow into the window. Based on this

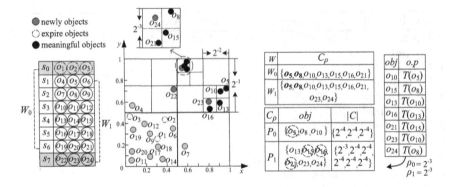

Fig. 3. Running Example of the partition-based Optimization($s = 3, N = 21, T_M = 9$)

observation, in order to enhance the query result quality, we propose a partition-based method that selects a group of ρ based on importance of objects contained in the ρ–CSS C_ρ.

This algorithm is not invoked until the first time $|T|$ achieves T_M after initialization. At that moment, we partition objects in C_ρ into $\{P_1, P_2, \cdots, P_t\}$ such that: (i) $|P_i| \geq 2|P_{i+1}|$; (ii) $\forall o \in P_i$ and $\forall o' \in P_{i+1}$, $o.p > o'.p$ is satisfied. In other words, we partition objects in C_ρ into a group of subsets based on their prediction moment. Then, we form a group of quad-trees $\{T_1, T_2, \cdots, T_t\}$ based on objects in $\{C_1, C_2, \cdots, C_m\}$ respectively.

After partition, for each T_i, we calculate a suitable parameter ρ_i for each T_i via ρ–selection and then remove meaningless objects in T_i respectively. From then on, we use $\{T_1, T_2, \cdots, T_t\}$ for supporting ρ–ACSQ. As the algorithm is simple, we only highlight three points.

Firstly, when processing a newly arrived object o, we scan each quad-tree $\{T_1, T_2, ...\}$, using the algorithm discussed in Sect. 3.3 to find objects that could be deleted. The scanning is terminated when finding an object o' that can dominate o. Then, we insert o into the proper quad-tree based on $o.p$. Secondly, given T_i, when elements in T_i have a chance to become query results, we should partition objects in T_i into a group of new partitions. Thirdly, when the number of objects in these quad-trees achieves T_M, we scan each T_i, adjust ρ_i, and remove meaningless objects based on ρ_i accordingly.

Figure 3 shows the result of $C_\rho = \{o_5, o_8, o_{10}, o_{13}, o_{15}, o_{16}, o_{21}\}$ under the window W_0 with $|C_\rho|$ being smaller than 9. When the window slides to W_1, C_ρ is updated to $\{o_5, o_8, o_{10}, o_{13}, o_{15}, o_{16}, o_{21}, o_{23}, o_{24}\}$, i.e., achieving to 9. In this case, we form the partition via the optimization algorithm (See Fig. 3). After partitioning, $P_0 = \{o_5, o_8, o_{10}\}$, $P_1 = \{o_{13}, o_{15}, o_{16}, o_{21}, o_{23}, o_{24}\}$. We then calculate the parameters ρ_1 and ρ_2, i.e., that are 2^{-3} and 2^{-3} respectively. Lastly, we update P_0 and P_1 based on ρ_1 and ρ_2. Finally, P_0 and P_1 are updated to $\{o_8, o_{10}\}$ and $\{o_{13}, o_{23}, o_{24}\}$ respectively.

The Cost Analysis. When processing a newly arrived object o, as the height of each T_i is bounded by $\log \frac{1}{\rho_i}$, and ρ_i is a constant, the cost of inserting an object o into T_i is bounded by $\mathcal{O}(1)$. When we try to find objects ρ−dominated by o, as each object will be deleted only once, this part of cost is $\mathcal{O}(1)$. As the partition amount is bounded by $\log \rho^{d-1}$, the overall cost of processing a newly arrived object is bounded by $\mathcal{O}(d)$. Besides, the cost of processing expired object is $\mathcal{O}(1)$. Therefore, the overall cost of processing one object is $\mathcal{O}(d)$.

4 Performance Evaluation

In this section, we first explain the experiment settings. Next, we conduct extensive experiments to demonstrate the efficiency of ρ−SEAK over sliding window.

4.1 Experiment Settings

Data sets. The experiments are based on two real data sets named STOCK and TRIP, and two synthetic data sets named NORMAL and RANDOM respectively. STOCK contains a set of 1G trading records of about 2300 stocks in 24 months. The second real data set is TRIP, which contains a set of 1G trip records in New York City, USA. Objects in the synthetic data sets obey the normal distribution and the random distribution respectively.

Parameter Settings. In our experiment, we evaluate algorithm performance differences under the following parameters, including the window size N, speed ratio s, the error threshold ρ, and the maximum number of objects the system could load M. The parameter settings are shown in Table 2.

Table 2. Parameter Settings.

Parameter	value
N	100KB, 500KB, **1MB**, 5MB,10MB
s	0.1%,0.5%,**1%**,5%, 10%($\times$ N)
ρ	0.001,0.005, **0.01**, 0.05 ,0.1
M	100, **200**, 400, 800, 1600

Competitors. In addition to ρ−SEAK, we also implement n-of-N [1], LAZY [5], and BBDS [6] for answering continuous skyline query as competitors. Note that, only the ρ−SEAK could self-adaptively adjust ρ. For the others, when the system cannot load all candidate objects, we randomly discard some objects contained in the candidate set. All algorithms are implemented in C++ with 6226R CPU @ 2.90GHz 16·2 core processor.

4.2   Experimental Evaluation

In this section, we first compare the performance of ρ−SEAK with its competitors. Next, we evaluate the impact of ρ to the algorithm performance.

Running Time and Space Cost Comparison. The first set of experiments evaluate the algorithm performance under different window length N, i.e., ranging from 100 K to 10 M, with the other parameters being defaulted. As shown in Fig. 4(a)-4(d), with the increase of N, we have the following findings. Firstly, ρ−SEAK performs the best of all. Its running time is about 0.65 times that of LAZY, and 0.55 times of BBDS. Secondly, the performance of these three competitors decreased slightly, while ρ−SEAK decreases the most slowly. The main reason is both n-of-N, and LAZY needs to maintain all potential skyline objects in the index, resulting in high maintenance cost. In contrast, ρ−SEAK only maintains a subset of candidates.

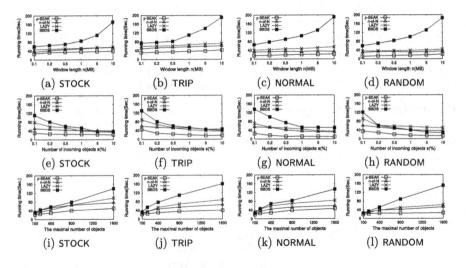

Fig. 4. Running time comparison of different algorithms under different data sets.

Next, we evaluate the algorithm performance under different s, i.e., ranging from 1K to 0.1M. The other parameters are set by default. As shown in Fig. 4(e)-4(h), ρ−SEAK still performs the best of all. With the increase of s, the running time of these four algorithms decreases slightly. The reason is, with the increasing of s, more objects could be deleted directly when they flow into the window.

We then evaluate the impact of M on the algorithm performance. As shown in Fig. 4(i)-4(l), the running time of these four algorithms is increasing with the increase of M. The reason behind it is, with the increase of M, these four algorithms all have to maintain more objects. However, it is important to note that ρ−SEAK efficiently prunes newly arrived objects in the window and deletes most objects contained in leaf nodes, retaining only one object with the highest arrived order. As a result, $\rho-$SEAK achieves the lowest running time among all methods, while using minimal resources.

In the following, we compare the space cost of $\rho-$SEAK with LAZY, BBDS and n-of-N under different window length N. As shown in Table 3. The space cost of $\rho-$SEAK is less than the other three algorithms. The reason is $\rho-$SEAK uses the concept of $\rho-$dominate to delete objects. In many cases, the size of the candidate set is approaching $\frac{M}{4}$. In addition, $\rho-$SEAK makes full use of the parameter s to delete newly arrived objects at the moment they arrive in the window. In this way, we can further reduce the space cost. By contrast, LAZY, BBDS and n-of-N do not use the parameter s for pruning newly arrived objects.

Table 3. Space cost size under different window length(MB)

Algorithm	STOCK						NORMAL					
	0.1	0.2	0.5	1	5	10	0.1	0.2	0.5	1	5	10
$\rho-$SEAK	0.209	0.232	0.246	0.278	0.302	0.311	0.243	0.266	0.28	0.312	0.323	0.332
LAZY	0.881	0.904	0.918	1.061	1.181	1.194	0.629	0.652	0.666	0.809	0.949	0.942
n -of-N	0.757	0.78	0.794	0.826	0.974	0.956	0489	0.512	0.526	0.584	0.705	0.714
BBDS	1.205	1.228	1.242	146	1.513	1.633	1.646	1.341	1.361	1.377	1.615	1.748

Table 4. Effect of the parameter ρ

Dataset		The Parameter(ρ)									
		0.001	0.002	0.003	0.005	0.01	0.02	0.03	0.04	0.05	0.1
STOCK	Running Time	38.25	36.16	35.99	34.33	33.53	30.15	29.46	27.49	26.33	25.01
	Accuracy	0.99	0.99	0.99	0.98	0.96	0.95	0.93	0.92	0.90	0.89
	cost	0.36	0.36	0.36	0.33	0.29	0.28	0.27	0.25	0.23	0.21
TRIP	Running Time	42.30	41.69	40.12	38.40	37.37	36.51	35.12	33.26	32.14	29.70
	Accuracy	0.99	0.99	0.99	0.97	0.96	0.94	0.92	0.91	0.90	0.88
	cost	0.33	0.33	0.30	0.27	0.25	0.24	0.23	0.22	0.21	0.19
RANDOM	Running Time	23.24	22.68	21.25	19.28	17.77	17.03	16.25	16.04	15.55	13.01
	Accuracy	0.99	0.99	0.99	0.96	0.95	0.93	0.92	0.91	0.91	0.90
	cost	0.36	0.34	0.33	0.32	0.30	0.28	0.27	0.25	0.24	0.20
NORMAL	Running Time	26.68	25.62	24.48	22.56	21.04	20.15	19.27	18.61	17.79	15.81
	Accuracy	0.99	0.99	0.98	0.98	0.97	0.96	0.96	0.95	0.94	0.91
	cost	0.34	0.34	0.33	0.31	0.29	0.28	0.27	0.25	0.23	0.23

Effect of the Parameter ρ. Last of this section, we evaluate the impact of ρ to the performance of $\rho-$SEAK. Under this experiment, the threshold ρ is set in advance, i.e., ranging from 0.001 to 0.1. The other parameters are set by default. As shown in Table 4. The running time of $\rho-$SEAK is decreased when ρ is increasing. The reason is that as ρ becomes larger, more objects in the window could be deleted. However, the accuracy of $\rho-$SEAK is decreased with the increasing of ρ. We also find that, when ρ reaches 0.005, the accuracy of $\rho-$SEAK tends to be stable. Therefore, it is very important to find an appropriate error threshold ρ based on the number of objects that the system can load, and the size of the candidate set.

5   Conclusion

In this paper, we propose a novel framework named $\rho-$SEAK for supporting $\rho-$ approximate continuous skyline query over data stream. It uses quad-tree for indexing streaming data, uses side-length of cubes corresponding to leaf nodes of the quad-tree for selecting a reasonable ρ, achieves the goal of supporting $\rho-$approximate skyline query under memory limitation-based data stream. Through extensive experiments on diverse datasets with various distributions, our results demonstrate the superior performance of $\rho-$SEAK.

Acknowledgement. This paper is partly supported by the National Key Research and Development Program of China(2020YFB1707901), the National Natural Science Foundation of Liao Ning(2022-MS-303, 2022-MS-302,and 2022-BS-218), the National Natural Science Foundation of China (62102271, 62072088, Nos. U22A2025, 62072088, 62232007, 61991404), and Ten Thousand Talent Program (No.ZX20200035).

References

1. Xuemin, L., Yidong, Y., Wei, W.: Stabbing the sky: efficient skyline computation over sliding windows. In: ICDE, pp. 502-513 (2005)
2. Soundararajan, R., Kumar, S.R., et al.: Retraction Note: skyline query optimization for preferable product selection and recommendation system (2023)
3. Nikos, S., Gautam, D., Nick, K.: Categorical skylines for streaming data. In: SIGMOD, pp. 239–250 (2008)
4. Lu, M., Sun, Y., Duan, H., et al.: A trend extraction method based on improved sliding window. In: Proceedings of the 9th International Conference on Computer Engineering and Networks, pp. 415–422 (2021)
5. Yufei, T., Papadias, D.: Maintaining sliding window skylines on data streams. In: IEEE Transactions on Knowledge and Data Engineering, pp. 377–391 (2006)
6. Xinjun, C., Bai, M., Dong, H., Wangguo, R.: An efficient processing algorithm for ρ-dominant skyline query. Chin. J. Comput. (2011)
7. Liang, S., Peng, Z., Yan,J.: Adaptive mining the approximate skyline over data stream. Int. Conf. Comput. Sci. (3), 742–745 (2007)
8. Xingxing, X., Jianzhong L.: Sampling-based approximate skyline calculation on big data. In: COCOA, pp. 32–46 (2020)
9. Tianyi, L., Yu, G., Xiangmin, Z., Qian, Ma., Ge, Yu.: An effective and efficient truth discovery framework over data streams. In: EDBT, pp. 180–191 (2017)
10. Borzsonyi, S., Kossmann, D.: The skyline operator. In: ICDE, pp. 421–430 (2001)
11. Donald, K., Frank, R., Steffen, R.: Shooting stars in the sky: an online algorithm for skyline queries. In: VLDB, pp. 275–286 (2002)
12. Dimitris, P., Yufei, T., Greg, F., Bernhard, S.: An optimal and progressive algorithm for skyline queries. In: SIGMOD, pp. 467–478 (2003)
13. Tianyi, L., Lu, C., Christian, S.: TRACE: real-time compression of streaming trajectories in road networks. In: PVLDB, pp. 1175–1187 (2021)

Identifying Backdoor Attacks in Federated Learning via Anomaly Detection

Yuxi Mi[1], Yiheng Sun[2], Jihong Guan[3], and Shuigeng Zhou[1(✉)]

[1] Fudan University, Shanghai 200438, China
{yxmi20,sgzhou}@fudan.edu.cn
[2] Tencent, Shenzhen 518000, China
elisun@tencent.com
[3] Tongji University, Shanghai 201804, China
jhguan@tongji.edu.cn

Abstract. Federated learning has seen increased adoption in recent years in response to the growing regulatory demand for data privacy. However, the opaque local training process of federated learning also sparks rising concerns about model faithfulness. For instance, studies have revealed that federated learning is vulnerable to backdoor attacks, whereby a compromised participant can stealthily modify the model's behavior in the presence of backdoor triggers. This paper proposes an effective defense against the attack by examining shared model updates. We begin with the observation that the embedding of backdoors influences the participants' local model weights in terms of the magnitude and orientation of their model gradients, which can manifest as distinguishable disparities. We enable a robust identification of backdoors by studying the statistical distribution of the models' subsets of gradients. Concretely, we first segment the model gradients into fragment vectors that represent small portions of model parameters. We then employ anomaly detection to locate the distributionally skewed fragments and prune the participants with the most outliers. We embody the findings in a novel defense method, ARIBA. We demonstrate through extensive analyses that our proposed methods effectively mitigate state-of-the-art backdoor attacks with minimal impact on task utility.

Keywords: Federated learning · Backdoor attack · Anomaly detection

1 Introduction

Federated learning (FL) [15,21] is a rapidly evolving machine learning paradigm that enables the collaborative training of a shared global model across multiple participants. The parameters of the shared model are iteratively updated under the orchestration of a central server by synchronizing the participants' *local model updates*. Federated learning offers effective protection of data privacy [14], as the sensitive training data is always retained on edge devices.

© The Author(s), under exclusive license to Springer Nature Singapore Pte Ltd. 2024
X. Song et al. (Eds.): APWeb-WAIM 2023, LNCS 14333, pp. 111–126, 2024.
https://doi.org/10.1007/978-981-97-2387-4_8

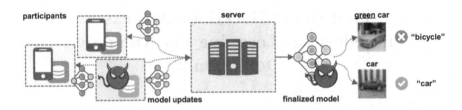

Fig. 1. Backdoor attack in FL systems. The attacker embeds a subtle modification in the shared model that changes model's behavior on inputs with backdoor triggers.

The fundamental aim of FL (as with any machine learning scheme) is to develop a faithful model that accurately represents and generalizes from the training data of all participants. However, recent studies show the faithfulness of FL models could be especially prone to malicious threats, as the distributed nature of FL hinders the server from auditing the training process, such as by purging contaminated data (Fig. 3(a)). Concretely, an attacker controlling some compromised participants can engage in *poisoning attacks* [8,16,20,31], by intentionally injecting malicious contributions to the shared model (*e.g.*, by training on contaminated data [3] or providing deceptive model updates [13,18]), to downgrade the predictions of the finalized model.

This paper investigates the targeted form of poisoning attacks known as the *backdoor attack* [8,16,20,31]. It differs from conventional poisoning in that the attack is both *targeted* and *stealthy*: The attacker embeds a subtle modification (*i.e.*, the *backdoor*) into model parameters, such that contaminated model behaves most time normally, yet in an incorrect and potentially destructive way when its input contains a specific trigger (Fig. 1). For instance, an undermined model may misclassify an image of *green* (the trigger) car as a bicycle while still classifying other car images correctly. Backdoor attacks are difficult to detect since as the backdoor is triggered only in rare cases, their negative (therefore, distinguishable) influence on model performance could be minimized.

We advocate an effective defense against backdoor attacks, to identify and prune compromised participants by examining their model gradients. We start with a key observation: as implanting backdoors involves changes in the attacker's data distribution and training objectives, *the presence of backdoors could be reflected as discernible disparities in terms of gradients' step sizes and directions*, due to the nature of gradient descent (Fig. 2). Therefore, one would be able to carry out defenses by examining the gradients.

However, we find it could be insufficient to discriminate gradients by a single solitary rudimentary metric, say, by examining the gradients' magnitude and orientation (Fig. 3(b)). Such a method is prone to blur the discriminative boundaries between malicious and normal gradients, thus impeding their effective separation. To reconcile the drawback, we propose to decouple the model gradients into subset vectors, *fragments*, and distinguish them by their statistical distribution (Fig. 3(c)). It turns out that backdoored gradients can be robustly and accurately identified by their distributional bias. We concretize our find-

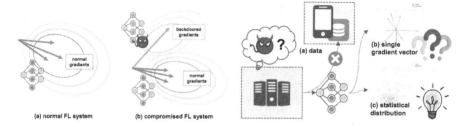

Fig. 2. Gradient Descent. (a) Normal participants usually produce similar gradients. (b) Gradients of backdoored clients bias in terms of magnitude and orientation, as they are obtained through skewed data distribution and different objectives.

Fig. 3. Paradigm of our idea. By defense: (a) Server cannot audit training data. (b) Directly examining gradient vectors could produce ambiguous results, as the decision boundary is unclear. (c) Distribution bias provides clear and robust disparity.

ings into a novel defense method, ARIBA, where the distributional disparity is leveraged by an anomaly-detection-based technique to reach pruning decisions.

Our contributions are three-fold: (1) We present an in-depth study on backdoor attacks by the attacker's threat model and techniques. (2) We advocate an effective defense, to identify compromised participants by the distributional bias of their subset gradient vectors. (3) We concretize our findings into the proposed ARIBA method and analyze its effectiveness through extensive experiments.

2 Related Work

2.1 Attacks on Model Faithfulness

The faithfulness of the FL models is prone to malicious threats. Poisoning attacks have long been explored in the context of centralized learning [5,9,11,20,22] and is extended to FL settings very recently. The attacker aims to manipulate the training process such that the trained model biases or downgrades its prediction in an attacker-desired specific way. For instance, a model compromised by *label flipping* [26] could misclassify all images of cats into "dogs". In FL, the attack can be engaged in by contaminating the data [24,26] or by tampering with the training process or finalized models [12,32].

Backdoor attacks are targeted and stealthy varieties of poisoning attacks. Specifically, the model's behavior is subtly modified by implanting a backdoor. Its performance downgrades only in the presence of a backdoor trigger, which could be manifested in various forms such as specific data [2], data with unique fingerprints [9], and data carrying certain semantic information [1]. In FL settings, attacks [1,2,25,27,29] commonly employ *scaling, i.e.,* multiplying the attacker's local model weights by a scaling factor σ, as means to survive from the aggregation, later elaborated on in Sect. 3.2.

2.2 Defenses Against Backdoor Attack

The distributed nature of FL and the stealthiness of backdoors both make the detection of backdoor attacks challenging: The server is unable to predicate the existence of backdoors by either examining training data or testing model performance. To this end, most prior arts focus on examining the model itself. We roughly categorize their means into three branches.

Attack-aware Aggregations. The server may replace FedAvg with *byzantine-resilient* aggregations such as Krum, coordinate-wise median (CooMed), and GeoMed [4,6,10,30], which prevent local model updates skewed in distribution from being aggregated. However, their effectiveness heavily relies on the specific distribution of local training data. Research [1] further suggests a nullifying of their defense if the attackers choose proper covert strategies.

Examination on Model Gradients. [2,7] are relevant to ours as we all differentiate malicious and benign updates by the magnitude or orientation of their model gradients. Prior arts examine general statistical traits such as the l2-norm of [2] or the cosine similarity between [7] model parameters. These methods mainly suffer two drawbacks: (1) They involve hyper-parameters to depict certain detection thresholds. Fine-tuning these hyper-parameters requires *a priori* knowledge about the attacker's capacity, which is not the case in the real world. (2) The coarse metrics they employed lack clues for detailed model behaviors, which could result in an ambiguous detection of some malicious updates.

Dedicated Defenses. Recent discovery [9] suggests a backdoor attack is engaged by activating certain model neurons. To this end, [28] proposes a pruning defense to identify and remove suspicious neurons by their activation. However, this defense only protects the inference phase against certain types of backdoors. [17] proposes a spectral anomaly detection technique that detects compromised model updates in their low-dimensional embeddings. However, their implementation relies on centralized training on auxiliary public datasets, which is unobtainable in a majority of FL settings.

3 Preliminaries and Attack Formulation

We first set up some basic notions. Let $\langle X, y \rangle$ denote a data sample and its corresponding label. $f(\cdot, \theta)$ denotes the model parameterized by θ. $l(\cdot, \cdot)$ denotes a generic loss function. $\mathbf{D} = \{D_1, \ldots, D_n\}$ denotes the training datasets.

3.1 Federated Learning

Federated learning [15,21] develops a shared global model $f(\cdot, \theta)$ by the collaborative efforts of n participants of edge devices $\mathbf{P} = \{P_1, \ldots, P_n\}$ under the coordination of a central server S. Each participant P_i possesses its own private training dataset $D_i \in \mathbf{D}$. To train the global model, rather than sharing the private data, participants train a copy of the model locally and synchronize the model updates with the server. Specifically, at initialization, S generates a model $f(\cdot, \theta^0)$ with initial parameters θ^0 and advertises it to all $\{P_i\}$. At each global round t, each P_i aligns its local model with received global weight $\theta_i^{t+1} = \theta^t$, and trains the model for several local iterations with $\arg\min_{\theta_i^{t+1}} l(f(X, \theta_i^{t+1}), y)$, where $\langle X, y \rangle \in D_i$. It then shares the updated θ_i^{t+1}. The server renews the global model by aggregating all received θ_i^{t+1} using the FedAvg [19] algorithm:

$$\theta^{t+1} = \theta^t + \frac{\eta}{n} \sum_{i=1}^{n} (\theta_i^{t+1} - \theta^t), \tag{1}$$

where η is the global learning rate. Note FedAvg can be replaced by attack-aware aggregations [4,6,10,30] discussed in Sect. 2.2. S then advertise θ^{t+1} again to all $\{P_i\}$ and the training continues iteratively, until the model reaches convergence at round r. The model is finalized as $f(\cdot, \theta^r)$.

3.2   Threat Model

The Attacker's Capability. We consider an attacker who gains control of a small subset of $k \ll n$ compromised participants, denoted as $\mathbf{P}_m = \{P_{m_1}, \ldots, P_{m_k}\}$. This could be achieved by injecting attacker-controlled edge devices into the FL system, or by deceiving some benign clients. We assume the attacker can develop malicious model updates by contaminating the participants' local training data or directly manipulating their model weights. We assume an honest server who endeavors to eliminate the attack.

The Attacker's Goal. The attacker wants to implant a backdoor in the global model weight (denote the manipulated weight as θ') such that the finalized model $f(\cdot, \theta'^r)$ produces attacker-desired incorrect outcomes only when the query $\langle X, y \rangle$ contains a backdoor trigger (see Sect. 3.3). Concretely for a classification model, it should predict $\tilde{y} \triangleq f(X, \theta'^r) \neq y$ for backdoored X, and $\tilde{y} = y$ otherwise. We assume the attacker takes two steps towards the objective: it first develops the backdoor in participants' local models by training on a mix of correct and backdoored data [9], then introduces it to the global model by *model replacement* [1].

Model Replacement. The attacker attempts to undermine the global model with backdoored weights $\{\theta_{m_i}\}$. However, we argue it cannot directly share $\{\theta_{m_i}\}$ as model updates. As $k \ll n$, the effect of backdoors can be diluted by other participants' updates during aggregation, and the global model forgets the backdoor quickly. Instead, it shall first wait until the model nearly converges at round t, where the updates of other participants start to cancel out, *i.e.*,

$$\sum_{P_i \in \mathbf{P} \setminus \mathbf{P}_m} (\theta_i^{t+1} - \theta^t) \approx 0. \tag{2}$$

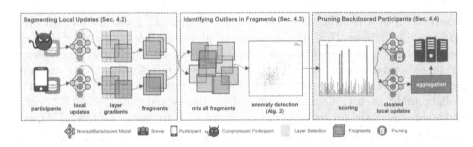

Fig. 4. Pipeline of ARIBA. (1) To earn a clear and robust disparity between backdoored and normal model gradients, we decouple them into fragments of subset gradient vectors. (2) We employ anomaly detection to identify the compromised participants by their skewed distribution of fragments. (3) We identify and prune the participants that suspiciously carry backdoors by scoring their outliers.

Then, to survive the averaging in FedAvg, it calculates a *scaled* local model update $\hat{\theta}_{m_i}^{t+1}$ for each compromised participant by multiplying its original updates with a scaling factor σ_{m_i}:

$$\hat{\theta}_{m_i}^{t+1} = \theta^t + \sigma_{m_i}(\theta_{m_i}^{t+1} - \theta^t), where \sum_{P_i \in \mathbf{P}_m} \sigma_{m_i} \triangleq \sigma = \frac{n}{\eta}. \tag{3}$$

Finally, we let the attacker share all $\{\hat{\theta}_{m_i}^{t+1}\}$ as model updates. As a result, the global model weight can be replaced by the attacker's updates in Eq. 1 as

$$\theta^{t+1} = \theta^t + \frac{\eta}{n}\left(\sum_{P_i \in \mathbf{P}_m}(\hat{\theta}_{m_i}^{t+1} - \theta^t) + \sum_{P_i \in \mathbf{P}\backslash\mathbf{P}_m}(\theta_i^{t+1} - \theta^t)\right)$$

$$\approx \theta^t + \frac{\eta}{n}\sum_{P_i \in \mathbf{P}_m}(\hat{\theta}_{m_i}^{t+1} - \theta^t) = \frac{\eta}{n}\sum_{P_i \in \mathbf{P}_m}\sigma_{m_i}\theta_{m_i}^{t+1} \triangleq \theta_m^{t+1}, \tag{4}$$

where θ_m^{t+1} denotes the collaborative efforts of all compromised participant. Therefore, the attacker conveniently implants the backdoor into the global model. It iterates the attack until the model $f(\cdot, \theta^{\prime r})$ is finalized with the backdoor.

3.3   Choice of Backdoor Triggers

The backdoor triggers can be manifested in various forms. We concretely study three types of state-of-the-art (SOTA) backdoors with different triggers in image classification: (1) **Targeted backdoors** [2]: The attacker possesses a collection of images with tampered labels. The finalized model is expected to misclassify the exact collection of images if they appear in inference queries. (2) **Pattern backdoors** [9]: The attacker endows arbitrary images from a certain class with a unique fingerprint, which is concretized as a bright pixel pattern at the corner of images. The model misclassifies images with the same pattern into an attacker-desired class. (3) **Semantic backdoors** [1]: The attacker chooses a naturally occurring semantic (*e.g.*, *green* car) rather than artificial fingerprints as the trigger. This makes the backdoor more stealthy as it requires no modification of images. Figure 5 exemplifies the three types of backdoors.

4   Methodology

We now discuss the motivation and technique details of our proposed method, ARIBA. The name comes from the core functionality of our defense, *i.e.*, enabling *a*ccurate and *r*obust *i*dentification of *b*ackdoor *a*ttacks.

4.1   Motivation

We propose to identify backdoor attacks in FL by letting the server examine the participant's shared model updates, concretely, by uncovering outliers in the segmented fragments of model gradients. To elucidate our motivation, we shall begin by revisiting the principle of gradient descent. Gradient descent is regarded as the most fundamental optimization method in machine learning. It iteratively adjusts the parameters of a model through certain *step sizes* in the *direction* of the steepest descent of a cost function, which gauges the disparity between the model's predicted output and ground truth under certain objectives. The step size and direction of adjustment can be reflected as the *magnitude* and *orientation* of the model's gradient vector, respectively.

In an uncompromised FL system, during each global round, the model gradients among normal participants' updates should possess similar magnitudes and orientations (Fig. 2(a)). This is due to they are trained under the exact same objectives and roughly consistent data distributions. However, it is not the case if a portion of participants are compromised by the attacker and submit backdoored model updates: We find that their model gradients deviate from the normal ones, as the contaminated data and backdoor influence their data distributions and optimization on cost functions, resulting in distinct gradients (Fig. 2(b)). One could leverage such disparity to identify backdoors.

However, research [1] has shown that directly distinguishing the disparity [2,7] could provide unsatisfactory protection. The drawbacks are two-fold: (1) It is a crude approach to depend on a single *coarse metric* (say, l2-norm or cosine similarity) to identify the gradient as a whole. The decision boundary between malicious and normal gradients in magnitude and orientation is not always clear (Fig. 3(b)), making it difficult to effectively separate them. (2) Attackers can easily conceal their intentions by optimizing the local model towards the elimination of corresponding metrical differences [1], which nullifies the defense.

Our goal is to find a clear and discernible disparity between gradients that allows accurate and robust identification. Studies in model perception [23] have widely shown that deep neural networks learn about the entirety from a fusion

---

Algorithm 1. ARIBA defense against backdoor attacks

Input: Local model updates $\{\theta_i^{t+1}\}_{P_i \in \mathbf{P}}$ at round t.
Parameter: A generous estimated number of malicious participants $\tilde{k}$.
Output: Local model updates of normal participants $\{\theta_i^{t+1}\}_{P_i \in \mathbf{P} \setminus \tilde{\mathbf{P}}_m}$.

1: **for all** $P_i \in \mathbf{P}$ **do**
2:     Obtain accumulated gradients: $\delta_i^{t+1} \leftarrow \theta_i^{t+1} - \theta^t$.
3:     Mean subtraction: $\hat{\delta}_i^{t+1} \leftarrow \delta_i^{t+1} - \frac{1}{n} \sum_{P_i \in \mathbf{P}} \delta_i^{t+1}$.
4:     Form segmented fragments: $\mathbf{M}_i = \{M_{i,1}, \ldots, M_{i,m}\} \leftarrow \hat{\delta}_i^{t+1}$
5: **end for**
6: $\mathbf{M} \leftarrow \{\mathbf{M}_1, \ldots, \mathbf{M}_n\}$
7: Obtain the scoring matrix: $\mathbf{S} = \{s_{i,j}\}^{n \times m} \leftarrow \text{Scoring}(\mathbf{M}, \tilde{k})$
8: **for all** $P_i \in \mathbf{P}$ **do**
9:     Calculate each participant's score: $S_i = \sum_j s_{i,j}$.
10: **end for**
11: Speculate malicious participants: $\tilde{\mathbf{P}}_m \leftarrow \{\tilde{P}_{m_1}, \ldots, \tilde{P}_{m_{\tilde{k}}}\}$,
     where $\tilde{P}_{m_i}$ is the participant with i-highest S_i ($i \leq \tilde{k}$).
12: **return** $\{\theta_i^{t+1}\}_{P_i \in \mathbf{P} \setminus \tilde{\mathbf{P}}_m}$

---

of local features. Therefore, we argue the disparity between malicious and normal model updates should also be reflected as the accumulated disparity in their local subsets of parameters. To testify, we segment the model gradients into *fragments*. Each fragment is a vector that represents the gradients of a small portion of model parameters (*e.g.*, kernels from models' convolutional layers). We visualize the distribution of fragments by principal component analysis (PCA). In Fig. 3(c), we find clearly a distinguishable statistical difference between malicious and normal model gradients, as their fragments form separate clusters. Such difference provides a more robust and distinguishable pattern than analyzing gradient magnitudes and orientations. Getting inspired, we propose to *identify the backdoored model updates by discerning the statistical bias among their subsets of model gradients, i.e., fragments.*

We concretize our findings in the proposed ARIBA defense, which can serve as a plug-in detection block in common FL systems. We first process the participants' model updates to obtain decoupled fragments (in our specific case, gradients of convolutional kernels). Then, we employ unsupervised anomaly detection as a means to uncover the fragments belonging to skewed distributions. We maintain a scoring matrix that counts the number of outliers for each participant. Finally, we prune the participants who scored highest, as their model updates are most suspicious in regard to the presence of backdoors. Figure 4 illustrates the framework of ARIBA. In Sect. 5, we demonstrate through extensive experiments that our method provides simple yet effective defenses.

4.2   Segmenting Local Updates

We start by segmenting fragments from the participants' updates. At any global round t, the server has on hand the shared model updates of all participants

---

Algorithm 2 Scoring by Mahalanobis distance

Input: Fragments of all participants $\mathbf{M}$ and the estimated $\tilde{k}$.
Output: The scoring matrix $\mathbf{S}$.
1: Initialize the scoring matrix: $\mathbf{S} = \{s_{i,j}\}^{n \times m}$.
2: Calculate the covariance of fragments: $\Sigma \leftarrow cov(\mathbf{M})$.
3: Calculate the mean of fragments: $\bar{M} \leftarrow mean(\mathbf{M})$.
4: **for all** $i \leq n, j \leq m$ **do**
5:    Calculate the Mahalanobis distance for each $M_{i,j} \in \mathbf{M}$:
      $d_{i,j} \leftarrow \sqrt{(M_{i,j} - \bar{M})^T \Sigma^{-1}(M_{i,j} - \bar{M})}$.
6: **end for**
7: Set: $s_{i,j} \leftarrow 1$, if $d_{i,j}$ is one of the top-$(\tilde{k}m)$ largest $\{d_{i,j}\}$; $s_{i,j} \leftarrow 0$, otherwise.
8: **return** $\mathbf{S}$

$\{\theta_i^{t+1}\}$ (some may be backdoored) and the global model weight of one round behind θ^t. The server obtains each participant's changes in model weights by

$$\delta_i^{t+1} \triangleq \theta_i^{t+1} - \theta^t. \tag{5}$$

Note δ_i^{t+1} actually represents the *accumulated* gradients of participants' local iterations during round t and is adjusted by its local learning rate. We here and later still call it gradients for simplicity. As we are interested in the *difference* between malicious and normal gradients, we further require the server to perform a mean subtraction on all $\{\delta_i^{t+1}\}$, as

$$\hat{\delta}_i^{t+1} \triangleq \delta_i^{t+1} - \frac{1}{n} \sum_{P_i \in \mathbf{P}} \delta_i^{t+1}, \tag{6}$$

which helps emphasize their disparity.

To segment fragments, concretely, we pick a convolutional layer from the model and extract the gradients from each of its kernels, as illustrated in Fig. 4. Recall the kernels are the local feature detectors of CNN composed of small matrices of weights. We choose kernels as their gradients are semantically meaningful, but such practice is not a must and one is free to segment the model gradients in arbitrary ways, as long as the outcome is favorable for statistical analyses. We denote the derived fragments of participant P_i as $\mathbf{M}_i \triangleq \{M_{i,1}, \ldots, M_{i,m}\} \subset \hat{\delta}_i^{t+1}$. Figure 6(b) visualizes some exemplar fragments of kernels from different P_i, where we can clearly observe the difference in pattern between malicious and normal fragments. This further testifies to our findings in Sect. 4.1.

4.3   Identifying Outliers in Fragments

Recall we try to discriminate the backdoored updates by the distributional bias of their fragments (Fig. 3(c)). However, how can the server make decisions based on the distributional disparity? We elucidate that *anomaly detection* can be leveraged as a convenient tool: Consider each fragment as an individual datum and

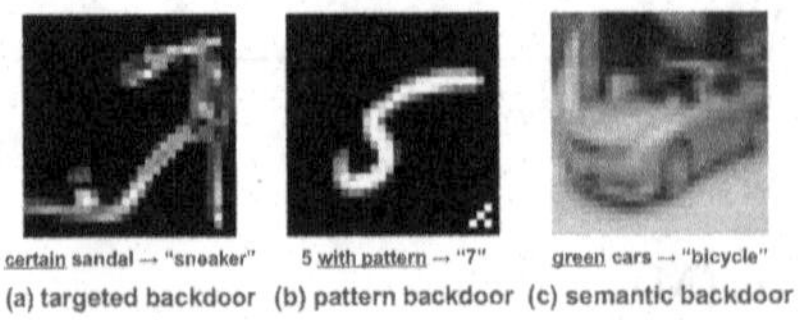

certain sandal → "sneaker"   5 with pattern → "7"   green cars → "bicycle"
(a) targeted backdoor   (b) pattern backdoor   (c) semantic backdoor

(a) distribution of score   (b) visualization of kernel gradients

Fig. 5. Three types of backdoors we studied. The trigger can manifest as (a) certain samples, (b) samples carrying specific patterns, (c) samples with certain semantics. The same settings of backdoors are adopted in our experiments.

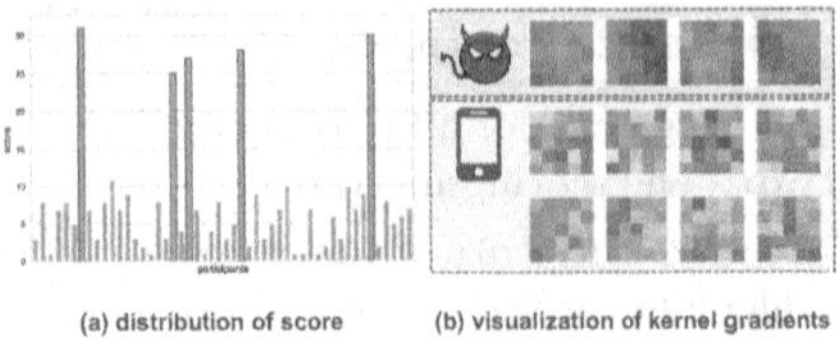

Fig. 6. (a) The scores of outliers of a specific experimental case. All backdoored participants (marked read) are clearly discerned. (b) A difference in pattern can be observed from the visualization between some malicious and normal fragments.

map all fragments onto a hyperspace. Since most normal participants exhibit similar statistical distributions, their fragments will densely populate the projected region. Conversely, backdoored participants' fragments are liable to project onto isolated and sparsely populated regions due to the observed skewed distribution. Thus, by performing anomaly detection in the hyperspace, the backdoored fragments are highly susceptible to being identified as outliers. As a result, *participants with significantly more outliers in their fragments are suspicious of providing backdoored updates.*

To embody the theory, we first gather the fragments of all participants $\mathbf{M} = \{\mathbf{M}_1, \ldots, \mathbf{M}_n\}$. We then feed them into a `Scoring` function parameterized by $\tilde{k}$ (Alg. 2), where unsupervised anomaly detection takes place to mark a portion $(\tilde{k}m)$ of fragments as outliers. Here, $\tilde{k}$ is the server's estimated number of backdoored participants. In practice, the server can choose a generous $\tilde{k}$ that ensures $\tilde{k} > k$. By $(\tilde{k}m)$, we note a *flawless* anomaly detection algorithm would classify m fragments of all $\tilde{k}$ participants as anomalous and classify the remaining as normal. Nevertheless, we anticipate (and tolerate) an approximation as some fragments are sure to be wrongfully classified due to their partial overlapping in distributions. We concretely choose a Mahalanobis-distance-based algorithm to elaborate our method, yet one is free to replace it with any unsupervised anomaly detection algorithms. `Scoring` returns a scoring matrix $\mathbf{S} = \{s_{i,j}\}^{n \times m}$, where $s_{i,j}=1$ if $M_{i,j}$ is marked as an outlier.

4.4   Pruning Backdoored Participants

The server now identifies backdoored participants $\tilde{\mathbf{P}}_m$ by counting the number of outliers. Conveniently, it calculates each participant's score as

$$S_i = \sum_{j=1}^{m} s_{i,j}, \tag{7}$$

and puts participants $\{\tilde{P}_{m_1}, \ldots, \tilde{P}_{m_{\tilde{k}}}\}$ with i-highest S_i $(i \leq \tilde{k})$ as $\tilde{\mathbf{P}}_m$. As the server speculates these $\tilde{k}$ participants to provide backdoored updates, their model updates are pruned from being aggregated. The FedAvg aggregation (Eq. 1) is therefore carried out on *cleaned* $\mathbf{P} \setminus \tilde{\mathbf{P}}_m$.

Note our defense is effective as long as $\mathbf{P}_m \subset \tilde{\mathbf{P}}_m$. Figure 6(a) exhibits the scores $\{S_i\}$ in a specific experimental case with $(n, k, \tilde{k}) = (50, 5, 10)$, from which we can observe (1) the defense is effective as all malicious participants are identified, and (2) the difference between malicious and normal S_i is salient, suggesting a clear disparity in identification thus providing robust defense. We summarize the proposed ARIBA method in Alg. 1.

5   Experiments

5.1   Experimental Settings

We leverage ARIBA to identify the three types of backdoor triggers discussed in Sect. 3.3. FL models are trained on 3 common image datasets, MNIST, Fashion-MNIST, and CIFAR-10. We apply a 4-layer toy model for MNIST and Fashion-MNIST, and a ResNet18 for CIFAR-10. We by default choose $(n, k, \tilde{k}) = (50, 5, 10)$, *i.e.*, the attacker compromises 5 out of 50 participants while the server generously estimates 10 of them as malicious. We later in Sect. 5.5 show over-estimation (choosing $\tilde{k} > k$) impacts model performance very slightly. We set the global learning rate $\eta = 1$, which is in favor of the attacker as a larger η eases model replacement (Eq. 4). We presume identical local learning rates among all participants, concretely, η_p=1e-3, 1e-4, 1e-4 for the three datasets, respectively. Experiments are carried out on an Nvidia 3090 GPU with PyTorch 1.10 and CUDA 11. The same random seed is sampled among all experiments.

5.2   Effectiveness of Our Defense

We first establish the backdoor attacks. Concretely, we train the FL model from scratch for $(t - 1)$ global rounds until it nearly converges. By the threat model in Sect. 3.2, the attacker waits (and behaves normally) till convergence. At round t, it begins with embedding the backdoor by letting each of its compromised participants train local updates on a mix of contaminated (that contains backdoor triggers) and normal data. It then scales the backdoored updates by σ (Eq. 3), where by $(n, \eta) = (50, 1)$ we naturally have $\sigma = 50$. We further presume each compromised participant P_{m_i} has an equal share of $\sigma_{m_i} = 10$.

Baselines. We launch the attacks without defense. We aggregate θ^{t+1} with local updates of all clients $\{\theta_i^{t+1}\}$ with FedAvg (Eq. 1). For each attack, we choose 30 images X' from the test dataset and implant them with the backdoor triggers, as exemplified in Fig. 5. We evaluate the model by test accuracy (*acc.*) and judge the attacker's performance by *backdoor accuracy*, *i.e.*, the proportion of X' the model misclassifies according to its objective ($X'\%$). Higher $X'\%$

Table 1. Summary of primary results. (a) The baseline FL models are compromised by backdoors. (b) Our proposed ARIBA provides an effective and robust defense in terms of confidence and proportion of attackers pruned. (c) Comparison with prior arts.

Backdoors	(a) baseline		(b) with ARIBA				(c) Prior arts						
	acc.	$X'\%$	acc.	$X'\%(\downarrow)$	C	$\mathbf{P}_m\%$	$[2]_1$	$[2]_2$	$[4]$	$[30]$	$[28]$	$[7]$	$[17]$
Targeted	88.87	0.90	90.01	0.13	0.79	1.0	✗	✓	✗	✗	–	–	–
Pattern	98.85	0.83	99.24	0.00	0.95	1.0	✗	✓	✗	✗	✓	✓	–
Semantic	71.30	0.67	72.94	0.23	0.94	1.0	✗	✓	✗	✗	–	–	✓

indicates successful attacks. As illustrated in Table 1(a), all three attacks succeed if without protection as the model wrongfully classifies most of the backdoored samples. Meanwhile, the compromised global model still performs well. This suggests one cannot identify the backdoor by examining model performances.

Effect of Our Defense. Now we plug in ARIBA before aggregation, *i.e.*, updates $\{\theta_i^{t+1}\}$ are first examined by our proposed technique in Sect. 4 where suspicious updates are pruned. Here, we introduce *confidence* $C = (\sum_{\{P_i \in \mathbf{P}_{m,j}\}} s_{i,j})/(\tilde{k}m)$ to measure how sure the server is about the pruning decision: Higher C indicates a better defense, as the server manages to classify more of the attacker's fragments as outliers, which contributes to the identification of backdoors. We illustrate the result of pruning, on the proportion of compromised participants pruned, by $\mathbf{P}_m\%$. We report test and backdoor accuracy as well. Results are summarized in Table 1(b). Note few X' may still be misclassified due to the model's wrong prediction, even without the presence of attacks. We highlight that (1) ARIBA effectively defends all three types of backdoor attacks, as *all* compromised clients are identified and pruned ($\mathbf{P}_m\%=1.0$); (2) We further provide robust defense as the confidence C is high, indicating a clear distinguishability between malicious and normal participants. (3) The model retains high performance regardless of the excessive pruning ($\tilde{k} > k$).

5.3 Comparison with Prior Arts

We compare ARIBA with prior defenses discussed in Sect. 2.2. Results in Table 1(c) are summarized from both previous literature and our experimental studies. Here, "✓, ✗, -" indicates effective defense, ineffective defense, and no result claimed, respectively. Specifically, [2] proposes to examine the model's test accuracy ($_1$) or the l2-norm of local updated weights ($_2$). We note though l2-norm effectively identifies baseline attacks, it can be easily bypassed by advanced attacks in Sect. 5.4. Krum [4] and CooMed [30] are byzantine-resilient aggregations, which defenses were nullified in [1]. [7,17,28] are attack-specific defenses that are not likely to generalize against other attacks. Some of these defenses require certain conditions such as auxiliary test datasets [17] or specific data distribution [28], which further constrains their practical use. Thereby, we argue ARIBA outperforms prior arts in terms of both generality and effectiveness.

5.4   Effectiveness on Advanced Attacks

We further study two varieties of advanced attackers, to illustrate ARIBA can provide effective protection even if against stealthy attack countermeasures.

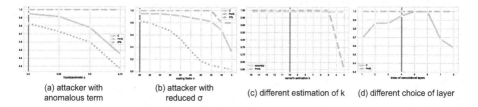

(a) attacker with anomalous term   (b) attacker with reduced σ (c) different estimation of k   (d) different choice of layer

Fig. 7. Experimental results. Note the blue vertical lines mark our default settings. Our proposed method provides robust protection with high C and $\mathbf{P}_m\%$ against two types of advanced attackers, which (a) use anomalous countermeasures to evade detection and (b) improve stealthiness with reduced σ. We further show (c) the excessive estimation of $\tilde{k}$ influences model performance very slightly, and (d) the confidence of defense could vary by the concrete choice of layers. (Color figure online)

Attacker with Anomalous Objectives. Studies [1,2] suggest an attacker that could intentionally evade some metric-based detection. Specifically, it appends an anomalous term $l_a(\cdot)$ to the compromised participants' training objective as $\arg\min_{\theta_{m_i}} ((1-\rho)l(f(X, \theta_{m_i}), y) + \rho l_a(g(\theta_{m_i})))$, where ρ is a hyperparameter and $g(\cdot)$ is the targeted metric. For example, by choosing $g(\theta_{m_i}) = \|\theta_{m_i} - \theta\|_2$ can the attacker deceive l2-norm bounding defense. Figure 7(a) presents the confidence C and $\mathbf{P}_m\%$ of ARIBA together with such attacker's *baseline* $X'\%$ under different ρ. We note the anomalous term *does* affect the confidence however at the cost of lowering backdoor accuracy(in regard to $X'\%$). Nevertheless, ARIBA still provides intact protection by pruning all the compromised clients ($\mathbf{P}_m\%$=1.0).

Attacker with Reduced σ. During model replacement, [1] suggest the attacker could reduce its capacity in exchange for better stealthiness, by choosing smaller $\sigma < \frac{n}{\eta}$ that *partially* replace the global model. We study attackers with different σ in Fig. 7(b). Results indicate ARIBA effectively identifies and prunes under most σ. Only in rare cases where σ is too small would ARIBA miss some participants. However, note the backdoor accuracy is concurrently impaired: At σ=5,10, we argue the attacker's $X'\%$ is too low to incur an effective threat.

5.5   Ablation Study

Server's Estimation on $\tilde{k}$. In Sect. 4.3, the server is let estimate generously on the number of compromised participants $\tilde{k} > k$. Excessive estimation can cause wrongful pruning of normal clients. In Fig. 7(c), we show a generous $\tilde{k}$ is acceptable as it affects model performance very slightly. On contrary, underestimation $\tilde{k} < k$ should be prohibited, since $\mathbf{P}_m\%$ reduces accordingly.

Choice of Convolutional Layers. By default, we choose one layer in the middle of the networks. We here alter the choice to observe its influence on defense. Results in Fig. 7(d) by C and $\mathbf{P}_m\%$ show though all choices provide effective protection, choosing in-the-middle layers mostly benefits robust identification. Note this seems to suggest *the attacker's influence on model weights differs by the stages of model components*, which may be leveraged as more detailed detection clues. We leave it as an interesting open problem due to our limits of space.

6   Conclusion

This paper discusses the backdoor attacks against model faithfulness in FL systems. We present an in-depth study on state-of-the-art attacks by the attacker's goal, capability, and possible attack approaches. By the observation of the magnitude and orientation disparity on the attacker's model gradients, we advocate an effective defense, to identify and prune compromised participants by the distributional bias of their fragments, *i.e.*, gradient vectors of subset model parameters. We concretize our findings into the proposed ARIBA defense and demonstrate through extensive experiments its effectiveness and robustness.

References

1. Bagdasaryan, E., Veit, A., Hua, Y., Estrin, D., Shmatikov, V.: How to backdoor federated learning. In: International Conference on Artificial Intelligence and Statistics, pp. 2938–2948. PMLR (2020)
2. Bhagoji, A.N., Chakraborty, S., Mittal, P., Calo, S.: Analyzing federated learning through an adversarial lens. In: International Conference on Machine Learning, pp. 634–643. PMLR (2019)
3. Biggio, B., Nelson, B., Laskov, P.: Poisoning attacks against support vector machines. arXiv preprint arXiv:1206.6389 (2012)
4. Blanchard, P., El Mhamdi, E.M., Guerraoui, R., Stainer, J.: Machine learning with adversaries: byzantine tolerant gradient descent. In: Proceedings of the 31st International Conference on Neural Information Processing Systems, pp. 118–128 (2017)
5. Chen, X., Liu, C., Li, B., Lu, K., Song, D.: Targeted backdoor attacks on deep learning systems using data poisoning. arXiv preprint arXiv:1712.05526 (2017)
6. Chen, Y., Su, L., Xu, J.: Distributed statistical machine learning in adversarial settings: byzantine gradient descent. Proc. ACM Measur. Anal. Comput. Syst. 1(2), 1–25 (2017)
7. Fung, C., Yoon, C.J., Beschastnikh, I.: Mitigating sybils in federated learning poisoning. arXiv preprint arXiv:1808.04866 (2018)
8. Goldblum, M., et al.: Dataset security for machine learning: data poisoning, backdoor attacks, and defenses. arXiv preprint arXiv:2012.10544 (2020)

9. Gu, T., Dolan-Gavitt, B., Garg, S.: BadNets: identifying vulnerabilities in the machine learning model supply chain. arXiv preprint arXiv:1708.06733 (2017)
10. Guerraoui, R., Rouault, S., et al.: The hidden vulnerability of distributed learning in byzantium. In: International Conference on Machine Learning, pp. 3521–3530. PMLR (2018)
11. Huang, L., Joseph, A.D., Nelson, B., Rubinstein, B.I., Tygar, J.D.: Adversarial machine learning. In: Proceedings of the 4th ACM Workshop on Security and Artificial Intelligence, pp. 43–58 (2011)
12. Ji, Y., Zhang, X., Ji, S., Luo, X., Wang, T.: Model-reuse attacks on deep learning systems. In: Proceedings of the 2018 ACM SIGSAC Conference on Computer and Communications Security, pp. 349–363 (2018)
13. Kairouz, P., et al.: Advances and open problems in federated learning. arXiv preprint arXiv:1912.04977 (2019)
14. Knaan, Y.: Under the hood of the pixel 2: how AI is supercharging hardware. Google AI (2017)
15. Konečný, J., McMahan, H.B., Yu, F.X., Richtárik, P., Suresh, A.T., Bacon, D.: Federated learning: strategies for improving communication efficiency. arXiv preprint arXiv:1610.05492 (2016)
16. Li, Q., et al.: A survey on federated learning systems: vision, hype and reality for data privacy and protection. arXiv preprint arXiv:1907.09693 (2019)
17. Li, S., Cheng, Y., Wang, W., Liu, Y., Chen, T.: Learning to detect malicious clients for robust federated learning. arXiv preprint arXiv:2002.00211 (2020)
18. Li, T., Sahu, A.K., Talwalkar, A., Smith, V.: Federated learning: challenges, methods, and future directions. IEEE Signal Process. Mag. **37**(3), 50–60 (2020)
19. Li, X., Huang, K., Yang, W., Wang, S., Zhang, Z.: On the convergence of FedAvg on Non-IID data. arXiv preprint arXiv:1907.02189 (2019)
20. Liu, Y., et al.: Trojaning attack on neural networks. NDSS (2018)
21. McMahan, B., Moore, E., Ramage, D., Hampson, S., y Arcas, B.A.: Communication-efficient learning of deep networks from decentralized data. In: Artificial Intelligence and Statistics, pp. 1273–1282. PMLR (2017)
22. Rubinstein, B.I., et al.: ANTIDOTE: understanding and defending against poisoning of anomaly detectors. In: Proceedings of the 9th ACM SIGCOMM Conference on Internet Measurement, pp. 1–14 (2009)
23. Selvaraju, R.R., Cogswell, M., Das, A., Vedantam, R., Parikh, D., Batra, D.: Grad-CAM: visual explanations from deep networks via gradient-based localization. In: IEEE International Conference on Computer Vision, ICCV 2017, Venice, Italy, October 22-29, 2017, pp. 618–626. IEEE Computer Society (2017). https://doi.org/10.1109/ICCV.2017.74
24. Steinhardt, J., Koh, P.W., Liang, P.: Certified defenses for data poisoning attacks. In: Proceedings of the 31st International Conference on Neural Information Processing Systems, pp. 3520–3532 (2017)
25. Sun, Z., Kairouz, P., Suresh, A.T., McMahan, H.B.: Can you really backdoor federated learning? arXiv preprint arXiv:1911.07963 (2019)
26. Tolpegin, V., Truex, S., Gursoy, M.E., Liu, L.: Data poisoning attacks against federated learning systems. In: Chen, L., Li, N., Liang, K., Schneider, S. (eds.) ESORICS 2020. LNCS, vol. 12308, pp. 480–501. Springer, Cham (2020). https://doi.org/10.1007/978-3-030-58951-6_24
27. Wang, H., et al.: Attack of the tails: yes, you really can backdoor federated learning. arXiv preprint arXiv:2007.05084 (2020)
28. Wu, C., Yang, X., Zhu, S., Mitra, P.: Mitigating backdoor attacks in federated learning. arXiv preprint arXiv:2011.01767 (2020)

29. Xie, C., Huang, K., Chen, P.Y., Li, B.: Dba: Distributed backdoor attacks against federated learning. In: International Conference on Learning Representations (2019)
30. Yin, D., Chen, Y., Kannan, R., Bartlett, P.: Byzantine-robust distributed learning: towards optimal statistical rates. In: International Conference on Machine Learning, pp. 5650–5659. PMLR (2018)
31. Zhang, J., Chen, J., Wu, D., Chen, B., Yu, S.: Poisoning attack in federated learning using generative adversarial nets. In: 2019 18th IEEE International Conference On Trust, Security And Privacy In Computing And Communications/13th IEEE International Conference On Big Data Science And Engineering (TrustCom/BigDataSE), pp. 374–380. IEEE (2019)
32. Zou, M., Shi, Y., Wang, C., Li, F., Song, W., Wang, Y.: PoTrojan: powerful neural-level trojan designs in deep learning models. arXiv preprint arXiv:1802.03043 (2018)

PaTraS: A Path-Preserving Trajectory Simplification Method for Low-Loss Map Matching

Ruoyu Leng, Chunhui Feng, Chenxi Hao, Pingfu Chao[✉], and Junhua Fang

School of Computer Science and Technology, Soochow University, Suzhou, China
{20214227061,chfeng99,20225227106}@stu.suda.edu.cn,
{pfchao,jhfang}@suda.edu.cn

Abstract. Massive and redundant vehicle trajectory data is being accumulated and recorded at an unprecedented speed and scale, incurring expensive cost for storage, transmission, and query processing. Trajectory simplification is a typical way to reduce the size of raw trajectory as well as maintaining its structural information. However, existing methods mainly focus on preserving the shape of the trajectory while ignoring its influence on downstream applications. Since most applications require trajectories to be map-matched into paths before further processing, in this paper, we propose PaTraS, a path-preserving trajectory simplification method that aims to minimize the accuracy loss on the map-matching results of the compressed trajectories. To achieve this objective, we build an index that materializes the road network connectivity, and propose a connectivity-based similarity function that measures the importance of a trajectory point with respect to how it contributes to the map-matching results. Extensive experiments show that, compared with state-of-the-art methods, our proposed solution can better preserve the path generated by trajectory map-matching at the cost of a slightly increased running time, and it works effectively in both online and offline modes.

Keywords: Trajectory simplification · trajectory compression · map-matching · candidate set similarity

1 Introduction

In the era of big data, the ubiquitous of GPS-equipped mobile devices enable the continuous generation of massive spatio-temporal trajectory data. However, due to the GPS measurement and sampling errors [2,6], the raw trajectory data is inherently inaccurate, which makes it hard to serve downstream applications directly. As an essential preprocessing step, a map-matching process reduces the uncertainty of a raw trajectory by aligning trajectory points to the road network and converting them into a smooth and reliable path [3]. For instance, despite a car moving along a coastal road, its GPS records may drift randomly. With the help of the map-matching technology, the actual position of the car can

X. Song et al. (Eds.): APWeb-WAIM 2023, LNCS 14333, pp. 127–144, 2024.
https://doi.org/10.1007/978-981-97-2387-4_9

be estimated more precisely, which is beneficial to downstream applications like urban traffic management, epidemic prevention and control, *etc.*

Meanwhile, as the lightweight GPS devices usually have limited storage space and network bandwidth, once the high-volume raw trajectories are captured, they can neither be transmitted to downstream processors directly nor stored locally. Hence, a trajectory simplification process is necessary to reduce the data size before data transmission and subsequent map-matching procedure.

Due to its high importance, the trajectory simplification problem has been extensively studied for decades. The main idea of trajectory simplification is to remove or/and adjust trajectory points to reduce the size of the trajectory while preserving its shape and structural information. From methodology perspectives, existing solutions can be categorized into two types: error-bounded line simplification and semantic-preserving trajectory simplification [1,4,11,12,16]. The line simplification methods [10] usually remove redundant points reckoning its contribution and the shape of a whole trajectory, while the semantic-preserving trajectory simplification methods [7] achieve data compression by encoding trajectories to a compact format with the help of external knowledge, like road network structure.

Unfortunately, existing works neglect the fact that simplified trajectories will inevitably go through a map-matching process before being used by downstream applications. Therefore, in most scenarios, the information loss caused by the simplification process does not matter as long as the map-matching result remains unchanged. In other words, the impact of a simplification process should be measured by its influence on the accuracy of subsequent map-matching. As exemplified in Fig. 1, following traditional line simplification methods, p_1 is more likely to be removed than p_3 as it does not affect the shape of the trajectory significantly. Notwithstanding, after the removal of p_1, the trajectory is more likely to be map-matched to road $A \rightarrow D \rightarrow E \rightarrow F \rightarrow K \rightarrow J \rightarrow I$ instead of the actual path $A \rightarrow B \rightarrow E \rightarrow F \rightarrow K \rightarrow J \rightarrow I$, whereas the possible removal of p_3 has no influence to the matching result.

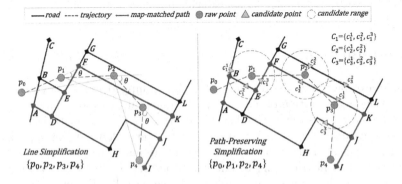

Fig. 1. Illustration of diverse features captured by different simplifications.

Based on our extensive research and analysis, few works consider the impact of trajectory simplification on the downstream map-matching [8]. Therefore, in this paper, we propose the **PaTraS** (Path-Preserving Trajectory Simplification) algorithm that aims to maintain the map-matching quality maximally while performing trajectory compression. The core idea of our algorithm is to measure the importance of each trajectory point in generating the map-matching result, and remove points that introduce the least uncertainty to the matching process, as depicted in Fig. 1. In other words, if the addition of a trajectory point (p_3) does not introduce new map-matching options, it is considered redundant. Note that such evaluation requires the participation of road network and shortest path calculation, which is of high space and time complexity, so we design an index structure, termed as $\mathcal{D}$, that selectively pre-computes and maintains shortest paths for fast online searches. In the meantime, we propose a connectivity-based similarity method that measures the similarity of a trajectory point to its neighbors in terms of the ability to impact the map-matching result. Overall, with the help of the pre-computed index, PaTraS can run in both online and offline modes. It outperforms all existing trajectory simplification methods when it comes to preserving the map-matching quality, notwithstanding a slightly increased time for the compression. In general, our contributions are listed as follows:

1. To the best of our knowledge, this is the first work that tackles the problem of accuracy loss on map-matching caused by trajectory simplification, and we propose *PaTraS* that maintains map-matching quality while effectively compressing the trajectories and reducing the complexity at the same time.
2. We enable PaTraS to run on online mode by designing a shortest path index $\mathcal{D}$. Besides, we propose a connectivity-based similarity measurement to better evaluate the importance of each trajectory point in map-matching process, and perform simplification accordingly.
3. We evaluate the effectiveness and efficiency of PaTraS against error-bounded methods, semantic-preserving methods, and other state-of-the-art methods. Extensive experiments on large-scale real datasets elucidate that our proposed model outperforms baseline methods on the map-matching quality.

2   Related Work

In this section, we overview the research related to our map-matching-based trajectory simplification work, grouped into two categories, namely error-bounded line simplification and semantic-preserving trajectory simplification.

2.1   Error-Bounded Line Simplification

Line simplification algorithms usually have a preset distance or direction constraint, points that violate the constraint are preserved as they contribute more to the shape of the trajectory. Most of the line simplification algorithms compress a continuous trajectory points into several segments, and they are commonly used in practice [10] due to their simplicity and efficiency.

The Douglas-Peucker(DP) algorithm is the most famous method which simplifies the raw trajectory by setting a distance threshold and dividing trajectories recursively. To further improve performance, DPhull [5] is proposed based on properties of convex hull and the greatest perpendicular Euclidean distance(PED), which returns the same point sets as DP algorithm but reduces the time complexity to $O(NlogN)$ (here N represents the total number of trajectory points). Muckell et al. proposes a method named SQUISH [12] by continuously adding trajectory points into a fixed-size buffer and removing those points with the smallest synchronized Euclidean distance(SED). They additionally put forward SQUISH-E [13] which strives to ensure the compression ratio. DOTS [1] alternatively uses local integral square SED and computes the shortest path in an incremental way. CISED [9] regularizes the start point, then employs synchronous circles as well as cone projection circles to abridge the redundant sequences. DPTS [11] measures error by angles rather than Euclidean distances with direction-aware distance(DAD).

It has been demonstrated that using different metrics has an influence on compression performance. Moreover, given the same distance metric, a proper value for the error bound varies significantly in different scenarios as it is closely related to the geometric properties of the dataset. Therefore, finding the best error bound is usually the main challenge of line simplification algorithms.

2.2   Semantic-Preserving Trajectory Simplification

Rather than relying solely on distance or direction error bounds, there are several semantic-preserving trajectory simplification algorithms aiming at discovering more latent knowledge, such as moving tendency, road network information, and others.

Li et al. [8] proposes three simplification algorithms, namely IS, SWS, and GS, based on different weighting functions. It considers spatio-temporal reliability including the density of trajectory points and the moving speed of objects. REST [17] maintains a prebuilt reference trajectory set and compresses trajectory by finding an optimal combination of matchable reference trajectories. In light of road networks, most existing works first conduct a map-matching process, then perform simplification based on the path sets. MMTC [7] designs a network-based trajectory similarity to compress the continuous path segments as final results. TrajCompressor [4] devises a light-weighted map matcher and makes great use of heading change at intersections. PRESS [15] calculates the spatial and temporal similarity, respectively, to format the mapped trajectory and build the compression framework based on frequent sub-segments.

2.3   Analysis of Existing Work

It is clear that the aforementioned two types of trajectory simplification methods have their respective application scenarios: 1) the line simplification is lightweight and efficient, which is good for the online scenario, but it does not consider its influence on downstream applications, like the map-matching process; 2) some semantic-preserving trajectory simplification methods utilize road network for lossless compression, but it has to perform the map-matching process directly, which is time-consuming thus only works in offline mode. Overall, none of the existing studies investigates the possibility of ensuring downstream map-matching quality during the simplification phase without performing the actual map-matching directly, which is the main focus of our work.

To achieve this goal, it is obvious that the road network information should be taken into account when simplifying trajectory, as the importance of a trajectory point to the map-matching result cannot be measured without the road network. In the meantime, since performing real-time map-matching involving road network information is infeasible, finding the map-matching similarity between trajectory points and removing redundant ones become challenging. There is little work that designs a feasible similarity calculation method for geographical points based on road network information, hence, it is necessary to design a proper calculation scheme to measure the influence of original trajectory points on the map-matching task.

3   Preliminaries

In this section, we describe the related concepts and definitions involved in this work. Subsequently, the problem statement of our task is formulated. Table 1 shows the frequently used notations in this paper.

Table 1. Notation Summary.

Notations	Description
$\mathcal{T}, T$	A trajectory set, a trajectory formalized as $T = \{p_1, p_2, \cdots\}$
$\mathcal{R}$	Candidate range used to determine the candidates of one trajectory point
$\mathcal{C}_i$	The candidate set of a trajectory point p_i
$P_s^{c_i, c_j}$	A shortest path between two candidates c_i, c_j in the road network
$Sim(p_i)$	Similarity value indicating the importance of p_i on map-matching
$S(P)$	Road segment set corresponding to path P, where $S(P) \subset G$
φ	A distance range when building the specific index $\mathcal{D}$ of shortest paths
β	Size of the online buffer window, with $\beta < T.size()$
δ	The number of points remained in β after each simplification process
CR	The compression ratio describes the proportion of retention points

3.1   Basic Definitions

Regarding our problem settings, a moving object travels along the road network and generates a raw trajectory. The trajectory is captured in real time, but can be processed in either online or offline manner. Here, we first list some basic definitions involved.

Definition 1 (Raw Trajectory). *A raw trajectory is a series of chronologically ordered points $T = \langle p_1 \rightarrow p_2 \rightarrow \cdots \rightarrow p_n \rangle$ sampled from a moving object. Each point $p_i = \langle lat, lon, t, speed, heading \rangle$ sampled by positioning devices indicates the location, speed and heading information at time $p_i.t$.*

Since our method also considers the underlying road network when simplifying trajectories, we give the definition of road network as follows:

Definition 2 (Road Network). *A road network $G(V, E)$ is formatted as a directed graph, where V is a set of vertices representing road intersections and road ends, and E is a set of edges, denoted as $e = \langle start, end, SEQ \rangle$, each of which is a **road segment** represented as a polyline that connects two vertices in V and has a sequence of intermediate points SEQ describing its shape.*

Definition 3 (Simplified Trajectory). *A simplified trajectory T_s is a subset of its corresponding raw trajectory T by removing points from T. Reviewing the Definition 1, we formulate the simplified trajectory as:*

$$T_s = \langle p_1 \rightarrow p_2 \rightarrow \cdots \rightarrow p_m \rangle, m \leq n$$

Note that m denotes the size of the simplified trajectory T_s and n is the size of the corresponding raw trajectory T, thus its compression ratio can be defined as $CR = \frac{n}{m}$.

Definition 4 (Map Matching). *The map matching is a process to identify the actual running path of a moving object by aligning its trajectory T to the underlying road network G. After the map-matching, the trajectory is converted to a sequence of road segments, which is defined as a path P.*

3.2   Methodology Analysis

The goal of map-matching is to estimate a path P_0 from a raw GPS trajectory T which is as similar as possible to its actual path P. Due to GPS measurement noise, the sampled GPS points randomly deviate from their actual positions on the true path P. Such uncertainty may lead to many errors in map-matching results. To address the above problem, a simplification process can first be applied to the raw trajectory T to remove noisy and stop points, which leads to a simplified trajectory T_0. The problem statement in this paper is as follows.

Problem Statement. Given a trajectory dataset $\mathcal{T}$ and the attached road network G, we aim to transfer $\forall T \in \mathcal{T}$ into T_s and finally achieve stable convincing map matching result. The whole simplification process is assisted by the candidate sets that extract more information about the characteristics of the road network where the trajectory points are located.

Our goal is to simplify the raw trajectories while guaranteeing the correctness of map-matching results. The existing simplification methods only consider the indicator of geometric error, such as PED, SED and DAD. However, their contribution to the accuracy of map-matching results is very limited. The quality of each trajectory point that reflects the surrounding environmental information will play a key role during map-matching. Thus, ensuring the quality of the trajectory points while simplifying the trajectory becomes a crucial challenge. To solve this problem, we creatively integrate road network information in the trajectory simplification process and propose PaTraS, a road-preserving trajectory simplification method for map matching. In a nutshell, to simplify raw trajectories, we detect the similarity between different raw points based on how it contributes to the map-matching results. To achieve this purpose, we generate a candidate set $\mathcal{C}_p$ for each trajectory point p, and remove points based on their similarity values which are calculated from the association relationship between neighbouring candidate sets. After going through this process, we will acquire similar map-matching results while largely reducing the scale of trajectories by trajectory simplification. This process effectively saves the consumption of computation and satisfies real-time scenarios. In the next section, we will describe the details of the simplification algorithm.

4   Path-Preserving Trajectory Simplification

Our method innovatively takes the road network information into consideration compared with Li *et al.* [8]. In order to lightly calculate the road network distance between candidate points and match the pairs with high connectivity, we store the shortest paths between different candidates corresponding to distinct trajectory points. It is unessential to prepare perfect matching results since there are several candidate pairs located at irrelevant roads. In the next part, we will introduce the main content of the *PaTraS* algorithm.

4.1   Overview of PaTraS

Figure 2 elucidates the framework of our simplification framework, PaTraS. Because the distance between the trajectory points is not too far away, based on different sampling rates and road characteristics, the storing range of these paths has different thresholds. Intuitively, people tend to pick up the shorter path to a destination. Consequently, we store the shortest path(s) between intersections or road endpoints on the road network. The component storing shortest paths is called SpDB (abbreviated as $\mathcal{D}$).

When the trajectory stream comes, to perceive the road network environment, each trajectory point corresponds to a set of candidate points in a specified candidate range. The candidate range is determined by a distance threshold ϵ. For different road network structures, we adjust their value to adapt to different road densities. By matching candidate pairs from distinct collections, we can calculate the similarity between trajectory points which brings in road network information to better perceive the similarity between them. In the process of simplifying the trajectory points, we will preferentially delete the trajectory point with the largest similarity value in the processing window and guarantee the requirement of the compression ratio at the same time. The pseudo-code of

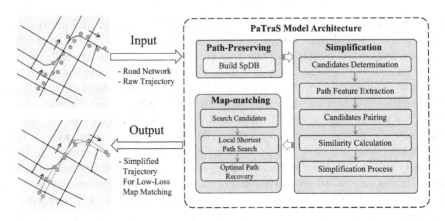

Fig. 2. Overview of PaTraS.

PaTraS is shown in Algorithm 1. By the way, our algorithm also has batch mode which receives complete trajectories before simplification.

4.2 Preserving Shortest Paths

In order to quickly find the shortest path in online scenarios, we pre-store the shortest paths in the road network. But it is not wise to store the paths between all the intersections on the road network, which will pay a large-scale storage space. Since the object does not move very far in the time period between the sampling interval, we only need to keep the paths around the intersection. Firstly, we divide the given road network into several parts. When storing the shortest path, we use $Dijkstra$ algorithm to calculate the path distance from one intersection to others. This operation will only be done once in each part of the road network. For intersection pairs across partitions, there is a certain overlap area between partitions. We end up storing the distance between all intersections so we can quickly query the path in need. When the road network structure is introduced, we calculate the path between the candidate points to evaluate the environmental information around the trajectory points in order to calculate the similarity score. Due to the location of candidates may not necessarily be an intersection, in addition to querying the distance between our pre-stored intersections from $\mathcal{D}$, we also need to add the distance between the two candidate points to their corresponding intersections.

The structure of path stored in $\mathcal{D}$ is as follow:

$$P_s^{c_i,c_j} = \{< R_{Idx}^i, R_{Idx}^j > | R_{Idx}^i, ..., R_{Idx}^k, ..., R_{Idx}^j\}(i < k < j)$$

In the above formula, candidates c_i and c_j are on road R^i and R^j respectively. R_{Idx} represents a specific road ID of road R. Path $P_s^{c_i,c_j}$ is a list of connected roads which is the shortest path between candidate c_i and c_j.

Algorithm 1: PaTraS

Input: $T = \{p_1, p_2, ...\}$: the continuous raw trajectory, G: the road network, CR: the compression ratio, δ: the number of points to be kept in sliding window β.

Output: T_s: Simplified trajectory at the current moment.

1 $\mathcal{D}$ = getDB(G); // The storage structure is described in subsection 4.2
2 Initialize $\beta = null$; // The maximum capacity of β is $CR*\delta$
3 Add the first trajectory point to T_s;
4 **for** p_i *in* T **do**
5   β.add(p_i);
6      $\mathcal{C}_i = getCandidates(p_i, \mathcal{R})$;
7      $CandidateList$.add($\mathcal{C}_i$);
8   **if** β *is full* **then**
        // Details can be found in subsection 4.4
9          $simValue$ = calSimilarity($\mathcal{D}$, $CandidateList$, φ);
10         $maxHeap$.add($simValue$);
11   **if** $maxHeap.size == \beta$ **then**
12   β.remove($maxHeap$, $\beta - \delta$);
13   Add remained δ points in β to T_s;
14             $maxHeap$.clear(); β.clear();

15 Add the last trajectory point to T_s;
    Result: Simplified trajectory T_s

4.3 Candidates Pairing

In order to capture the similarity between the original trajectory points, road network information is creatively combined in similarity calculation. Our main idea is to gain insight into the road structure around the trajectory points. Therefore, for each trajectory point, we construct candidate points around it under a certain candidate range to capture road features. According to this insight, when calculating the similarity value of one trajectory point, we take the connection between its adjacent trajectory points into account.

As mentioned above, in order to capture the similarity between contextual points, the similarity value can be calculated for every three trajectory points, which is namely the similarity value of the middle point. By introducing road networks, we can generate some candidate points around the corresponding trajectory points. These candidate points are calculated based on the preset candidate range R and road network G. All roads covered within this range generate candidates related to its trajectory point, as shown on the right of Fig. 1. Usually, a trajectory point may have multiple candidate points in urban areas with complicated traffic conditions. Let's use $\mathcal{C}_{i-1}, \mathcal{C}_i, \mathcal{C}_{i+1}$ to denote the sets of candidate points corresponding to three adjacent raw trajectory points. Firstly, we respectively sort the candidate points in different candidate sets by their projection distance. Then, we complete the candidates pairing between $\mathcal{C}_{i-1}$ and

$\mathcal{C}_{i+1}$ confirming to their distance priorities. Noting that the candidates which have been successfully matched will be excluded from the subsequent pairing. Considering the quantities of different sets may be different, we will eventually identify $min(M, N)$ pairs(M is the size of $\mathcal{C}_{i-1}$, N is the size of $\mathcal{C}_{i+1}$).

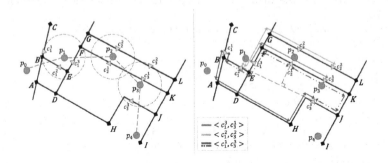

Fig. 3. Example of similarity calculation.

4.4   Similarity Computation

In this section, we introduce how to calculate the similarity value of trajectory points and use it to help us effectively simplify raw trajectories.

The similarity represents the road connectivity captured by the point. The larger similarity value indicates the higher redundancy of the information that can be captured by the corresponding trajectory point. After the pairing process is completed, we obtain the shortest paths between these candidate pairs from $\mathcal{D}$. In order to perceive the connectivity characteristics between candidates of different candidate sets, the positional distribution of the intermediate set $\mathcal{C}_i$ counts. For ease of expression, we denote the corresponding candidate points in these three candidate sets as c_{i-1}^x, c_i^y, and c_{i+1}^z(x, y, and z correspond to the indices of the candidates in different candidate sets, respectively). For each obtained path, we sequentially check whether the candidate c_i^y of $\mathcal{C}_i$ falls on the path. If so, the candidate is on the shortest path between the preceding candidate c_{i-1}^x and succeeding candidate c_{i+1}^z. Thus, due to its positional information is already included in the shortest path, we consider this candidate to be of low importance. For the other case, if it comes out that no candidate falls on the path, we take the candidate which is closest to the line connecting c_{i-1}^x and c_{i+1}^z as a reference. Then, we can obtain the shortest path between this candidate and the preceding candidate c_{i-1}^x and the shortest path between this candidate and the latter candidate c_{i+1}^z from $\mathcal{D}$. In general, we concatenate these two paths, denote as $P_s^{con_i}$, and use the intersection and union roads between the concatenation path $P_s^{con_i}$ and the shortest path $P_s^{c_{i-1}^x, c_{i+1}^z}$ between c_{i-1}^x and c_{i+1}^z to calculate the similarity value. Specific details are described below.

$$P_s^{con_i} = P_s^{c_{i-1},c_i} \odot P_s^{c_i,c_{i+1}} \tag{1}$$

First of all, we need to pay attention to how to link roads. To avoid duplicate calculation, the road where c_i^y is located will only be counted once. Equation (1) shows how to obtain the spliced result $P_s^{con_i}$ and the symbol $\odot$ indicates the way of roads concatenation.

$$Sim(p_i|p_{i-1}, p_{i+1}) = \sum_{j=0}^{min(m,n)} \frac{len(P_{s_j}^{con_i} \cap P_{s_j}^{c_{i-1},c_{i+1}})}{len(P_{s_j}^{con_i} \cup P_{s_j}^{c_{i-1},c_{i+1}})} \tag{2}$$

Then we can start calculating the similarity of the middle point p_i. We gather the results for $min(M, N)$ pairs in total. As described in Subsect. 4.3, for each pair, we calculate the intersection and union parts of the roads based on the reference candidate and record the length information for each road. Then, as shown in Equation (2), we combine the results of all matched pairs and obtain the final similarity value of the intermediate point p_i. The larger this value is, the more similar the road network information that different candidate sets can represent. Therefore, we first simplify the trajectory points with larger similarity values in our simplification process.

Taking an easy-to-understand example shown in Fig. 3. When the candidates of three trajectory points p_1, p_2 and p_3 are obtained, we sort the candidates in candidate sets separately according to their projection distances. Then we pair the sorted results in sequence and obtain 3 pairs. For the first pairing $< c_1^3, c_3^1 >$, c_2^1 falls on the shortest path $\{c_1^3 \to F \to c_3^1\}$ between c_1^3 and c_3^1, so its intersection and union roads are the same, namely $P_s^{\cap_1} = P_s^{\cup_1} = \{c_1^3 \to F \to c_3^1\}$. The same principle applies to the next pair $< c_1^2, c_3^2 >$ as well, with the corresponding intersection and union roads $\{c_1^2 \to E \to F \to G \to c_3^2\}$. The shortest path of $< c_1^1, c_3^3 >$ is $\{c_1^1 \to B \to A \to D \to H \to c_3^3\}$. Since there is no candidate falling on the shortest path of $< c_1^1, c_3^3 >$ in C_2, we find the candidate c_2^1 closest to the line connecting c_1^1 and c_3^3 as the reference candidate. Thus, the intersection roads is $P_s^{\cap_3} = \{c_1^1 \to B\}$, the union parts is $P_s^{\cup_3} = \{c_1^1 \to B \to A \to D \to H \to c_3^3, B \to E \to F \to K \to J \to c_3^3\}$

$$Sim(p_2|p_1, p_3) = \frac{len(P_s^{\cap_1}) + len(P_s^{\cap_2}) + len(P_s^{\cap_3})}{len(P_s^{\cup_1}) + len(P_s^{\cup_2}) + len(P_s^{\cup_3})}$$

4.5 Pairing Optimization

In the process of pairing, simply matching candidates by projection distance may lead to biased results because the candidate points may not be on the same path or may be moving in opposite directions. However, attempting to pair each candidate with all other candidate points comes at a cost of $O(MN)$ which requires inevitable large calculation.

Lemma 1. *Assuming that there are three continuous trajectory points p_1, p_2 and p_3, corresponding to candidate sets C_1, C_2 and C_3. Given $P_s^{c_1^x, c_2^y}$, $P_s^{c_2^y, c_3^z}$ and*

$P_s^{c_1^x, c_3^z}$ which are paths between candidates c_1^x, c_2^y and c_3^z respectively from $\mathcal{C}_1$, $\mathcal{C}_2$ and $\mathcal{C}_3$, if the concatenation of $S(P_s^{c_1^x, c_2^y})$ and $S(P_s^{c_2^y, c_3^z})$ is equal or nearly equal to $S(P_s^{c_1^x, c_3^z})$, we then decide that the middle point p_2 is not very necessary for map matching, which means it can be removed.

Proof. Let $P_s^{c_1^x, c_3^z}$ be the shortest path from c_1^x to c_3^z, if c_i is on the $P_s^{c_1^x, c_3^z}$, the shortest path $P_s^{c_1^x, c_i}$ from c_1^x to c_i is definitely the subset of $P_s^{c_1^x, c_3^z}$, otherwise, there is a shorter path existing, which contradicts our previous assumption.

To solve this problem, we also adopt a reliable optimization solution. We further consider the connectivity characteristics of the roads where candidates are located and the orientation information of the trajectory points. On the one hand, because the road network is a directed graph, the roads where candidate points are located have limited driving directions. If we ignore this feature we will inevitably introduce many wrong candidate points. Considering this feature, we first determine whether the trend of road extension is in the same direction when pairing. According to common sense, people tend to take shorter paths. If a candidate point corresponding to an intermediate point falls on the shortest path, then the paths between this candidate and its adjacent ones are also the shortest, illustrated as Lemma 1. If there is a significant deviation or even a situation where the road doubles back on itself, then these two candidates cannot be paired. On the other, the direction we are heading is generally the same as the road extends intuitively. Since the collected original trajectory points have speed and orientation information, we assign the heading direction of all original trajectory points to their corresponding candidate sets. When we are pairing candidates between two different collections, we no longer pair candidates just based on projection distance. The angular difference between the extension direction of road segment and the heading of candidate is calculated. If this angle difference is greater than a certain value, we then consider them deviating too much and the pairing of these two candidates fails. In practice, we filter out candidates whose angle difference exceeds 135 °C, and according to statistics, about 60% of the points are filtered. For convenience, we call the basic version PaTraS-B and the optimized version PaTraS-O. Next, we will describe our experiments details.

5   Experiments

In this section, we conduct multiple experiments to verify the precision and efficiency of our proposed algorithm, analyzing the effects of each sub-procedure.

5.1   Experimental Setup

In this subsection, we elaborate on the experimental setup. Referring to the ultimate objective of our proposed PaTraS (Sect. 1), we empirically evaluate the effectiveness and efficiency on a large-scale dataset.

Datasets. Our experiments are conducted on real-world commercial datasets encompassing taxi trajectories and road network datasets located in Beijing. Due to the distinct regional characteristics of the road network structure, we classify the maps into urban and rural areas based on the map density. For conducting large-scale performance tests, we also extract a larger area of the urban area. In a nutshell, we extract three sub-areas, namely Beijing-U, Beijing-R and Beijing-L, as listed in Table 2.

Table 2. Summary of experiment datasets

Name	Input Trajectory			Road Network			
	Trajectory Count	Trajectory Point Count	Sampling Rate(sec)	# ofvertices + mini nodes	# of edges	Map Size(km²)	Map Density (km/km²)
Beijing-U	7902	243964	11.7	5358	4484	9.9	27.3
Beijing-R	645	20666	13.7	929	282	9.9	4.98
Beijing-L	36666	1673671	10.8	28313	17909	47.5	24.7

Through the evaluation results between Beijing-U and Beijing-R, we find that the trajectory distribution have a significant influence on the efficiency. Compared with the distribution of rural trajectories, trajectories in complex urban road networks are more valuable to study. Therefore, we select Beijing-L as a representative to analyze.

Baselines. In order to compare the differences between the algorithms more comprehensively, we selected the more classical methods such as DP(Douglas Peucker) and SQUISH [12]. Besides, the state-of-the-art methods CISED-S and CISED-W [9] in the field of trajectory compression is also included.

Environment. All algorithms are implemented in Java. Our experiments are conducted on a single server with Intel(R) Xeon(R) Gold 6258R CPU 2.70GHz, 498GB memory and Ubuntu 20.04.1. To ensure the result reliability, all tests are repeated over 3 times and recorded the average.

5.2   Evaluation Oriented to Map-Matching

In our experiments, we choose HMM model [14] to match the simplified trajectories. The evaluation metrics are introduced as follows.

Metric. We measure our results against these metrics which universally used to evaluate map-matching results. They are F_1-score, RMF (Route Mismatch Fraction), and Accuracy. The detailed definitions are presented below.

– F_1-score: This metric comprehensively evaluates the accuracy and completeness of map-matching results, is formulated as:

$$F_1 - score = 2 \cdot \frac{Precision \cdot Recall}{Precision + Recall}$$

We use *Precision* and *Recall* to estimate the performance, whose $Precision = length(P \cap P')/length(P')$, $Recall = length(P \cap P')/length(P)$

- *RMF*: Route Mismatch Fraction computes the total length of the false positive road segments and false negative road segments, denoted as d_+ and d_-. Let d_0 represent the total length of the real path. A small RMF value indicates the map-matching results are more similar to the real path.

$$RMF = (d_+ + d_-)/d_0$$

- *Accuracy*: Evaluate the accuracy of the point matching results. It use the quantity of candidates that correctly matched on their segments compared with corresponding ones in ground truth at the same timestamp.

$$Accuracy = \sum_{i=0}^{n} \frac{N(seg|seg \in P^i \cap P_{gt}^i)}{N(seg_{gt}|seg_{gt} \in P_{gt}^i)}$$

Performance. In this part, we compare the scalability and dependency of PaTraS with other baselines. Then, we show the necessity of our optimization about connectivity-based similarity.

First, we compare PaTraS with other approaches from two metrics, which are respectively F_1-score and *Accuracy*. The compression ratio range is set from 2 to 9, comprehensively showing the F_1-score and accuracy performances of different methods at both low and high compression ratios.

Table 3. Comparison of different Compression Ratio

Method	Metric	Compresion Ratio							
		2	3	4	5	6	7	8	9
DP	F1-score	91.07	88.11	86.36	86.17	84.19	81.06	80.15	78.68
SQUISH		87.83	87.27	83.11	81.18	79.85	77.07	74.09	68.06
CISED-S		90.52	89.12	87.52	85.71	84.10	83.63	77.48	75.23
CISED-W		90.23	87.06	78.77	75.11	70.15	67.37	66.05	62.60
PaTraS		**92.85**	**90.30**	**90.07**	**86.96**	**86.36**	**85.42**	**81.94**	**82.21**
DP	Accuracy	78.28	69.80	63.24	61.18	58.99	59.33	58.57	55.94
SQUISH		84.58	83.37	80.40	79.23	75.94	74.10	67.44	65.85
CISED-S		79.12	77.07	75.03	73.01	70.83	68.12	62.22	57.97
CISED-W		76.93	66.32	56.79	51.32	43.36	40.95	41.06	40.43
PaTraS		**85.88**	**83.97**	**81.37**	77.68	75.28	72.82	**71.10**	**69.04**

As shown in Table 3, we compare F_1-score and Accuracy after trajectory simplification at different compression ratios. The upper part shows PaTraS outperforms other four methods under F_1-score. This is because DP is a global scheme based on geometry simplification strategy. It may inevitably loses semantic information of local trajectory points. CISED is a local geometry-based simplification scheme, which detects whether the movement trend of continuous trajectory points has a large change, but is easily affected by abnormal deviation points. SQUISH simply removes trajectory points in the fixed window, it is difficult to guarantee whether the distribution and connectivity of the simplified trajectory points is helpful for road network matching. Although the lower part of Table 3 shows SQUISH's Accuracy is higher than PaTraS under compression ratio 5–7, its Accuracy starts to decline rapidly at 8 or higher, which shows PaTraS has better stability at different compression ratios. In terms of overall performance, PaTraS is the best solution of all methods.

Then, we compare the performance of PaTraS-B and PaTraS-O, the former represents unoptimized, the latter represents the optimized. The differences in the experimental results, as proposed in Table 4, strongly reflect the importance of orientation analysis and path correlations in the process of matching candidates. After introducing angle difference, the simplification time greatly reduced. This phenomenon indicates that the amount of computation during pairing process decrease very much, because the optimization for road connectivity and heading information filters out many completely wrong pairings, such as a route with a loop-back. Meanwhile, it also improves the accuracy of matching candidates pairs, making the calculation of similarity between trajectory points more reliable.

Table 4. Comparison after Optimization

Methods	CR	F_1-Score	RMF	Accuracy	Simplification Time
PaTraS-B	2	90.800	0.186	83.299	22.5 min
	4	88.609	0.231	78.491	17.5 min
	6	85.722	0.268	72.404	14.7 min
	8	80.189	0.398	68.718	13.2 min
PaTraS-O	2	92.855	0.144	85.882	5.6 min
	4	90.074	0.168	81.370	4.5 min
	6	86.361	0.245	75.282	3.8 min
	8	81.946	0.358	71.103	3.6 min

5.3 Parameter Sensitivity Study

The influence of different parameters involved in our model as follows:

- $\mathcal{R}$: Candidate Range. We take the original trajectory point as the center of candidate range. Roads covered by this range generate counterpart candidate points which mean the road network information.
- CR(or k): Compression Ratio. When we have simplified the trajectory, we need this ratio to indicate the degree of trajectory simplification. The compression ratio is obtained by dividing the number of original trajectory points

by the number of simplified trajectory points. In the following charts, we also use k to represent CR, so hopefully there will be no confusion about this mixed use.

– φ: Distance Threshold. When we build the specific index $\mathcal{D}$, we should set a certain range to guarantee the shortest paths we store between two candidates are not particularly long.

First of all, we focus on the impact of $\mathcal{R}$ determining the range of candidates we extract. As shown in Fig. 4(a), regardless of the compression ratio, the simplification time grows as the candidate range $\mathcal{R}$ increases. A larger range covers more candidates on the road network, so it takes longer to perform the pairing process. In Fig. 4(c), when candidate range $\mathcal{R}$ equals to 10 m, the F_1-score is relatively low and the performance is not stable at different compression ratios. The number of candidates obtained from a small range is very limited, and the reliability of these points is relative low due to the presence of GPS errors. As the compression ratio k increases, the time consumed by the whole process gradually decreases. This is because the larger k is, the larger the window β we construct and therefore, for the same trajectory, the fewer times we need to process. The score performs better when $\mathcal{R}$ is 70 m, while larger ranges introduce more noise and lead to an increase in unnecessary computation.

Then, we analyze the impact of φ which decides the max path length preserving in $\mathcal{D}$. In this part, we unify the candidate range to 70m. We can observe

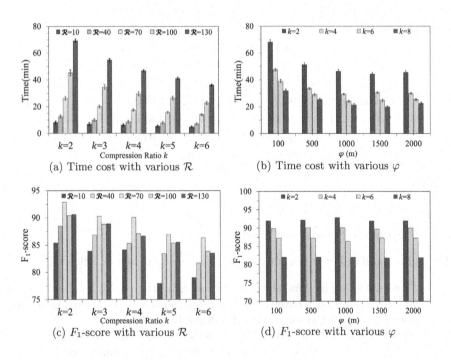

(a) Time cost with various $\mathcal{R}$ (b) Time cost with various φ

(c) F_1-score with various $\mathcal{R}$ (d) F_1-score with various φ

Fig. 4. Parameter Sensitivity

the simplification time between different compression ratios under different φ in Fig. 4(b). As reported by statistics, 64.3% of the trajectory points have a distance interval of 50–70 m owing to the relationship between the speed and the sampling interval. Therefore, when the pre-stored distance of $\mathcal{D}$ is changed from 100 m to 500 m, the simplification time has changed significantly. Most pairings fail because paths between two sampling intervals could not be found under φ of 100 m. In this case, an online shortest path search is performed. The simplification time tends to stabilize when φ getting larger because almost all candidate pairs are able to get the shortest path directly in $\mathcal{D}$. Although the larger φ captures the paths with longer length, the storage of longer distance paths does not contribute to the calculation of similarity value. As presented in Fig. 4(d), the map-matching performance of the simplified trajectory is stable at different distances. Thus, the setting of φ should be adapted to both sampling interval and road network characteristics for better performance. However, since the distance between adjacent trajectory points is not too far, storing a longer path has little improvement to F_1-score.

6 Conclusion

To conclude, under the premise of ensuring the correctness of map-matching results, we propose a novel simplification algorithm that compresses raw trajectories data with different compression ratios. Compared with other models, our model PaTraS can retain valuable trajectory points in the process of simplifying the trajectory, which makes the simplified trajectory have outstanding advantages in map-matching. Trajectory simplification is of great significance to storage, communication or downstream applications. For future work, our model could be extended to discovery further knowledge of trajectories and other temporal-spatial data. Also, we will investigate broader simplification strategy for downstream applications.

References

1. Cao, W., Li, Y.: DOTS: an online and near-optimal trajectory simplification algorithm. J. Syst. Softw. **126**, 34–44 (2017)
2. Chao, P., Hua, W., Mao, R., Xu, J., Zhou, X.: A survey and quantitative study on map inference algorithms from GPS trajectories. IEEE Trans. Knowl. Data Eng. **34**(1), 15–28 (2020)
3. Chao, P., Xu, Y., Hua, W., Zhou, X.: A survey on map-matching algorithms. In: Borovica-Gajic, R., Qi, J., Wang, W. (eds.) ADC 2020. LNCS, vol. 12008, pp. 121–133. Springer, Cham (2020). https://doi.org/10.1007/978-3-030-39469-1_10
4. Chen, C., Ding, Y., Xie, X., Zhang, S., Wang, Z., Feng, L.: TrajCompressor: an online map-matching-based trajectory compression framework leveraging vehicle heading direction and change. IEEE Trans. Intell. Transp. Syst. **21**(5), 2012–2028 (2019)
5. Hershberger, J.E., Snoeyink, J.: Speeding up the douglas-peucker line-simplification algorithm (1992)

6. Jiang, L., Chen, C.X., Chen, C.: L2MM: learning to map matching with deep models for low-quality GPS trajectory data. ACM Trans. Knowl. Discov. Data (TKDD) (2022)
7. Kellaris, G., Pelekis, N., Theodoridis, Y.: Map-matched trajectory compression. J. Syst. Softw. **86**(6), 1566–1579 (2013)
8. Li, H., Kulik, L., Ramamohanarao, K.: Spatio-temporal trajectory simplification for inferring travel paths. In: Proceedings of the 22nd ACM SIGSPATIAL International Conference on Advances in Geographic Information Systems, pp. 63–72 (2014)
9. Lin, X., Jiang, J., Ma, S., Zuo, Y., Hu, C.: One-pass trajectory simplification using the synchronous Euclidean distance. VLDB J. **28**(6), 897–921 (2019)
10. Lin, X., Ma, S., Jiang, J., Hou, Y., Wo, T.: Error bounded line simplification algorithms for trajectory compression: an experimental evaluation. ACM Trans. Database Syst. (TODS) **46**(3), 1–44 (2021)
11. Long, C., Wong, R.C.W., Jagadish, H.: Direction-preserving trajectory simplification. Proc. VLDB Endowment **6**(10), 949–960 (2013)
12. Muckell, J., Hwang, J.H., Patil, V., Lawson, C.T., Ping, F., Ravi, S.: Squish: an online approach for GPS trajectory compression. In: Proceedings of the 2nd International Conference on Computing for Geospatial Research & Applications, pp. 1–8 (2011)
13. Muckell, J., Olsen, P.W., Hwang, J.H., Lawson, C.T., Ravi, S.: Compression of trajectory data: a comprehensive evaluation and new approach. GeoInformatica **18**(3), 435–460 (2014)
14. Newson, P., Krumm, J.: Hidden Markov map matching through noise and sparseness. In: Proceedings of the 17th ACM SIGSPATIAL International Conference on Advances in Geographic Information Systems, pp. 336–343 (2009)
15. Song, R., Sun, W., Zheng, B., Zheng, Y.: Press: a novel framework of trajectory compression in road networks. arXiv preprint arXiv:1402.1546 (2014)
16. Zhang, D., Ding, M., Yang, D., Liu, Y., Fan, J., Shen, H.T.: Trajectory simplification: an experimental study and quality analysis. Proc. VLDB Endowment **11**(9), 934–946 (2018)
17. Zheng, K., Zhao, Y., Lian, D., Zheng, B., Liu, G., Zhou, X.: Reference-based framework for spatio-temporal trajectory compression and query processing. IEEE Trans. Knowl. Data Eng. **32**(11), 2227–2240 (2019)

Coordinate Descent for k-Means
with Differential Privacy

Yuchen Xie, Yi-Jun Yang$^{(\boxtimes)}$, and Wei Zeng

School of Computer Science and Technology, Xi'an Jiaotong University, Xi'an, China
xieyuchen@stu.xjtu.edu.cn, {yangyijun,wz}@xjtu.edu.cn

Abstract. In recent years, Lloyd's heuristic has become one of the most useful methods to solve k-means problem due to its simplicity. However, Lloyd's heuristic suffers from the bad local minimum and the privacy issues which make it not proper to be used in the privacy-preserving scenarios. In this paper, we propose a differentially private framework for k-means clustering by using the coordinate descent method. Firstly, we propose an approximate version of the updating functions of the indicator matrix which claims each point's assignment. Then we ensure differential privacy for k-means clustering by using exponential mechanism to perturb the indicator matrix. Finally, we conduct several experiments based on multiple real-world datasets. Our experimental results show that our algorithm outperforms state of the art in terms of the trade-off between utility and privacy.

Keywords: k-means · Coordinate descent · Differential privacy · Exponential mechanism

1 Introduction

k-means, as one of the most famous clustering methods, has been applied to multiple fields such as data mining [5,28], signal process [22] and pattern recognition [10,21]. The purpose of k-means is to give a partition of the data points into k clusters such that the sum of the squared errors (SSE) of each point to the center is minimized. Because of its simplicity and practicability, k-means has aroused broad attention in AI community.

However, existing algorithms for solving k-means clustering problem still have privacy issues. For example, when financial institutions make a clustering with respect to their customers, the models may leakage sensitive information of customers such as gender and outcome. Existing privacy-preserving methods such as multi-party computation (MPC) [29] and anonymity [11] cannot be used in this scenario when the adversaries have the maximum background knowledge. Aiming at this issue, differential privacy (DP) was proposed by Dwork et al. [6] against inference attacks with maximum background knowledge and has been widely used by Google [1], Apple [4] and Samsung [23] as the mathematical definition of privacy.

© The Author(s), under exclusive license to Springer Nature Singapore Pte Ltd. 2024
X. Song et al. (Eds.): APWeb-WAIM 2023, LNCS 14333, pp. 145–158, 2024.
https://doi.org/10.1007/978-981-97-2387-4_10

To ensure DP for k-means clustering, several works [3,7,13,16,19,24–26,30] consider to combine DP with Lloyd's heuristic [12]. Some works [3,7,13,19,26, 30] use Laplace mechanism [8] and exponential mechanism [14] to perturb the outputs of Lloyd's heuristic. Mohan et al. [16] proposed a sampling framework to ensure DP for k-means. However, these works have a common issue, that is the bad convergence of Lloyd's heuristic. Several works [17,18,20,27] have found that Lloyd's heuristic usually converges to a bad local minimum. Moreover, existing differentially private methods k-means approaches mostly add Laplace noise to perturb the outputs and use exponential mechanism to perturb the assignment of each point. These issues will make the utility of clustering much worse.

To address the bad local minimum issue, Nie et al. [18] proposed a coordinate descent method for k-means (CDKM) which can reach a lower objective value than Lloyd's heuristic. Moreover, this method just outputs the indicator matrix (a matrix which can imply the assignment of each point) instead of the final centers. Therefore, the adversaries cannot get sensitive information from the centers, which makes the CDKM a better method for clustering while ensuring strong privacy guarantee.

However, there are still two critical issues of CDKM. Firstly, we find that the time complexity of CDKM is $O(ndk)$ at each iteration, where n, d and k represent the number of points, features and clusters, respectively. Note that this complexity is still too large. Secondly, the adversaries can still get sensitive information from the exported indicator matrix of CDKM. To solve these issues, we propose an efficient and privacy-preserving framework for CDKM based on DP. We elaborate the contributions as follows:

- We propose an approximate coordinate descent method for k-means (ACDKM) which can reduce the complexity to $O(nk)$ in each iteration.
- We propose a differentially private framework for ACDKM (DP-ACDKM) based on exponential mechanism. Our DP-ACDKM just perturbs the objective function and does not output the centers.
- Our experimental results show that DP-ACDKM outperforms existing methods in terms of the trade-off between utility and privacy.

2   Related Work

In this section, we review several popular methods with respect to the differentially private k-means. Blum first proposed the DP-Lloyd algorithm based on Lloyd's heuristic and DP in 2005 [3]. The main idea of their algorithm is to add Laplace noise to each step of the original Lloyd's heuristic. Zhang et al. [30] proposed PrivGene, a differentially private k-means framework based on genetic algorithm in 2013. Given a dataset and scoring function, the goal is to find the optimal parameters to maximize the scoring function. Balcan et al. [2] proposed a privacy-preserving algorithm by combining local search and exponential mechanism in 2017. The main idea of their algorithm refers to the privacy-preserving algorithm of the k-median proposed by Gupta et al. [9]. In the traditional local search algorithm, they exchange the pair-wise elements with the largest improvement of the current solution in each step to generate a new solution. While in the privacy-preserving algorithm, they exchange the pair-wise elements with the largest increase in each step of the algorithm by using the exponential mechanism. Along with the other differentially private frameworks [7,13,16,19,24–26,30], these works are all based on Lloyd's heuristic which has bad a local minimum. Nie et al. [18] proposed a coordinate method for k-means which has a smaller local minimum than Lloyd's heuristic. This method use a coordinate method to update an indicator matrix and omit the update of the centers. Note that omitting this update can preserve privacy of users since the update of centers will lead to privacy leakage. Therefore, CDKM is a proper and generic method for applying differential privacy to k-means clustering.

3   Preliminary

In this section, we elaborate the coordinate descent (CD) method for k-means clustering problem proposed by Nie et al. [18] and the definition of differential privacy (DP) proposed by Dwork et al. [6].

3.1 k-Means

We first review the k-means clustering. Let $X = \{x_1, x_2, \cdots, x_n\} \in \mathrm{R}^{d \times n}$ be a dataset composed of n points where each point contains d features, i.e. $x_i \in \mathrm{R}^d$, let $D = \{D_1, D_2, \cdots, D_k\}$ be a set of k clusters, let $C = \{c_1, c_2, \cdots, c_k\} \in \mathrm{R}^{d \times k}$ be a set of k centers, which are the mean values of points in $D_j, j = 1, 2, \cdots, k$, then the purpose of k-means is to minimize the sum of squared errors (SSE)

$$\min_C \sum_{j=1}^{k} \sum_{x \in D_j} \|x - c_j\|_2^2. \tag{1}$$

Lloyd's heuristic is one of the most famous algorithms for solving k-means problem. It iteratively assigns each point to its nearest center and refines the centers by computing the mean value in each cluster. However, Lloyd's heuristic has several issues: 1) it has a bad local minimum; 2) it will leakage the privacy of users. We next introduce a novel method named CDKM for solving k-means problem which can address the above issues.

3.2   Coordinate Descent for k-Means

Let $U \in \mathrm{R}^{n \times k}$ denote an indicator matrix composed of 0 and 1, where each row of U has only one 1, and other elements are all 0. Let u_{ij} be the element in the i-th row and j-th column of U, when $x_i \in D_j$, then $u_{ij} = 1$, otherwise $u_{ij} = 0$. Now we can formally define the SSE of k-means clustering problem as

$$\min_{U \in Ind, C} \sum_{j=1}^{k} \sum_{i=1}^{n} \|x - c_j\|_2^2 u_{ij} = \min_{U \in Ind, C} \|X - CU^T\|_F^2, \tag{2}$$

Algorithm 1: CDKM

Input: X, k
1 Initialize k centers by k-means++ to get $U \in \mathrm{R}^{n \times k}$;
2 Compute and store Xu_j and $u_j^T u_j$;
3 **while** *not converge* **do**
4 $\quad$ **for** $i = 1$ *to* n **do**
5 $\quad\quad$ Compute $\phi(U^{(j)})$, $j = 1, 2, \cdots, k$ by (8);
6 $\quad\quad$ Update the i-th row of U by (9);
7 **return** U;

---

where $\| \cdot \|_F$ denotes the Frobenius norm.

We can know that the loss function (2) is a NP-hard problem due to the highly non-convex property. One can iteratively update the indicator matrix U and the center matrix C by using Lloyd's heuristic [12]. Nie et al. propose to use an elimination method to omit the center matrix C. When fixing the indicator matrix F, the objective function (2) can be rewritten as follows:

$$\min_{U \in Ind, C} \|X - CU^T\|_F^2 = \min_{U \in Ind, C} \mathrm{Tr}((X - CU^T)(X - CU^T)^T), \qquad (3)$$

where $\mathrm{Tr}(\cdot)$ is the trace of a matrix. Then by taking the derivative with respect to C and making it equal to zero, we have

$$C = XU(U^T U)^{-1} \qquad (4)$$

Plug (4) into (3), we have

$$\min_{U \in Ind} \mathrm{Tr}(XX^T) - \mathrm{Tr}(XU(U^T U)^{-1} U^T X^T). \qquad (5)$$

As XX^T is a constant, then the optimization problem (5) can be simplified as

$$\min_{U \in Ind} -\mathrm{Tr}(XU(U^T U)^{-1} U^T X^T). \qquad (6)$$

Note that the (h, h)-element of $U^T U$ is equal to $u_h^T u_h$, where u_h is the h-th column of U, then the optimization problem (6) can be rewritten as

$$\min_{U \in Ind} -\sum_{h=1}^{k} \frac{u_h^T X^T X u_h}{u_h^T u_h}. \qquad (7)$$

Nie et al. [18] proposed to use a coordinate method to solve the optimization problem (7) by updating row by row of U. They define k indicator matrices $\{U^{(1)}, U^{(2)}, \ldots, U^{(k)}\}$, where each $U^j \in \mathrm{R}^{n \times k}$. When updating the i-th row of U, then the (i, j)-element of $U^{(j)}$ is 1, and other elements in the i-th row are 0. The optimization problem of updating the i-th row of U is

$$\min_{j \in \{1,2,\ldots,k\}} \phi(U^{(j)}) = \min_{j \in \{1,2,\ldots,k\}} -\sum_{h=1}^{k} \frac{(u_h^{(j)})^T X^T X u_h^{(j)}}{(u_h^{(j)})^T u_h^{(j)}}, \qquad (8)$$

where $u_h^{(j)}$ is the h-th column of $U^{(j)}$. Then, according to CD and the nature of indicator matrix U, we give the updating formulation for the i-th row of U:

$$u_{il} = \begin{cases} 1, & l = \arg\min_j \phi(U^{(j)}) \\ 0, & \text{otherwise,} \end{cases} \tag{9}$$

By computing (8) and (9) iteratively, then the k-means clustering problem can be solved. Algorithm 1 gives the details of this process.

3.3 A Fast Version of CDKM

Note that the computation of optimization problem (8) is too heavy. Nie et al. [18] proposed a fast version of CDKM. To simplify the optimization problem (8), they introduce $U^{(0)}$ whose elements in i-th row are all 0, and other elements are same with $U^{(j)}$. Then the optimization problem (8) is equivalent to

$$\begin{aligned}
&\min_{j\in\{1,2,\dots,k\}} \quad obj(U^{(j)}) \\
=&\min_{j\in\{1,2,\dots,k\}} -\sum_{h=1}^{k} \frac{(u_h^{(j)})^T X^T X u_h^{(j)}}{(u_h^{(j)})^T u_h^{(j)}} + \sum_{h=1}^{k} \frac{(u_h^{(0)})^T X^T X u_h^{(0)}}{(u_h^{(0)})^T u_h^{(0)}} \\
=&\min_{j\in\{1,2,\dots,k\}} -\frac{(u_j^{(j)})^T X^T X u_j^{(j)}}{(u_j^{(j)})^T u_j^{(j)}} + \frac{(u_j^{(0)})^T X^T X u_j^{(0)}}{(u_j^{(0)})^T u_j^{(0)}}.
\end{aligned} \tag{10}$$

Then the i-th row of U is updated by the following equation,

$$u_{il} = \begin{cases} 1, & l = \arg\min_j obj(U^{(j)}) \\ 0, & \text{otherwise,} \end{cases} \tag{11}$$

where l is the updated position of 1 in the i-th row of the indicator matrix. Let U denote the current indicator matrix before the next iteration, let p denote the location of 1 in the i-th row of U, let u_j denote the j-the column of U, let $u_j^{(j)}$ denote the j-the column of $U^{(j)}$, let $u_j^{(0)}$ denote the j-th column of $U^{(0)}$, then we can rewrite the optimization problem (10) as the following two situations:

- Situation 1: When $j = p$, then we have :

$$u_j^{(j)} = u^{(j)}, X u_j^{(0)} = X u_j - x_i, (u_j^{(0)})^T u_j^{(0)} = u_j^T u_j - 1. \tag{12}$$

- Situation 2: When $j \neq p$, then we have:

$$u_j^{(0)} = u^{(j)}, X u_j^{(j)} = X u_j + x_i, (u_j^{(j)})^T u_j^{(j)} = u_j^T u_j + 1. \tag{13}$$

The second and the third equations in (12) and (13) hold for $\delta_j = u_j^{(j)} - u_j^{(0)}$ and $\delta_j = u_j^{(0)} - u_j^{(j)}$ respectively, where $\delta_j \in R^n$ denotes a vector whose i-th element equals to 1, and other elements are all 0. Plug (12) and (13) into the optimization problem (10), we have

$$
\min_{j \in \{1,2,\ldots,k\}} obj(U^{(j)}) =
\begin{cases}
\dfrac{u_j^T X^T X u_j}{u_j^T u_j} - \dfrac{u_j^T X^T X u_j - 2x_i^T X u_j + x_i^T x_i}{u_j^T u_j - 1}, & j = p \\[2ex]
-\dfrac{u_j^T X^T X u_j}{u_j^T u_j} + \dfrac{u_j^T X^T X u_j + 2x_i^T X u_j + x_i^T x_i}{u_j^T u_j + 1}, & j \neq p.
\end{cases}
\tag{14}
$$

Then by computing (14) iteratively, the k-means clustering problem can be solved efficiently.

3.4 Differential Privacy

Differential privacy (DP) has been one of the most useful privacy models for data mining. Intuitively, a mechanism satisfies DP if its outputs are approximately same with the outputs when a single data is changed. Then the adversary cannot infer any information with respect to the data owners from the outputs. Here we review the definition of DP.

Definition 1 (Differential Privacy [6,8]). *A randomized mechanism $\mathcal{M}$: $\mathcal{X} \to \mathcal{R}$ satisfies ϵ-differential privacy, where $\epsilon > 0$, if for any adjacent datasets X, X' and for any subsets of $S \subseteq \mathcal{R}$, it holds that*

$$
\Pr[\mathcal{M}(X) \in S] \leq exp(\epsilon) \cdot \Pr[\mathcal{M}(X') \in S].
\tag{15}
$$

Exponential mechanism is one of the most commonly used mechanisms that ensure ϵ-DP for the queries [14]. Given a utility function over the outputs, then the exponential mechanism will sample the high utility output with higher probability. The formal definition of exponential mechanism is given as follows.

Definition 2 (Exponential Mechanism [14]). *Let $f : \mathcal{X} \times \mathcal{R} \to \mathrm{R}$ denote an utility function with sensitivity $\Delta(f) = \max_{X,X'} |f(X, S) - f(X', S)|$, the exponential mechanism $\mathcal{M}$ ensures ϵ-DP with the outputs distribution*

$$
\Pr[\mathcal{M}(X) = S] \propto exp(\frac{\epsilon f(X, S)}{2\Delta(f)}).
\tag{16}
$$

We will propose a differentially private k-means clustering method based on exponential mechanism and CDKM in Sect. 4.

Definition 3 (Sequential Composition [8]). *Suppose a set of mechanisms $\mathcal{M}_1, \mathcal{M}_2, \cdots, \mathcal{M}_m$ sequentially acts on a dataset X, and each mechanism M_i ensures ϵ_i-DP, then their compositions ensure ϵ-DP, where $\epsilon = \sum_i^m \epsilon_i$.*

4   Proposed Our Method

This section elaborates the main contents of our method toward differentially private k-means clustering with coordinate descent. Firstly, we propose an approximate CDKM algorithm which can effectively reduce the time complexity. Then we show the sensitivity of CDKM, and by applying exponential mechanism to the CDKM, we can ensure ϵ-DP for k-means clustering.

Algorithm 2: DP-ACDKM

Input: X, k, T, ϵ

1 Initialize k centers by k-means++ to get $U \in \mathrm{R}^{n \times k}$;
2 Compute and store Xu_j and $u_j^T u_j$;
3 **for** $t = 1$ *to* T **do**
4   **for** $i = 1$ *to* n **do**
5   Compute $obj(U^{(j)})$, $j = 1, 2, \cdots, k$ by (17);
6   Update the i-th row of U by (18);
7   **if** $l \neq p$ **then**
8   Update Xu_l, Xu_p, $u_l^T u_l$ and $u_p^T u_p$ by (19);
9 **return** U;

4.1 Approximate CDKM

In Sect. 3.3, we review the fast version of CDKM. However, this method is still time-consuming due to the high time complexity. In this section, we will propose an approximate CDKM (ACDKM) to simplify the computation of updating.

By observing the form of (14), we find that these two terms

$$\frac{u_j^T X^T X u_j}{u_j^T u_j} - \frac{u_j^T X^T X u_j}{u_j^T u_j - 1}, \quad -\frac{u_j^T X^T X u_j}{u_j^T u_j} + \frac{u_j^T X^T X u_j}{u_j^T u_j + 1}$$

are approximately equal to 0, and $x_i^T x_i$ does not influence the final result. Then we simplify the objective function as

$$\min_{j \in \{1, 2, \ldots, k\}} obj(U^{(j)}) = \begin{cases} \frac{x_i^T X u_j}{u_j^T u_j - 1}, & j = p \\ \frac{x_i^T X u_j}{u_j^T u_j + 1}, & j \neq p. \end{cases} \tag{17}$$

We next propose our differentially private framework for ACDKM based on exponential mechanism.

4.2 Proposed DP-ACDKM

Algorithm 2 shows our DP-ACDKM framework. To ensure ϵ-DP for each updating of the i-th row of U in each iteration (the fifth and sixth steps in Algorithm 2), we give the updating function as follows:

$$u_{il} = \begin{cases} 1, & \Pr[u_{il} = 1] \propto exp(\frac{\epsilon obj(U^{(l)})}{2\Delta(f)}) \\ 0, & \text{otherwise.} \end{cases} \tag{18}$$

To update the i-th row of U, we need to compute $\frac{x_i^T X u_j}{u_j^T u_j - 1}$ and $\frac{x_i^T X u_j}{u_j^T u_j + 1}$ in each iteration. The time complexity of this step is $O(ndk)$. To reduce the time complexity, we use an acceleration strategy (the eighth step in Algorithm 2). Consider these two situations:

- When $l = p$, then U, Xu_j and $u_j^T u_j (j = 1, 2, \cdots, k)$ are unchanged.
- When $l \neq p$, then we have $Xu_p = Xu_p^{(0)}$, $Xu_l = Xu_l^{(l)}$, $u_p^T u_p = (u_p^{(0)})^T u_p^{(0)}$ and $u_l^T u_l = (u_l^{(l)})^T u_l^{(l)}$. Then we can use the following equations to update Xu_j and $u_j^T u_j, j = l, p$:

$$
\begin{aligned}
Xu_l &\leftarrow Xu_l + x_i, & u_l^T u_l &\leftarrow u_l^T u_l + 1 \\
Xu_p &\leftarrow Xu_p - x_i, & u_p^T u_p &\leftarrow u_p^T u_p - 1.
\end{aligned}
\tag{19}
$$

Note that (19) can enormously reduce the time complexity since computing Xu_j needs nd additions, while computing $Xu_j + x_i$ just needs d additions. Our algorithm can reduce the time complexity to $O(nk)$, while the time complexity of vanilla CDKM is $O(ndk)$.

4.3   Privacy Analysis

We consider to use exponential mechanism to ensure DP for ACDKM by perturbing the objective function in (17). We next prove our algorithm ensures ϵ-DP for k-means clustering.

Theorem 1. *The sensitivity of the objective function is 1, and Algorithm 2 ensures $t \cdot \epsilon$-DP, where t is the number of iterations.*

Proof. We first prove the sensitivity. Without loss of generality, we discuss the utility function of $obj(U^{(j)})$ when $j \neq p$. Now we give the utility function as follows

$$
f(X, S) = \frac{x_i^T X u_j}{u_j^T u_j + 1}
\tag{20}
$$

Let X and X' denote a pair of adjacent datasets, and without loss of generality, we let x_{11} and $x_{11'}$ be the different element in X and X', then the sensitivity of the output is

$$
\begin{aligned}
\max_{X, X'} |f(X, S) - f(X', S)| &\leq |\frac{x_i^T X u_j - (x_i')^T X' u_j}{u_j^T u_j + 1}| \\
&= |\frac{[x_{11}^2 - (x_{11}')^2, x_{21}(x_{11} - x_{11}'), \cdots, x_{n1}(x_{11} - x_{11}')] u_j}{u_j^T u_j + 1}| \\
&\leq |\frac{[x_{11}^2 - (x_{11}')^2, x_{21}(x_{11} - x_{11}'), \cdots, x_{n1}(x_{11} - x_{11}')] u_j}{u_j^T u_j}| \\
&= 1,
\end{aligned}
\tag{21}
$$

where the first inequality holds for (20), the second inequality holds for $u_j^T u_j + 1 \geq u_j^T u_j$, the last equality is holds for normalizing each x_{ij} to $[0, 1]$. When $j \neq p$, the proof is similar with $j = q$. The only difference is $u_j^T u_j - 1 \leq u_j^T u_j$.

When n is large enough, we have $u_j^T u_j \gg 1$. Then we can approximately give $u_j^T u_j - 1 \approx u_j^T u_j$. Therefore, the sensitivity is proved and each iteration of Algorithm 2 can ensure ϵ-DP.

Then by using the sequential composition [8], we can easily know the Algorithm 2 ensures $t \cdot \epsilon$-DP.

Remark 1. Note that the privacy parameters can be bounded tightly by using strong composition [6] or r-DP [15]. Since this is not the point we discuss in this paper, we use the sequential composition for intuitiveness.

5 Experiments

We evaluate our DP-ACDKM in terms of the trade-off between utility and privacy by varying the privacy parameter ϵ among several real-world datasets[1] in Sect. 5.1. To evaluate the convergence of DP-ACDKM, we show the objective function value as a function of the iteration t. We give a brief description of the datasets in Table 1. Our experiments are all implemented on Windows 10 with i9-12900K CPU and 64G RAM. In each experiment, we perform 100 times for each algorithm and compute the mean value of the results. For each dataset, we normalize each element in x_i to $[0, 1]$ by using the following equation:

$$x'_{ij} = \frac{x_{ij} - \min_j(x_i)}{\max_j(x_i) - \min_j(x_i)}, \tag{22}$$

where $i \in \{1, 2, \cdots, n\}$ and $j \in \{1, 2, \cdots, d\}$.

Table 1. A description of datasets.

Datasets	# of points	# of features	# of clusters
Binalpha	1,404	320	$\{5, 10\}$
Mpeg	1,400	6,000	$\{5, 10\}$
Palm	2,000	256	$\{5, 10\}$
Protein	694	30	$\{5, 10\}$

5.1 Privacy-Utility Trade-Off

To evaluate the effectiveness of DP-ACDKM, we show the SSE as a function of the privacy level. Our empirical results compare DP-ACDKM algorithm against DP-Lloyd [3], ACDKM and CDKM [18] with $k = 5$ and $k = 10$. Note that we give two values of k for generality. Figure 1 quantifies the trade-off between privacy and utility for different privacy levels, where the ϵ varies among 0.1, 0.2,

---

[1] http://archive.ics.uci.edu/ml/index.php.

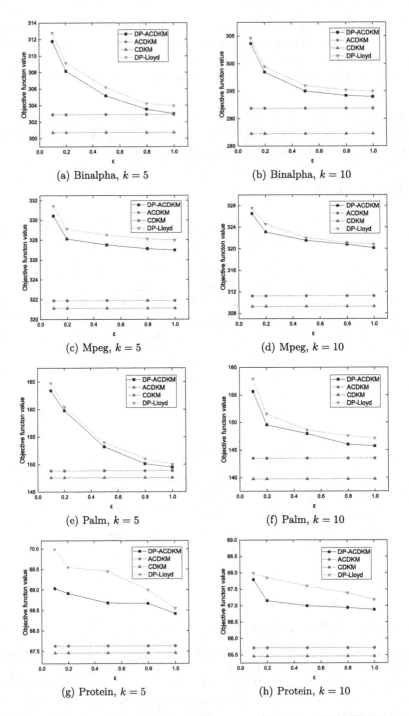

Fig. 1. Privacy-utility trade-off. Better utility means lower loss. DP-ACDKM outperforms DP-Lloyd for the k-means clustering.

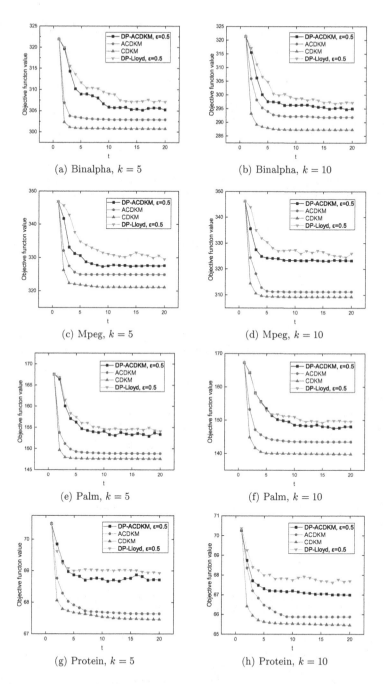

(a) Binalpha, $k = 5$

(b) Binalpha, $k = 10$

(c) Mpeg, $k = 5$

(d) Mpeg, $k = 10$

(e) Palm, $k = 5$

(f) Palm, $k = 10$

(g) Protein, $k = 5$

(h) Protein, $k = 10$

Fig. 2. The impact of differential privacy on convergence. Differential privacy does not prevent convergence but makes the loss reduction more noisy and also make the loss larger than the optimum.

0.5, 0.8, 1. It is obvious that our DP-ACDKM outperforms DP-Lloyd. With the increase of ϵ, the utility of DP-ACDKM becomes better. The loss of our ACDKM is larger than CDKM since ACDKM is an approximate algorithm. However, this increase of loss is not larger than 2%. Therefore, DP-ACDKM is an effective algorithm which can preserve privacy with remarkable trade-off.

5.2   Convergence

We evaluate the convergence of DP-ACDKM in this section. In Fig. 2, we choose the privacy level $\epsilon = 0.5$ for both DP-ACDKM and DP-Lloyd to show the results of objective function value as a function of the number of iterations. Here we set the number of maximum iteration to 20. It is obvious that DP-ACDKM converges to smaller values than DP-Lloyd, and DP-ACDKM converges faster than DP-Lloyd among all datasets. Both DP-ACDKM and DP-Lloyd have noisy reduction of objective function values due to the differential privacy, but DP-ACDKM is smoother than DP-Lloyd.

6   Conclusion

In this paper, we proposed a differentially private framework called DP-ACDKM for k-means clustering by using the novel CDKM and exponential mechanism. We first used an approximate method called ACDKM to speed up the convergence of CDKM. This method can reduce the amount of redundant calculation while maintaining a small difference with the CDKM in terms of the objective function value. By using the exponential mechanism to perturb the objective function, our algorithm can ensure ϵ-DP for k-means. Our results showed that DP-ACDKM outperforms the state of the art in terms of the trade-off between utility and privacy.

Acknowledgments. This work was supported by the National Key R&D Program of China (2021YFA1003002), the China National Natural Science Foundation (12090021, 61872224).

References

1. Abadi, M., et al.: Deep learning with differential privacy. In: Proceedings of the 2016 ACM SIGSAC Conference on Computer and Communications Security, pp. 308–318 (2016)
2. Balcan, M.F., Dick, T., Liang, Y., Mou, W., Zhang, H.: Differentially private clustering in high-dimensional Euclidean spaces. In: International Conference on Machine Learning, pp. 322–331. PMLR (2017)
3. Blum, A., Dwork, C., McSherry, F., Nissim, K.: Practical privacy: the SULQ framework. In: Proceedings of the Twenty-Fourth ACM SIGMOD-SIGACT-SIGART Symposium on Principles of Database Systems, pp. 128–138 (2005)

4. Cormode, G., Jha, S., Kulkarni, T., Li, N., Srivastava, D., Wang, T.: Privacy at scale: local differential privacy in practice. In: Proceedings of the 2018 International Conference on Management of Data, pp. 1655–1658 (2018)
5. Dennis, D.K., Li, T., Smith, V.: Heterogeneity for the win: one-shot federated clustering. In: International Conference on Machine Learning, pp. 2611–2620. PMLR (2021)
6. Dwork, C.: Differential privacy. In: Bugliesi, M., Preneel, B., Sassone, V., Wegener, I. (eds.) Automata, Languages and Programming. ICALP 2006. LNCS, vol. 4052pp. 1–12. Springer, Heidelberg (2006). https://doi.org/10.1007/11787006_1
7. Dwork, C.: A firm foundation for private data analysis. Commun. ACM **54**(1), 86–95 (2011)
8. Dwork, C., Roth, A., et al.: The algorithmic foundations of differential privacy. Found. Trends® Theor. Comput. Sci. **9**(3–4), 211–407 (2014)
9. Gupta, A., Ligett, K., McSherry, F., Roth, A., Talwar, K.: Differentially private combinatorial optimization. In: Proceedings of the Twenty-First Annual ACM-SIAM Symposium on Discrete Algorithms, pp. 1106–1125. SIAM (2010)
10. Horn, D., Gottlieb, A.: Algorithm for data clustering in pattern recognition problems based on quantum mechanics. Phys. Rev. Lett. **88**(1), 018702 (2001)
11. Li, N., Li, T., Venkatasubramanian, S.: t-closeness: privacy beyond k-anonymity and l-diversity. In: 2007 IEEE 23rd International Conference on Data Engineering, pp. 106–115. IEEE (2006)
12. Lloyd, S.: Least squares quantization in PCM. IEEE Trans. Inf. Theory **28**(2), 129–137 (1982)
13. Lu, Z., Shen, H.: Differentially private k-means clustering with convergence guarantee. IEEE Trans. Depend. Secure Comput. **18**(4), 1541–1552 (2020)
14. McSherry, F., Talwar, K.: Mechanism design via differential privacy. In: 48th Annual IEEE Symposium on Foundations of Computer Science (FOCS 2007), pp. 94–103. IEEE (2007)
15. Mironov, I.: Rényi differential privacy. In: 2017 IEEE 30th Computer Security Foundations Symposium (CSF), pp. 263–275. IEEE (2017)
16. Mohan, P., Thakurta, A., Shi, E., Song, D., Culler, D.: Gupt: privacy preserving data analysis made easy. In: Proceedings of the 2012 ACM SIGMOD International Conference on Management of Data, pp. 349–360 (2012)
17. Nie, F., Li, Z., Wang, R., Li, X.: An effective and efficient algorithm for K-means clustering with new formulation. IEEE Trans. Knowl. Data Eng. **35**(4), 3433–3443 (2023)
18. Nie, F., Xue, J., Wu, D., Wang, R., Li, H., Li, X.: Coordinate descent method for k-means. IEEE Trans. Pattern Anal. Mach. Intell. **44**(5), 2371–2385 (2021)
19. Park, M., Foulds, J., Choudhary, K., Welling, M.: Dp-em: differentially private expectation maximization. In: Artificial Intelligence and Statistics, pp. 896–904. PMLR (2017)
20. Pei, S., Chen, H., Nie, F., Wang, R., Li, X.: Centerless clustering. IEEE Trans. Pattern Anal. Mach. Intell. **45**(1), 167–181 (2022)
21. Peng, X., Zhou, C., Hepburn, D.M., Judd, M.D., Siew, W.H.: Application of k-means method to pattern recognition in on-line cable partial discharge monitoring. IEEE Trans. Dielectr. Electr. Insul. **20**(3), 754–761 (2013)
22. Sahoo, S.K., Makur, A.: Dictionary training for sparse representation as generalization of k-means clustering. IEEE Signal Process. Lett. **20**(6), 587–590 (2013)
23. Shin, H., Kim, S., Shin, J., Xiao, X.: Privacy enhanced matrix factorization for recommendation with local differential privacy. IEEE Trans. Knowl. Data Eng. **30**(9), 1770–1782 (2018)

24. Stemmer, U.: Locally private k-means clustering. J. Mach. Learn. Res. **22**(1), 7964–7993 (2021)
25. Stemmer, U., Kaplan, H.: Differentially private k-means with constant multiplicative error. Adv. Neural Inf. Process. Syst. **31** (2018)
26. Su, D., Cao, J., Li, N., Bertino, E., Lyu, M., Jin, H.: Differentially private K-means clustering and a hybrid approach to private optimization. ACM Trans. Privacy Secur. **20**(4), 1–33 (2017)
27. Wang, R., Lu, J., Lu, Y., Nie, F., Li, X.: Discrete and parameter-free multiple kernel k-means. IEEE Trans. Image Process. **31**, 2796–2808 (2022)
28. Wu, J.: Advances in K-Means Clustering: A Data Mining Thinking. Springer, Heidelberg (2012). https://doi.org/10.1007/978-3-642-29807-3
29. Yao, A.C.: Protocols for secure computations. In: 23rd Annual Symposium on Foundations of Computer Science (SFCS 1982), pp. 160–164. IEEE (1982)
30. Zhang, J., Xiao, X., Yang, Y., Zhang, Z., Winslett, M.: Privgene: differentially private model fitting using genetic algorithms. In: Proceedings of the 2013 ACM SIGMOD International Conference on Management of Data, pp. 665–676 (2013)

DADR: A Denoising Approach for Dense Retrieval Model Training

Mengxue Du[1,2], Shasha Li[1,2], Jie Yu[1,2(✉)], Jun Ma[1,2], Huijun Liu[1,2],
Miaomiao Li[1,2], and Bin Ji[1,2]

[1] College of Computer, National University of Defense Technology, Changsha, China
{dumengxuenudt,shashali,majun,liuhuijun,limiaomiao21}@nudt.edu.cn,
jibin@nus.edu.sg
[2] Institute of Data Science, National University of Singapore, Singapore, Singapore
yj@nudt.edu.cn

Abstract. With the development of representation learning techniques, Dense Retrieval (DR) has become a new paradigm to retrieve relevant texts for better ranking performance. Although current DR models have achieved encouraging results, their performance is highly affected by the noise level in training samples. In particular, a large number of examples that were not labeled as positives (which were used as negative samples by default) were found to actually be positive or highly relevant. As such, it is of critical importance to account for the inevitable noises in DR model training. However, little work on dense retrieval has taken the noisy nature into consideration. In this work, we intensely investigate the serious negative impacts of noisy training samples and propose a new denoising approach, i.e., A **D**enoising **A**pproach based on dynamic weights for **D**ense **R**etrieval model training (DADR), which reduces the effects of noise on model performance by assigning diverse weights to the different samples during the training process. We incorporate the proposed DADR approach with three representative kinds of sampling methods and different loss functions. Experimental results on two publicly available retrieval benchmark datasets show that our approach significantly improves the performance of the DR model over normal training.

Keywords: sample denoising · dense retrieval · dynamical weights

1 Introduction

In recent years, with the development of deep learning, dense retrieval models have been a promising solution for IR tasks such as question answering [16, 26,27], web search [7,9,13,18] and cross-modal retrieval [5,6]. However, there are some unique noise issues with DR models that have not been well addressed.

This work was supported by Hunan Provincial Natural Science Foundation Project (No. 2022JJ30668) and (No. 2022JJ30046).

Specifically, totally manually annotating all the candidate texts for a given query is infeasible due to the huge size of the candidate data.

As a consequence of the sparsity of the manual annotations and the large demand for negative samples, researchers supplement negative samples by randomly sampling from the unlabeled corpus or adopting the evaluation results of additional models. Unfortunately, quite a few unlabeled positives might be mistakenly added to the training set as negative samples, which introduces noise to model training. For instance, as the work of predecessors [26] shows, 70% of the top-ranked unlabeled passages are found to actually be positives or highly relevant in the original MSMARCO dataset [7].

In this work, we argue that false-negative samples would impose error penalties on the model parameters during training, leading to low-quality retrieval rankings. Table 1 shows the negative effects of false-negative samples when training a typical BERT-based DR model on two benchmark datasets. We construct a "noisy" testing set by adding some false-negative samples for retriever evaluation. As Table 1 shows, compared to the original dataset, training the DR model with false-negative noise results in a significant performance drop over two datasets with respect to Recall@10 and NDCG@10. The above results provide empirical evidence that it is crucial to consider the impact of false-negative noise in retrieval tasks.

Table 1. Results of the normal training and noisy training under 10% additional label noise on MS MARCO. # Drop denotes the relative performance drop of noisy training as compared to normal training.

Dataset	MARCO Dev Passage		MARCO Dev Doc	
	MRR@10	R@100	MRR@10	R@100
Normal training	0.273	0.851	0.319	0.805
Noisy training	0.224	0.748	0.207	0.725
#Drop	23.02%	6.97%	35.11%	9.94%

To study how false-negative noises affect the performance of DR models during training, we compare the loss of true-negative and false-negative samples along the normal training process. We observe that the noisy data are significantly different in two dimensions compared to the clean samples, *i.e.*, the gradient direction in the early training phase and the probability density function of loss over the entire training period (see Fig. 1). Based on the empirical evidence provided by our observations, we design two noisy evaluation signals, which are as follows:

1) Inconsistency of the gradients. As Fig. 1(a) shows, the loss of false-negative noises remained high in the early period of training and began to decrease sharply in the late training period, which means the model tends to fit the clean samples first, while the fit to the noisy samples (which is harder for

the model to understand) often occurs in the late training stage. Therefore, one can infer from the inconsistency between the gradient directions of the sample and the model optimization in the early period of training whether a sample is more likely to be clean or noisy.

2) Noisy estimation by probabilistic modeling. As Fig. 1(b) shows, there are significant differences between the probability density function distribution of noisy and clear samples loss over the entire training period. Inspired by previous work [1], we leverage the mixture distribution model to distinguish the clean and noisy samples from the loss distribution alone.

Combining these two noisy signals above, we propose a novel denoising strategy for Dense Retrieval that dynamically weights the sample loss along the training process. We simultaneously consider the two noisy signals above as the final noisy estimation, and the samples with a higher noisy probability are assigned a smaller loss weight. In detail, we put the first noisy signal (the inconsistency of the gradients) in the dominant position in the early stage of training due to the empirical evidence that noise could barely be fitted by the model at first. On the contrary, we gradually shift dominance to the second noisy signal because the performance of the probability estimation signal tends to be stable in the later stages. On two benchmarks, we test the proposed DADR over three representative sampling methods: random negative sampling [12,15,17], static hard negative sampling [10] and mixed sampling [30]. The results demonstrate the superiority and universality of DADR.

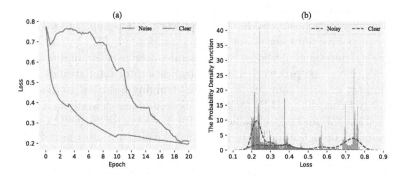

Fig. 1. The observations on MS MARCO under 50% label noise for clean and noisy samples. (a) Pointwise loss over the entire training period. (b) The probability density function distribution of loss after 20 epochs.

In summary, this work makes the following contributions:

1) We propose a novel denoising approach for dense retrieval model training that does not introduce any extra data or models for the first time.

2) Compared to previous work, our approach denoises the samples with negligible overhead at the same time as model training instead of adding an additional denoising stage.

3) We applied the proposed paradigm to three representative kinds of sampling strategies, *i.e.*, random negative sampling, static negative sampling, and mixed sampling, with two typical loss functions, *i.e.*, pointwise loss and pairwise loss. Extensive experiments on two benchmarks validate and demonstrate the effectiveness of DADR in improving the performance of DR models over normal training.

2   Related Work

In recent years, with the development of deep learning and the reliance of the deep neural networks on large-scale labeled databases, many innovative denoising approaches have been proposed [21]. [22] proposed a noise identification approach based on confidence learning, which identifies erroneous labels by estimating the conditional probability between a given sample and a potentially correct label. Mariya, T. et al. [28] argue that noisy samples tend to experience more forgetting events than normal samples during model training. Based on the heuristic, they further identified noisy data by noting the total number of forgotten events experienced by each sample during model training. It is worth mentioning that a method to identify noise based on the relative magnitude of loss values of samples during the training process was proposed in [14]. This work adopts the recurrent learning strategy to make the neural network switch between under-fitting and over-fitting several times and tracks the loss of models with different parameters on samples at different stages. The probability that the sample belongs to the noise sample is estimated by counting the mean and variance of the loss of each sample. In the latest work, a denoising method based on learning contrastive representations has been proposed in [21]. In addition, some other classical denoising methods, such as semi-supervised [20] and unsupervised [2] methods, have also provided valuable inspiration for later studies.

The impact of noisy samples [3] in model training has been studied in Natural Language Processing (NLP) tasks such as recommender systems [11,29], text summarization [23] and text classificaton [19]. However, little attention has been paid to research on denoising for retrieval tasks, which are substantially different in the noisy distribution of training samples from traditional tasks. Specifically, as with other deep models, the performance of retrieval models strongly depends on the size and quality of training samples. In addition, however, retrieval has a large demand for negative samples in particular. In general, retrieval performance can be effectively improved by focusing on higher-quality negative samples, especially hard negative samples. However, due to the high cost of manual labeling and the difficulty of obtaining hard negative samples accurately, the supplement of negative samples is always accompanied by obvious noise, *i.e.*, false-negative samples.

Qu, Y. et al. [26] proposed an advanced denoising approach for dense passage retrieval that uses a pre-trained cross-encoder to identify false-negative samples. Specifically, the top-K paragraphs recalled by the retrieval model are given to the cross-encoder for noisy evaluation except the labeled positive samples, and the paragraphs with high relevance scores will be identified as false-negative samples in the above work [26]. However, the above approach needs an additional pre-trained noise recognizer, which must have better performance than the training retrieval model, which requires a high level of resources.

Different from the above methods, we propose a novel Denoising Approach for Dense Retrieval (DADR), which doesn't introduce any additional data or model resources. Since the denoising task of DADR is performed synchronously during model training, it does not add an additional denoising step to the model training pipeline. Moreover, since our method mainly employs two simple computational signals, the time overhead incurred by denoising DADR is a negligible constant at each training step.

3   Method

In this section, we detail the proposed denoising approach based on dynamical weights for dense retrieval model training.

3.1   Task Formulation

Given a query q and a corpus $\mathcal{C}$, the target of information retrieval is to find the relevant texts $\mathcal{D}^+$ with parameters θ. DR model encodes queries and documents separately into embeddings, denoted as $e_{q;\theta}$ and $e_{d;\theta}$, respectively, and then uses the similarity function $f(q, d|\theta)$ to predict relevance score.

$$f(q, d|\theta) = \langle e_{q;\theta}, e_{d;\theta} \rangle \tag{1}$$

where $\langle , \rangle$ denotes the similarity function. Ideally, the setting of DR model training is to learn θ from a reliable training dataset $\mathcal{D}^*$. The parameters θ is learning by minimizing the $\mathcal{L}_P$ loss over $\mathcal{D}^*$ as follows:

$$\theta^* = \min \mathcal{L}_P(\mathcal{D}^*) \tag{2}$$

However, due to the widespread occurrence of false-negative samples in retrieval set $\widehat{\mathcal{D}}$, dense retrieval training might be misled and result in poor model. Thus, this denoising work aim to train a DR model by denoising $\widetilde{\mathcal{D}}$ as:

$$\theta^* = \min \mathcal{L}_P(denoise(\widetilde{\mathcal{D}})) \tag{3}$$

In this work, we perform denoising by dynamically weighting the samples in the training set $\widetilde{\mathcal{D}}$. The denoising process happens at training time rather than in an extra stage.

3.2   Denoising Approach Based on Dynamical Weight

We propose the DADR approach for the DR model, which mainly adopts the strategy of dynamic weighting for sample loss. To reduce the impact of false-negative samples, DADR considers two noise signals. We detail the two signals as follows.

Inconsistency of the Gradients. As shown in Fig. 1(a), the inconsistency between the gradient direction of the sample and the model optimization implies that the sample is more difficult to fit and is generally more likely to be a noisy sample. Based on the above assumption, this work adopts the inconsistency of gradients as an important noisy signal, which is obtained by the following equation:

$$c_x = \frac{\sum_{y \in \mathcal{B}} \langle g_x, g_y \rangle}{|g_x| |\sum_{y \in \mathcal{B}} g_y|} \tag{4}$$

where c_x is the consistency of sample x and model optimization, g is the gradient contribution of a single sample, and $\mathcal{B}$ is the sample set from one training batch. Noted c_x is actually equivalent to the cosine similarity between the gradient update vector g_x and the average gradient of all samples in the batch. A higher similarity indicates that this sample is more consistent with the direction of model optimization. The inconsistency, inc_x is obtained by normalizing the inverse of c_x:

$$inc_x = \frac{-c_x + 1}{2} \tag{5}$$

Noisy Estimation by Probabilistic Modeling. The observation in Fig. 1(b) reveals a clear difference in the distribution between the noisy samples and the clean samples. We record the loss values of samples over a period of training, which is set to be 20 epochs in this work. We find that the clean and noisy data conform to a bimodal distribution, which can be modeled using a mixture of probabilistic models.

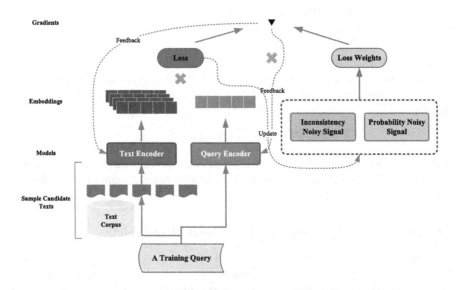

Fig. 2. The overall framework of DADR.

In our work, we fit them with the Gaussian Mixture Model(GMM) after normalizing all losses. Specifically, the losses are assumed to follow a distribution of GMM, and the density estimation on the loss l can be defined as:

$$p(l|\pi, \mu, \Sigma) = \sum_{k=1}^{K} \pi_k \mathcal{N}(l|\mu_k, \Sigma_k) \qquad (6)$$

where $\mathcal{N}(l|\mu_k, \Sigma_k)$ is the kth component of the GMM model, π_k is the mixture coefficient, and K is the number of components. We fit a GMM with two components to model the distribution of clean and noisy samples. Note that the EM algorithm is adopted to calculate the parameters π, μ, Σ of the distribution iteratively. Finally, we obtain the probability of noise based on the posterior probability of the loss distribution:

$$p(k|l) = \frac{p(k)p(l|k)}{p(l)} \qquad (7)$$

where $k = 1$ denotes a noisy class and $k = 0$ denotes a clean class. Thus, $p(1|\bar{l}_x)$ is the probability of a sample x being noisy. In this work, we adopt the average of losses for example x in the previous training epoch, which can be defined as:

$$p(\bar{l}_x^E) = \frac{1}{E} \sum_{e=1}^{E} l_x^e \qquad (8)$$

Final Noisy Estimation. We dynamically weight the sample loss using both of the above noise signals, c_x and $p(1|\tilde{l}_x)$ simultaneously, but there are different adoption weights for the two in different training stages. We find that inconsistency signals c_x are more suggestive of noise in the early stage of model training but gradually lose their effect in the later stage of training. On the contrary, the performance of the probability estimation signal $p(1|\tilde{l}_x)$ gradually improves in the early stages of training and tends to be stable in the later stages. Given this observation, we dynamically adjust the weights of the two signal contributions to ensure that the inconsistent signal c_x is used as the dominant signal in the early stage of model training and that the dominant signal is post-converted into probability estimation $p(1|\tilde{l}_x)$ in the later stage. The final noisy estimate for sample x is given in the following equation:

$$noise_x = \epsilon(E) \cdot inc_x + (1 - \epsilon(E)) \cdot p(k = 1|l_x) \qquad (9)$$

where inc_x and $p(k = 1|l_x)$ are mentioned in Equation (5) and (8) as inconsistency signals and probability estimation signal, respectively. $\epsilon(E)$ is defined as:

$$\epsilon(E) = max(\epsilon_{max} - \alpha E, \epsilon_{min}) \qquad (10)$$

where E is the epoch number, α, ϵ_{max} and ϵ_{min} are hyper-parameters. ϵ_{max} and ϵ_{min} are upper bound and lower bound respectively, and α adjusts the pace to reach the minimum. Note that we use a simple linear function rather than a more complex function to adjust the proportion of the two signals, which has the benefit of reducing the cost of model tuning.

During model training, $noise_x$ is updated on each epoch and applied to dynamic weight loss in the next epoch. The weight of loss for sample x is assigned as:

$$w_x = 1 - noise_x \qquad (11)$$

Summary. DADR dynamically weights the sample loss using both of the above noise signals while adjusting the weights of the two signal contributions to ensure that the inconsistent signal is used as the dominant signal in the early stage of model training and that the dominant signal is post-converted into probability estimation in the later stage.

As Fig. 2 shows, the DR model consists of two representation encoders, $i.e.$, a Query Encoder and Text Encoder. The vector similarity between the query representation and the candidate text representation is used as the relevance score, and the loss of this candidate text is calculated according to the relevance score. DADR computes the proposed two noisy signal estimates based on the loss of the samples. It is worth noting that the probability noise model is accumulated from previous epochs and updated every epoch. According to the above two noise signals, the sample noise is estimated and the loss weight is obtained. The sample loss is fed back to update the model parameters after reweighting the loss value. The whole algorithm is explained in Algorithm 1.

Algorithm 1: Denoising Approach Based on Dynamical Weight

Input: trainable parameters θ, training samples $\widetilde{\mathcal{D}}$, the maximum epoch number E_{max}, α, ϵ_{max}, ϵ_{min}, learning rate η, parameter optimization function ∇.

1   **Initialize** $\mathcal{W}_0 = \{w_x = 1 | x \in \widetilde{\mathcal{D}}\}$;

2   **Initialize** $\epsilon_0 = \epsilon_{max}$;

3   **for** $E = 1 \rightarrow E_{max}$ **do**

4   **Fetch** the loss weights $\mathcal{W}_{E-1} = \{w_x | x \in \widetilde{\mathcal{D}}\}$;

5   **Update** $\theta_E = \nabla(\theta_{E-1}, \eta, \mathcal{W}_{E-1}, \widetilde{\mathcal{D}}\})$;

6   **Fetch** gradient contributions $\mathcal{G}_E = \{g_x | x \in \widetilde{\mathcal{D}}\}$;

7   **Fetch** sample loss set $\mathcal{L}_E = \{l_x | x \in \widetilde{\mathcal{D}}\}$;

8   **Obtain** inconsistency noisy signal $\mathcal{INC}_E = \{inc_x | x \in \widetilde{\mathcal{D}}\} \leftarrow \mathcal{G}_E$;

9   **Obtain** probability noisy signal $\mathcal{P}_E = \{p(k=1|l_x) | x \in \widetilde{\mathcal{D}}\} \leftarrow \mathcal{L}_E$;

10   **Update** the loss weights
     $\mathcal{W}_E = \{\epsilon_{E-1} \cdot inc_x + (1 - \epsilon_{E-1}) \cdot p(k=1|l_x) | x \in \widetilde{\mathcal{D}}, inc \in \mathcal{INC}_E, p \in \mathcal{P}_E\}$;

11   **Update** $\epsilon_E = max(\epsilon_{max} - \alpha E, \epsilon_{min})$;

12 **end**

    Output: the optimized parameters $\theta_{E_{max}}$ of DR model

---

4 Experiment

4.1 Dataset and Metrics

Dataset. To evaluate the effectiveness of the proposed DADR on DR model training, we conducted experiments on the TREC 2019 Deep Learning (DL) Track [7]. The track focuses on ad-hoc retrieval and consists of passage retrieval and document retrieval tasks.

1) Passage Retrieval: The passage retrieval task has a corpus of 8.8 million passages with about 500 thousand training queries, 7 thousand development queries, and 43 test queries.
2) Document Retrieval: The document retrieval task has a corpus of 3.2 million documents with 367 thousand training queries, 5 thousand development queries, and 43 test queries [30].

Evaluation Protocols. We use the official metrics to evaluate the retrieval performance, such as MRR@10, MRR@100, and R@100.

4.2 Experiment Settings

Baseline Methods. To test the performance of this work on DR models, we apply DADR to three kinds of sampling strategies with two kinds of loss functions.

We tested the generalization of our method using three popular sampling methods:

1) Random negative sampling: For random negative sampling baselines, we adopt Rand Neg [15]. Rand Neg randomly draws negative samples from the entire corpus.
2) Static hard negative sampling: For the static hard negative sampling baseline, we adopt BM25 [10]. BM25 Neg uses the BM25 top candidates as the negative samples.
3) Mixed sampling: We adopt the Stable Training Algorithm for dense Retrieval (STAR) [30]. STAR aims to employ static hard negatives to improve top-ranking performance and random negatives to stabilize the training process. It can be viewed as a combination of random and static sampling.

DR models are typically trained with pointwise loss and pairwise loss. For each of the above sampling strategies, we use the two kinds of loss functions to test the effectiveness of the proposed method:

1) Pointwise loss: For the pointwise loss baseline, We adopt cross-entropy loss.

$$l_x = -q_x * log(p_x) - (1 - q_x) * log(1 - p_x) \tag{12}$$

where q is the relevant label and p is the predicted score.
2) Pairwise loss: We adopts RankNet [4] as pairwise loss. Given a query q, let t^+ and t^- be a relevant text and a negative text. The loss function is formulated as follows:

$$\mathcal{L}_P(t^+, t^-) = \log(1 + e^{f(q,t^-)-f(q,t^+)}) \tag{13}$$

Implementation Details

- We fit Gaussian Mixture Moderated by using the Scikit-Learn machine learning library [25].
- The α, ϵ_{max} and ϵ_{min} are hyper-parameters in Sect. 3, and we set ϵ_{max} and ϵ_{min} to 0.9 and 0.1 by default. As we discuss later in more detail, we typically fix α range as (0.01, 0.03, 0.05, 0.07, 0.09). We set α=0.03 by default.
- Noisy dataset (mentioned in Sect. 1) synthesis method used for the prior observations in this paper. We control the proportion of noise by partially extracting the positive samples in the benchmark dataset, reversing their labels and using them as negative samples. We use the noise added dataset to observe the loss characteristics of the noise samples.
- We choose the max query length as 32 and the max doc length as 128 in passage retrieval task while set max doc length 180 in document retrieval task on the TREC 2019 Deep Learning (DL) Track.
- We use BERT$_{base}$ [8] as pre-trained embedding encoder of the DR model to embed the queries and candidate texts with a dimension of 768 and a vocab size of 30522.
- Our model is implemented using Python 3 and PyTorch 1.8. We use the popular transformers library for the pre-trained BERT model [24]. All models are trained with the AdamW optimizer.

4.3   Experiment Results

Table 2 summarizes the retrieval performance of the DR models trained with normal or DADR strategies over two datasets using three negative sampling methods.

Main Results

Performance Comparison of Different Loss Functions. By comparing the performance on pointwise loss and pairwise loss, we found that the proposed method contributes more to pairwise loss than to pointwise loss. This phenomenon may be caused by the fact that pairwise is more sensitive to noisy samples; that is, one mislabeled sample will cause multiple pair errors. This also implies that our proposed method effectively alleviates the negative impact of noisy data on retrieval performance and improves the robustness of DR model training.

Performance Comparison of Different Sampling Strategies. By comparing the performances of the three negative sampling methods, we found that the proposed method has the most contribution to the static hard negative sampling while having the least contribution to the randon negative sampling. This may suggest that the static hard negative sampling is more susceptible to noise than random negative sampling. We suspect the higher sensitive to noise of the static hard negative sampling is affected by hard negative samples, *i.e.*, hard negative samples are more likely to be false negatives than normal negative samples. On the contrary, the improvement in random negative sampling is relatively small, showing that the design of random sampling can improve the robustness against false-negative noise to some extent. Nevertheless, applying DADR still leads to a performance gain, which further validates the rationality of denoising negative samples.

Summary. In all cases, the proposed method effectively improves retrieval performance. Among them, the effect of DADR is particularly obvious in the static hard negative sampling, which increases by 5.17% and 6.29% in the cases of pointwise and pairwise loss, respectively. The significant improvement in performance indicates that adopting DADR makes the DR model have better generalization ability.

Table 2. Overall performance of DR models trained with DADR strategies and normal training over two datasets. We adopt two loss functions and apply them to three negative sampling methods. Note that Recall@K is shorted as R@K to save space, respectively, and "RI" in the last column denotes the relative improvement of DADR over normal training on average. The best results are highlighted in bold.

(a) Performance Comparisons on Pointwise Loss.

Strategy	MARCO Passage		MARCO Doc		RI
	MRR@10	R@100	MRR@100	R@100	
Rand negative sampling					
Rand Neg	0.273	0.851	0.316	0.848	–
Rand Neg + DADR	**0.280**	**0.854**	**0.328**	**0.851**	1.77%
Static hard negative sampling					
BM25 Neg	0.279	0.797	0.306	0.783	–
BM25 Neg + DADR	**0.301**	**0.816**	**0.325**	**0.816**	**5.17%**
Mixed sampling					
STAR	0.327	0.869	0.384	0.879	–
STAR + DADR	**0.332**	**0.878**	**0.391**	**0.904**	1.81%

(b) Performance Comparisons on Pairwise Loss.

Strategy	MARCO Passage		MARCO Doc		RI
	MRR@10	R@100	MRR@100	R@100	
Rand negative sampling					
Rand Neg	0.297	0.860	0.329	0.856	–
Rand Neg + DADR	**0.310**	**0.867**	**0.339**	**0.863**	2.26%
Static hard negative sampling					
BM25 Neg	0.291	0.804	0.319	0.805	–
BM25 Neg + DADR	**0.317**	**0.833**	**0.341**	**0.851**	**6.29%**
Mixed sampling					
STAR	0.336	0.870	0.393	0.905	–
STAR + DADR	**0.347**	**0.891**	**0.417**	**0.912**	3.14%

Ablation Study. In Fig. 3, we study the effect of the proposed denoising signals on noisy samples fitting during dense retrieval model training. As Fig. 3 shows, noisy samples are eventually fitted by the retrieval model under normal training. On the contrary, by applying the proposed denoising with respect, the loss values of noisy samples also show a decreasing trend, showing that the retrieval model still fits such noise. However, when the training gradually converges, the noise loss with applying the denoising strategy is significantly larger than that over normal training, indicating that both denoise signals reduce the effect of noisy samples on retrieval model training, which can explain their improvement over normal training.

Furthermore, we notice that inconsistency denoising signal performs better than probability estimation denoising signal. We believe that the possible reason is that the gradient inconsistency between the noise samples and the clean samples is more obvious in the early training, and the adoption of inconsistency denoising signal further delays the fitting of the noise. However, the probability estimation denoising signal has a poor prediction effect in the early stage of training, but gradually plays a certain role in the later stage of training with the accumulation of sample loss data.

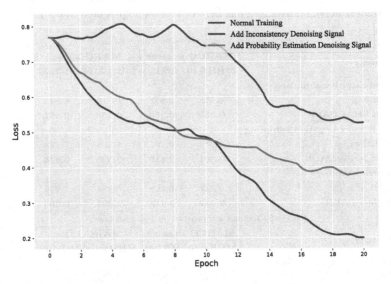

Fig. 3. The ablation observation on MS MARCO under 30% label noise. The figure shows the loss of information from noisy samples in three situations, *i.e.*, normal training and using the two kinds of noisy signals proposed in this paper, respectively. We smoothed the data with a smoothing factor of 0.7.

In Table 3, we report the performance of the proposed denoising signals with respective. Different from Fig. 3 (with addition noise), we further do ablation study on the official benchmark (without addition noise). Note that we adopt the average of performances on different loss functions and sampling strategies as the reported performance on Table 3, where we can conclude that both denoising signals contribute to the performance of DADR.

Table 3. The ablation study of DADR on MSMARCO.

Strategy	MARCO Passage		MARCO Doc		RI
	MRR@10	R@100	MRR@100	R@100	
Normal training without denoising	0.301	0.842	0.341	0.846	–
+ Inconsistency denoising signal	0.312	**0.857**	0.350	0.864	2.55%
+ Probability estimation denoising signal	0.305	0.843	0.346	0.853	0.94%
+ Both denoising signals (DADR)	**0.315**	0.856	**0.353**	**0.866**	**3.05%**

5 Conclusions

In this paper, we introduce DADR, a novel denoising approach based on dynamical weights for dense retrieval model training. By using two noisy evaluation signals to dynamically weight samples during the model training, DADR can improve the performance of the DR model without requiring extra resources, e.g, additional data, auxiliary models, or extra time-consuming stages. Extensive experiments conducted on two benchmark datasets with three representative sampling strategies and two typical loss functions demonstrate the superiority and universality of DADR.

References

1. Arazo, E., Ortego, D., Albert, P., O'Connor, N.E., McGuinness, K.: Unsupervised label noise modeling and loss correction (2019)
2. Arazo, E., Ortego, D., Albert, P., O'Connor, N., McGuinness, K.: Unsupervised label noise modeling and loss correction. In: International Conference on Machine Learning, pp. 312–321. PMLR (2019)
3. Brodley, C.E., Friedl, M.A.: Identifying mislabeled training data (1999)
4. Burges, C.J.C.: From ranknet to lambdarank to lambdamart: an overview (2010)
5. Chen, Y., Zhou, D., Li, L., Han, J.M.: Multimodal encoders for food-oriented cross-modal retrieval. In: U, L.H., Spaniol, M., Sakurai, Y., Chen, J. (eds.) Web and Big Data. APWeb-WAIM 2021. LNCS, vol. 12859, pp. 253–266. Springer, Cham (2021). https://doi.org/10.1007/978-3-030-85899-5_19
6. Cheng, M., et al.: Vista: vision and scene text aggregation for cross-modal retrieval. In: Proceedings of the IEEE/CVF Conference on Computer Vision and Pattern Recognition, pp. 5184–5193 (2022)
7. Craswell, N., Mitra, B., Yilmaz, E., Campos, D., Voorhees, E.M.: Overview of the Trec 2019 deep learning track. Text REtrieval Conference (2020)

8. Devlin, J., Chang, M.W., Lee, K., Toutanova, K.: Bert: pre-training of deep bidirectional transformers for language understanding. In: North American Chapter of the Association for Computational Linguistics (2018)
9. Du, M., et al.: Topic-grained text representation-based model for document retrieval. In: Pimenidis, E., Angelov, P., Jayne, C., Papaleonidas, A., Aydin, M. (eds.) Artificial Neural Networks and Machine Learning. ICANN 2022. LNCS, vol. 13531, pp. 776–788. Springer, Cham (2022). https://doi.org/10.1007/978-3-031-15934-3_64
10. Gao, L., Dai, Z., Fan, Z., Callan, J.: Complementing lexical retrieval with semantic residual embedding. Cornell University - arXiv (2020)
11. Gao, Y., et al.: Self-guided learning to denoise for robust recommendation (2022)
12. Gutmann, M.U., Hyvärinen, A.: Noise-contrastive estimation: a new estimation principle for unnormalized statistical models. In: International Conference on Artificial Intelligence and Statistics (2010)
13. Guu, K., Lee, K., Tung, Z., Pasupat, P., Chang, M.W.: Realm: retrieval-augmented language model pre-training. arXiv : Computation and Language (2020)
14. Huang, J., Qu, L., Jia, R., Zhao, B.: O2u-net: a simple noisy label detection approach for deep neural networks (2019)
15. Huang, J.T., et al.: Embedding-based retrieval in Facebook search (2020)
16. Karpukhin, V., et al.: Dense passage retrieval for open-domain question answering. In: Proceedings of the 2020 Conference on Empirical Methods in Natural Language Processing (EMNLP), pp. 6769–6781 (2020)
17. Karpukhin, V., et al.: Dense passage retrieval for open-domain question answering. arXiv : Computation and Language (2020)
18. Khattab, O., Zaharia, M.: Colbert: efficient and effective passage search via contextualized late interaction over bert. In: Proceedings of the 43rd International ACM SIGIR Conference on Research and Development in Information Retrieval, pp. 39–48 (2020)
19. Kim, D., Koo, J., Kim, U.M.: Osp-class: open set pseudo-labeling with noise robust training for text classification. In: 2022 IEEE International Conference on Big Data (Big Data), pp. 5520–5529. IEEE (2022)
20. Li, J., Socher, R., Hoi, S.C.H.: Dividemix: learning with noisy labels as semi-supervised learning. arXiv : Computer Vision and Pattern Recognition (2020)
21. Li, Y., Liu, S., She, Q., Mcleod, A., Wang, B.: On learning contrastive representations for learning with noisy labels (2023)
22. Northcutt, C.G., Jiang, L., Chuang, I.L.: Confident learning: estimating uncertainty in dataset labels. arXiv : Machine Learning (2019)
23. Parker, B., Sokolov, A., Ahmed, M., Kalebic, M., Akinli Kocak, S., Shai, O.: Domain specific fine-tuning of denoising sequence-to-sequence models for natural language summarization. arXiv e-prints pp. arXiv–2204 (2022)
24. Paszke, A., et al.: Pytorch: an imperative style, high-performance deep learning library. Neural Inf. Process. Syst. (2019)
25. Pedregosa, F., et al.: Scikit-learn: machine learning in python. J. Mach. Learn. Res. **12**, 2825–2830 (2011)
26. Qu, Y., et al.: Rocketqa: an optimized training approach to dense passage retrieval for open-domain question answering (2020)
27. Ren, R., et al.: Rocketqav2: a joint training method for dense passage retrieval and passage re-ranking. In: Proceedings of the 2021 Conference on Empirical Methods in Natural Language Processing, pp. 2825–2835 (2021)

28. Toneva, M., Sordoni, A., des Combes, R.T., Trischler, A., Bengio, Y., Gordon, G.J.: An empirical study of example forgetting during deep neural network learning. In: International Conference on Learning Representations (2018)
29. Wang, W., Feng, F., He, X., Nie, L., Chua, T.S.: Denoising implicit feedback for recommendation. arXiv : Information Retrieval (2020)
30. Zhan, J., Mao, J., Liu, Y., Guo, J., Zhang, M., Ma, S.: Optimizing dense retrieval model training with hard negatives. Cornell University - arXiv (2021)

Multi-pair Contrastive Learning Based on Same-Timestamp Data Augmentation for Sequential Recommendation

Shun Zheng, Shaoqing Wang$^{(\boxtimes)}$, Lijie Zhang, Yao Zhang, and Fuzhen Sun

Shandong University of Technology, Zibo 255091, China
wsq0533@163.com

Abstract. The core of sequential recommendations is to model users' dynamic preferences from their sequential historical behaviors. Bidirectional representation models can make better sequential recommendations because each item in user's historical behaviors fuses information from both left and right sides. Despite their effectiveness, we argue that such bidirectional models are sub-optimal due to the limitations including: a) items with the same timestamp interactions have adverse effect on user modeling; b) the random masking process often produces noises. To address these limitations, we propose Multi-pair Contrastive Learning based on same-timestamp data augmentation for Sequential Recommendation (MCL4SR). Specifically, we firstly modify the masking strategies of BERT encoder. Then we propose a multi-pair contrastive learning framework by exploring data augmentation of the same timestamp interactions. During the training and testing process, we design three types of samples so as to imitate human learning. Extensive experiments on two benchmark datasets show that our model outperforms state-of-the-art sequential models.

Keywords: Self-attention · Sequential recommendation · Dynamic masking · Contrastive learning

1 Introduction

With the development of Web 2.0, people are active in various applications. The users' behaviors forms many historical interaction sequences which reflect the users' dynamic preferences. The goal of sequential recommendation is to predict the subsequent items that users will probably interact with in the future.

In recent years, BERT [1] has become a research focus in the areas of sequential recommendation [2,3]. The bidirectional self-attention mechanism in BERT can combine contextual information from two directions to make predictions, and obtain better performance. Despite their effectiveness, we argue that such bidirectional models are sub-optimal due to the following limitations.

This work was supported by Shandong Provincial Natural Science Foundation, China (ZR2020MF147, ZR2021MF017).

First, the random masking process often produces noises. The mask proportion of many BERT-based models is 15%. However, in recent years, researchers in the field of language models propose different opinions: a larger mask proportion and masking without substitution can bring better performance [4]. Are the above conclusions, which are drawn from the language model, effective in sequential recommendation? Therefore, this paper raises following questions: 1) What is the appropriate masking proportion? 2) Is it necessary to replace the items randomly during the masking process?

Second, items with the same timestamp interactions have adverse effect on user modeling. For example, on a e-commerce site, a user added many items to a cart and bought them. Then many same timestamp interactive data were created. Existing sequential recommendation researches sort construct sequence according to timestamps from earliest to latest. But they ignore the influence of items that are interacting with the same timestamp on user modeling. We find that this factor plays an important role on learning user preferences, especially when considering the problem of interest drift. In the sequences of many datasets, same timestamp data accounts for a large proportion of the total interactions. It is inadequate to simply sort items according to timestamps. So, how can this issue be addressed to ensure that the user's interaction sequence reflects their interest drift?

Aiming at the above problems, this paper proposes a BERT-based Multi-pair Contrastive Learning model for Sequential Recommendation (MCL4SR). Specifically, we propose dynamic masking strategies which is different from the static masking strategies in BERT. BERT model masks some percentage of the input tokens at random, and then predict those masked tokens. It is often referred to as a *Cloze* task. It should be pointed out that, BERT only executes the masking operation once on a sequence and generates a static masked sequence. Inspired by multi-pair contrastive learning method [3], we propose that multiple masked sequences can be generated in advance for a sequence, and different masked sequences provide corresponding samples to train model as it goes deeper. To be specific, we propose to generate many masked sequences using different masking proportion and name these masked sequences as **study sample, exercise samples** and **quiz samples**, respectively. The training process of our model is similar to the process of human knowledge acquisition. By learning from the study sample, the model learns the representation of the sequence. It can then predict each masked item in the sequence by practicing with the exercise samples, and finally verify what it has learned by testing with the quiz samples. Both the exercise samples and the quiz samples require assistance of the real-world sample, and the multi-pair contrastive learning method provides the framework for the learning approach.

The effectiveness of the proposed model was verified through experiments on two public benchmark datasets. The contributions of this paper can be summarized as follows:

- We design a method of same-timestamp data augmentation which generates three kinds of samples: study sample, exercise samples and quiz samples.

- We propose dynamic masking strategies and feed multiple masked sequences into multi-pair contrastive learning framework.
- We conduct extensive experiments on two public benchmark datasets to verify the effectiveness of the proposed model.

2   Related Work

2.1   Self-supervised Learning

Self-supervised learning aims to extract contextual features from unlabelled data. Early self-supervised models for sequential recommendation, such as S3-Rec [5], incorporate contextual information into sequential recommendation by maximizing mutual information. In recent years, contrastive learning has shown great potential in sequential recommendation. For example, CL4SRec [6] adopts three kinds of data augmentation methods applicable to generate positive samples. CoSeRec [7] improves CL4SRec by introducing robust data augmentation methods. DuoRec [8] performs contrastive learning at the model level by introducing supervised and unsupervised objectives. CBiT [3] constructs multiple positive pairs through unsupervised dropout [9] and sliding window technique. Different from their works, this paper proposes a contrastive learning method based on the BERT architecture, where the masking strategies are modified and three kinds of samples are designed.

2.2   Sequential Recommendation

Early sequential recommendation works take advantage of Markov chains to capture the dynamic transitions of user interactions, such as MDP [10], FPMC [11], and Fossil [12]. Later, recurrent neural networks (RNNs), such as gated recurrent units, were adopted to model user interactions in sequential recommendation. Convolutional neural networks (CNNs) were proven to be effective on modeling short-term user preference. Recently, the success of Transformers in natural language processing (NLP) and computer vision (CV) has led researchers to explore in sequential recommendation. SASRec [13] is one of pioneer works, which adopt Transformers for sequential recommendation. BERT4Rec [2] improves SASRec by using a bidirectional Transformer, i.e., BERT. LSSA [14] proposes a long and short term self-attention networks which models both the user's long-term preferences and short-term interests. SR-GNN [15] and GC-SAN [16] combine graph neural networks with self-attention to capture local and global features of the sequences. FDSA [17] and S3-Rec incorporate mutual information into Transformers to capture context information. CL4SRec, CoSeRec, and DuoRec fuse Transformers to contrastive learning framework to improve the performance of sequential recommendations.

3   The Proposed Model

3.1   Problem Definition

We denote $\mathcal{U} = \{u_1, u_2, \cdots, u_{|\mathcal{U}|}\}$ as a set of users, $\mathcal{V} = \{v_1, v_2, \cdots, v_{|\mathcal{V}|}\}$ be a set of items, and $S_u = \left[v_1^{(u)}, \cdots, v_t^{(u)}, \cdots v_{|S_u|}^{(u)} \right]$ be a list of user u interaction sequence sorted in chronological order, where $v_t^{(u)} \in \mathcal{V}$ represents the item that user u interacts with at t-th timestamp. The task of sequential recommendation is to predict the next item for user u at the time step $|S_u| + 1$:

$$p\left(v_{|S_u|+1}^{(u)} = v \mid S_u\right) \tag{1}$$

3.2   Model Framework

The framework of the proposed model is shown in Fig. 1. First, each sequence is copied m ($m \geq 3$) times in data augmentation module. Then two kinds of masked samples are obtained through static masking layer, the one kind of masked samples are obtained through dynamic masking layer. Dynamic masking operation is executed multiple times. Next, the masked samples are sent into the embedding layer in turn during training. The embedding layer transforms these samples into embedding vectors, which are then passed through BERT. This paper adopts the output of the last layer as the hidden representation of the sequences. The *Cloze* task is introduced as the main training objective, which requires the model to reconstruct the masked items based on their corresponding hidden representations. These hidden representations of the sequences are treated as a positive set for multi-pair contrastive learning and compared with other sequences. A dynamic loss re-weighting strategy is introduced to balance the main *Cloze* task loss and the multi-pair contrastive loss [3].

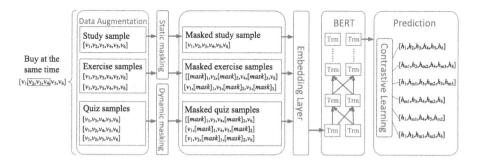

Fig. 1. Model framework

3.3   Data Augmentaion

When the length of users' interaction sequences is greater than a threshold value T, the sequences are truncated according to the sliding window technique. Each sequence is copied m $(m \geq 3)$ times.

- We name the original sequence as **study sample**.
- Two of the copied sequences are labeled as **exercise samples**.
- The remaining $(m-2)$ copies are **quiz samples**.
  For each sequence in quiz samples, only some of same-timestamp items should be reserved and the other be removed. To be specific, we firstly identify the number of same-timestamp subsequences n. Then their lengths $L_i^{sub}(i \in \{1,..,n\})$ are obtained. Lastly, we filter out interaction sequences whose length are shorter than a threshold value T due to remove the same-timestamp interactions. Therefore, only the sequences that meet the condition showed in Eq. 2 can be executed data augmentation of same-timestamp separation.

$$T \leqslant L_{seq} - \sum_{i=1}^{n} L_i^{sub} + n, \qquad (2)$$

where L_{seq} is the length of the original sequence.

The above three kinds of samples serve the training process of our model. It is similar to the process of human knowledge acquisition. By learning from the study sample, we can obtain the representation of the sequence. It can then predict each masked item in the sequence by practicing with the exercise samples, and finally verify what it has learned by testing with the quiz samples.

3.4   Masking Operation

We apply different masking operations to each kind of sample. Assuming that the length of original sequence of user u is 6, and the k-th round training is coming out soon.

After the masking operation, the study sample remains unchanged. Let $S_u^{k,0}$ denote the masked study sample of user u.

$$S_u^{k,0} = [v_1^{(u)}, v_2^{(u)}, v_3^{(u)}, v_4^{(u)}, v_5^{(u)}, v_6^{(u)}].$$

The exercise samples are only masked at the odd or even positions of the items. The result of masking operation is shown as follows:

$$S_u^{k,1} = [[mask]_1, v_2^{(u)}, [mask]_2, v_4^{(u)}, [mask]_3, v_6^{(u)}],$$

$$S_u^{k,2} = [v_1^{(u)}, [mask]_1, v_3^{(u)}, [mask]_2, v_5^{(u)}, [mask]_3].$$

The masked quiz samples, i.e., $S_u^{k,3}, S_u^{k,4}, \cdots, S_u^{k,m}$, are obtained by randomly mask the quiz samples according to a given proportion. Finally, $(m+1)$ positive samples S_u^k of user u are obtained at the k-th round:

$$S_u^k = \{S_u^{k,0}, S_u^{k,1}, S_u^{k,2}, \cdots, S_u^{k,m}\}.$$

3.5 Embedding Layer

Project embedding matrix $\boldsymbol{E} \in \mathbb{R}^{|\mathcal{V}| \times d}$ and position embedding matrix $\boldsymbol{P} \in \mathbb{R}^{T \times d}$ are added together to construct hidden representations of sequences, where T represents the maximum sequence length supported by the model, and d represents the hidden dimension of the model. Therefore, given an item v_i, its input can be expressed as:

$$h_i^0 = e_i + p_t, \qquad 1 \le t \le T, \tag{3}$$

where $e_i \in \boldsymbol{E}$ and $p_t \in \boldsymbol{P}$ are the embedding vector of the item v_i and its location, respectively.

After the embedding layer, a matrix $\boldsymbol{H}^0 \in \mathbb{R}^{T \times d}$ stacked by h_i^0 is generated as a hidden representation of the entire sequence:

$$\boldsymbol{H}^0 = [h_1^0, h_2^0, h_3^0, \cdots h_T^0]. \tag{4}$$

3.6 BERT Encoder

After feeding the matrix $\boldsymbol{H}^0 \in \mathbb{R}^{T \times d}$ shown in Eq. 4 into BERT, it passes through the L-layer bidirectional Transformer encoder, which can be expressed as:

$$\boldsymbol{H}^l = \mathrm{Trm}\left(\boldsymbol{H}^{l-1}\right), \quad \forall l \in \{1, 2, \cdots L\}. \tag{5}$$

The output $\boldsymbol{H}^L = [h_1^L, h_2^L, h_3^L, \cdots, h_T^L]$ passing through the last layer of BERT is used as the final hidden representation of the sequence.

3.7 Prediction Layer

Let i denote the position of the item in the sequence, a linear layer is used to transform the final hidden representation h_i^L into a probability distribution of candidate items.

$$p(v) = \boldsymbol{W}^p h_i^L + b^p, \tag{6}$$

where $\boldsymbol{W}^p \in \mathbf{R}^{|v| \times d}$ is the weight matrix, $b^p \in \mathbf{R}^{|v|}$ is the bias term of the prediction layer.

3.8   Multi-pair Contrastive Learning

The goal of contrastive learning is to pull the distance of positive pair closer in the vector space, and push the distance between positive samples and negative samples far away.

At k-th round, given a batch of samples $\{S_u^k\}_{u=1}^N$ where N is the batch size. A pair of hidden features $\boldsymbol{H}_u^x$ and $\boldsymbol{H}_u^y$, which come from the same original sequence, are combined together and considered as a positive pair, while the other $2(N-1)$ samples in the same batch are considered as negative samples. Based on InfoNCE, the contrastive loss for a pair can be defined as follows [3]:

$$\ell\left(\boldsymbol{H}_u^x, \boldsymbol{H}_u^y\right) = -\log \frac{e^{<H_u^x, H_u^y>/\tau}}{e^{<H_u^x, H_u^y>/\tau} + \Sigma_{i=1, i\neq u}^N \Sigma_{c\in\{x,y\}} e^{<H_u^x, H_i^c>/\tau}}, \tag{7}$$

where τ is a temperature hyperparameter, and $< \cdot, \cdot >$ represents the cosine function to calculate the similarity between the two hidden representations.

Only one positive pair for contrastive learning is not enough to fully take advantage of the huge potential of BERT, because the difference between them may be small. It is difficult to learn much useful information from one simple positive pair. Moreover, false positive samples may exist in a batch and without many positive samples to balance them out, the performance of the learnt model will be sub-optimal. Therefore, this paper adopts multi-pair contrastive learning. Since the connections between items in sequential recommendation are weak, the model may have difficulty on predicting missing items when a large proportion of the sequence is masked. Therefore, study samples and exercise samples are provided within the positive pairs for contrastive learning. The hidden representations obtained from Eq. 5 are combined together as positive pairs. The multi-pair contrastive loss for m positive pairs is defined as follows [3]:

$$\mathcal{L}_{\mathrm{CL}} = \sum_{x=1}^{m} \sum_{y=1}^{m} \mathbb{1}_{[x\neq y]} \ell\left(\boldsymbol{H}_u^x, \boldsymbol{H}_u^y\right), \tag{8}$$

where $\mathbb{1}_{[x\neq y]} \in \{0,1\}$ is an indicator function which equals 1 only when $x \neq y$.

4   Experiments

4.1   Experimental Setup

Dataset. Experiments are conducted on two public benchmark datasets. The Amazon dataset [18] includes user reviews on different domains. In Amazon dataset, the length of sequence is short. We select Video domain as the experimental dataset from the Amazon dataset. Another dataset, MovieLens-1M (ML-1M) [19], includes user ratings of movies with longer sequences.

All interactions are considered implicit feedback. Duplicate interactions are removed, and interactions for each user are sorted in chronological order to construct sequences. Following the existing works [3,7,8], users with less than 5

Table 1. Statistics of the datasets

Dataset	Users	Items	Interaction	Sparsity
Amazon	826,767	50,210	1,324,753	99.99%
ML-1M	6,040	3,952	1,000,209	99.99%

interactions and items related with less than 5 users are filtered out. The leave-one-out evaluation strategy is used, where the last item in the sequence is used for testing, the second-to-last item is used for validation, and the rest of the items are used for training. The statistics of the datasets are shown in Table 1.

Evaluation Metrics. To make a fair comparison [20], all item predictions on the dataset are ranked. This paper calculates scores for the top-k hit rate (HR@K) and the normalized discounted cumulative gain (NDCG@K), where $K \in \{5, 10, 20\}$.

Baseline Methods.

- BERT4Rec [2]. It uses BERT as the sequence encoder and employs the *Cloze* task to train the model.
- CBiT [3]. It uses BERT as the sequence encoder and employs multi-contrastive learning to reduce the impact of false positive samples.
- SASRe [13]. It uses Transformer as the sequence encoder, serving as the base model for contrastive methods.
- CL4SRec [6]. It applies contrastive learning to sequential recommendation and uses data augmentation to generate positive examples.
- DuoRec [8]. This is a model that uses a Transformer encoder and contrastive learning, which alleviates the representation degradation problem in contrastive learning.

4.2 Experimental Results

The training and testing of all models are conducted on the NVIDIA GeForce RTX 3090.

For CBiT and DuoRec, this paper uses the codes provided by their authors. For SASRec, CL4Rec, and BERT4Rec, this paper uses the codes reproduced by the RecBole [21] framework.

The proposed model is implemented with PyTorch [22]. The number of Transformer encoding blocks is 2 for ML-1M and Amazon. The number of attention heads is 4 for ML-1M and Amazon. The step of sliding window is 10 for ML-1M and 1 for Amazon, respectively. The masking proportion of the quiz samples is 0.3 for ML-1M and 0.35 for Amazon, respectively. The dynamic masking rounds are 3 for ML-1M and Amazon. The number of positive samples for contrastive learning is 8 for ML-1M and Amazon. The dropout was set to 0.3 for ML-1M and 0.35 for Amazon, respectively. The threshold value for augmenting sequence length could be 5 for Amazon, and 40 for ML-1M, respectively. The model optimizer used is Adam [23], with a learning rate of 0.001, which is reduced by a factor of 0.1 after each 75 epochs for ML-1M or 250 epochs for Amazon. The model was trained for a maximum of 500 epochs with early stopping, and if the model validation accuracy did not improve after 50 epochs, the training is stopped. The checkpoint with the best NDCG@10 score on the validation set is selected for testing.

Table 2. Overall performance of different methods

Dataset	Metric	SASRec	CL4SRec	BERT4Rec	DuoRec	CBiT	Ours	Improve(%)
Amazon	Recall@5	0.0699	0.0568	0.0260	0.0756	0.0665	**0.0796**	5.29
	Recall@10	0.1139	0.0939	0.0456	0.1188	0.1032	**0.1246**	4.88
	Recall@20	0.1698	0.1470	0.0745	0.1812	0.1489	**0.1834**	1.21
	NDCG@5	0.0412	0.0352	0.0159	0.0479	0.0438	**0.0525**	9.60
	NDCG@10	0.0553	0.0470	0.0222	0.0618	0.0555	**0.0669**	8.25
	NDCG@20	0.0694	0.0604	0.0294	0.0775	0.0670	**0.0816**	5.29
ML-1M	Recall@5	0.1078	0.1142	0.1308	0.1930	0.2024	**0.2129**	5.19
	Recall@10	0.1810	0.1815	0.2219	0.2865	0.2891	**0.3041**	5.19
	Recall@20	0.2745	0.2818	0.3354	0.3901	0.3997	**0.4123**	3.15
	NDCG@5	0.0681	0.0705	0.0804	0.1327	0.1357	**0.1443**	6.34
	NDCG@10	0.0918	0.0920	0.1097	0.1586	0.1636	**0.1736**	6.11
	NDCG@20	0.1156	0.1170	0.1384	0.1843	0.1915	**0.2009**	4.91

The performance comparison results are shown in Table 2, and it can be observed that the experimental results of the proposed model are generally better than those of other models. For the BERT encoder, the masking strategies of this paper are better than that of BERT4Rec and CBiT. Furthermore, a larger masking proportion has a good effect, because that there are many short sequences, and a small masking proportion often leads to some sequences not being masked, which affects the performance of the model.

For comparisons of contrastive learning methods, DuoRec and CBiT, which adopt mixed data augmentation, outperform CL4SRec. Our model with three kinds of samples performs better than CBiT. Existing researches show inappropriate data augmentation may break the original semantic of the sequence. Therefore, we don't randomly replace items during masking, as it may introduce noise into the sequences. The data augmentation method we used was better than the one proposed in CBiT, which may masks important items reflecting user preferences during data augmentation.

Our model also demonstrates good adaptability and achieves the best performance on both short and long sequence datasets. The performance improvement of our proposed model owes to three parts: the masking strategies, the method of data augmentation, and multi-pair constrastive learing.

4.3   Hyperparameter Experiments

In this section, the influence of four hyperparameters is studied. The experiments are conducted on the Amazon-video dataset, and the evaluation metric is NDCG@10.

Threshold of the Length of the Sequences. The threshold of sequence length for data augmentation determines how long a user sequence should be. If the length of the augmented sequences is too short, it may not reflect the user's preferences well. When they are used as positive samples for model training, short sequences will reduce the accuracy of the model. If the length of the augmented sequence is too long, many short sequences contains many items purchased at the same timestamp, which is of no benefit to model user preferences. Therefore, it is crucial to select a suitable threshold. As shown in Fig. 2, the model performs best when the threshold is 7.

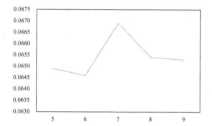

Fig. 2. Threshold of the length of the sequences.

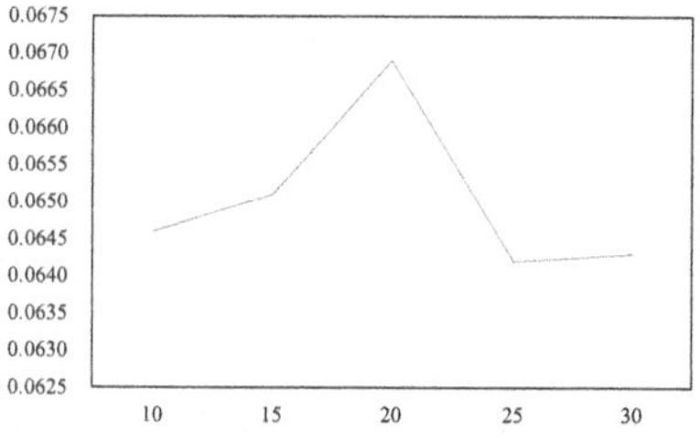

Fig. 3. Number of dynamic masking rounds.

Number of Dynamic Masking Rounds. The number of dynamic masking rounds determines value that the sequences are sent into the model to train before replacing with new masked sequences. After a certain number of training batches, the model will extract another masked sequence for learning. Multiple different masked sequences perform better than a fixed masked sequence, because the model can use more different testing samples to make predictions and learn sentence semantics at different positions. However, when there are too many masked sequences, it not only greatly increases the time and memory required for model training but also reduces the accuracy of model training because different masked sequences need a certain number of training batches to learn. Therefore, it is important to find an appropriate value. As shown in Fig. 3, when the number of rounds is 20, the model performs the best.

Masking Proportion. Masking proportion determines how many items of a sequence will be masked and sent to the model for learning. If the proportion is too small, most short sequences may not have any masked item, and the model cannot learn anything useful from them. If the proportion is too large, too many masked items in the sequence may make it difficult for the model to rely on sufficient contextual information to understand the semantics of the masked content. As shown in Fig. 4, the model performs best when the masking proportion is 0.35. Meanwhile, we have incorporated the masking strategy recommended by BERT into our model, and conducted experiments using masking rates of 15% (the same rate used in BERT, BERT4Rec, and CBiT) and 35% respectively, as shown in Fig. 5, the masking strategy we employed outperforms the one used by BERT.

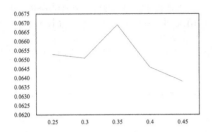

Fig. 4. Masking proportion.   **Fig. 5.** Masking methods.

The Number of Positive Samples. As shown in Fig. 6, the performance can be improved with the increasing the number of positive samples. Contrastive learning with multiple positive samples performs better than that with only one positive sample, due to the various semantic patterns brought by multiple positive samples. However, when there are enough positive samples, the performance will reach a peak. This is because that there are small errors between the predicted values of BERT and the representations, and too many samples with errors will reduce the weight of learning samples in sample pairs, then lead to a decrease of accuracy. The model performs best when the number of positive samples is 8.

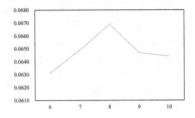

Fig. 6. The number of positive samples.

4.4 Ablation Study

In this section, the influence of three kinds of samples is studied. The experiments are conducted on the Amazon-video dataset, and the evaluation metric is NDCG@10.

First, experiments are conducted by removing the study samples from the positive pairs and replacing them with quiz samples. The results are shown in Fig. 7, where the performance on the test sets decreased. We observe that as a denoising encoder, BERT requires a hidden representation during the training process to learn the right things. Study sample provides a reference sequence for the model to learn from.

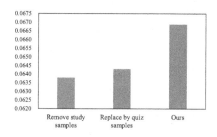

Fig. 7. Ablation study about study samples.

Second, ablation study is conducted by removing the exercise samples and replacing them with quiz samples to understand their effectiveness. The results of the experiments, shown in Fig. 8, indicates that removing or replacing the exercise samples leads to a certain decrease of the performance. We observe that during training, the model needs to predict all items in the sequence, and the exercise samples provide the model with ample opportunities to do so through a large number of masked items. Moreover, by learning from the study sample, the model is able to learn more fine-grained representation at different positions in the sequences.

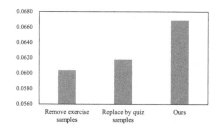

Fig. 8. Ablation study about the exercise samples.

Finally, ablation studies were conducted by removing both the study and exercise samples, and the results are shown in Fig. 9. It can be observed that the model's accuracy decreases to some extent when the masking proportion is unchanged. The possible reason is that the connections between items in sequence are not as closely related to NLP tasks. After random masking, the model has difficulty on predicting items in the sequence based on context alone. However, after adding study and exercise samples, the model has a reference for prediction, which makes it easier to predict at the right direction.

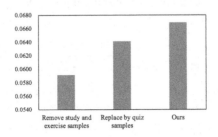

Fig. 9. Ablation study about both the study and exercise samples.

5   Conclusion

This paper proposes a multi-pair contrastive learning method for sequential recommendation based on same-timestamp data augmentation. The proposed model fully utilizes BERT's ability to learn contextual representations and obtains more accurate hidden representations for each item in the sequence, and ultimately achieves more accurate semantics for the sequence with good interpretability. Experimental results on two public datasets show that the proposed model have best performance comparing to SOTA baselines.

In the future, we will focus on lightweight model, so that it can learn from a greater number of different masked sequences and positive samples while using the same amount of memory, as well as reducing the training time.

References

1. Devlin, J., Chang, M., Lee, K., Toutanova, K.: BERT: pre-training of deep bidirectional transformers for language understanding. In: NAACL-HLT 2019, Minneapolis, 2–7 June 2019, vol. 1, pp. 4171–4186 (2019)
2. Sun, F., et al.: Bert4rec: sequential recommendation with bidirectional encoder representations from transformer. In: CIKM 2019, Beijing, 3–7 November 2019, pp. 1441–1450 (2019)
3. Du, H., et al.: Contrastive learning with bidirectional transformers for sequential recommendation. In: CIKM, Atlanta, 17–21 October 2022, pp. 396–405 (2022)

4. Wettig, A., Gao, T., Zhong, Z., Chen, D.: Should you mask 15% in masked language modeling? In: EACL 2023, Dubrovnik, 2–6 May 2023, pp. 2977–2992 (2023)
5. Zhou, K., et al.: S3-rec: self-supervised learning for sequential recommendation with mutual information maximization. In: CIKM 2020, Virtual Event, 19–23 October 2020, pp. 1893–1902 (2020)
6. Xie, X., et al.: Contrastive learning for sequential recommendation. In: ICDE 2022, Kuala Lumpur, 9–12 May 2022, pp. 1259–1273 (2022)
7. Liu, Z., Chen, Y., Li, J., Yu, P.S., McAuley, J.J., Xiong, C.: Contrastive self-supervised sequential recommendation with robust augmentation. arXiv preprint arXiv:2108.06479 (2021)
8. Qiu, R., Huang, Z., Yin, H., Wang, Z.: Contrastive learning for representation degeneration problem in sequential recommendation. In: WSDM 2022 Virtual Event/Tempe, 21-25 February 2022, pp. 813–823 (2022)
9. Srivastava, N., Hinton, G.E., Krizhevsky, A., Sutskever, I., Salakhutdinov, R.: Dropout: a simple way to prevent neural networks from overfitting. J. Mach. Learn. Res. **15**(1), 1929–1958 (2014)
10. Shani, G., Heckerman, D., Brafman, R.I.: An MDP-based recommender system. J. Mach. Learn. Res. **6**, 1265–1295 (2005)
11. Rendle, S., Freudenthaler, C., Schmidt-Thieme, L.: Factorizing personalized Markov chains for next-basket recommendation. In: WWW 2010, Raleigh, 26–30 April 2010, pp. 811–820 (2010)
12. He, R., McAuley, J.J.: Fusing similarity models with Markov chains for sparse sequential recommendation. In: ICDM 2016, 12–15 December 2016, Barcelona, pp. 191–200 (2016)
13. Kang, W., McAuley, J.J.: Self-attentive sequential recommendation. In: ICDM 2018, Singapore, 17–20 November 2018, pp. 197–206 (2018)
14. Xu, C., et al.: Long- and short-term self-attention network for sequential recommendation. Neurocomputing **423**, 580–589 (2021)
15. Wu, S., Tang, Y., Zhu, Y., Wang, L., Xie, X., Tan, T.: Session-based recommendation with graph neural networks. In: EAAI 2019, Honolulu, 27 January–1 February 2019, pp. 346–353 (2019)
16. Xu, C., et al.: Graph contextualized self-attention network for session-based recommendation. In: IJCAI 2019, Macao, 10–16 August 2019, pp. 3940–3946 (2019)
17. Zhang, T., et al.: Feature-level deeper self-attention network for sequential recommendation. In: IJCAI 2019, Macao, 10–16 August 2019, pp. 4320–4326 (2019)
18. McAuley, J.J., Targett, C., Shi, Q., van den Hengel, A.: Image-based recommendations on styles and substitutes. In: SIGIR 2015, Santiago, 9–13 August 2015, pp. 43–52 (2015)
19. Harper, F.M., Konstan, J.A.: The movielens datasets: history and context. ACM Trans. Interact. Intell. Syst. **5**(4), 19:1–19:19 (2016)
20. Krichene, W., Rendle, S.: On sampled metrics for item recommendation. Commun. ACM **65**(7), 75–83 (2022)
21. Zhao, W.X., et al.: Recbole: towards a unified, comprehensive and efficient framework for recommendation algorithms. In: CIKM 2021, Virtual Event, Queensland, 1–5 November 2021, pp. 4653–4664 (2021)
22. Paszke, A., et al.: Pytorch: an imperative style, high-performance deep learning library. In: NeurIPS 2019, 8–14 December 2019, Vancouver, pp. 8024–8035 (2019)
23. Kingma, D.P., Ba, J.: Adam: a method for stochastic optimization. In: ICLR 2015, San Diego, 7–9, May 2015, Conference Track Proceedings (2015)

Enhancing Collaborative Features with Knowledge Graph for Recommendation

Lingang Zhu, Yi Zhang, and Gang Li(✉)

College of Intelligence and Computing, Tianjin University, Tianjin, China
{zlg122,yizhang,ligang}@tju.edu.cn

Abstract. Knowledge Graph (KG) is of great help in improving the performance of recommendation systems. Graph neural networks (GNNs) based model has gradually become the mainstream of knowledge-aware recommendation (KGR). However, existing GNN-based KGR models underutilize the semantic information in KG to enhance collaborative features. Therefore, we propose a Collaborative Knowledge Graph-Aware framework (CKGA). In general, we first use the knowledge graph to obtain the semantic representation of items and users, and then feed these representations into the Collaborative Filtering (CF) model to obtain better collaborative features. Specifically, (1) we design a novel CF model to learn the collaborative features of items and users, which partitions the interaction graph into different subgraphs of similar interest and performs high-order graph convolution inside subgraphs. (2) For learning important semantic information in KG, we design an attribute aggregation scheme and an inference mechanism for GNN which directly propagates further attributes and inference information to the central node. Extensive experiments conducted on three public datasets demonstrate the superior performance of CKGA over the state-of-the-arts.

Keywords: Knowledge Graph · Recommendation · Collaborative Filtering · Graph Neural Networks

1 Introduction

Recommender system has increasingly become an integral part of many Internet services to reduce information overload and improve users' experience, which ranges from E-commerce platform, movie website, and music software. Collaborative Filtering (CF) based models [1,4] are arguably one of the most successful methods in multiple scenarios because they effectively model the users' preferences and are easy to implement. With the development of collaborative filtering paradigms, matrix factorization (MF) modeling the user-item interactions with inner product [1] has been gradually replaced by neural CF models, which use neural networks to learn latent user and item embeddings [4].

However, CF-based models rely heavily on interaction data which suffers from data sparsity and cold start problems [13]. To address these issues, Knowledge Graph (KG) has been introduced into the recommender system to enrich the user and item representations. In the early stages, studies utilize knowledge graph embedding (KGE) [5,8]

© The Author(s), under exclusive license to Springer Nature Singapore Pte Ltd. 2024
X. Song et al. (Eds.): APWeb-WAIM 2023, LNCS 14333, pp. 188–203, 2024.
https://doi.org/10.1007/978-981-97-2387-4_13

algorithms (e.g., TransD [5], TransR [8]) to encode entity and relation for item representation learning. To overcome the poor performance in capturing high-order KG connectivity, path-based models [20, 22] are proposed to utilize the connectivity similarity of users and items for recommendation. Nevertheless, most path-based methods depend on effective meta paths, which require domain knowledge and manual labor. Because GNNs show superiority in modeling KG, more studies focus on leveraging information of multi-hop neighbors to capture node features and high-order connectivity, such as KGCN-LS [15], KGAT [16], and KGIN [17].

Despite the success of these KG-aware recommendation models in recent years, we have noticed that previous models only use KG to learn the semantic features of items and users. KG semantics are added with the user-item interaction information to encode the final representation, but seldom help to enhance the user-item collaborative features. As shown in Fig. 1, most methods only utilize a single dimension information from different graphs. Besides, they simply sum the collaboration features learned from the user-item graph with the semantic information from the global graph.

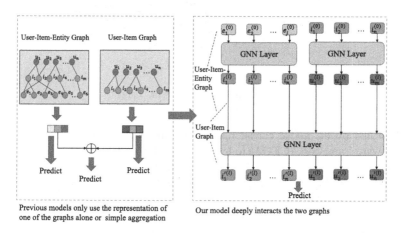

Fig. 1. Differences between previous models and our model. Our models utilizes the semantic features to enhance the collaborative representations.

In the stage of obtaining collaborative signals, previous models suffer from over-smoothing. Node embeddings become more similar and indistinguishable when stacking more layers. Recently, some GCN models, such as LR-GCN [11] and LightGCN [3], are proposed to alleviate the over-smoothing problem. Liu et al. [9] argues that these models overlook an important factor for the over-smoothing problem in the recommendation. That is, high-order neighboring users with no common interests can be involved in the information aggregation and passing in the graph convolution operation. Aggregating the information from all those neighbors indistinguishably violates the core idea behind collaborative filtering in which similar users like similar items.

For learning the semantic representations of users and items in KG, most models [17, 19] introduce noise of irrelevant relations and entities. They use path-aware aggregation which recursively integrates the relation sequences of long-range connectivity.

However, entity vector will be multiplied by the relationship vector of the different hops when aggregating to the central node. Meanwhile, implicit inference information in KG has not been fully explored. We can usually infer latent information about an item based on the connections of relations and entities in KG. As shown in Fig. 2, we can infer "The Avengers" is a science fiction movie from the KG triplet (The Avengers, star, Robert Downey Jr) and the triplet (The Avengers, star, Mark Alan Ruffalo). This is because the two actors have worked in many science fiction films so that the relation-entity pairs: (star, Rober Doweny Jr) and (star, Mark Alan Ruffalo) have science fiction tags.

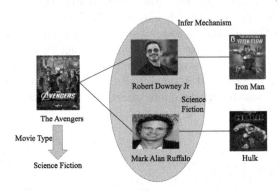

Fig. 2. Examples of inference mechanism

Motivated by the above considerations, we propose a Collaborative Knowledge Graph-Aware framework (CKGA) to learn high-quality collaborative information with the help of the KG. Overall, CKGA first utilizes the KG to learn the semantic representations of users and items on the global graph, and then feeds them into the CF model to enhance collaborative features. Specially, to alleviate the over-smoothing problem in the previous CF models, we propose Multi-IMP-GCN based on the IMP-GCN [9], which groups users and their interactive items into different interest subgraphs and operates high-order graph convolutions inside subgraphs. Our proposed GCN module has an advantage in considering the diversity of user interests compared to IMP-GCN [9]. In order to learn the discriminative semantic representation in the KG, we design a KG Explore module to encode the attributes and inference information in the KG. This module uses the attribute aggregator to directly aggregate the multi-hop neighbor entities and relations to the center node in KG. Furthermore, it utilizes an inference mechanism to capture implicit information based on these attributes of the center node. We conduct extensive experiments on three real-world datasets. Experimental results show that our CKGA outperforms the state-of-the-art methods.

We summarize the contributions of this work as:

- We propose a CKGA framework which utilizes the semantic features in KG to explore richer collaborative information.
- We design the Multi-IMP-GCN to group users into different subgraphs according to the users' interests. It is proved to be more effective in capturing collaborative signals and alleviating the over-smoothing problem.

- We design a KG Explore Module to explore attributes and inference information behind KG and have experimentally demonstrated that it could improve the accuracy of recommendations.
- We conduct empirical studies on three benchmark datasets to demonstrate the superiority of CKGA.

2 Related Work

Based on existing methods of KG-aware recommendation, we group them into three categories: embedding-based methods, path-based methods and GNN-based methods.

Embedding-Based Methods. [14,21] leverage knowledge graph embedding (KGE) [5,8] algorithms to encode entities and relations embeddings and then integrate these side information into the recommendation. For example, CKE [21] incorporates side information into the collaborative filtering model. The item's structural knowledge, which is utilized by TransR [8], is combined to the item embeddings. In the news recommendation system, DKN [14] models the news by combining the knowledge-level embedding of entities in news content via TransD [5]. Embedded-based approaches show high flexibility in using KG, but the KGE algorithm is more concerned with modeling strict semantic relatedness, which is better suited for link prediction than the recommendation.

Path-Based Methods. [20,22] leverage the connectivity similarity of users and/or items for recommendation. For instance HeteRec [20] uses the meta-path similarity to enrich the user-item interaction matrix, thus extracting more comprehensive embeddings of users and items. To overcome the limitation of the meta-path's representation ability, FMG [22] replaces the meta-path with the meta-graph to capture similarities between entities. Path-based methods use KG more naturally but rely heavily on manually designed meta-path.

GNN-Based Methods. [15–17,19] work with the information aggregation mechanism of graph neural networks (GNNs). Typically it aggregates multi-hop neighbors' information to the node representation to capture node features and graph structure so that long-range connections can be modeled. KGAT [16] combines the user-item graph and KG as a heterogeneous graph, which is then recursively aggregated using GCN. KGIN [17] leverages KG to explore the user intent behind the user-item graph. GNN-based methods combine the advantages of the above two methods. It not only learns the semantic information contained in the KG but also captures the connectivity information between entities.

However, previous GNN-based models only use KG to learn the semantic features of item and user, and do not utilize these semantic features to learn better quality collaborative information. Our work utilizes the KG to learn the semantic representations of users and items on the global graph, and then feeds them into the CF model to enhance collaborative features. Specifically, we design a KG Explore module for learning semantic representations and a novel CF model to alleviate the over-smoothing problem.

3   Preliminaries

This section first introduces the structural data: user-item interactions and knowledge graph, and then present the task of knowledge-aware recommendation.

Interaction Data. We denote $\mathcal{U} = \{u_1, u_2, ..., u_n\}$ as the set of N users and $\mathcal{I} = \{i_1, i_2, ..., i_m\}$ as the set of M items respectively. Let user-item interaction matrix be $Y \in R^{M \times N}$, where $y_{ui} = 1$ denotes that the user u has implicitly interacted with the item i (e.g., view, click, purchase); otherwise $y_{ui} = 0$.

Knowledge Graph. We let $\mathcal{G} = \{(h, r, t) | h, t \in \mathcal{E}, r \in \mathcal{R}\}$ be the knowledge graph, where entity-relation-entity triplet (h, r, t) denotes the head and tail entity h and t with the relation r; $\mathcal{E}$ and $\mathcal{R}$ refer to the sets of entities and relations in KG. The knowledge graph of datasets in this paper only contains item attributes with different types of entities and corresponding relationships without constructing the knowledge graph of the user attributes. So an item $i \in \mathcal{I}$ corresponds to one entity $v \in \mathcal{E}$ ($\mathcal{I} \subset \mathcal{E}$). This side information in KG could be complementary to the interaction data and improve the performance and interpretability of recommendation.

Task Description. Given the user-item interaction matrix Y and item knowledge graph data $\mathcal{G}$, the task of knowledge-aware recommendation is to learn a function that can predict how likely a user would adopt an item.

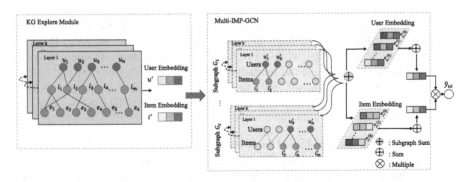

Fig. 3. An overview of CKGA framework consisting of two main parts as illustration. The KG Explore Module get the semantic representation after global grpah view, details in Fig. 4; Multi-IMP-GCN presents the propagation detail of subgraphs.

4   Methodology

Figure 3 shows the working flow of CKGA, which comprises two main components: (1) KG Explore module, which learns the semantic representation by capturing the semantic attributes and deducing implicit information from this explicit information; and (2) Multi-IMP-GCN, which alleviates the over-smoothing by performing graph convolution on interaction subgraphs. Technical details are discussed in the following sub-sections.

4.1   KG Explore Module

Figure 4 shows the framework of the KG Explore module. Unlike the relation path neighborhood aggregation, the kg explore module consists of two key parts: (1) attribute aggregator; and (2) inference module. The attributes and inference information of multiple hops are directly aggregated in the central node.

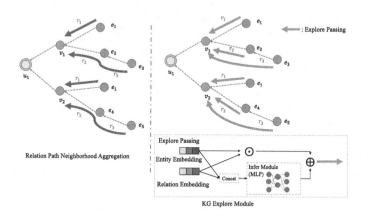

Fig. 4. An example of relation path neighborhood aggregation and KG Explore module schemes, where a (dashed or solid) arrow is an information flow among nodes. Best viewed in color. And the detail of KG Explore module is in right subgraph. (Color figure online)

Attribute Aggregator. Previous models [17,19] use path-aware aggregation, which recursively integrates the relation sequences of long-range connectivity. Path-aware aggregation aggregates neighboring information for K times meanwhile preserving the path information and distinguishing which paths they originate from. As shown in the left of Fig. 4, in the k-th layer ($\forall k \in K'$), the aggregation process can be formulated as:

$$e_u^k = \frac{1}{\sqrt{N_u}} \sum_{i \in N_u} e_i^{k-1}$$

$$e_i^k = \frac{1}{\sqrt{N_i}} \sum_{(r,v) \in N_i} e_r \odot e_v^{k-1} \tag{1}$$

where e_i^k and e_v^k denote the representations of item i and entity v respectively, which memorize the relational signals after $k-1$ layers propagation. e_u^k represent embeddings of user u at the k-th layer. N_i denotes the item i's set of neighboring entities and N_u means the set of items that interact with user u; $\odot$ is the elementwise product. However, the entity vector will be multiplied by the relationship vector of the different hops when aggregating to the central node. Inspired by the design of FFM [6], we regard each relationship as an attribute. As shown in the right of Fig. 4, different layers' neighbor entities directly propagate relational signals as attributes to the center node. The aggregation process can be formulated as:

$$e_u^k = \frac{1}{\sqrt{N_u}} \sum_{i \in N_u} e_i^{k-1}$$

$$e_i^1 = \frac{1}{\sqrt{N_i}} \sum_{(r,v) \in N_i} e_r \odot e_v^0$$

$$e_i^k = \frac{1}{\sqrt{N_i}} \sum_{(r,v) \in N_i} e_v^{k-1} \tag{2}$$

where e_v^0 is the ID embedding of entity v. Attribute aggregation divides the long-range path into multiple multi-hop paths, which only memorize relations and entities of the same layer, thus it solves the confusion of the multiplication of relationship vectors and entities at different layers and filters the noise.

Inference Module. We propose the inference module, which infers latent information according to each relation and entity. For example, we can deduce from the actor Jackie Chan that this movie is most likely an action movie. So we assign each latent tag $t \in T$ representing a kind of inference classification information such as comedy, science fiction, horror, and so on in movie recommendations. Technically, the inference procedure and latent information aggregator are shown as follows:

$$e_{ii}^0 = \frac{1}{\sqrt{N_i}} \sum_{(r,v) \in N_i} \sum_{t \in T} \beta(r,v)e_t \tag{3}$$

where e_t is the ID embedding of latent tag t and N_i is the neighboring entities of item i based on different types of relations in the knowledge graph. e_{ii}^0 is the latent information embedding of an entity or item at the initial layer.

And aiming to weigh the importance of each latent tag to the relation and entity, the attention weights $\beta(r,v)$ are calculated as follows:

$$\beta(r,v) = softmax(W(e_r \parallel e_v)) \tag{4}$$

$W \in \mathbb{R}^{2d \times tag\_num}$ are the trainable weight matrices and $\parallel$ denotes concat operation. The tag_num is a hyper-parameter to set the total number of the latent tag.

Then the inference information is transmitted to the central node like the attribute information:

$$e_{ui}^k = \frac{1}{\sqrt{N_u}} \sum_{i \in N_u} e_{ii}^{k-1}$$

$$e_{ii}^k = \frac{1}{\sqrt{N_i}} \sum_{(v) \in N_i} e_{vi}^{k-1} \tag{5}$$

where e_{vi}^k denotes the latent information embedding of the entity at k-th layer and e_{ii}^k represents the latent information embedding of entity or item at the k-th layer. e_{ui}^k denotes the inference information embedding of the users.

Finally, we sum all layers' basic attribute and infer representations up to have the global view representation u' and i':

$$u' = e_u^0 + e_{ui}^0 + \cdots + e_u^k + e_{ui}^k$$
$$i' = e_i^0 + e_{ii}^0 + \cdots + e_i^k + e_{ii}^k \tag{6}$$

e_u^k and e_i^k represent the user and item basic attribute representation at k-th layer after attribute aggregation. And e_{ui}^k and e_{ii}^k represent the user and item infer representation at k-th layer after infer model.

In this way, KG Explore module obtains the semantic representations consisting of basic and hidden information. However, we notice that semantic features could help CF models learn the higher quality collaborative representations. And we feed the semantic representation obtained by KG Explore module into the next Multi-IMP-GCN to enhance collaborative features.

4.2 Multi-IMP-GCN

The framework of the Multi-IMP-GCN is shown on the right of Fig. 3. It consists of two main components: (1) subgraph generation module, which divides user-item graph into different subgraphs according to the user's different interests; and (2) collaborative signal aggregator, which aggregates the collaborative signals from the same layer of different subgraphs.

Subgraph Generation Module. In this section, we introduce the subgraph generation module, which is designed to construct the user-item subgraph G_s with $s \in \{1, \cdots, N_s\}$ from the given user-item interaction graph. Users in a subgraph have similar interests. We formulate the user grouping as a classification task, i.e., users are classified into different groups according to their different interests. If a user has multiple interests and may be divided into multiple subgraphs. Specifically, each user is represented by a feature vector:

$$F_u = LeakyReLU(W_1 e_{u'}^{(0)} + b_1) \tag{7}$$

where F_u is the obtained user feature. $e_{u'}^{(0)}$ is the embedding of the user after KG Explore module propagation, which contains more veracious characteristics than the initial embedding of user ID. $W_1 \in \mathbb{R}^{d \times d}$ and $b_1 \in \mathbb{R}^{1 \times d}$ are respectively the weight matrix and bias vector. LeakyReLU [10] is the activation function that encodes positive and small negative signals. To classify the users into different subgraphs, we utilize the obtained user feature to a prediction vector through neural networks:

$$U_h = LeakyReLU(W_2 F_u + b_2)$$
$$U_o = W_3 U_h + b_3 \tag{8}$$

where U_o is the prediction vector representing the probability of the user belonging to each group/subgraph. Meanwhile, we can set top_n as a hyper-parameter, which represents the number of groups the user can belong to. $W_2 \in \mathbb{R}^{d \times d}$, $W_2 \in \mathbb{R}^{d \times N_s}$ and

$b_2 \in \mathbb{R}^{1 \times d}$, $b_3 \in \mathbb{R}^{1 \times N_s}$ are respectively the weight matrices and bias vectors of the two layers. For users with similar interests, Eq. 7 will generate the similar group classification. Compared to the IMP-GCN [9], our model can classify one user into different groups, for one user have various interest.

Collaborative Signal Aggregator. Through the subgraph generation module above, users who share a common interest are grouped into a subgraph, and items that are directly linked to those users also belong to this subgraph. Therefore, each user belongs to top_n multiple subgraphs, and an item can be associated with multiple subgraphs. Meanwhile, we set the N_s as the total number of subgraphs and let G_s with $s \in \{1, \cdots, N_s\}$ denotes a subgraph. To avoid noisy information aggregation, the neighbor information aggregation of each subgraph does not interfere with each other. The high-order propagation in the module is defined as follows:

$$
\begin{aligned}
e_{u'_s}^k &= \sum_{i' \in N_u^s} \frac{1}{\sqrt{|N_u|}\sqrt{|N_i|}} e_{i'}^{k-1} \\
e_{i'_s}^k &= \sum_{u' \in N_i^s} \frac{1}{\sqrt{|N_u|}\sqrt{|N_i|}} e_{u'}^{k-1}
\end{aligned}
\tag{9}
$$

where u' and i' denote the embedding of the user and item after KG Explore module propagation respectively. $e_{u'_s}^k$ and $e_{u'_s}^k$ denote the embedding of user u' and item i' respectively in subgraph s after k layers graph convolution. Through the way that the node in the subgraph only aggregates the neighboring information in that subgraph, noisy information could effectively be avoided. $e_{i'}^{(\cdot)}$ and $e_{u'}^{(\cdot)}$ can be regarded as the features of one interest learned from the interest subgraph G_s. The final representation of node after k layers graph convolution is a combination of its embeddings learned in different subgraphs:

$$
\begin{aligned}
e_{u'}^k &= \sum_{s \in S} e_{u'_s}^k \\
e_{i'}^k &= \sum_{s \in S} e_{i'_s}^k
\end{aligned}
\tag{10}
$$

where S is the subgraph the node belongs to. In this way, users with more similar interests are grouped into a subgraph and the noise propagation in the graph convolution operation is excluded through using subgraphs.

4.3   Model Prediction

After Multi-IMP-GCN performing k layers aggregation for every subgraph, we obtain the different layers representations of user u and item i and then sum them up as the final representations:

$$
\begin{aligned}
e_{u'} &= e_{u'}^{(0)} + \cdots + e_{u'}^{(k)} \\
e_{i'} &= e_{i'}^{(0)} + \cdots + e_{i'}^{(k)}
\end{aligned}
\tag{11}
$$

Like previous models [16, 17], we predict their matching score through inner product, as follows:

$$\hat{y}_{ui} = e_{u'}{}^\top e_{i'} \tag{12}$$

4.4 Model Optimization

For the KG-aware recommendation task, we adopt the pairwise BPR loss [12] to reconstruct the historical data, which user's historical items should be assigned with higher prediction scores than the unobserved items:

$$\mathcal{L}_{BPR} = \sum_{(u,i,j)\in\mathcal{O}} -ln\sigma(\hat{y}_{ui} - \hat{y}_{uj}) \tag{13}$$

where $\mathcal{O} = \{(u,i,j)|(u,i) \in \mathcal{O}^+, (u,j) \in \mathcal{O}^-\}$ is the training dataset consisting of the observed interactions $\mathcal{O}^+$ and unobserved counterparts $\mathcal{O}^-$; σ is the sigmoid function. By combining the independence loss and BPR loss, we minimize the following objective function to learn the model parameter:

$$\mathcal{L}_{CKGA} = \mathcal{L}_{BPR} + \lambda \|\theta\|_2^2 \tag{14}$$

where θ is the model parameter set, λ is the hyper parameter to control the L_2 regularization term.

5 Experiments

We provide empirical results to demonstrate the effectiveness of our proposed CKGA.

5.1 Experimental Settings

Datasets. To evaluate the effectiveness of CKGA, we use three public datasets collected from different real-life platforms: Amazon-Book for book recommendation, Yelp2018 for business venue recommendation, Alibaba-iFashion for outfit recommendation. Table 1 presents the statistical information of our experimented datasets with interaction information and knowledge graph characteristics.

Table 1. Statistics of the datasets.

		Yelp2018	Amazon-book	Alibaba-iFashion
User-Item Interaction	#User	45,919	70,679	114,737
	#Items	45,538	24,915	30,040
	#Interactions	1,183,610	846,434	1,781,093
Knowledge Graph	#Entities	47,472	29,714	59,156
	#Relations	42	39	51
	#Triplets	869,603	686,516	279,155

Evaluation Metrics. For fair evaluation, we adopt the all-ranking strategy rather than sampling strategies [15, 18]. Specifically, for each user, the items that he/she has not interacted are viewed as negative samples, and while the items that are relevant to the user in the testing set are viewed as positive samples. For the top-K recommendation evaluation, recall@K and ndcg@K are the two representative metrics. The average metrics result for all users in the testing set are reported with $K = 20$ by default.

Baselines. We compare CKGA with various lines of recommender systems for performance evaluation, including conventional method(MF), embedding-based(CKE), and GNN-based(KGAT, KGNN-LS, CKAN, and KGIN) methods:

- **MF** [12] (matrix factorization) directly projects the ID of a user (or an item) into an embedding vector through the user-item interactions.
- **CKE** [21] adopts TransR to encode the items' semantic information and incorporates the side information into the CF model.
- **KGNN – LS** [15] is a GNN-based model which enriches item embeddings with GNN and label smoothness regularization.
- **KGAT** [16] utilizes an attentive neighborhood aggregation mechanism to exploit high-order connectivities on the heterogeneous graph consisting of the user-item graph and knowledge graph.
- **CKAN** [18] explicitly encodes the collaborative signals by heterogeneous collaboration propagation and proposes a natural way of combining collaborative signals with knowledge associations together.
- **KGIN** [17] reveals user intents behind the interactions, and performs GNNs on the proposed user-intent-item-entity graph.

Parameter Settings. We implement our CKGA model in PyTorch. For a fair comparison, we fix the embedding dimensionality as 64 for all models, and the embedding parameters are initialized with the Xavier method [2]. We set the optimizer as Adam [7] and batch size as 1024. The global graph aggregation layer K is set to 3, and the aggregation layer K' in Multi-IMP-GCN is set to 5. In the inference module, we set the number of the latent tags tag_num to the 16. And we search the num of the interest groups (subgraph) N_s in the Multi-IMP-GCN in $\{1, 3, 5\}$.

5.2 Performance Comparison

We compare all methods and report the overall performance in Table 2. From the results, we find that:

- CKGA consistently outperforms the other baselines across three datasets. We attribute the improvements obtained by CKGA to two aspects: i) Benefiting from the KG Explore module, CKGA filters the noise of irrelevant relations and entities and get the rich semantic representation. ii) Depending on the Multi-IMP-GCN, CKGA alleviates the over-smoothing in CF.

Table 2. Overall performance comparison.

	Amazon-book		Yelp2018		Alibaba-iFashion	
	recall	ndcg	recall	ndcg	recall	ndcg
MF	0.1244	0.0658	0.0555	0.0375	0.1095	0.0670
CKE	0.1375	0.0685	0.0686	0.0431	0.1103	0.0676
KGAT	0.1390	0.0739	0.0675	0.0432	0.1030	0.0627
KGCN	0.1111	0.0569	0.0532	0.0338	0.1039	0.0557
CKAN	0.1380	0.0726	0.0689	0.0441	0.0970	0.0509
KGIN	0.1436	0.0748	0.0712	0.0462	0.1147	0.0716
CKGA	**0.1590**	**0.0852**	**0.7450**	**0.0483**	**0.1188**	**0.0742**

- By comparing the MF and other methods, we can find that introducing knowledge graph into the recommender system can effectively improve the performance of recommendation and tackle the sparsity issue. However, by comparison of the performance of our CKGA and other knowledge-aware models, we conclude that semantic features in the global graph enhance the collaborative information. .

5.3 Ablation Studies

We investigate the contributions of the main components in our model to the final performance by comparing CKGA with the following two variants:

- $CKGAw/o\ K$: In this variant, the KG explore module is replaced by the Path relation module, which was proposed by KGIN [17]. Path relation module uses the relation path neighborhood aggregation, which aggregates neighboring information meanwhile preserving the path information.
- $CKGAw/o\ M$: This variant removes the Multi-IMP-GCN, and let the representation after KG explore module as the final representation.

The results of the two variants and CKGA are shown in Table 3, from which we have the following observations: 1) Replacing the KG explore module by path relation module degrades the model's performance, which reflects that KG Explore module could filter the noise of irrelevant relations and entities and learn discriminative semantic representation to improve recommendation performance. 2) Removing the Multi-IMP-GCN model degrades the model performance, which demonstrates semantic representation in KG could help models learn the richer collaborative representation and improve the model performance.

Table 3. Effect of ablation study.

Model	Amazon-book		Yelp2018		Alibaba-iFashion	
	recall	ndcg	recall	ndcg	recall	ndcg
CKGA	**0.1590**	**0.0852**	**0.0745**	**0.0483**	**0.1188**	**0.0742**
$CKGA_{w/o\ K}$	0.1536	0.0827	0.0736	0.0477	0.1180	0.0737
$CKGA_{w/o\ M}$	0.1474	0.0780	0.0693	0.0445	0.1141	0.0709

5.4   Impacts of Multi-IMP-GCN

As shown in Fig. 5, dividing the users of interest groups could improve the recommendation performance. This also reflects that indistinguishably aggregating the information from all neighbors will make a lot of noise, IMP-GCN [9] and Multi-IMP-GCN can alleviate the over-smoothing problem by performing high-order graph convolution inside interest subgraphs. At the same time, we found in the Alibaba-iFashion experiment that the number of groups should be reasonably determined; otherwise, the performance would be affected.

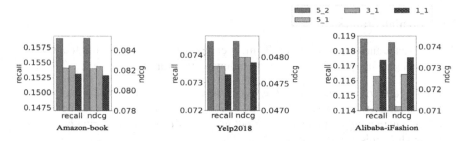

Fig. 5. Effect of Multi-IMP-GCN Model study. 5_2 means user interests can be divided into five broad categories, and a user can belong to two groups, and the same goes for other representations. 1_1 means no grouping, just the LightGCN.

When we add the interest group nums and specify that users can belong to two interest groups; we find that recommendation performance is much better than if users only belonged to one interest group. This proves that the Multi-IMP-GCN has an advantage in considering the diversity of user interests compared to IMP-GCN [9].

As shown in Fig. 6, the performance of Multi-IMP-GCN is always better than the performance of LightGCN [3] in different layers, which demonstrates Multi-IMP-GCN could alleviate the over-smoothing problem comparing to the LightGCN [3].

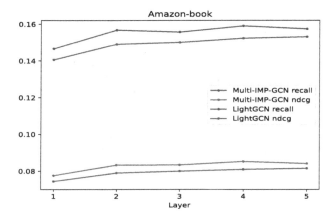

Fig. 6. Impact of the number of layers L.

6 Conclusion and Future Work

In this work, we focus on utilizing the semantic features to enhance the user-item collaborative features. We proposed a novel framework, CKGA, which generates more discriminative representation from two stages: (1) Multi-IMP-GCN utilizes the semantic features to enhance the collaborative representations and alleviates the over-smoothing problem in CF by performing high-order graph convolution inside interest subgraphs. (2) KG Explore module learns the semantic representations by filtering the noise and inferring the implicit knowledge. In summary, We provide a novel approach to enhance collaborative signaling using semantic features. In future work, we will explore other auxiliary information such as the social network to enhance the collaborative features.

Acknowledgments. This work was supported in part by National Key Research and Development Program of China (2022YFF0904301).

References

1. Bell, R.M., Koren, Y.: Lessons from the Netflix prize challenge. ACM SIGKDD Explor. Newsl. **9**(2), 75–79 (2007). https://doi.org/10.1145/1345448.1345465
2. Glorot, X., Bengio, Y.: Understanding the difficulty of training deep feedforward neural networks. In: Proceedings of the Thirteenth International Conference on Artificial Intelligence and Statistics, JMLR Workshop and Conference Proceedings, pp. 249–256 (2010)
3. He, X., Deng, K., Wang, X., Li, Y., Zhang, Y., Wang, M.: LightGCN: Simplifying and Powering Graph Convolution Network for Recommendation. arXiv preprint arXiv:2002.02126 (2020)

4. He, X., Liao, L., Zhang, H., Nie, L., Hu, X., Chua, T.S.: Neural Collaborative Filtering. arXiv preprint arXiv:1708.05031 (2017)

5. Ji, G., He, S., Xu, L., Liu, K., Zhao, J.: Knowledge graph embedding via dynamic mapping matrix. In: Proceedings of the 53rd Annual Meeting of the Association for Computational Linguistics and the 7th International Joint Conference on Natural Language Processing (Volume 1: Long Papers), pp. 687–696. Association for Computational Linguistics, Beijing (2015). https://doi.org/10.3115/v1/P15-1067

6. Juan, Y., Zhuang, Y., Chin, W.S., Lin, C.J.: Field-aware factorization machines for CTR prediction. In: Proceedings of the 10th ACM Conference on Recommender Systems, pp. 43–50. ACM, Boston (2016). https://doi.org/10.1145/2959100.2959134

7. Kingma, D.P., Ba, J.: Adam: A Method for Stochastic Optimization (2017). https://doi.org/10.48550/arXiv.1412.6980

8. Lin, Y., Liu, Z., Sun, M., Liu, Y., Zhu, X.: Learning entity and relation embeddings for knowledge graph completion. Proc. AAAI Conf. Artif. Intell. **29**(1) (2015). https://doi.org/10.1609/aaai.v29i1.9491

9. Liu, F., Cheng, Z., Zhu, L., Gao, Z., Nie, L.: Interest-aware Message-Passing GCN for Recommendation (2021)

10. Maas, A.L.: Rectifier Nonlinearities Improve Neural Network Acoustic Models (2013)

11. Ren, M., Huang, X., Li, W., Song, D., Nie, W.: LR-GCN: latent relation-aware graph convolutional network for conversational emotion recognition. IEEE Trans. Multimedia **24**, 4422–4432 (2022). https://doi.org/10.1109/TMM.2021.3117062

12. Rendle, S., Freudenthaler, C., Gantner, Z., Schmidt-Thieme, L.: BPR: Bayesian Personalized Ranking from Implicit Feedback (2012)

13. Sun, Z., et al.: Research commentary on recommendations with side information: a survey and research directions. Electron. Commer. Res. Appl. **37**, 100879 (2019). https://doi.org/10.1016/j.elerap.2019.100879

14. Wang, H., Zhang, F., Xie, X., Guo, M.: DKN: Deep Knowledge-Aware Network for News Recommendation (2018)

15. Wang, H., et al.: Knowledge-aware Graph Neural Networks with Label Smoothness Regularization for Recommender Systems (2019). https://doi.org/10.48550/arXiv.1905.04413

16. Wang, X., He, X., Cao, Y., Liu, M., Chua, T.S.: KGAT: knowledge graph attention network for recommendation. In: Proceedings of the 25th ACM SIGKDD International Conference on Knowledge Discovery & Data Mining, pp. 950–958. ACM, Anchorage (2019). https://doi.org/10.1145/3292500.3330989

17. Wang, X., et al.: Learning intents behind interactions with knowledge graph for recommendation. In: Proceedings of the Web Conference 2021, pp. 878–887 (2021). https://doi.org/10.1145/3442381.3450133

18. Wang, Z., Lin, G., Tan, H., Chen, Q., Liu, X.: CKAN: collaborative knowledge-aware attentive network for recommender systems. In: Proceedings of the 43rd International ACM SIGIR Conference on Research and Development in Information Retrieval (SIGIR 2020), pp. 219–228. Association for Computing Machinery, New York (2020). https://doi.org/10.1145/3397271.3401141

19. Yang, Y., Huang, C., Xia, L., Li, C.: Knowledge graph contrastive learning for recommendation. In: Proceedings of the 45th International ACM SIGIR Conference on Research and Development in Information Retrieval, pp. 1434–1443 (2022). https://doi.org/10.1145/3477495.3532009

20. Yu, X., et al.: Recommendation in heterogeneous information networks with implicit user feedback. In: Proceedings of the 7th ACM Conference on Recommender Systems, pp. 347–350. ACM, Hong Kong (2013). https://doi.org/10.1145/2507157.2507230

21. Zhang, F., Yuan, N.J., Lian, D., Xie, X., Ma, W.Y.: Collaborative knowledge base embedding for recommender systems. In: Proceedings of the 22nd ACM SIGKDD International Conference on Knowledge Discovery and Data Mining, pp. 353–362. ACM, San Francisco (2016). https://doi.org/10.1145/2939672.2939673
22. Zhao, H., Yao, Q., Li, J., Song, Y., Lee, D.L.: Meta-graph based recommendation fusion over heterogeneous information networks. In: Proceedings of the 23rd ACM SIGKDD International Conference on Knowledge Discovery and Data Mining, pp. 635–644. ACM, Halifax (2017). https://doi.org/10.1145/3097983.3098063

PageCNNs: Convolutional Neural Networks for Multi-label Chinese Webpage Classification with Multi-information Fusion

Jiawei Zheng[1,2], Junying Chen[1,2(✉)], and Yi Cai[1,2]

[1] School of Software Engineering, South China University of Technology, Guangzhou, China
[2] Key Laboratory of Big Data and Intelligent Robot (SCUT), Ministry of Education, Guangzhou, China
jychense@scut.edu.cn

Abstract. Along with the popularity and development of the Internet in China, Chinese webpage classification has become an important research topic. As the webpage text is a kind of text, webpage classification is constructed based on text classification. But due the particularity of the webpage composition, the external linked webpages can leverage helpful information to improve the webpage classification performance. The goal of this work is to design accurate multi-label Chinese webpage classification models by effectively fusing the information extracted from current webpage and external linked webpages, including the text information and label information of external linked webpages. A convolutional neural network for webpage classification (PageCNN) model and its two variants (PageCNN-CLL and PageCNN-WLL) are proposed to effectively fuse the text and label information extracted from multiple Chinese webpages. The proposed PageCNN models are compared with two modified traditional machine learning models, the modified TextCNN model, and three state-of-the-art deep learning based multi-label text classification models. The experimental results demonstrate that the PageCNN models perform better than the compared models in terms of subset accuracy, Hamming loss, macro F1, and micro F1. Moreover, the in-depth analysis of the effectiveness of the external linked webpages on current webpage classification is conducted by analyzing the error correction rate and hit rate of the proposed models and preliminary prediction variables. As demonstrated in the experiments, the multi-information fusion methods developed in the PageCNN models can effectively manipulate the input data from multiple webpages to enhance the multi-label Chinese webpage classification performance.

Keywords: Chinese Webpage Classification · Multi-information Fusion · Multi-label Classification · Convolutional Neural Network

1 Introduction

Internet technology has brought China into the rapid development of the information era. Alongside the popularization and development of the Internet, millions of webpages have been created with tremendous commercial value to impulse the information spread on the Internet. At the very beginning, webpage classification was done manually. However, the number of webpages on the Internet has been exploding, which makes it more difficult to classify the webpages manually. Therefore, automatic webpage classification is required. Besides, China has the largest population around the world, where a large number of Chinese webpages have been developed to promote the Internet economy in China. The classification of Chinese webpages is beneficial to the subsequent data mining of Chinese webpages, such as recommendation systems and user profiling. As a result, Chinese webpage classification is an important research topic.

As the webpage text is a kind of text, webpage classification is basically constructed based on text classification. Text classification is a classic problem and basic task in the field of natural language processing. According to the number of classification categories, text classification problems can be divided into binary classification and multi-classification. Multi-label text classification problem is further derived from the text multi-classification problem. Each sample in the multi-classification problem can only belong to one category, while there are multiple categories. On the other hand, each sample in the multi-label classification problem can belong to multiple categories. For example, the movie *"The Avengers"* can be labeled as an action movie and a science fiction movie. Recently, multi-label text classification problem has been actively investigated with the development of deep learning models, such as convolutional neural network (CNN) [7,11], recurrent neural network (RNN) [13], ensemble of CNN and RNN [1], canonical correlated autoencoder [21], sequence generation model (SGM) [19], multi-level dilated convolution and hybrid attention (MDC-HA) [10], sequence-to-set model [18], combination of self-attention and contextualized embedding [23], multi-step co-attention model [12], label-specific attention network (LSAN) [17], hierarchical sequence-to-sequence model [20], combination of SGM and CNN [9], and graph-based methods [2,14].

In terms of multi-label webpage classification problem, a webpage can be labeled with multiple tags. For example, a news webpage covering a sports event can be labeled with two tags: sports and news, and a webpage for a TV variety show can be labeled with two tags: video and entertainment. The existing multi-label webpage classification methods treat the multi-label webpage classification problem as the multi-label text classification problem, which is beneficial to solve subsequent problems such as automated tag suggestion and bid phrase recommendation. Traditional machine learning algorithms are utilized in the multi-label webpage classification methods, such as weight-in-order construction algorithm (WOCA) [15], which transformed the multi-label classification problem into a series of single-label multi-class classification problems. Unlike English and other Indo-European languages, there is no natural separator between Chinese words [3,16]. Chinese word segmentation makes Chinese webpage processing more difficult than English webpage processing. As a result, Chinese webpage classification is not a straightforward application of the webpage classification methods developed for other languages. Existing Chinese webpage classification methods also take advantage of machine learning models, such as support vector machine (SVM) [3] and neural networks [4]. Upon literature review, there are no studies on multi-label Chinese webpage classification yet, and applying deep learning models to multi-label Chinese webpage classification should be feasible and meaningful.

Due to the particularity of the webpage composition, there are some differences between webpage contents and plain text, which bring difficulties but also richer infor-

Fig. 1. An example webpage with two external linked webpages.

mation into webpage classification. A webpage contains information other than plain text, such as uniform resource locator, webpage structure, linked webpages, etc. Using extra information of webpages like structural information and hyperlinks can improve the webpage classification performance. Among lots of the extra information of a webpage, we find that the external linked webpages of the current webpage are very helpful for determining the tags of the current webpage. For example, the current webpage in Fig. 1 is a news webpage reporting a student won the national track and field champion. If the current webpage is classified only based on the webpage text, the current webpage may be labeled as a sports and news webpage. But if its external linked webpages are considered, such as linked webpage 1 and linked webpage 2 in Fig. 1, the tags of the current webpage will be different. The linked webpage 1 and 2 are the homepage of the school and the homepage of the administrative office of the school, respectively. Considering the information obtained from these two externally linked webpages, the current webpage shows a close connection with the school, and as a result, an additional education tag is added to the current webpage. Hence, the external linked webpages leverage helpful information to improve the multi-label Chinese webpage classification.

To the best of our knowledge, this is the first work to solve multi-label Chinese webpage classification problem. The goal of this work is to design accurate multi-label Chinese webpage classification models by effectively merging the information extracted from the current webpage and external linked webpages. The CNN for webpage classification (PageCNN) model and its two variants are proposed to effectively fuse the information extracted from multiple webpages so as to enhance the multi-label Chinese webpage classification accuracy. The proposed PageCNN models are summarized as follows:

1. *PageCNN*: PageCNN model fuses the text information extracted from the current webpage and external linked webpages to expand the scope of the feature representation of the current webpage.
2. *PageCNN-CLL*: PageCNN with concatenated link labels (PageCNN-CLL) fuses the label information of the external linked webpages and the text information of the current webpage and external linked webpages together, which further expands the scope of the feature representation.
3. *PageCNN-WLL*: PageCNN with weighted link labels (PageCNN-WLL) fuses the label and text information of the current webpage and external linked webpages all together, with a weight coefficient adjusting the influence of the external linked webpages on the classification of the current webpage. The weighted method realizes better multi-information fusion for the current webpage classification.

2 Multi-label Chinese Webpage Classification Models

The input webpages are first preprocessed to obtain a corpus after Chinese word segmentation, and the corpus is furthered represented as word vector matrices. The initial set of word vectors in this work is obtained by training a *Baidubaike* corpus [8] using word2vec model. The pre-trained initial word vectors are further fine-tuned and updated during the training process of the multi-label Chinese webpage classification model. Each word extracted from the input webpage is represented as a fixed-sized K-dimensional word vector x_m by looking up the pre-trained initial set of word vectors. Hence, a webpage text T containing M words can be converted into a $K \times M$ word vector matrix, which is represented as $T = [x_1, x_2, \cdots, x_M]$. The word vector matrix T of the webpage is used as the input of the multi-label Chinese webpage classification model.

The proposed multi-label Chinese webpage classification models are built upon the CNN model [5] which is developed to solve the binary text classification problem, *a.k.a.*, TextCNN. The structure of the TextCNN model is divided into four layers, which are: input layer, convolutional layer, pooling layer, and output layer. The input layer normalizes the input word vector matrices for various webpages, by truncating the matrix when the corresponding webpage length exceeds the set maximum length and filling the matrix when the corresponding webpage length is insufficient. The convolutional layer performs convolutional operations, and the pooling layer conducts the max pooling operations. The output layer is a fully connected layer with a softmax activation function. In this work, the activation function of the output layer is modified to sigmoid function so as to realize multi-label text classification. As a result, the modified TextCNN model can be used for multi-label Chinese webpage classification when only considering the current webpage text for classification.

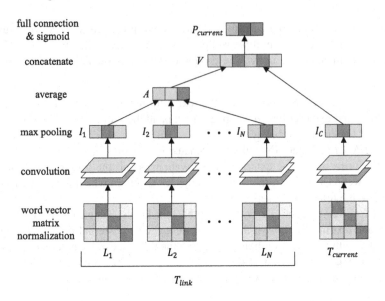

Fig. 2. Structure of the proposed PageCNN model combining the external linked webpage text for multi-label Chinese webpage classification.

PageCNN Model Combining Linked Webpage Text. In the modified TextCNN model, the resulting vector after convolutional and max pooling operations is the final feature vector extracted from the input word vector matrix to predict the labels of current webpage. The same network structure is used to extract the feature vectors of the external linked webpages, which are combined with the feature vector of current webpage to construct the PageCNN model. The structure of the proposed PageCNN model combining the external linked webpage text for multi-label Chinese webpage classification is shown in Fig. 2. As seen from Fig. 2, the input of the proposed PageCNN model includes the word vector matrices of current webpage $T_{current}$ and external linked webpages T_{link}. Assuming that there are N external links for current webpage, T_{link} is represented as a set of word vector matrices $\{L_1, L_2, ..., L_N\}$.

In the convolutional layer, Q convolutional kernels with the size of $K \times H$ are used to extract local features of the input word vector matrix. The parameter H means performing the convolutional operation on H consecutive word vectors. For the ith convolutional kernel, the local feature f_i is computed as $f_i = ReLU(W_i \cdot T_{i:(i+H-1)} + b_i)$, where W_i is a $K \times H$ weight matrix for the ith convolutional kernel, and $T_{i:(i+H-1)}$ is a segmented sequence containing the word vectors from the ith to $(i+H-1)$th columns in the input word vector matrix, i.e., $T_{i:(i+H-1)} = [x_i, x_{(i+1)}, \cdots, x_{(i+H-1)}]$. The dot product of W_i and $T_{i:(i+H-1)}$ is then adjusted by a bias term b_i. The convolutional kernel starts from the first segmented sequence and slides down to the last segmented sequence of the input word vector matrix. For the set of segmented sequences $\{T_{1:H}, T_{2:(H+1)}, ..., T_{(M-H+1):M}\}$, a feature vector $f = [f_1, f_2, ..., f_{(M-H+1)}]$ is finally obtained after convolutional operations.

The subsequent pooling layer applies max pooling on the feature vector to obtain the value of the maximum element in the feature vector, which is $\hat{f} = max\{f\}$. The parameter H can be set as multiple values to construct various convolutional kernels of different sizes, so as to extract various local features from the input word vector matrix. Supposing that H is set as E different values, the total number of convolutional kernels is $Q \times E$. Hence, the max pooling result of a webpage text T is represented as: $I = [\hat{f}^1, \hat{f}^2, ..., \hat{f}^{Q \times E}]$.

The max pooling result of current webpage $T_{current}$ is denoted as I_C, and those of the external linked webpages $\{L_1, L_2, ..., L_N\}$ are denoted as $\{I_1, I_2, ..., I_N\}$. For the sake of taking the information of every external linked webpage into consideration, an average operation is conducted to calculate the average feature vector of the external linked webpages, which is $A = \sum_{n=1}^{N} I_n / N$. A is then concatenated to I_C to obtain the final feature vector for current webpage classification: $V = A \parallel I_C$, where $\parallel$ is the concatenation operator. The final feature vector V fuses the text information extracted from current webpage and external linked webpages together, which expands the scope of the feature representation for current webpage.

Finally, the sigmoid activation function is utilized to calculate the probability of each label class for current webpage: $P_{current} = sigmoid(W_p V + b_p)$, where $P_{current} = [P_C^1, P_C^2, ..., P_C^J]$ is the predicted probability result, J is the total number of label classes, and W_p and b_p are the weight matrix and bias term of the fully connected layer. The multiplication result of W_p and V is adjusted by b_p. The final classification result $C_{current}$ is obtained based on the predicted probability result $P_{current}$. The label class elements in $P_{current}$ which are larger than 0.5 are first selected, and then the classes with top 3 probability values are labeled as 1 while the other classes are labeled as 0 in $C_{current}$. If the number of elements in $P_{current}$ which are larger than 0.5 is less than 3, the number of elements in $C_{current}$ which are labeled as 1 is less than 3, with 1 or 2 labels for current webpage.

Two Variant PageCNN Models with Additional Linked Webpage Labels. Apart from the text contents of the external linked webpages, the labels of the external linked webpages are also valuable information for the classification of current webpage. Therefore, the PageCNN model is further improved by considering additional linked webpage labels. Two variant models are proposed, which are: PageCNN with concatenated link labels (PageCNN-CLL) and PageCNN with weighted link labels (PageCNN-WLL).

The PageCNN-CLL model is illustrated in Fig. 3. In this model, the first fully connected layer applies a sigmoid activation function on the average feature vector of the external linked webpages to obtain a predicted label distribution of external linked webpages: $P_{link} = sigmoid(W_l A + b_l)$, where W_l and b_l are the weight matrix and bias term of the first fully connected layer, and $P_{link} = [P_L^1, P_L^2, ..., P_L^J]$ contains the predicted probability of each label class for the external linked webpages. P_{link} is regarded as the high-level feature of the external linked webpages, which is concatenated to the feature vectors of current webpage and its external linked webpages to obtain a modified final feature vector V_{CLL} for current webpage classification: $V_{CLL} = P_{link} \parallel A \parallel I_C$. The modified final feature vector V_{CLL} fuses the label information of the external linked webpages and text information of current webpage and external linked webpages together, which further expands the scope of the feature representation. The modified final feature vector V_{CLL} then substitutes V to calculate the predicted probability result $P_{current}$ of the PageCNN-CLL model as: $P_{current} = sigmoid(W_p V_{CLL} + b_p)$.

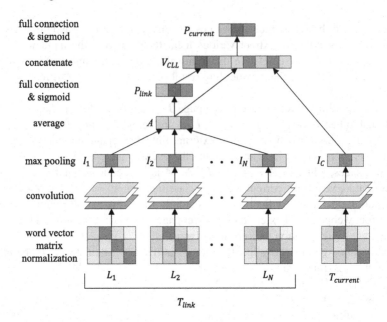

Fig. 3. Structure of the proposed PageCNN-CLL model with additional linked webpage labels.

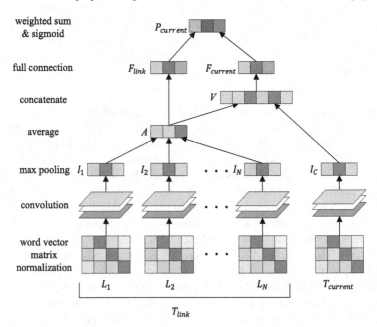

Fig. 4. Structure of the proposed PageCNN-WLL model with additional linked webpage labels.

The PageCNN-WLL model is presented in Fig. 4. In this model, the average linked webpage feature vector A and final feature vector V are both processed in the output layer using full connection. The role of full connection is to connect particular features to the correlated label classes and perform a linear transformation from the feature space to label space. The outputs of the full connection are J-dimensional vectors, denoted as F_{link} and $F_{current}$. The calculations of the two full connection outputs are: $F_{link} = W_l A + b_l$, and $F_{current} = W_p V + b_p$. Then, a weighted sum of F_{link} and $F_{current}$ is calculated before sigmoid activation. Hence, the predicted probability result of the PageCNN-WLL model is $P_{current} = sigmoid(\lambda F_{link} + (1 - \lambda) F_{current})$, where λ is the weight coefficient of link labels. Experiments will be conducted in Sect. 3 to determine the suitable weight coefficient of link labels. The PageCNN-WLL model fuses the label information and text information of current webpage and external linked webpages all together, with a weight coefficient adjusting the influence of the external linked webpages on the classification of current webpage. The weighted method realizes better multi-information fusion for current webpage classification.

Relationships of the Three Proposed Models. On the basis of the modified TextCNN, PageCNN introduces the text information of external linked webpages into the classification of current webpage. If the average text feature vector of external linked webpages A is removed from the concatenated final feature vector V (which combines the text information from external linked webpages and current webpage), the elements remained in the final feature vector of the PageCNN model are actually the elements of the max pooling result of current webpage I_C, then the PageCNN model becomes the modified TextCNN model. The PageCNN-CLL model adds the label information of external linked webpages to PageCNN model by concatenating the predicted label distribution of external linked webpages P_{link} to obtain the modified final feature vector V_{CLL}. If P_{link} is removed from V_{CLL}, the PageCNN-CLL model becomes PageCNN model. Moreover, the PageCNN-WLL model incorporates the label information of external linked webpages into PageCNN model in another way. It uses the full connection operation to connect the feature vectors (A and V) to the correlated label classes while transforming the feature space to label space, obtaining two output vectors (F_{link} and $F_{current}$). The sigmoid activation is performed on the weighted sum of F_{link} and $F_{current}$ to get the final prediction result. If F_{link} and $F_{current}$ are removed, the final prediction result is obtained from the concatenated feature vector V, then the PageCNN-WLL model becomes PageCNN model.

3   Experimental Results and Discussions

3.1   Multi-label Chinese Webpage Dataset

In this work, 7550 Chinese webpages from 228 websites are collected by a distributed Python crawler. Each of the webpages is tagged with 1 to 3 labels, depending on the contents of the webpage. There are totally 16 categories for the webpage classification, which are shown in Table 1. In order to make the sample distribution of the label categories as balanced as possible, some samples belonging to the categories with a larger number of webpage samples are discarded, thereby generating a dataset with 6650 webpage samples. The number of webpage samples in each category is also shown

Table 1. Label Categories for Chinese Webpage Classification and The Number of Webpage Samples in Corresponding Label Categories

Label Code	Category Name	# of Samples	Label Code	Category Name	# of Samples
1	Gaming	776	9	Life service	1631
2	Reading	436	10	Education or Culture	744
3	Shopping	494	11	Internet Technology	1115
4	Health care	798	12	Sports or Fitness	504
5	Audio or Video	724	13	Leisure or Amusement	1362
6	Blog or BBS	1370	14	Comprehensive Portal	733
7	News	1305	15	Public Organization	591
8	Traffic or Tour	590	16	Financial or Business	1116

Table 2. The Distribution of Webpage Samples with Different Number of Labels

# of Webpage Labels	# of Webpage Samples
1	774
2	4113
3	1763

in Table 1, and the distribution of webpage samples referring to different number of webpage labels is shown in Table 2. In this Chinese webpage dataset, each webpage has an average of 8.36 external linked webpages.

3.2 Implementation Details and Evaluation Metrics

In the experiments, the word vector size K is set as 300, the convolutional kernel parameter H is set as three values (3, 4, and 5), and the number of convolutional kernels Q for each H value is 128. During the training process, the batch size is 32, and the dropout rate is 0.5. Furthermore, to reduce the influence of the distribution of Chinese webpage dataset on the experimental results, a 10-fold cross validation is conducted. During the training process, a dropout operation is conducted before entering the output layer, which randomly hides a part of the neurons according to a certain ratio. The dropout operation can effectively prevent data overfitting. Furthermore, the cross-entropy loss function is applied to minimize the error between the predicted labels and ground truth labels, and Adam optimizer [6] is adopted to learn the parameters in the network model during backpropagation.

We modify the WOCA in this work. We sort the ground truth labels of the training sample with their weights, and supplement labels to the training sample according to the order of weights of the ground truth labels. The modified WOCA is more reasonable than original WOCA. The classifier applied after WOCA is an SVM as presented in [15]. In this work, we adopt naive Bayes (NB) as another choice of classifier. Hence, the classification results of original and modified WOCA-SVM and WOCA-NB will be evaluated in the experiments. Various PageCNN models and the modified TextCNN model are implemented based on Tensorflow framework, the compared SGM, MDC-HA and LSAN models are implemented using PyTorch framework, and the compared WOCA models are implemented with Sklearn toolkit. Moreover, an Nvidia GTX1080 Ti GPU is used to train the deep learning models.

In multi-label Chinese webpage classification, each webpage sample may have more than one labels. Hence, the methods to evaluate multi-label classification models are different from those commonly used in single-label classification models. In this work, it is assumed that each label is independent of other labels, and there is no hierarchy between the labels. As a result, subset accuracy, Hamming loss, macro-averaging (macro F1), and micro-averaging (micro F1) are used as the evaluation metrics [22].

3.3   Multi-label Chinese Webpage Classification Results

Analysis of Evaluation Metrics. The multi-label Chinese webpage classification results of the proposed PageCNN models and the compared models in terms of the evaluation metrics are shown in Table 3. As a 10-fold cross validation is conducted in the experiment, the average and standard deviation values of the evaluation metrics are listed in the table. For subset accuracy, macro F1, and micro F1, the higher the value, the better the classification performance. But for Hamming loss, the lower the metric value, the better the classification performance. As seen from Table 3, most deep learning models perform better than the traditional machine learning models, but the modified WOCA-SVM model demonstrates better performance than the MDC-HA model. Among all the deep learning models, the proposed PageCNN models achieve the best results in all evaluation metrics, especially the PageCNN-WLL model. The subset accuracy, Hamming loss, macro F1, and micro F1 of the PageCNN-WLL model are 86.14%, 0.01427, 94.41%, and 94.59%, respectively. It is worth noted that the subset accuracy and Hamming loss of the PageCNN models are higher than 85% and less than 0.015, respectively, while none of the compared models achieves a subset accuracy higher than 85% and a Hamming loss less than 0.015.

As seen from the experimental results of the traditional machine learning models, the modified WOCA improves the model performance as compared with the original WOCA with 15.35~25.45% subset accuracy improvement. In addition, the adopted NB algorithm combined with the original WOCA performs better than the SVM (utilized in [15]) combined with the original WOCA, but the NB algorithm with modified WOCA is worse than the SVM with modified WOCA. This is because SVM can clearly distinguish the difference between categories, while NB algorithm has a higher misjudgment rate when there are similar categories and the feature similarity between categories is high.

Without information extracted from external linked webpages, the performance of several representative deep learning based multi-label text classification models is worse than the proposed PageCNN models which incorporate information from both current webpage and external linked webpages, as demonstrated in Table 3. As for the deep learning models which only extract information from current webpage, the performance of SGM and MDC-HA are worse than that of the modified TextCNN. The reason may be that the webpage text is more disorganized than usual text with fluent semantics, and these two sequence generation models based on LSTM sequence model

Table 3. Experimental Results of Multi-label Chinese Webpage Classification

Model	Subset Accuracy	Hamming Loss	Macro F1	Micro F1
original WOCA-SVM	0.5696	0.06261	0.8985	0.7940
	(± 0.0197)	(± 0.00173)	(± 0.0039)	(± 0.0054)
original WOCA-NB	0.6257	0.05328	0.8840	0.8194
	(± 0.0146)	(± 0.00212)	(± 0.0075)	(± 0.0061)
modified WOCA-SVM	0.8241	0.02114	0.9232	0.9210
	(± 0.0094)	(± 0.00164)	(± 0.0059)	(± 0.0060)
modified WOCA-NB	0.7792	0.02358	0.9087	0.9098
	(± 0.0131)	(± 0.00162)	(± 0.0058)	(± 0.0060)
modified TextCNN	0.8385	0.01637	0.9318	0.9349
	(± 0.0088)	(± 0.0091)	(± 0.0059)	(± 0.0061)
SGM	0.8315	0.01887	0.9239	0.9293
	(± 0.0151)	(± 0.00181)	(± 0.0076)	(± 0.0067)
MDC-HA	0.7959	0.02427	0.9067	0.9097
	(± 0.0105)	(± 0.00186)	(± 0.0084)	(± 0.0070)
LSAN	0.8487	0.01637	0.9355	0.9381
	(± 0.0151)	(± 0.00113)	(± 0.0054)	(± 0.0044)
PageCNN	0.8580	0.01474	0.9416	0.9441
	(± 0.0110)	(± 0.00151)	(± 0.0060)	(± 0.0056)
PageCNN-CLL	0.8591	0.01443	0.9432	0.9458
	(± 0.0111)	(± 0.00130)	(± 0.0054)	(± 0.0052)
PageCNN-WLL	**0.8614**	**0.01427**	**0.9442**	**0.9459**
	(± 0.0126)	**(± 0.00131)**	**(± 0.0050)**	**(± 0.0048)**

are more focused on extracting contextual semantic information, which makes them perform poorer than the TextCNN model that focuses on extracting local feature information. The LSAN model, which combines textual information with label semantic information, achieves the best results among the models extracting information only from current webpage. However, because it is only work on the text and label information of current webpage, its classification results are worse than the PageCNN models which fuse the external linked webpage information.

The experimental results demonstrate that the external linked webpages leverage helpful information to improve the multi-label Chinese webpage classification. The PageCNN models can effectively merge the information extracted from the current webpage and external linked webpages, including the text information and label information of the external linked webpages. By using a weight coefficient to adjust the influence of the external linked webpages on the classification of current webpage, the PageCNN-WLL model realizes better multi-information fusion for current webpage classification than the PageCNN and PageCNN-CLL models.

Analysis of Weight Coefficient λ in PageCNN-WLL. In the PageCNN-WLL model, a weight coefficient λ is used to adjust the influence of the external linked webpage labels

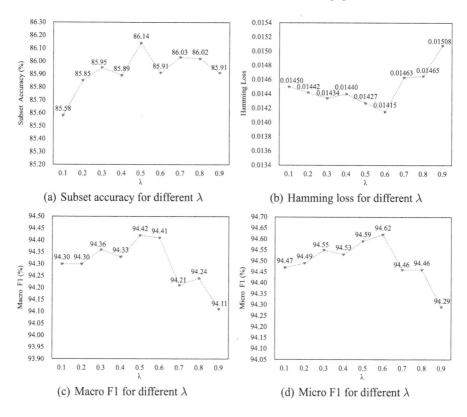

Fig. 5. Classification performance curves of the PageCNN-WLL model with different λ values.

on the current webpage classification result. The coefficient λ represents the weight of the external linked webpage label features in the calculation of the final classification result. The best practice of λ is obtained via experimentation, by successively increasing the λ value from 0.1 to 0.9 with a stride of 0.1 in the experiment. The classification performance curves of the PageCNN-WLL model in terms of subset accuracy, Hamming loss, macro F1 and micro F1 with different weight coefficient λ values are shown in Fig. 5. It can be seen from Fig. 5 that, when λ increases from 0.1 to 0.9, the classification performance curves basically show a trend of decreasing after increasing, while Hamming loss shows a trend of increasing after decreasing. When λ exceeds 0.5 or 0.6, the classification results of the model become worse. As presented in Fig. 5, when λ is 0.5, 0.6, 0.5, and 0.6, the PageCNN-WLL model obtains the best subset accuracy, Hamming loss, macro F1, and micro F1, respectively. When λ is 0.5 and 0.6, the differences between the corresponding Hamming loss, macro F1, and micro F1 values are not obvious, but the difference between the corresponding subset accuracy values is relatively larger. Moreover, since subset accuracy is the strictest metric to measure the performance of multi-label classification, and the subset accuracy curve appears a concave-point when λ is 0.6, it is more appropriate to set the λ value in the PageCNN-WLL model as 0.5. This experimental result indicates that the information extracted from

Table 4. Error Correction Effects of External Linked Webpages

Model	PageCNN	PageCNN-CLL		PageCNN-WLL			
Metric	Correction Rate against TextCNN	Correction Rate against PageCNN	Hit Rate of P_{link}	Correction Rate against PageCNN	Hit Rate of $F_{current}$	Hit Rate of F_{link}	Correction Rate of F_{link} against $F_{current}$
Score	2.27%	1.03%	21.15%	1.44%	90.86%	84.75%	3.23%

external linked webpages is basically equally important as the information extracted from current webpage.

Analysis of Error Correction Effects of External Linked Webpages. As discussed in Sect. 2, the PageCNN model improves the modified TextCNN model by introducing text information of external linked webpages into current webpage classification, and the PageCNN-CLL and PageCNN-WLL models further improves the PageCNN model by further incorporating label information of external linked webpages. In PageCNN-CLL and PageCNN-WLL models, P_{link}, F_{link} and $F_{current}$ are important preliminary prediction variables related to the classification improvements. Therefore, we will analyze the effectiveness of the text and label information of external linked webpages by analyzing the error correction effects of the proposed models and related preliminary prediction variables.

We first define a correction rate. The correction rate of Model A against Model B (or Variable A against Variable B) is defined as: $Correction\ Rate_{B|A} = \#\ of\ corrected\ labels/\#\ of\ all\ labels$, where $\#\ of\ corrected\ labels$ is the number of labels which are predicted wrong by Model B (or Variable B) but correctly predicted by Model A (or Variable A), and $\#\ of\ all\ labels$ is the total number of labels in the test set. Besides, to further analyze the effects of the preliminary prediction variables (P_{link}, F_{link} and $F_{current}$) on the classification results, we define a hit rate. The hit rate of the preliminary prediction variables is defined as: $Hit\ Rate = \#\ of\ hit\ labels/\#\ of\ all\ labels$, where $\#\ of\ hit\ labels$ is the number of labels which are correctly predicted by the preliminary prediction variable. It should be noted that the preliminary prediction results of F_{link} and $F_{current}$ are obtained by adding individual sigmoid activation operations to them. The correction rate of the proposed models and the hit rate of the preliminary prediction variables are presented in Table 4.

As seen from the error correction rates in Table 4, it is verified that the PageCNN model improves the modified TextCNN model, and the PageCNN-CLL and PageCNN-WLL models further improves the PageCNN model. Besides, it is illustrated in Table 4 that the hit rate of P_{link} is relatively low, while the hit rate of F_{link} is much higher. As discussed in Sect. 2, from the architectural point of view (comparing Fig. 3 and Fig. 4), adding a sigmoid operation to F_{link} makes it have the same functionality as P_{link}. Hence, adding a sigmoid operation to F_{link} is supposed to output the same results as P_{link}. However, the network parameters trained in PageCNN-CLL and PageCNN-WLL models are different. In the PageCNN-CLL model, P_{link} is concatenated to V as a part of the final feature vector V_{CLL}, which weakens the influence of P_{link} to the final prediction results. As a result, the network parameters for P_{link} need not train towards precise preliminary prediction results. But in the PageCNN-WLL model, the label information extracted from external linked webpages is basically equally important as the label information extracted from current webpage, as demonstrated in Sect. 3.3. Hence, the network parameters for F_{link} should be trained towards precise preliminary prediction results. This is why the hit rate of F_{link} is much higher than that of P_{link}. Furthermore, the hit rate of $F_{current}$ is higher than that of F_{link}, which indicates that the text information extracted from current webpage and external linked webpages is more relevant to the classification of current webpage than the text information extracted solely from external linked webpages. However, the text information extracted solely from external linked webpages is still very useful for the classification of current webpage, as illustrated in the correction rate of F_{link} against $F_{current}$.

4 Conclusion

In this work, the PageCNN model and its two variants (PageCNN-CLL and PageCNN-WLL) are proposed to effectively fuse the text and label information extracted from the current Chinese webpage and external linked Chinese webpages. The proposed PageCNN models are compared with two traditional machine learning models, the modified TextCNN model, and three state-of-the-art deep learning based multi-label text classification models. The experimental results demonstrate that the PageCNN models perform better than the compared models in terms of subset accuracy, Hamming loss, macro F1, and micro F1. By using a weight coefficient to adjust the influence of the external linked webpages on the classification of current webpage, the PageCNN-WLL model realizes better multi-information fusion for the current webpage classification than the PageCNN and PageCNN-CLL models. This work may be extended by further considering label ranking and label correlation in the proposed PageCNN models. Besides, the mechanism to remove the noise in the webpages such as advertising text and links may be helpful to the classification results. Moreover, the data in other modalities, such as images and videos on the webpages, may provide useful information for the multi-label Chinese webpage classification, so the integration of multi-modal information can be considered.

Acknowledgements. This work was supported by Guangdong Natural Science Foundation (2021A1515012651), National Natural Science Foundation of China (62076100), Science and Technology Planning Project of Guangdong Province (2020B0101100002), Fundamental Research Funds for the Central Universities (SCUT) (x2rjD2220050), and CAAI-Huawei Mind-Spore Open Fund.

References

1. Chen, G., Ye, D., Xing, Z., Chen, J., Cambria, E.: Ensemble application of convolutional and recurrent neural networks for multi-label text categorization. In: Proceedings of International Joint Conference on Neural Network, pp. 2377–2383 (2017)
2. Fang, L., Zhang, L., Wu, H., Xu, T., Zhou, D., Chen, E.: Patent2vec: Multi-view representation learning on patent-graphs for patent classification. World Wide Web **24**, 1791–1812 (2021)

3. Zou, J.-q., Chen, G.-l., Guo, W.-z.: Chinese web page classification using noise-tolerant support vector machines. In: Proceedings of International Conference on Natural Language Processing and Knowledge Engeering, pp. 785–790 (2005)
4. Liang, J.-z.: Chinese web page classification based on self-organizing mapping neural networks. In: Prodings of International Conference on Computer Intelligent and Multimedia Application, pp. 96–101 (2003)
5. Kim, Y.: Convolutional neural networks for sentence classification. In: Proceedings of Conference on Empirical Methods in Natural Language Process, pp. 1746–1751 (2014)
6. Kingma, D., Ba, J.: Adam: a method for stochastic optimization. In: Proceedings of International Conference on Learning Representations (2015)
7. Kurata, G., Xiang, B., Zhou, B.: Improved neural network-based multi-label classification with better initialization leveraging label co-occurrence. In: Proceeedings of the Conference of the North American Chapter of the Association for Computer Linguistics: Human Language Technology, pp. 521–526 (2016)
8. Li, S., Zhao, Z., Hu, R., Li, W., Liu, T., Du, X.: Analogical reasoning on Chinese morphological and semantic relations. In: Proceedings of the Annual Meeting of the Association for Computer Linguistics, vol. 2, pp. 138–143 (2018)
9. Liao, W., Wang, Y., Yin, Y., Zhang, X., Ma, P.: Improved sequence generation model for multi-label classification via CNN and initialized fully connection. Neurocomputing **382**, 188–195 (2020)
10. Lin, J., Su, Q., Yang, P., Ma, S., Sun, X.: Semantic-unit-based dilated convolution for multi-label text classification. In: Proceedings of the Conference on Empirical Methods in Natural Language Process, pp. 4554–4564 (2018)
11. Liu, J., Chang, W.C., Wu, Y., Yang, Y.: Deep learning for extreme multi-label text classification. In: Proceedings of the International ACM SIGIR Conference on Research and Development in Information Retrieval, pp. 115–124 (2017)
12. Ma, H., Li, Y., Ji, X., Han, J., Li, Z.: Mscoa: multi-step co-attention model for multi-label classification **7**, 109635–109645 (2019)
13. Nam, J., Loza Mencía, E., Kim, H.J., Fürnkranz, J.: Maximizing subset accuracy with recurrent neural networks in multi-label classification. In: Proceedings of Advance in Neural Information Processing System, pp. 5413–5423 (2017)
14. Sang, J., Wang, Y., Yuan, L., Li, H., Jiang, X.: Multi-label transfer learning via latent graph alignment. World Wide Web **25**, 879–898 (2022)
15. Wang, X., et al.: Research and implementation of a multi-label learning algorithm for Chinese text classification. In: Proceedings of International Conference on Big Data Computation and Communication, pp. 68–76 (2017)
16. Wu, Y., Pei, C., Ruan, C., Wang, R., Yang, Y., Zhang, Y.: Bayesian networks and chained classifiers based on svm for traditional chinese medical prescription generation. World Wide Web **25**, 1447–1468 (2022)
17. Xiao, L., Huang, X., Chen, B., Jing, L.: Label-specific document representation for multi-label text classification. In: Proceedings of Conference on Empirical Methods in Natural Language Processing International Joint Conference on Natural Language Process, pp. 466–475 (2019)
18. Yang, P., Luo, F., Ma, S., Lin, J., Sun, X.: A deep reinforced sequence-to-set model for multi-label classification. In: Proceedings of the Annual Meeting of the Association for Computer and Linguistics, pp. 5252–5258 (2019)
19. Yang, P., Sun, X., Li, W., Ma, S., Wu, W., Wang, H.: SGM: sequence generation model for multi-label classification. In: Proceedings of the International Conference on Computer Linguistics, pp. 3915–3926 (2018)
20. Yang, Z., Liu, G.: Hierarchical sequence-to-sequence model for multi-label text classification **7**, 153012–153020 (2019)

21. Yeh, C.K., Wu, W.C., Ko, W.J., Wang, Y.C.F.: Learning deep latent spaces for multi-label classification. Proc. AAAI Conf. Artif. Intell. **31**(1), 838–2844 (2017)
22. Zhang, M.L., Zhou, Z.H.: A review on multi-label learning algorithms **26**(8), 1819–1837 (2013)
23. Zhong, T., Liu, F., Zhou, F., Trajcevski, G., Zhang, K.: Motion based inference of social circles via self-attention and contextualized embedding **7**, 61934–61948 (2019)

MFF-Trans: Multi-level Feature Fusion Transformer for Fine-Grained Visual Classification

Qi Hang[(✉)], Xuefeng Yan, and Lina Gong

Nanjing University of Aeronautics and Astronautics, Nanjing, China
1161075573@qq.com

Abstract. In fine-grained visual classification, fusing both local and global information is crucial. However, current methods based on vision transformer tend to just focus on selecting discriminative patch tokens, which ignore the variation of rich global and semantic information in classification tokens at different layers. To address this limitation, we propose a novel framework dubbed MFF-Trans that considers the mutual relationships between all tokens. Specifically, we put forward the important token election module (ITEM) which utilizes multi-headed self-attention mechanism in vision transformer to evaluate the importance of all tokens. This module will guide the model to select tokens which contain discriminative local information and global information with different semantics at each ViT layer. Meanwhile, to enhance the model's perception of semantic connection between selected patch tokens, we further introduce the semantic connection enhancing module (SCEM) which use the graph convolutional network to mine the structural information between them in deep layers of vision transformer. Extensive experimental results on three benchmark datasets indicate that MFF-Trans achieves satisfactory performance compared with other methods. We achieve good results in CUB (92.1%), Stanford Cars (95.4%), and Stanford Dogs (92.3%).

Keywords: fine-grained visual classification · vision transformer · graph convolutional network

1 Introduction

In the past few years, neural networks based on convolutional neural network (CNN) have made profound achievements in visual classification. Many classical neural networks such as VGG [13], GoogLeNet [16], ResNet [6] have been proven to be effective for coarse-grained classification. However, due to small inter-class differences, large intra-class differences, and insufficient annotations, fine-grained categories cannot be effectively distinguished. Unlike traditional

This work is supported by the Basic Research for National Defense under Grant Nos. JCKY2020605C003.

coarse-grained visual classification, fine-grained visual classification (FGVC) refers to the grouping of subordinate categories under basic categories, such as the types of birds and dogs. In FGVC tasks, objects belonging to the same class often look different. On the contrary, objects of different classes may look similar to each other. For example, the images from the CUB-200-2011 dataset [18] are shown in Fig. 1, they belong to three subclasses of Cuckoo and almost look the same in shape and color. This highlights the demand for fine-grained visual classification models to discriminate tiny differences between them.   In

Black_Billed_Cuckoo   Mangrove_Cuckoo   Yellow_Billed_Cuckoo

Fig. 1. The images are selected from three sub-classes of Cuckoo, which have very similar shapes and colors.

addition, on account of the limitations of training data, annotating fine-grained datasets usually requires a great deal of expertise. Therefore, fine-grained visual classification is considered to be a very challenging task.

Up to date, thanks to tremendous well-designed networks, FGVC methods have gained steady progress in recent years. A number of works have revealed that the mining of discriminative regions is essential for processing FGVC tasks. Some early works [5,9] localized discriminative regions with the help of additional hand-marked annotations. Some other authors used RPN [12] or attention mechanism [7,15,23] to find discerning areas. However, the relationship between discriminant regions was often ignored by them which could result in a sharp decline in model performance.

Recently, the successful application of vision transformer (ViT) [2] in the field of image processing has promoted the development of FGVC methods. ViT was proposed by Google in 2020 to apply transformer [17] framework to image classification. Because of its simple structure and excellent performance, it has attracted many researchers' attention. The multi-headed self-attention (MSA) mechanism in ViT fuses the patch tokens' features (local information) into the classification token (global information), indicating its great potential in the field of FGVC. However, ViT cannot directly leverage its advantages over FGVC tasks because the fixed number of tokens per layer is not conducive to the network capturing critical regional concerns [4]. Therefore, the emerging ViT-based FGVC approaches try to filter important tokens from previous layers and regard them as the input of the final layer by conducting studies on the attention weights matrix in MSA. For example, in TransFG [4], the authors fuse all the attention weights matrixes in each layer and select discriminative tokens from the penultimate layer according to the relationship between the classification token and other tokens. In FFVT [19], the authors assert tokens in deep layers of ViT are more focused on high-level information. So they aggregate important tokens from each transformer layer rather than just one layer to compensate for low-level and inter-level features. In AFTrans [21], the authors also take advantage of the attention weights matrix in the ViT and adaptively filter tokens features with multi scales according to their relative importance. In SIM-Trans [14], the authors incorporate object structure information into ViT to strengthen the discriminative feature representation.

In the shallow layers of ViT, the patch tokens' features are focused on surface information of images such as color and texture. The classification (cls) tokens in these layers don't necessarily correlate highly with the label, so the attention weights between patch tokens and cls tokens may not effectively represent the deep-seated semantic connection between them. Therefore, relying on the cls token alone to filter important tokens is not comprehensive enough. Additionally, current ViT-based FGVC methods tend to ignore selecting cls tokens which contain global information with different semantics of various layers. In FGVC tasks, it is crucial to fuse both local and global information, especially in the early stages of model learning. Therefore, the cls token shouldn't solely determine which patch token to be selected. We ought to pay attention to the mutual relationships between all tokens, and let them choose the significant ones by themselves. To achieve this, we introduce the important token election module (ITEM), which utilizes the self-attention mechanism in MSA to evaluate the importance of all tokens. This module will guide the model to gather crucial patch and cls tokens to mine discriminative local information and global information with different semantics at each ViT layer.

As mentioned above, the semantic connection between the cls token and patch tokens has not been established well in the shallow layers of ViT. On the contrary, in the deep layers of ViT, such a relation becomes much clearer to the model. The attention weights between the cls token and patch tokens in deep layers are highly related to whether or not this token contains the object's information. Enhancing the model's perception of this relation is unquestionably effective. Therefore, we introduce the semantic connection enhancing module (SCEM) to mine the structural information between selected important patch tokens in deep layers. We first take the relative polar coordinates of selected patch tokens as positional information and add it to the token feature. We then use relative attention weights between them to build the adjacent matrix of a graph structure. Finally, we apply a two-layer graph convolution network on them to mine their semantic connection and inject it into the cls token for model improvement. We don't do this enhancement at every layer of ViT. Because if we carry out this enhancement too early, the model will not be able to perceive it. While applying it too late, the model will get mostly redundant information, which also does not have much effect on the performance. The beginning layer of executing the enhancement is 9 in this paper. Specific ablation experiments are shown in Fig. 4.

To fully verify the validity of our approach, we explored the performance of the model on three widely used FGVC datasets. Extensive experiments have shown that MFF-Trans achieves good results compared with previous ViT-based FGVC methods.

Our main contributions can be summarized as follows:

- We explore a novel ViT-based framework dubbed MFF-Trans for FGVC tasks, which can automatically identify discriminant regions and enhance their semantic connection to improve the model performance.
- We propose a novel token selection approach named important token election module (ITEM), which aims to select important tokens for fusing discriminative local features and global information with different semantics in each layer of the ViT. We also introduce the semantic connection enhancing module (SCEM) to enhance the semantic connection between selected patch tokens, which will improve the model's understanding ability of semantic comprehension. The proposed two modules are lightly weighted and can be easily plugged into any ViT-based method.
- We conduct extensive experiments on three widely used FGVC datasets to validate the effectiveness of the proposed method. The results have shown that our method can achieve satisfactory performance with previous ViT-based methods for fine-grained visual classification. We achieve good results in CUB (92.1%), Stanford Cars (95.4%), and Stanford Dogs (92.3%).

2 Related Works

2.1 CNN-Based FGVC Methods

CNN-based FGVC methods can be divided into the following two parts, location-identification sub-networks, and attention mechanism. For the first kind of method, some works [5,9] mainly used existing partial annotations of discriminant parts of the object to capture visual details. WSCPM [3] used region proposal networks (RPN) to raise candidate bounding boxes, which contain different regions with different sizes. After selecting the informative regions of the image, the cropped images were discriminative a predefined size and feed into the encoder network to obtain local features. Mask-cnn [20] uses FCN to localize key components and generate masks. However, collecting the annotations recorded in txt files is labourious. Moreover, it requires skilled experts and a lot of cost. Therefore, using such methods is impractical for real world applications. For the second kind of method, MAMC [15] and CAL [11] used an attention mechanism for the rough location of objects. Then, a particular position of the feature map was extracted by the attention maps. In addition to learning location information, attention mechanisms could also improve the expressiveness of features. However, it is easy to overfit in the case of small-scale datasets.

2.2   ViT-Based FGVC Methods

In recent years, many ViT-based methods have been proposed for computer vision tasks. They were widely used in many fields and had made remarkable achievements. ViT [2] was the first work to introduce a pure transformer into image classification. It first divided the image into a series of flat patches and mapped them into tokens, and then continuously calculated the relationship between them to obtain the description of the image. TransFG [4] designed a token selection module to select discriminant tokens. FFVT [19] aggregates important tokens from every layer of the transformer to generate multi-level features. RAMS-Trans [8] learned to resolve regional attention in a multi-scale way. AFTrans [21] also takes advantage of the attention weights matrix in the ViT and adaptively filter tokens features with multi scales according to their relative importance. SIM-Trans [14] incorporated object structure information into the transformer to strengthen the discriminative feature representation. Although these ViT-based methods have achieved good performance, most of them focused on mining discriminative regions while ignoring the semantic relations between image patches. Different from their methods, we propose a new ViT-based model for FGVC named MFF-Trans, which considers both the identification of discriminant regions and the enhancement of the semantic relations between image patches.

3   Proposed Method

An overview of our network is shown in Fig. 2. The vision transformer encoder is in the middle, we split the image into a series of patch tokens and put them into it for feature extraction. The proposed important token election module (ITEM) is on the right side, which makes the use of the self-attention mechanism in MSA to collect important tokens of each ViT layer. The introduced semantic connection enhancing module (SCEM) is on the left side, which enhances semantic connection of selected patch tokens to improve the model performance. In the following subsections, we first give a brief description of the vision transformer encoder and then show the details of our proposed modules.

3.1   Vision Transformer Encoder

In the field of NLP, methods based on transformer [17] usually treat each word of a sentence as an element and input them into linear projection to generate tokens. However, an image is different from a sentence because it is composed of pixels. If treating pixels in a picture as words and using a method similar to NLP, the sequence length of a 224×224 image will be 50176. This number will far exceed the bearing capacity of the transformer. Therefore, based on this problem, ViT [2] proposes a strategy of image processing that divides the image into small patches rather than pixels to reduce the number. Especially, given an image $I \in R^{H \times W \times 3}$ and a partition parameter P, ViT first reshapes I into

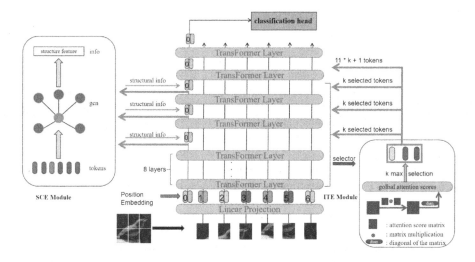

Fig. 2. The overall framework of MFF-Trans. An image is split into patches and sent to the ViT encoder for feature extraction. The important token election module is used to select important tokens for fusing discriminative local features and global features with different-level semantics. The semantic connection enhancing module is used to benefit the semantic comprehension of the model.

N small patches $I_p \in R^{P \times P \times 3}$, where $N = \lfloor \frac{H}{P} \rfloor * \lfloor \frac{W}{P} \rfloor$, H and W represent the height and width of I and 3 is the number of channels in the original RGB image. The patch I_p is then linearly projected into the D-dimensional embedding space through operation $LP(\cdot)$. To keep the same paradigm with previous nlp methods, a cls token is put at the beginning of the input sequence. To cover position influences, a learnable position embedding $E_{pos} \in R^{(N+1) \times D}$ is added to the patch embeddings. The formula is shown in Eq. 1 where z_0 is the input of the first transformer layer.

$$z_0 = [cls; LP(I_{p1}); LP(I_{p2}); ...; LP(I_{pN})] + E_{pos} \tag{1}$$

The transformer encoder consists of L transformer layers. Each layer contains a multi-head self-attention (MSA) module and a feed-forward network (FFN) module. The outputs of the m_{th} layer are expressed by 2 and 3, where $LN(\cdot)$ represents the layer normalization.

$$z_m = MSA(LN(z_{m-1})) + z_{m-1} \tag{2}$$

$$z_m = FFN(LN(z_m)) + z_m \tag{3}$$

Finally, the classification token in the final layer z_f^0 is fed into a multi-layer perceptron to predict the image's category.

3.2   Important Token Election Module

As mentioned above, the most important challenge in FGVC is precisely recognizing the discriminative regions that help to explain subtle differences between similar subordinate categories. Among many CNN-based FGVC approaches, they usually accomplish this requirement by making use of manual annotations, region proposal networks, or the attention mechanism. However, they consume a lot of labor and redundant calculation.

The multi-head self-attention mechanism in ViT naturally meets the above requirement well for it fuses the patch tokens' features (local information) into the classification token (global information). The attention weights matrix in each layer of ViT is a square matrix of order $N + 1$. The element at a particular position (i, j) in it represents the attention weight between the ith token and jth token from the perspective of the ith token. In addition, the classification token aggregates information from other tokens, which can be regarded as a description of the whole image. Some ViT-based methods take advantage of this property and replace the raw input before the final layer of ViT. TransFG [4] first generates a weights matrix by multiplying the attention weights matrix of all layers to integrate all the information. Then, the first row of the weights matrix, which represents the correlation between the cls token and the other tokens from the perspective of the cls token, is extracted and the tokens with the higher value are selected. FFVT [19] not only concerns the first row of the weights matrix, but also pays attention to the first column of the weights matrix, which represents the above relations from the perspective of other tokens. It applies the hadamard product on the two extracted vectors and sorts it as the basis for selecting important token. In addition, TransFG selects tokens from only one layer while FFVT from all layers.

The shallow layers of ViT extract relatively simple feature information such as color and texture. With the deepening of the layer, ViT will extract more complex feature information like semantic relations between patches tokens and cls tokens, where it can make a more accurate judgment. Therefore, just using the relationship between the cls token and other tokens in shallow layers of ViT is a lack of consideration. Moreover, cls tokens in different layers fuse different levels of local information, which is also important for the model. Since the great importance of cls tokens from different layers, then why not select them as the input of the final layer? A simple way is to regard all the cls tokens from previous layers as the input of the final layer of ViT. However, due to the fact that cls tokens are more focused on global information, ignoring other tokens will lead to the lack of a local description of the object, which is important in FGVC tasks. Another straightforward idea is to use the first column or the first row of the attention weights matrix as the judge for selecting important tokens(including the cls token). But the value in position $(0, 0)$ of the weights matrix represents the self-correlation of the cls token, which gives it an unfair role compared with other tokens. Therefore, inspired by FFVT [19], we also identify the importance of tokens from their mutual perspective of them. Specifically, assume that the model has N_h self-attention heads, the query Q and key K of all tokens are

D-dimensional vectors, then the attention weights matrix can be calculated by Eq. 4.

$$A_M = softmax(\frac{QK^T}{\sqrt{D/N_h}}) \tag{4}$$

$$A_h = \frac{1}{N_h} \sum_{M=1}^{N_h} A_M \tag{5}$$

As is shown in Eq. 5, we then average the attention weights matrix of all heads per layer to get A_h. The details of averaged attention weights matrix $A_h \in R^{(N+1)\times(N+1)}$ are shown in Eqs. 6 and 7:

$$A_h = [b^0; b^1; b^2; \ldots; b^i; \ldots; b^N] \tag{6}$$

$$b_i = [b_{i,0}, b_{i,1}, b_{i,2}, \ldots, b_{i,j}, \ldots, b_{i,N}] \tag{7}$$

where $b_{i,j}$ represents the relations between token i and token j, which is calculated by token i's query and token j's key. We multiply it by itself and get the attention map A_{map}. As shown in Eq. 8, we multiply it by itself, and the ith diagonal element of the calculated matrix represents the correlation between token i and all tokens from mutual views of each other. Its value of it can determine whether the corresponding token is important or not. We select K tokens of each layer for the input of the final layer.

$$\begin{pmatrix} b_{00} & \ldots & b_{0N} \\ \ldots & b_{ii} & \ldots \\ b_{N0} & \ldots & b_{NN} \end{pmatrix} \begin{pmatrix} b_{00} & \ldots & b_{0N} \\ \ldots & b_{ii} & \ldots \\ b_{N0} & \ldots & b_{NN} \end{pmatrix} = \begin{pmatrix} x_0 & \ldots & \ldots \\ \ldots & x_i & \ldots \\ \ldots & \ldots & x_N \end{pmatrix} \tag{8}$$
$$x_i = b_{i0}b_{0i} + \ldots + b_{iN}b_{Ni} \qquad i = 0 \ldots N$$

After that, we intercepts all the elements of the diagonal of the matrix as $g \in R^{N+1}$, which is shown in Eq. 9. We then sort them and selects default number of tokens with higher value.

$$g = diag(A_h A_h) = [g_0, g_1, g_2, \ldots, g_i, \ldots, g_N] \tag{9}$$

The process is shown in the right part of Fig 2. By applying this module, important tokens corresponding to discriminative regions will be picked out and the description of pictorial representation information with different semantics in all layers will be supplemented.

3.3   Semantic Connection Enhancing Module

In the methods based on the ViT encoder, a token can be seen as a description of an image patch. The selected important tokens are usually encoded from the image patches to which key areas (such as the head or the claw of a bird) of the object belong. Most of the ViT-based FGVC methods focus on mining discriminative regions. However, they ignore the spatial semantic relations between them, which is also important for the model's performance. Although the semantic relations of patch tokens in deep layers of ViT is much clearer to cls token than in shallow layers, we still can enhance it with some physical positional information. Therefore, inspired by SIM-Trans [14], we introduce a semantic connection enhancing module to help the model capture their interaction information of them.

Before exploring the spatial semantic relations between discriminative regions of the object, we need to get the important patch tokens. Thanks to the previously proposed token selection approach, they are on standby. We will establish a graph structure for the selected tokens of deep layers of ViT to describe their relations and further use a graph convolution network to study their semantic connection. Specially, we choose the polar coordinates to measure the spatial relationship between the selected tokens from the same layer to excavate their structural information. Among them, we set the patch token with the highest attention score in g as the reference patch P_{ref}. If the reference patch token is the cls token, we will change it into the patch token with the second-highest attention score. This is because the cls token is not corresponding to a certain patch in the raw image, which lacks the information of physical coordinates.

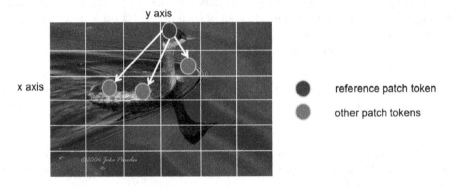

Fig. 3. The description of calculating relative polar coordinates. We set the coordinates of the reference patch as the base coordinates and calculate other important tokens' relative coordinates. The positional information will be added to the token feature.

Let's say the coordinates of reference patch P_{ref} is (x_{ref}, y_{ref}), and the polar coordinates of patch $P_{x,y}$ are calculated by Eq. 10, where P is the patch size, H and W are the height and width of the input image, $\rho_{x,y}$ is the relative distance between P_{ref} and $P_{x,y}$ and $\theta_{x,y}$ is their relative direction. We add the relative position information into the token feature to get the contextual correlation patch feature s_p.

$$\rho_{x,y} = \sqrt{\left(\frac{x-x_{ref}}{W/P}\right)^2 + \left(\frac{y-y_{ref}}{H/P}\right)^2}$$
$$\theta_{x,y} = \frac{(\arctan 2(y-y_{ref}, x-x_{ref}) + \pi)}{2\pi} \tag{10}$$

To cover the above relations in the model, we decide to use the graph convolutional network to encode a structural feature. The input of GCN has two parts, one is the token features containing the information of calculated relative positions above, and the other is the adjacent matrix of a graph. To satisfy the network second input requirement, we determine to use the attention weight between the patch token and the cls token. Specially, we first remove the element of b_0 at position 0 to get $b_{attn} \in R^{N \times 1}$ because we will ignore cls token when calculating important tokens' relations. The designed constant number which represents the number of selected tokens of each layer is $N_{selected}$. We sort g in Eq. 9 from largest to smallest to get a guiding vector g_{sorted}. We choose the $N_{selected}th$ element named as g_{TH} of the vector and treat it as a threshold to filter out unimportant tokens' attention scores in b_{attn}.

$$b_{attn} = \begin{cases} b_{attn(x,y)}, & g_{sorted(x,y)} >= g_{TH} \\ 0, & otherwise \end{cases} \quad (11)$$

Then we reshape the dimensions of b_{attn} and g_{sorted} into $R^{H/P \times W/P}$. After that, as shown in Eq. 11, We keep elements that are greater than or equal to the threshold g_{TH} and set those that are less than it to 0, where (x, y) on behalf of (x, y) position (shown in Fig. 3) in the raw image. Finally, we restore the dimension of $b_{attn} \in R^{N \times 1}$, the adjacent matrix $adj \in R^{N \times N}$ can be computed by $b_{attn} \times b_{attn}{}^T$. After all the preparatory work is done, we use a two-layer GCN to mine the semantic relations between important tokens and inject it into the ViT encoder. The process of obtaining structural feature s_f is shown in 12, where $relu$ is the activation function, W_1 and W_2 are learnable parameters.

$$s_f = relu(adj \times relu(adj \times s_p \times W_1) \times W_2) \quad (12)$$

The reference patch P_{ref}'s feature is then taken as the description of semantic relations and fused into the cls token's feature. The semantic relations between discriminative regions of the object will be clearer to the model during the training process, which definitely improve its performance.

All in all, we summarize the operation flow of our proposed model. Firstly, we apply a series of operations in Sect. 3.1 to get z_0. Supposing ViT has L transformer layers, we then apply the global attention selection module to select important tokens of the first $bl - 1$ layers. After that, starting at the blth layer, we not only select important tokens but also use a semantic connection enhancing module to enhance the semantic connection information of them to the cls token. Further, we gather these important tokens selected from previous layers and concatenate the cls token from the penultimate layer as the final layer input. Finally, we send the cls token from the final layer to the classification head for category prediction.

4   Experiments

In this section, we analyze the performance of our method MFF-Trans on three widely used fine-grained visual categorization datasets CUB-200-2011 (CUB),

Stanford Cars (CAR), and Stanford Dogs (DOG). We will first introduce the datasets and details of experimental hyperparameters. Then, we compare our method with some advanced methods according to classification accuracy in benchmark datasets. Finally, we conduct ablation studies and analysis on the specific proposed modules and hyperparameters.

4.1 DataSets and Implement Details

In this paper, we use three widely accepted fine-grained visual classification datasets. They are CUB-200-2011 (CUB), Stanford Cars (CAR), and Stanford Dogs (DOG). Cub is proposed by the California Institute of Technology in 2010, and is also the benchmark image dataset for current fine-grained classification research. It contains 11788 bird images including 5994 training images and 5794 testing images. It has 200 bird subcategories and each of which provides image labeling, a bounding box, and key part information about birds. The Stanford Cars dataset has 16185 images of different types of cars. The Stanford Dogs dataset contains images of 120 dog species from around the world. This dataset is built using images and annotations in ImageNet for fine-grained image classification tasks. The details are shown in Table 1.

Table 1. The details of three benchmark datasets. There are three distinct datasets whose meta categories are bird, car, and dog.

DataSet	Meta Category	Number of Subcategories	Total Images
CUB-200-2011	Bird	200	11788
Standford Cars	Car	196	16185
Standford Dogs	Dog	120	20580

In our experiments, the encoder ViT-B/16 is pre-trained on the ImageNet21K dataset, which is the same as most of the ViT-based FGVC methods. We resize the input image into 600×600 and randomly cropped it into 448×448 before feeding it to the network. The bl which represents the beginning layer of applying SCEM is 9. The number of attention heads K is 12 for CUB and 16 for DOG. We adopt the stochastic gradient descent (SGD) optimizer with a momentum of 0.9 to optimize the model. The learning rate is initialized as 0.04 and is updated by the cosine annealing schedule. The batch size is set as 16 for CUB and 8 for DOG and CAR. We train the model 150 epochs for CUB and 200 epochs for DOG and CAR. Experiments are conducted on one RTX 3090 (24 GB) with PyTorch 1.8.1 version.

4.2 Comparisons with Advanced Methods

We compare our methods with some advanced methods on the three widely used datasets mentioned above. The results are shown in Table 2. The first column of the table represents the method name, the second column represents the backbone used in the method and the third and fifth column refers to the performance of the method with top-1 classification accuracy metric on the certain dataset. The character string "(*)" in the table means the authors of the original article don't do the experiment corresponding to the specific dataset while we conduct it locally and report their results. From the Table 2, we can see that in CUB dataset, our method achieve good results compared with other ViT-based methods, especially using the ViT-B/16 backbone.

Table 2. Comparison experiments with advanced methods on CUB, CAR and DOG.

Method	backbone	Accuracy (CUB)	Accuracy (CAR)	Accuracy (DOG)
API-Net [23]	DenseNet161	90.0	95.3	90.3
PIM [1]	Swin-Trans	92.8	88.6(*)	89.1(*)
TransFG [4]	ViT-B/16	91.7	94.1	90.6
FFVT [19]	ViT-B/16	91.6	93.5(*)	91.5
DCAL [22]	ViT/DeiT	92.0	95.3	91.1(*)
SIM-Trans [14]	ViT-B/16	91.8	92.6(*)	91.4
AF-Trans [21]	ViT-B/16	91.5	95.0	91.6
TPSKG [10]	ViT-B/16	91.3	94.7	92.5
RAMS-Trans [8]	ViT-B/16	91.3	94.2(*)	92.4
MFF-Trans	ViT-B/16	92.1	95.4	92.3

In the Car dataset, we achieve the best results among ViT-based methods, which definitely illustrates the effectiveness of our method. In the DOG dataset, our method achieves better results than some ViT-based methods such as FFVT and AF-Trans, and approximate performance with TPSKG and RAMS-Trans.

4.3 Ablation Studies

In this paper, we propose two light weighted and effective modules. We first conduct the effectiveness analysis on them in three datasets to figure out the importance of them to the model performance.

Table 3. The ablation studies on two proposed modules.

Method	Accuracy (CUB)	Accuracy (CAR)	Accuracy (DOG)
ViT	90.4	93.5	91.2
ViT + ITEM	91.7	94.8	91.9
ViT + ITEM + SCEM	92.1	95.4	92.3

As shown in Table 3, take the results of CUB as an example, we can see that our ITEM gets 1.3% improvement over baseline accuracy. Since SCEM needs to cooperate with the token selection module, so we don't test it separately. We can see that SCEM works well with the first module, which further improves the model accuracy with a margin of 0.4%.

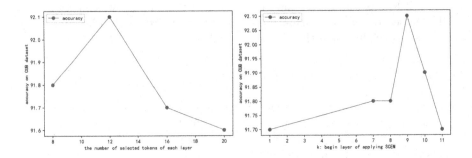

Fig. 4. The ablation studies on bl and k.

We then conduct ablation studies on the two hyperparameters bl and K, which represent the beginning layer where apply SCEM and the number of tokens selected for each layer. Since we use ViT-B/16 as the backbone which has 12 layers, our SCEM module is only possibly applied in the first 11 layers for enhancing the semantic relations of important tokens. As shown in the left part of Fig. 4, we can see that the best choice of bl is 9. It is because, in the shallow layers of ViT, semantic relations between patch tokens have not been established well, so the enhancement of semantic associations is not well perceived by the model. While in the deep layers of ViT, the enhancement of it may provide some redundant information, which is not that effective for the model.

We also evaluate the influences of K on our model. As shown in the right part of Fig. 4, we can see that when K is 12, the model achieves the best accuracy. The performance degradation is due to the fusion of noise tokens with a large K value, while a small K value leads to the loss of discriminant local information for classification.

5 Conclusion

In this paper, we propose a novel ViT-based model dubbed MFF-Trans for fine-grained visual classification. We first filter important tokens (including cls tokens) according to the mutual relationships between tokens to enrich the local and global information with different semantics to the model. We then mine the semantic relation of selected patch tokens in deep layers of ViT to enhance the model ability of semantic comprehension. We conduct experiments on benchmark datasets with good results, which indicate the effectiveness of our model. We also conduct elaborate ablations on our proposed modules and hyperparameters. According to the results, we believe the ViT-based methods have a great future on FGVC tasks.

References

1. Chou, P.Y., Lin, C.H., Kao, W.C.: A novel plug-in module for fine-grained visual classification. arXiv e-prints pp. arXiv–2202 (2022)
2. Dosovitskiy, A., et al.: An image is worth 16x16 words: transformers for image recognition at scale. In: International Conference on Learning Representations (2021)
3. Ge, W., Lin, X., Yu, Y.: Weakly supervised complementary parts models for fine-grained image classification from the bottom up. In: IEEE Conference on Computer Vision & Pattern Recognition (2019)
4. He, J., et al.: TransFG: a transformer architecture for fine-grained recognition. In: Proceedings of the AAAI Conference on Artificial Intelligence, vol. 36, pp. 852–860 (2022)
5. He, K., Gkioxari, G., Dollár, P., Girshick, R.: Mask R-CNN. IEEE Trans. Pattern Anal. Mach. Intell. **42**, 386–397 (2017)
6. He, K., Zhang, X., Ren, S., Sun, J.: Deep residual learning for image recognition. In: Proceedings of the IEEE Conference on Computer Vision and Pattern Recognition, pp. 770–778 (2016)
7. Hu, T., Qi, H., Huang, Q., Lu, Y.: See better before looking closer: weakly supervised data augmentation network for fine-grained visual classification. arXiv preprint arXiv:1901.09891 (2019)
8. Hu, Y., et al.: RAMS-TRANS: recurrent attention multi-scale transformer for fine-grained image recognition. In: Proceedings of the 29th ACM International Conference on Multimedia, pp. 4239–4248 (2021)
9. Liu, C., Xie, H., Zha, Z.J., Ma, L., Zhang, Y.: Filtration and distillation: enhancing region attention for fine-grained visual categorization. In: Proceedings of the AAAI Conference on Artificial Intelligence, vol. 34, no. 7, pp. 11555–11562 (2020)
10. Liu, X., Wang, L., Han, X.: Transformer with peak suppression and knowledge guidance for fine-grained image recognition. Neurocomputing **492**, 137–149 (2022)
11. Rao, Y., Chen, G., Lu, J., Zhou, J.: Counterfactual attention learning for fine-grained visual categorization and re-identification. In: 2021 IEEE/CVF International Conference on Computer Vision (ICCV), pp. 1005–1014. IEEE Computer Society (2021)
12. Ren, S., He, K., Girshick, R., Sun, J.: Faster R-CNN: towards real-time object detection with region proposal networks. IEEE Trans. Pattern Anal. Mach. Intell. **39**(6), 1137–1149 (2017)
13. Simonyan, K., Zisserman, A.: Very deep convolutional networks for large-scale image recognition. Computer Science (2014)
14. Sun, H., He, X., Peng, Y.: SIM-Trans: structure information modeling transformer for fine-grained visual categorization. In: Proceedings of the 30th ACM International Conference on Multimedia, pp. 5853–5861 (2022)
15. Sun, M., Yuan, Y., Zhou, F., Ding, E.: Multi-attention multi-class constraint for fine-grained image recognition. In: Ferrari, V., Hebert, M., Sminchisescu, C., Weiss, Y. (eds.) ECCV 2018. LNCS, vol. 11220, pp. 834–850. Springer, Cham (2018). https://doi.org/10.1007/978-3-030-01270-0_49
16. Szegedy, C., Liu, W., Jia, Y., Sermanet, P., Rabinovich, A.: Going deeper with convolutions. IEEE Computer Society (2014)
17. Vaswani, A., et al.: Attention is all you need. Adv. Neural Inf. Process. Syst. **30** (2017)

18. Wah, C., Branson, S., Welinder, P., Perona, P., Belongie, S.: The Caltech-UCSD birds-200-2011 dataset. California Institute of Technology (2011)
19. Wang, J., Yu, X., Gao, Y.: Feature fusion vision transformer for fine-grained visual categorization. In: BMVC 2021 (2021)
20. Wei, X.S., Xie, C.W., Wu, J.: Mask-CNN: localizing parts and selecting descriptors for fine-grained image recognition. arXiv preprint arXiv:1605.06878 (2016)
21. Zhang, Y., et al.: A free lunch from ViT: adaptive attention multi-scale fusion transformer for fine-grained visual recognition. In: ICASSP 2022-2022 IEEE International Conference on Acoustics, Speech and Signal Processing (ICASSP), pp. 3234–3238. IEEE (2022)
22. Zhu, H., Ke, W., Li, D., Liu, J., Tian, L., Shan, Y.: Dual cross-attention learning for fine-grained visual categorization and object re-identification. In: Proceedings of the IEEE/CVF Conference on Computer Vision and Pattern Recognition, pp. 4692–4702 (2022)
23. Zhuang, P., Wang, Y., Qiao, Y.: Learning attentive pairwise interaction for fine-grained classification. In: Proceedings of the AAAI Conference on Artificial Intelligence, vol. 34, pp. 13130–13137 (2020)

Summarizing Doctor's Diagnoses and Suggestions from Medical Dialogues

Tianbao Zhang[1], Yuan Cui[2], Zhenfei Yang[1], Shi Feng[1(✉)], and Daling Wang[1]

[1] Northeastern University, Shenyang, China
2001851@stu.neu.edu.cn, {fengshi,wangdaling}@cse.neu.edu.cn
[2] Jeonbuk National University, Jeonju, South Korea

Abstract. Nowadays, doctors can provide consultation services to patients by dialogues on the online medical platforms, and need to summarize their diagnoses and suggestions according to the regulations of the platform, which will play an important guiding role in the follow-up treatments. The essential challenges of automatic summarization lie in the high overlap between summaries and doctors' original utterances and the adaption to the specific structure of medical dialogues, which are overlooked by the majority of existing work. In response to this problem, we propose a pointer generator network model, dubbed as PMDS, to generate accurate and concise summaries for doctors' diagnoses and suggestions. PMDS takes the pointer generator network as the basic architecture and uses multi-level enhanced input feature representation, the latter of which helps to effectively distinguish speakers and better focus on key information. We evaluate our proposed model on a Chinese medical dialogue summarization dataset, and the experimental results exceeded several strong baselines in previous studies.

Keywords: Medical dialogue summarization · PMDS · Pointer generator network · RoBERTa · Speaker-level embedding

1 Introduction

The safety and convenience of online medical consultation have become increasingly prominent under the COVID-19 pandemic [11]. Through video, audio or text ways of communication [5,25], patients can elaborate their illness, and then doctors can provide diagnoses and suggestions. After each dialogue, some online platforms require the doctor to write summaries for key information, including disease diagnoses and treatment suggestions [12,14,18]. These summaries not only give important medical advice for the current patients, but also provide powerful references for the follow-up treatments and patients with the same diseases. However, due to the lengthy and highly professional characteristics of medical dialogues, the summarization is a repetitive and onerous task for doctors [32,35], and greatly reduces the efficiency of online medical service.

X. Song et al. (Eds.): APWeb-WAIM 2023, LNCS 14333, pp. 235–249, 2024.
https://doi.org/10.1007/978-981-97-2387-4_16

Chinese original texts	English translation
患者：如何治疗胃窦炎？（男，31 岁）	PT: How to treat gastric sinusitis?(male, 31)
医生：现在有什么症状？	DR: What are the symptoms now?
患者：早上总是胃痛。	PT: Always have a stomachache in the morning.
医生：有恶心或呕吐吗？	DR: Any nausea or vomiting?
患者：没有恶心，但总是烧心。	PT: No nausea, but always heartburn.
医生：之前吃过什么药吗？	DR: Have you taken any medicine before?
患者：泮托拉唑吃了一周。	PT: I took pantoprazole for a week.
患者：烧心减轻了，但还是胃痛。	PT: Less heartburn, but still stomachache.
医生：好的，那我给个方案：奥美拉唑肠溶片服用两周，西沙必利片服用一周。	DR: *Alright, I'll give you a plan: omeprazole enteric-coated tablet for two weeks and cisapride tablet for a week.*
患者：还有其他需要注意的吗？	PT: Is there anything else to note?
医生：忌烟酒和辛辣油腻生冷的食物。	DR: *Avoid tobacco, alcohol, spicy, greasy, raw and cold food.*
患者：好的，谢谢大夫。	PT: Okay, thank you, doctor.
医生：不客气，希望你早日康复。	DR: You're welcome, hope you get well soo
诊疗建议摘要 奥美拉唑肠溶片服用两周，西沙必利片服用一周。 忌烟酒和刺激性食物。	**Diagnose and suggestion summary** Omeprazole enteric-coated tablet for two weeks and cisapride tablet for a week. Avoid tobacco, alcohol and stimulating food.

Fig. 1. An example of medical dialogue summarization. The English example is translated from Chinese to illustrate the details and is not part of the original dataset. PT represents the patient, and DR represents the doctor. Italics in the dialogue represent utterances that have the value of generating summaries.

Compared with text summarization and other dialogue summarization tasks, medical dialogue summarization has unique characteristics and challenges. Firstly, as shown in Fig. 1, the summaries usually copy from doctors' original utterances, and should be more in line with the facts rather than pursue creativity [10]. Meanwhile, valuable medical terms need not be briefly summarized, but should be completely retained. Although summaries highly overlap with original dialogues, extractive methods are not completely appropriate because the extracted sentences may contain redundant and non critical information. Secondly, both patients and doctors can enter multiple utterances before another speaker responds, so the model should distinguish whether the utterances are said by patients or by doctors. Finally, the key information is scattered in dialogues, and not all the utterances of doctors have the value of generating summaries. The meaningless part includes questions and answers unrelated to the diagnosis and treatment suggestions, as well as polite words such as greetings and thanks. Thus, the model should have the ability to identify specific utterance semantics to focus on valuable information. Pointer generator network [31] takes into account both replication and generation, but the traditional text summarization model does not well adapt to the structure of two speakers and the fragmentation of key information in medical dialogues.

In order to tackle these challenges, we propose an improved pointer generator network, **P**ointer **M**edical **D**ialogue **S**ummarization network (PMDS), to generate diagnosis and suggestion summaries from medical dialogues. In order to make our model adapt to the structure and specific characteristics of medical dialogues, we propose three major improvements.

First, in order to cope with the high duplicability, we employ the pointer generator network [31] as the basic architecture, because it can selectively copy from original texts and retain the ability of abstractive generation. In order to prevent the repeated generation of the same words, we introduce the intra-attention mechanism [29] instead of the original coverage loss mechanism [31].

Second, PMDS sets a speaker embedding vector for the roles of patient and doctor to distinguish the speakers of utterances. PMDS adds the speaker embedding vector and the token embedding vector directly and inputs the composite embedding vector to the encoder.

Third, we leverage the pre-trained language model RoBERTa [7] to identify valuable utterances that contain key information. We input each utterance into RoBERTa separately and take the output of the position of [CLS] as the semantic representation of the corresponding utterance. Each token is attached with the semantic representation vector of its sentence to participate in attention calculation. Attention calculation takes into account the hidden state and sentence semantics, so that the model can focus on the key information.

2   Related Work

Text summarization is an important research issue in the field of natural language processing , which is designed to convert text into a short summary containing key information and is an important way to give the reader a quick overview of the main idea and key messages of the text [28]. According to the number of input documents, text summarization can be divided into single-document summarization and multi-document summarization. Single-document summarization generate a summary from a given document, while multi-document summarization generate a comprehensive summary from a given set of subject-related documents [8,19]. According to the classification of the method of generating summaries, text summarization can be divided into extractive summarization and generative summarization. Extractive summarization consist of key sentences and keywords extracted from the source document, so the summarization text is all derived from the original text. Generative summarization, on the other hand, is based on the original text and allow the use of new words that do not exist in the original text to generate the summary. The early works are mainly extractive summarization methods [4,26,27], which extract important sentences from original documents, but extractive methods may result in poor flexibility and readability. Later studies mostly focus on abstractive summarization methods [3,6,31], that is, documents are summarized into new sentences. Abstractive summaries are more flexible, but they tend to contain factual errors. Recent work has introduced large-scale pre-training language models [17,21,36] into summarization research.

Dialogue summarization is a special case of text summarization, but unlike the input data of text summarization, it is oriented to dialogue data in multiple fields. However, the basic purpose of dialogue summarization and text summarization is the same, both are to capture the key information in the dialogue and help readers quickly understand the core content of the dialogue. Dialogue summarization is characterized by scattered key information, low information density and multiple speakers, so it is not entirely appropriate to apply the original text summarization method directly to the conversation summarization problem [23]. There have been many deep learning methods for dialogue summarization in different fields [10], such as meeting summarization [15,39], chat summarization [2,38], email threads summarization [13,37] and customer service summarization [40]. Aiming at medical dialogue summarization, [22] set topics according to symptoms and designed a topic-level attention. [12] focused on the negative words. These two methods require additional manual annotations. [16] attempts to generate long semi-structured SOAP summaries in a variety of pipeline ways. However, the two-stage method makes the model optimization difficult.

Different from previous studies, aiming at summarizing doctors' diagnoses and suggestions, our model integrates pointer mechanism, speaker embedding and utterance semantics to adapt to scenes of medical dialogue summarization.

3   Model

As shown in Fig. 2, Our PMDS model is based on the pointer generator network architecture [31]. The encoder is a single-layer bidirectional LSTM, and the decoder is a single-layer unidirectional LSTM. Its goal is to maximize the conditional probability of the summary Y given the medical dialogue X and the network parameters θ: $P(Y|X, \theta)$.

3.1   Pointer Generator Network as Backbone

We use the pointer generator network [31] as the basic model, which uses a probability variable P_{copy} to adjust between replication and generation.

Our model uses the intra-attention [29] to replace the coverage loss of the pointer generator network, which can better reduce the generation of repeated words. We define e_{ti} as the attention score of the encoder hidden state h_i^e at decoding time step t. PMDS penalizes input tokens that have obtained high attention scores in previous decoding steps. We define new encoder attention scores e_{ti}':

$$e_{ti}' = \begin{cases} exp(e_{ti}) & t = 1 \\ \frac{exp(e_{ti})}{\sum_{j=1}^{t-1} exp(e_{ji})} & t > 1 \end{cases} \tag{1}$$

Then, we normalize the encoder attention scores and use these scores to obtain the encoder context vector c_t^e. For each decoding step t, PMDS computes a new decoder attention scores $e_{tt'}^d$ to reduce the generation of the previously generated words and computes a decoder context vector c_t^d.

$$e_{tt'}^d = h_t^{d^T} W_{attn}^d h_{t'}^d \tag{2}$$

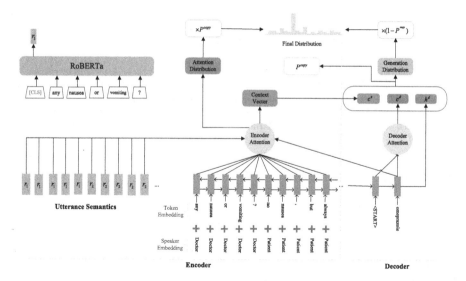

Fig. 2. Pointer Medical Dialogue Summarization network (PMDS) model structure. The attention distribution and the generation distribution are multiplied by the corresponding weights and added together to get the final distribution.

$$\alpha_{tt'}^d = \frac{exp(e_{tt'}^d)}{\sum_{j=1}^{t-1} exp(e_{tj}^d)} \tag{3}$$

$$c_t^d = \sum_{j=1}^{t-1} \alpha_{tj}^d h_j^d \tag{4}$$

The abstractive generation layer generates the generation probability distribution:

$$p_t^{gen}(y_t) = \text{softmax}(W_{gen}[h_t^d||c_t^e||c_t^d] + b_{gen}) \tag{5}$$

Meanwhile, the pointer mechanism uses the encoder attention scores α_{ti}^e as the probability to copy the original input token w_i .

$$p_t^{attn}(w_i) = \alpha_{ti}^e \tag{6}$$

We compute probabilities of using the copy mechanism for the decoding step t:

$$p_t^{copy} = \sigma(W_{copy}[h_t^d||c_t^e||c_t^d] + b_{copy}) \tag{7}$$

We use the weighted sum of the attention probability distribution for the original dialogue and the abstractive generation probability distribution to obtain final probability distribution for the output token y_t.

$$p(w) = p_t^{copy} p_t^{attn}(w) + (1 - p_t^{copy}) p_t^{gen}(w) \tag{8}$$

Finally, we use the cross entropy loss as the loss function for training. w_t^* is the target word at time t.

$$\text{loss} = \frac{1}{T} \sum_{t=0}^{T} -\log P(w_t^*) \tag{9}$$

3.2   Input Token Enhancement by Speaker-level Embedding

In order to adapt to the scenes of medical dialogues, we introduce the speaker-level embedding to distinguish speakers. We set up a trainable speaker embedding vector $E_{speaker}$, which includes two roles of doctor and patient. The speaker embedding vector $E_{speaker}$ and the token embedding vector E_{token} of the corresponding speaker are added to get the final input E_{input}:

$$E_{input} = E_{speaker} + E_{token} \tag{10}$$

Each input integrates both token information and speaker information. According to the results of experiments, the speaker embedding plays an important role in the performance of the model.

3.3   Input Token Enhancement by Utterance-level Embedding

Since the token-level embedding alone is not enough to locate the key words, we introduce utterance semantics to enhance input feature representation from the utterance level. In order to obtain the semantic representation of each utterance, we introduce the Chinese pre-training language model RoBERTa[1] [7].

All utterances in a medical dialogue are represented as $U = (u_1, u_2, ..., u_n)$, $u_i = (w_{i1}, w_{i2}, ..., w_{il})$. We separately input each utterance u_i into RoBERTa [7] and insert a [CLS] in front of each utterance. Finally, we take the corresponding output of [CLS] as the semantic representation r_i of the utterance u_i:

$$r_i = \text{RoBERTa}([CLS], w_{i1}, w_{i2}, ..., w_{il}) \tag{11}$$

Then, we give each token a semantic vector r_i of the utterance that the token is in and use the utterance semantics vector to calculate the encoder attention score of each input token:

$$e_t^{w_{il}} = v^T \tanh(W_e h_{w_{il}}^e + W_d h_t^d + W_r r_i) \tag{12}$$

The parameters of RoBERTa [7] are always fixed during training and testing, which ensure that each sentence has a fixed semantic representation. The results of experiments show that utterance semantics enables the model to focus on important information.

---

[1] https://huggingface.co/hfl/chinese- roberta-wwm-ext

4 Experiment

4.1 Dataset

We use the Chinese medical dialogue summarization dataset[2] provided by [33]. The original data is downloaded from Doctor Chunyu [3], the well-known online medical service provider in China. All the data are real online medical consultation data. In these medical dialogues, patients consult registered doctors for medical health questions, and doctors ask about specific disease conditions and give advice on diagnosis and treatment. In addition to medical dialogues and summaries, the dataset also contains some meta information, such as disease names and departments.

Table 1. Long dialogues represents the number of dialogues whose total number of tokens exceeds 512.

Number of data	Train	Valid	Test	All
Total	32631	4079	4079	32631
Long dialogues	4380	523	506	5409

Table 2. Average number of tokens for medical dialogues and summaries.

Average number of tokens	Train	Valid	Test
Dialogue	293.0	291.9	288.1
Summarization	95.4	94.2	93.6

As shown in Table 1, the dataset contains a total of 40789 pieces of data, and the train/valid/test random division: 80%/10%/10%. We define the dialogues with a total token number of more than 512 as long dialogues, and according to the statistical results, the proportion of long dialogues in each division is about 8%. As shown in Table 2, we count the average number of tokens of dialogues and summaries in each dataset partition, and it can be seen that the average token number difference of each partition is very small.

4.2 Baseline Models

In order to prove the effectiveness of our model, we select a variety of strong baselines from previous studies: **Lead-3** [26], which extracts doctors' first three utterances as summaries; **Random** [30], which randomly extracts doctors' three utterances as summaries; **Extractive Oracle** [39], which extracts doctors' first

---

[2] https://github.com/cuhksz-nlp/HET-MC
[3] https://www.chunyuyisheng.com/

three utterances with the highest ROUGE-1 scores as summaries in the valid dataset and the test dataset; the ranking extractive method **TextRank** [24]; the pointer Generator Network **PGNet** [31]; **ML+RL** [29], which combines maximum-likelihood training and reinforcement learning; the hierarchical extractive model **HET** [33] using the pre-training language model ZEN [9], which is dedicated to medical dialogue summarization; the transformer based [34] model **Longformer** [1], which focuses on dealing with long input text; the pre-training language model **BART** [17].

4.3 Settings

By counting the average number of utterances in summaries, Lead-3 [26], Random [30], Extractive Oracle [39] and TextRank [24] extract doctors' three utterances from dialogues. We used the Chinese implementation of BART[4] as our baseline. In order to converge as soon as possible, Longformer [1] is initialized with the weights of BART [17].

As shown in Table 1, although the average number of tokens in dialogues does not exceed 300, the number of long dialogues with more than 512 tokens accounts for about 8% of the dataset. Due to the input length limitation of BART, We keep the first 512 tokens for truncation.

In order to prevent error accumulation, we use the beam search algorithm when decoding on the valid set and the test set and set *beamsize* = 4.

4.4 Evaluation Metrics

We use ROUGE-1,ROUGE-2 and ROUGE-L scores [20] as automatic evaluation metrics. These three metrics separately represent the accuracy of unigrams, bigrams and the longest common subsequence. We use a universally accepted third-party library[5] to calculate the ROUGE scores and get the means as the final results.

In addition to ROUGE automatic evaluation, we also invite three doctors to evaluate the summaries generated by BART, PGNet, HET and PMDS. Given the complete medical dialogue, we asked them to evaluate the factual correctness, information integrity and correlation between the generated summary and the real summary. Taking the degree of correctness of facts as an example, we divide it into four levels: 3 points for completely correct, 2 points for most correct, 1 point for some correct, and 0 point for completely incorrect. Finally, we average the scores of all the evaluated samples as the final result.

4.5 Automatic Evaluation

As shown in Table 3, PMDS performs better than all the baseline models on all the ROUGE metrics. The two results of Lead-3 [26] and Random [30] are similar,

---

[4] https://huggingface.co/fnlp/bart-base-chinese
[5] https://github.com/pltrdy/rouge

which shows that the position of doctors' diagnoses and suggestions is not fixed. The high performance of Extractive Oracle shows that summaries and doctors' original utterances are highly duplicate. Longformer [1] and BART [17] perform poorly in all metrics. This proves that direct abstractive generation does not work. PGNet [31] performs at a medium level, which indicates the replication mechanism is effective and traditional text summarization models need to adapt to the scenes of medical dialogues. For HET [33], the results indicate that the extractive method also takes redundant words as summaries when extracting valuable utterances. ML+RL [29] uses replication mechanism and uses a variety of optimization techniques, so it performs well in all metrics.

Table 3. ROUGE-1,ROUGE-2,ROUGE-L scores of generated summaries in the test dataset.Numbers in bold are the overall best results.

Model	R-1	R-2	R-L
Lead-3	56.21	42.77	53.32
Random	59.08	45.82	51.08
Extractive Oracle	82.62	69.47	69.16
TextRank	74.66	60.25	67.94
Longformer	79.82	68.62	78.57
BART	79.98	68.60	78.76
PGNet	80.29	70.49	78.83
HET	81.59	73.48	81.57
ML+RL	83.70	75.31	82.61
PMDS	**89.52**	**82.86**	**88.79**

Ablation Experiments. As shown in Table 4, after removing intra-attention, speaker embedding and utterance semantics separately, the ROUGE-1 scores decrease by 2.33, 4.04 and 2.88 respectively. On the ROUGE-2 evaluation metric, the average scores decreased by 2.74, 4.81 and 4.04 respectively. On the ROUGE-L evaluation metric, the average scores decreased by 2.04, 4.30 and 3.03 respectively. By comparing the improvement effect of the ablation experiment, the improvement effect of the speaker encoding is the most obvious, which shows that identifying the speaker role can indeed greatly improve the performance of the model. Moreover, the results show that all our modules can effectively improve the performance of our model.

4.6 Doctor Evaluation

As shown in Table 5, each metric of PMDS under the actual evaluation of doctors has achieved the best. Since HET is a extractive method, we do not take into account the evaluation score of its factual correctness. HET may extract some useless words, such as greeting words, and omit utterances with important

Table 4. Ablation experiments of PMDS. Numbers in bold are the overall best results.

Model	R-1	R-2	R-L
no intra-attention	87.19	80.12	86.75
no speaker embedding	85.48	78.05	84.49
no utterance semantics	86.64	78.82	85.76
PMDS	**89.52**	**82.86**	**88.79**

information, which reduces the performance of the model. Since HET will connect sentences with commas to generate summaries after extracting important sentences, this will lead to redundancy of punctuation and affect the reading experience. BART performs poorly on the two metrics of factual correctness and information integrity, but performs well on the correlation indicators.

Table 5. Doctor evaluation scores for BART, PGNet, HET and PMDS. Numbers in bold are the overall best results.

Model	Correctness	Integrity	Correlation
BART	2.41	2.37	2.56
PGNet	2.50	2.49	2.44
HET	–	2.52	2.53
PMDS	**2.68**	**2.66**	**2.61**

4.7   Case Study

We selecte two medical dialogue ummarization data with typical characteristics in the test set and carefully studied the problems exhibited by the summaries generated by the different models. As shown in Table 6, One of the problems with BART is that it will lose some of the doctor's diagnosis and treatment recommendations. This experimental result also proves that the completely abstractive method is not applicable to medical dialogue summarization, because it will cause factual errors and omit important information. Although the summaries generated by BART are concise, the loss of key medical concepts will greatly reduce the quality and credibility of the summaries. Original PGNet performs poorly on all the metrics.

According to the example in Table 7, PGNet still has the problem of repeated generation and the ability of accurate replication is weak, which will result in greatly reduced readability of the summaries. The HET model generates the summary with redundant salutations such as "Hello" and "Welcome". Also, since the HET model adds a comma after the extracted sentences and then splices them together to form the summary, this creates some redundancy in punctuation and reduces the reading fluency. Through the two examples in Table 7 and Table 6, although PMDS can roughly express complete and correct diagnosis and treatment suggestions, there is still some redundancy in the generated summaries.

Table 6. The first comparison of summaries generated by several baselines and PMDS. The colored part represents the main difference between the reference summary and the generated summary. Text crossed out indicates the diagnosis and treatment information that appears in the reference summary but is missing from the model-generated summary.

Dialogue	**PT:** Diagnosed as glaucoma, I want to consult (female, 28 years old) **DC:** Glaucoma is a kind of lifelong disease once diagnosed, which cannot be completely cured. **PT:** The picture cannot be displayed due to privacy issues. **PT:** Would you please take a look at the extent of glaucoma and the probability of blindness? **PT:** Hello, it looks normal after taking CT. Do you suggest another MRI? **DC:** It is recommended that you do an MRI of the sella turcica area. If there is no problem, you can determine that it is glaucoma. **PT:** How can I describe MRI in sella turcica area to the hospital? Too professional. **DC:** It's a magnetic resonance imaging of the pituitary gland. **PT:** Okay, thank you. **DC:** You're welcome. If you have any questions, please contact me again.
Reference summary	Glaucoma is a kind of lifelong disease once diagnosed, which cannot be completely cured. It is recommended that you do an MRI of the sella turcica area. If there is no problem, you can determine that it is glaucoma. It's a magnetic resonance imaging of the pituitary gland.
BART	Glaucoma is a kind of ~~lifelong disease once diagnosed, which cannot be completely cured~~. It is recommended that you do it in the sella turcica area. If there is no problem, you can determine that it is glaucoma. It's a magnetic resonance imaging of the pituitary gland.
PGNet	Glaucoma is a kind of lifelong disease once diagnosed, which cannot be completely cured. It is suggested that you make one in the sella turcica area. If there is no problem, you can determine that it is glaucoma. ~~It's a magnetic resonance imaging of the pituitary gland.~~
PMDS	Glaucoma is a kind of lifelong disease once diagnosed , which cannot be completely cured. It is recommended that you do an MRI in the sella turcica area. If there is no problem, you can determine that it is glaucoma. It's a magnetic resonance imaging of the pituitary gland.

Table 7. The second comparison of summaries generated by several baselines and PMDS. The colored part represents the main difference between the reference summary and the generated summary.

Dialogue	**PT:** Papular urticaria, can my baby take a bath or swim? (female, 5 years old)) **DC:** Hello, welcome to consult, for papular urticaria, frequent bathing is also good. **DC:** That is to say, your baby can take a bath and swim. **PT:** Thank you. Can my baby with pimples and urticaria heal themselves without medication? **PT:** Look at the picture. Is it papule urticaria? It has lasted for ten days and itches occasionally. **DC:** The baby can heal herself without medicine, but will feel itchy. **PT:** She didn't scratch, and she should not itch. I don't know if it's papular urticaria. **DC:** From the picture, it is considered to be papular urticaria. **PT:** If so, you can swim, right? **DC:** Yes, your baby can swim.
Reference summary	For papular urticaria, frequent bathing is also good. The baby can heal herself without medicine, but will feel itchy. From the picture, it is considered to be papular urticaria. Your baby can swim.
PGNet	Welcome to consult, for papular urticaria, regularly consult, is good. That is to say, you can take a bath and swim. The baby can heal herself without medicine, but will feel itchy. From the picture, it is considered to be papular urticaria. Yes, your baby can swim.
HET	Hello, welcome to consult, for papular urticaria, frequent bathing is also good., That is to say, the baby can take a bath and swim,, The baby can heal herself without medicine, but will feel itchy., It is considered to be papular urticaria.
PMDS	Welcome to consult, for papular urticaria, frequent bathing is also good. The baby can take a bath and swim. The baby can heal herself without medicine, but will feel itchy. From the picture, it is considered to be papular urticaria. Yes, your baby can swim.

5   Conclusion

This paper proposes a pointer generator network model PMDS for medical dialogue summarization. By integrating intra-attention, speaker embedding and utterance semantics, we use multilevel enhanced input feature representations to adapt to the scenes of medical dialogues. According to the results of automatic evaluation metrics, we find that PMDS outperforms all the baselines and all the modules contribute to the performance. The professional evaluation of three doctors also confirms the actual effectiveness and performance superiority of PMDS. Meanwhile, we conduct a case study on the test set and found that PMDS can indeed compensate for the problems of missing critical information and redundancy of useless information that exist in other models. In the future, we will continue to explore and practice how to generate more concise and accurate medical dialogue summaries from the perspective of integrating medical knowledge.

Acknowledgements. This work is supported by the National Natural Science Foundation of China under Grant (No. 62272092, No. 62172086) and the Fundamental Research Funds for the Central Universities of China under Grant (No. N2116008).

References

1. Beltagy, I., Peters, M.E., Cohan, A.: Longformer: the long-document transformer. arXiv preprint arXiv:2004.05150 (2020)
2. Chen, J., Yang, D.: Structure-aware abstractive conversation summarization via discourse and action graphs. In: Proceedings of the 2021 Conference of the North American Chapter of the Association for Computational Linguistics: Human Language Technologies, pp. 1380–1391 (2021)
3. Chen, Y.C., Bansal, M.: Fast abstractive summarization with reinforce-selected sentence rewriting. In: Proceedings of the 56th Annual Meeting of the Association for Computational Linguistics (Volume 1: Long Papers), pp. 675–686 (2018)
4. Cheng, J., Lapata, M.: Neural summarization by extracting sentences and words. In: ACL, vol. 1 (2016)
5. Cheng, L., Shi, Y., Zhang, K.: Medical treatment migration behavior prediction and recommendation based on health insurance data. World Wide Web **23**, 2023–2042 (2020)
6. Chopra, S., Auli, M., Rush, A.M.: Abstractive sentence summarization with attentive recurrent neural networks. In: Proceedings of the 2016 Conference of the North American Chapter of the Association for Computational Linguistics: Human Language Technologies, pp. 93–98 (2016)
7. Cui, Y., Che, W., Liu, T., Qin, B., Wang, S., Hu, G.: Revisiting pre-trained models for Chinese natural language processing. In: Proceedings of the 2020 Conference on Empirical Methods in Natural Language Processing: Findings, pp. 657–668. Association for Computational Linguistics, November 2020. https://www.aclweb.org/anthology/2020.findings-emnlp.58
8. DeYoung, J., Beltagy, I., van Zuylen, M., Kuehl, B., Wang, L.L.: Ms2: multi-document summarization of medical studies. arXiv preprint arXiv:2104.06486 (2021)
9. Diao, S., Bai, J., Song, Y., Zhang, T., Wang, Y.: Zen: Pre-training Chinese text encoder enhanced by n-gram representations. In: Findings of the Association for Computational Linguistics: EMNLP 2020, pp. 4729–4740 (2020)
10. Feng, X., Feng, X., Qin, B.: A survey on dialogue summarization: Recent advances and new frontiers. arXiv preprint arXiv:2107.03175 (2021)
11. Gachabayov, M., Latifi, L.A., Parsikia, A., Latifi, R.: The role of telemedicine in surgical specialties during the covid-19 pandemic: a scoping review. World J. Surg. **46**(1), 10–18 (2022)
12. Joshi, A., Katariya, N., Amatriain, X., Kannan, A.: Dr. summarize: global summarization of medical dialogue by exploiting local structures. In: Findings of the Association for Computational Linguistics: EMNLP 2020, pp. 3755–3763 (2020)

13. Kano, R., Miura, Y., Taniguchi, T., Ohkuma, T.: Identifying implicit quotes for unsupervised extractive summarization of conversations. In: Proceedings of the 1st Conference of the Asia-Pacific Chapter of the Association for Computational Linguistics and the 10th International Joint Conference on Natural Language Processing, pp. 291–302 (2020)
14. Kim, E., Rubinstein, S.M., Nead, K.T., Wojcieszynski, A.P., Gabriel, P.E., Warner, J.L.: The evolving use of electronic health records (EHR) for research. In: Seminars in radiation oncology, vol. 29, pp. 354–361. Elsevier (2019)
15. Koay, J.J., Roustai, A., Dai, X., Liu, F.: A sliding-window approach to automatic creation of meeting minutes. In: Proceedings of the 2021 Conference of the North American Chapter of the Association for Computational Linguistics: Student Research Workshop, pp. 68–75 (2021)
16. Krishna, K., Khosla, S., Bigham, J.P., Lipton, Z.C.: Generating soap notes from doctor-patient conversations using modular summarization techniques. In: Proceedings of the 59th Annual Meeting of the Association for Computational Linguistics and the 11th International Joint Conference on Natural Language Processing (Volume 1: Long Papers), pp. 4958–4972 (2021)
17. Lewis, M., et al.: Bart: denoising sequence-to-sequence pre-training for natural language generation, translation, and comprehension. In: Proceedings of the 58th Annual Meeting of the Association for Computational Linguistics, pp. 7871–7880 (2020)
18. Li, M., Liu, R., Wang, F., Chang, X., Liang, X.: Auxiliary signal-guided knowledge encoder-decoder for medical report generation. World Wide Web 26(1), 253–270 (2023)
19. Li, W., Zhuge, H.: Abstractive multi-document summarization based on semantic link network. IEEE Trans. Knowl. Data Eng. 33(1), 43–54 (2019)
20. Lin, C.Y.: Rouge: a package for automatic evaluation of summaries. In: Text summarization branches out, pp. 74–81 (2004)
21. Liu, Y., Lapata, M.: Text summarization with pretrained encoders. In: Proceedings of the 2019 Conference on Empirical Methods in Natural Language Processing and the 9th International Joint Conference on Natural Language Processing (EMNLP-IJCNLP), pp. 3730–3740 (2019)
22. Liu, Z., Ng, A., Lee, S., Aw, A.T., Chen, N.F.: Topic-aware pointer-generator networks for summarizing spoken conversations. In: 2019 IEEE Automatic Speech Recognition and Understanding Workshop (ASRU), pp. 814–821. IEEE (2019)
23. Ma, B., et al.: Distant supervision based machine reading comprehension for extractive summarization in customer service. In: Proceedings of the 44th International ACM SIGIR Conference on Research and Development in Information Retrieval, pp. 1895–1899 (2021)
24. Mihalcea, R., Tarau, P.: Textrank: bringing order into text. In: Proceedings of the 2004 Conference on Empirical Methods in Natural Language Processing, pp. 404–411 (2004)
25. Morgenstern-Kaplan, D., Rocha-Haro, A., Canales-Albarrán, S.J., Núñez-García, E., León-Mayorga, Y.: An app-based telemedicine program for primary care and specialist video consultations during the covid-19 pandemic in Mexico. Telemedicine e-Health 28(1), 60–65 (2022)
26. Nallapati, R., Zhai, F., Zhou, B.: Summarunner: a recurrent neural network based sequence model for extractive summarization of documents. In: Thirty-first AAAI Conference on Artificial Intelligence (2017)

27. Narayan, S., Cohen, S.B., Lapata, M.: Ranking sentences for extractive summarization with reinforcement learning. In: Proceedings of the 2018 Conference of the North American Chapter of the Association for Computational Linguistics: Human Language Technologies, Volume 1 (Long Papers), pp. 1747–1759 (2018)
28. Nenkova, A., McKeown, K.: Automatic Summarization. Now Publishers Inc, Norwell (2011)
29. Paulus, R., Xiong, C., Socher, R.: A deep reinforced model for abstractive summarization. In: International Conference on Learning Representations (2018)
30. Riedhammer, K., Gillick, D., Favre, B., Hakkani-Tür, D.: Packing the meeting summarization knapsack. In: Ninth Annual Conference of the International Speech Communication Association (2008)
31. See, A., Liu, P.J., Manning, C.D.: Get to the point: summarization with pointer-generator networks. In: Proceedings of the 55th Annual Meeting of the Association for Computational Linguistics (Volume 1: Long Papers), pp. 1073–1083 (2017)
32. Shanafelt, T.D., et al.: Relationship between clerical burden and characteristics of the electronic environment with physician burnout and professional satisfaction. In: Mayo Clinic Proceedings, vol. 91, pp. 836–848. Elsevier (2016)
33. Song, Y., Tian, Y., Wang, N., Xia, F.: Summarizing medical conversations via identifying important utterances. In: Proceedings of the 28th International Conference on Computational Linguistics, pp. 717–729 (2020)
34. Vaswani, A., et al.: Attention is all you need. In: Advances in Neural Information Processing Systems, vol. 30 (2017)
35. Wu, Y., Pei, C., Ruan, C., Wang, R., Yang, Y., Zhang, Y.: Bayesian networks and chained classifiers based on SVM for traditional Chinese medical prescription generation. In: World Wide Web, pp. 1–22 (2022)
36. Zhang, J., Zhao, Y., Saleh, M., Liu, P.: Pegasus: pre-training with extracted gap-sentences for abstractive summarization. In: International Conference on Machine Learning, pp. 11328–11339. PMLR (2020)
37. Zhang, R., Tetreault, J.: This email could save your life: introducing the task of email subject line generation. In: Proceedings of the 57th Annual Meeting of the Association for Computational Linguistics, pp. 446–456 (2019)
38. Zhao, L., Xu, W., Guo, J.: Improving abstractive dialogue summarization with graph structures and topic words. In: Proceedings of the 28th International Conference on Computational Linguistics, pp. 437–449 (2020)
39. Zhu, C., Xu, R., Zeng, M., Huang, X.: A hierarchical network for abstractive meeting summarization with cross-domain pretraining. In: Findings of the Association for Computational Linguistics: EMNLP 2020, pp. 194–203 (2020)
40. Zou, Y., et al.: Unsupervised summarization for chat logs with topic-oriented ranking and context-aware auto-encoders. In: Proceedings of the AAAI Conference on Artificial Intelligence, vol. 35, pp. 14674–14682 (2021)

HSA: Hyperbolic Self-attention for Sequential Recommendation

Peizhong Hou[1], Haiyang Wang[2], Tianming Li[2], and Junchi Yan[1(✉)]

[1] MOE Key Lab of Artificial Intelligence & Student Innovation Center,
Shanghai Jiao Tong University, Shanghai, China
{houpzh,yanjunchi}@sjtu.edu.cn
[2] Ant Group, Shanghai, China
{chixi.why,william.ltm}@antgroup.com

Abstract. Recently, researchers apply various deep neural networks to the task of sequential recommendation, which captures dynamics of user preference from user behavior data to make accurate recommendation. Self-attention based approaches have been proposed to effectively identify relevant items and better capture long-term dependencies, achieving competitive results in sequential recommendation domain. However, most existing methods perform in the Euclidean space, which expands only polynomially, limiting the capacity of models. Besides, methods typically do not consider and leverage latent hierarchical structures existing in real-world datasets. To this end, we propose to learn representations in hyperbolic space for sequential recommendation, bringing two advantages. First, hyperbolic space expands exponentially and thus provides higher representation ability. Second, it is able to effectively model the latent hierarchical structures, which are indicated by the power-law distributions of user behavior sequences. Specifically, we propose a novel hyperbolic self-attention model, which learns item embeddings in hyperbolic space and adopts self-attention to model user sequence representation, using hyperbolic distance to measure preference and make recommendation. Extensive experiments conducted on three real-world datasets demonstrate the superiority of our proposed hyperbolic embedding approach over various competitive baselines, including Euclidean self-attention counterpart. We apply the proposed hyperbolic embedding method to classic sequential recommendation models and observe improvement, showing it a general technique which can boost other models.

Keywords: Sequential recommendation · Hyperbolic embedding learning · Self-attention mechanism

1 Introduction

Recommender systems [29,30] play an important role in online platforms by helping users find their interests among many item candidates. In real-world scenes, users' behaviors are consecutive and dynamic. Therefore, capturing the sequential patterns of user behaviors is significant for making effective recommendations, which is the basic concept of sequential recommendation [7,8,22,25].

X. Song et al. (Eds.): APWeb-WAIM 2023, LNCS 14333, pp. 250–264, 2024.
https://doi.org/10.1007/978-981-97-2387-4_17

To model the evolving properties of users' historical behaviors, many sequential recommendation models have been proposed based on deep neural networks [29]. Models based on Recurrent Neural Networks (RNNs) [7,15] and CNN [25,28] have made a success. Furthermore, self-attention mechanism is adopted [8,14,23] to improve model capacity to better capture dynamic user interests. Among those models, SASRec [8] has achieved strong and competitive results in this task by stacking self-attention blocks. However, all the mentioned models learn and predict in the Euclidean space. They face a fundamental limitation: the capability to model complex patterns (e.g. user behavior sequences, hierarchical structures in recommendation datasets) is intrinsically limited by the dimensionality of the Euclidean embedding space [18].

To better uncover and model the characteristics of user interaction sequences, we examine the distribution of the number of interaction for both users and items. We observe that they follow power-law distributions: a majority of nodes only have few interactions while a few nodes interact for a huge number of times. Besides, it has been shown that many real-world datasets display the typical property of complex networks, for example the power-law degree distribution [20], including many recommendation datasets [4,13,17]. Power-law distributions often indicate underlying hierarchical structure, as studied in [18,20]. These observations motivate how can one can model underlying hierarchical structures in user interaction sequences to improve sequential recommendation result.

Recently, hyperbolic representation learning approaches [18,19] are proposed to model latent hierarchical structures. This inspires researchers to learn representation in hyperbolic space to better capture users' preferences. Hyperbolic-based methods were proposed for different recommendation tasks, such as collaborative filtering [13,24], next-POI recommendation [3] and session-based recommendation [4], by adapting powerful architectures like graph convolutional networks (GCNs) and metric learning to hyperbolic settings, achieving better performance. Moreover, the learned hyperbolic representations could naturally capture both similarity and hierarchy through hyperbolic distance and their norm [18].

Focusing on the sequential recommendation problem, the number of possible user behavior sequences grows exponentially with its length theoretically. However, the Euclidean space only grows polynomially [18], which might limits the capacity of models to represent user behavior sequences. In contrast, the hyperbolic space expands exponentially with the radius [12,18], which is equipped with more powerful representation ability and might be a better choice than the Euclidean space to model user sequences.

Inspired by these, we propose a **H**yperbolic **S**elf-**A**ttention approach for sequential recommendation, named HSA. HSA learns item embeddings and user sequence representations in the hyperbolic space by attending to all the previous items through stacked self-attention blocks in the tangent space of a reference point. With the advantage of hyperbolic space, the proposed model can effectively learn with the cross entropy loss based on hyperbolic distances optimized with the Riemannian Adam. For inference, HSA computes the hyperbolic distance between users and items as the preference score, ranks those scores and

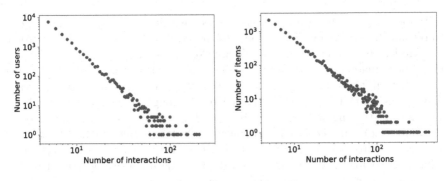

(a) Distribution of user interaction   (b) Distribution of item interaction

Fig. 1. Distributions of the user-item interaction on Amazon Beauty. X-axis: number of interactions of a user or item; Y-axis: number of such users/items.

makes recommendations. Moreover, our proposed hyperbolic embedding learning is a general technique which can be applied to existing sequential recommendation models. We adapt existing models to hyperbolic version and observe improvement, confirming the generality and effectiveness of proposed hyperbolic embedding.

The main contributions of this paper are summarized as follows. The source code will be made publicly available.

1) We develop the hyperbolic self-attention for sequential recommendation.
2) We develop a plugin for equipping existing sequential recommendation baselines with hyperbolic embedding learning.
3) Extensive experiments on public benchmarks show that the proposed HSA outperforms strong baselines including the Euclidean counterpart SASRec [8]. The hyperbolic versions of existing sequential recommendation models show consistently improvement, confirming our proposed hyperbolic embedding learning as a general technique which can boost existing models.

2 Preliminaries and Related Work

2.1 Empirical Analysis of Datasets

To understand the properties of user-item interactions and better model them, we present the distribution of the number of interactions for users and items in Fig. 1. We only show the results of preprocessed Amazon Beauty dataset, because of the space limitation. We observe the power-law distribution: a majority of users/items only have few interactions while a few users/items have a large number of interactions. Similar power-law distributions are observed from Amazon Video and Steam datasets. The findings empirically demonstrate user-item interactions follow power-law distribution. Researches show that hyperbolic space can naturally and effectively model data with power-law structures [12,18], inspiring us that hyperbolic space might be suitable for recommendation datasets and tasks.

2.2   Lorentz Model of Hyperbolic Space

To learn d-dimensional item embeddings in the hyperbolic space. There are several equivalent mathematical models for the hyperbolic space. Recently, [2] propose to learn GCNs in the hyperbolic space and show superior results on graph datasets where graphs show hierarchical structure. It adopts the Lorentz model of hyperbolic space and utilize the tangent space of points to perform deep learning operations. We build on this work to bring the benefits of hyperbolic spaces to the sequential recommendation domain. Here, we adopt the Lorentz model for its numerical stability and efficient optimization [19]. We briefly introduce the Lorentz model of hyperbolic space.

Recall that a d-dimensional hyperbolic space is a Riemannian Manifold $\mathcal{M}$ with a constant negative curvature, denoted by c. Use $k = -1/c$ to denote the negative reciprocal of the curvature such that $k > 0$. The tangent space $\mathcal{T}_{\mathbf{x}}\mathcal{M}$ centered at point $\mathbf{x}$ on $\mathcal{M}$ is a d-dimensional Euclidean space which is a local, first-order approximation of the hyperbolic manifold $\mathcal{M}$ at $\mathbf{x}$. The Lorentz formulation is defined by an underlying set and a metric tensor. The Lorentz model is defined as $\mathcal{L}^d = (\mathcal{H}^d, g_{\mathcal{L}})$:

$$\mathcal{H}^d = \{\mathbf{x} \in \mathbb{R}^{d+1} : \langle \mathbf{x}, \mathbf{x} \rangle_{\mathcal{L}} = -k, x_0 > 0\} \tag{1}$$

where $\langle \mathbf{x}, \mathbf{y} \rangle_{\mathcal{L}}$ is the Lorentzian inner product that $\langle \mathbf{x}, \mathbf{y} \rangle_{\mathcal{L}} := -x_0 y_0 + x_1 y_1 + \ldots + x_d y_d$ for $\mathbf{x}, \mathbf{y} \in \mathbb{R}^{d+1}$, and $g_{\mathcal{L}} = \text{diag} [-1, 1, 1, \cdots, 1]$ is the metric tensor. The metric $g_{\mathcal{L}}$ induces a distance between a pair of points $\mathbf{x}, \mathbf{y} \in \mathcal{H}^d$ by:

$$d_{\mathcal{L}}(\mathbf{x}, \mathbf{y}) = \sqrt{k} \operatorname{arcosh} \left(-\frac{\langle \mathbf{x}, \mathbf{y} \rangle_{\mathcal{L}}}{k} \right) \tag{2}$$

The tangent space centered at point $\mathbf{x}$ in the Lorentz manifold is defined as:

$$\mathcal{T}_{\mathbf{x}}\mathcal{H}^d = \{\mathbf{v} \in \mathbb{R}^{d+1} : \langle \mathbf{v}, \mathbf{x} \rangle_{\mathcal{L}} = 0\} \tag{3}$$

2.3   Self-attention Mechanism for Sequential Recommendation

We briefly introduce a competitive sequential recommendation model **SASRec** [8], which uses self-attention mechanism to model user sequences.

Embedding Layer. The input is a fixed-length user behavior sequence $s = (s_1, s_2, \ldots, s_n)$, where n is the length. It creates an item embedding matrix $\mathbf{M} \in \mathbb{R}^{|\mathcal{I}| \times d}$ where $|\mathcal{I}|$ is the number of items and d is the embedding dimensionality. It retrieves the input embedding matrix $\mathbf{E} \in \mathbb{R}^{n \times d}$, where $\mathbf{E}_i = \mathbf{M}_{s_i}$.

Positional Embedding: The self-attention is not aware of the positions of previous items in user sequence, thus it can't make use of the information of item position and the order of sequence. Therefore it injects a learnable positional embedding $\mathbf{P} \in \mathbb{R}^{n \times d}$ into the input embedding.

$$\widehat{\mathbf{E}} = \begin{bmatrix} \mathbf{M}_{s_1} + \mathbf{P}_1 \\ \mathbf{M}_{s_2} + \mathbf{P}_2 \\ \cdots \\ \mathbf{M}_{s_n} + \mathbf{P}_n \end{bmatrix} \tag{4}$$

Self-attention Block. The scaled dot-product attention [26] is defined as:

$$\text{Attention}(\mathbf{Q}, \mathbf{K}, \mathbf{V}) = \text{softmax}\left(\frac{\mathbf{QK}^{\top}}{\sqrt{d}}\right)\mathbf{V} \tag{5}$$

where $\mathbf{Q}$ represents the queries, $\mathbf{K}$ the keys and $\mathbf{V}$ the values (an item is represented by a row). Intuitively, the attention layer calculates a weighted sum of all values, where the weight of query i and value j reflects the importance of key j to query i. The scale factor $\sqrt{d}$ aims to avoid overly large results of the inner product, especially when the dimensionality is high.

Self-attention Layer: [26] proposed a self-attention method which takes the same objects as queries, keys and values. In this case, the self-attention operation takes the embedding $\widehat{\mathbf{E}}$ as input, first linear transforms it to three matrices and then sends them into an attention layer:

$$\mathbf{S} = \text{SA}(\widehat{\mathbf{E}}) = \text{Attention}(\widehat{\mathbf{E}}\mathbf{W}^Q, \widehat{\mathbf{E}}\mathbf{W}^K, \widehat{\mathbf{E}}\mathbf{W}^V) \tag{6}$$

where the linear transformation matrices $\mathbf{W}^Q, \mathbf{W}^K, \mathbf{W}^V \in \mathbb{R}^{d \times d}$. The transformations enable the model to be more flexible.

Causality: Because of the nature of sequences, the model should only consider the first t items when predicting the $(t+1)$-st item. However, the t-th output of the self-attention layer ($\mathbf{S}_t$) includes embeddings of subsequent items, making the model ill-posed. Therefore, it modifies the attention by disconnecting all links between $\mathbf{Q}_i$ and $\mathbf{K}_j (j > i)$.

Point-Wise Feed-Forward Network: Though the self-attention mechanism is able to aggregate embeddings of all previous items with adaptive weights based on their importance, ultimately it is still a linear model. To enable the model to learn nonlinearity and to consider effects between different latent dimensions, it applies a two-layer point-wise feed-forward network to all $\mathbf{S}_i$ identically (sharing parameters):

$$\mathbf{F}_i = \text{FFN}(\mathbf{S}_i) = \text{ReLU}(\mathbf{S}_i\mathbf{W}^{(1)} + \mathbf{b}^{(1)})\mathbf{W}^{(2)} + \mathbf{b}^{(2)} \tag{7}$$

where $\mathbf{W}^{(1)}, \mathbf{W}^{(2)}$ are $d \times d$ matrices and $\mathbf{b}^{(1)}, \mathbf{b}^{(2)}$ are d-dimensional vectors. Note that there is no interaction between $\mathbf{S}_i$ and $\mathbf{S}_j$ $(i \neq j)$, demonstrating that it still prevent information leaks (from later to earlier).

Stacking Self-attention Blocks. After the first self-attention block, $\mathbf{F}_i$ essentially aggregates embeddings of all the previous items (i.e. $\widehat{\mathbf{E}}_j, j \leq i$). However, it is probably effective to learn more complex item transitions and higher order relationships via another self-attention block based on $\mathbf{F}$. Specifically, it stacks self-attention blocks (i.e. a self-attention layer and a feed-forward network), and the b-th $(b > 1)$ block is defined as:

$$\begin{aligned} \mathbf{S}^{(b)} &= \text{SA}(\mathbf{F}^{(b-1)}) \\ \mathbf{F}_i^{(b)} &= \text{FFN}(\mathbf{S}_i^{(b)}), \quad \forall i \in \{1, 2, \ldots, n\} \end{aligned} \tag{8}$$

and the 1-st block is defined as $\mathbf{S}^{(1)} = \mathbf{S}$ and $\mathbf{F}^{(1)} = \mathbf{F}$.

However, as the network goes deeper, the increased model capacity tends to overfitting, the training process becomes unstable (affected by vanishing gradients etc.) and models usually consumes more training time. Based on [26], it adopts residual connection, dropout and layer normalization to alleviate these problems:

$$f'(x) = x + \text{Dropout}\left(f(\text{LayerNorm}(x))\right)$$

where $f(x)$ represents the self-attention layer or the feed-forward network.

Prediction Layer. It uses dot product to predict the preference of item i:

$$r_{i,t} = \mathbf{F}_t^{(b)} \mathbf{M}_i^\top$$

where $r_{i,t}$ is the relevance of item i being the next item given the first t items. Thus, a high interaction score $r_{i,t}$ means a high preference, then it can generate recommendations by ranking the scores.

3 Proposed Approach

3.1 Problem Formulation and Approach Overview

Denote a set of users and items by $\mathcal{U}$ and $\mathcal{I}$ respectively, where $u \in \mathcal{U}$ denotes a user and $i \in \mathcal{I}$ stands for an item. The numbers of users and items are denoted as $|\mathcal{U}|$ and $|\mathcal{I}|$. For sequential recommendation with implicit feedback, each user has an interaction sequence $\mathcal{S} = (\mathcal{S}_1, \mathcal{S}_2, \ldots, \mathcal{S}_{|\mathcal{S}|})$, where $\mathcal{S}_t$ is the t-th item that the user u has interacted with. We attempt to predict the next item that the user is likely to interact with, depending on the previous items.

We present an overview of the proposed HSA (Hyperbolic Self-attention) approach, a generalization of self-attention mechanism in hyperbolic geometry for sequential recommendation that benefits from the capabilities of both self-attention and hyperbolic embeddings. To apply HSA, we first initialize embeddings for all items in hyperbolic space $\mathcal{H}^d$. Then, given a user u with interaction sequence history, we map items in the sequence to the tangent space of a reference point where the sequence is passed through several self-attention blocks and we get the representation of the sequence. The sequence representation, are mapped back to $\mathcal{H}^d$ where a hyperbolic distance based cross-entropy loss is applied to the sequence representation and item embeddings. The loss is back-propagated to update parameters. We then describe each component in detail.

3.2 Item Embeddings in Hyperbolic Space

We transform the training sequence $(\mathcal{S}_1, \mathcal{S}_2, \ldots, \mathcal{S}_{|\mathcal{S}|-1})$ into a fixed-length sequence $s = (s_1, s_2, \ldots, s_n)$, for each position $p \in \{1, 2, \ldots, |\mathcal{S}| - 1\}$, where n represents the maximum length that the proposed model can process. For each position, if the sequence length up to p is longer than n, we consider the most recent n interactions. If it is shorter than n, we repeatedly add a 'padding' item to the left to make the length reach n.

We use the Lorentz representation for item embeddings, due to its numerical stability and efficient optimization [19]. We fix the origin $\mathbf{o} = (\sqrt{k}, 0, \cdots, 0) \in \mathcal{H}^d$ and use it as a reference point. Note that $k = -1/c$ is the reciprocal of the curvature c that is considered as a hyper-parameter and set empirically. The embeddings are initialized by sampling from the normal distribution in the tangent space $\mathcal{T}_\mathbf{o}\mathcal{H}^d$ of the reference point $\mathbf{o}$. Formally, given an item i, we first sample from the multivariate normal:

$$\boldsymbol{\theta}'_i \sim \mathcal{N}(\mathbf{0}, \sigma\mathbf{I}_{d \times d})$$

and then set:

$$\boldsymbol{\theta}''_i = [0; \boldsymbol{\theta}'_i]$$

where $[;]$ denotes concatenation. Since $\boldsymbol{\theta}''_i$ satisfies $\langle \boldsymbol{\theta}''_i, \mathbf{o} \rangle_{\mathcal{L}} = 0$ and therefore it belong to $\mathcal{T}_\mathbf{o}\mathcal{H}^d$. To obtain the corresponding embeddings in $\mathcal{H}^d$ we project them into the hyperbolic space by the exponential map $\exp_\mathbf{o} : \mathcal{T}_0\mathcal{H}^d \to \mathcal{H}^d$ [2]:

$$\exp_\mathbf{o}(\mathbf{v}) = \cosh\left(\frac{\|\mathbf{v}\|_{\mathcal{L}}}{\sqrt{k}}\right)\mathbf{o} + \sqrt{k}\sinh\left(\frac{\|\mathbf{v}\|_{\mathcal{L}}}{\sqrt{k}}\right)\frac{\mathbf{v}}{\|\mathbf{v}\|_{\mathcal{L}}} \tag{9}$$

where $\mathbf{v} \in \mathcal{T}_\mathbf{o}\mathcal{H}^d$ and $\|\mathbf{v}\|_{\mathcal{L}} = \sqrt{\langle \mathbf{v}, \mathbf{v} \rangle_{\mathcal{L}}}$. Then we get item embeddings in $\mathcal{H}^d$:

$$\boldsymbol{\theta}_i = \exp_\mathbf{o}(\boldsymbol{\theta}''_i) \tag{10}$$

We use these hyperbolic embeddings to initialize all items $\{\boldsymbol{\theta}_i\}_{i \in \mathcal{I}}$ of the proposed model. We create an item embedding matrix $\mathbf{M} \in \mathbb{R}^{|\mathcal{I}| \times d}$ where d is the embedding dimensionality, and retrieve the input embedding $\mathbf{E} \in \mathbb{R}^{n \times d}$, where $\mathbf{E}_i = \mathbf{M}_{s_i}$. A constant zero vector $\mathbf{0}$ is used as the embedding for the padding item.

Positional Embedding: We inject a learnable positional embedding $\mathbf{P} \in \mathbb{R}^{n \times d}$ into the input embedding as in Sect. 2.3. However, there is no notion of vector space structure in hyperbolic space. Thus we can't sum up two vectors directly. To deal with this problem, the main idea is to leverage the exp and log maps so that we can use the tangent space $\mathcal{T}_\mathbf{o}\mathcal{H}^d$ to perform Euclidean operations (e.g. vector addition).

To this end, we first project the item embeddings $\boldsymbol{\theta}_i$ to $\mathcal{T}_\mathbf{o}\mathcal{H}^d$ via the logarithmic map $\log_\mathbf{o} : \mathcal{H}^d \to \mathcal{T}_\mathbf{o}\mathcal{H}^d$. For the Lorentz representation, the logarithmic map is defined as [2]:

$$\log_\mathbf{o}(\mathbf{x}) = \sqrt{k}\,\text{arcosh}\left(-\frac{\langle \mathbf{o}, \mathbf{x} \rangle_{\mathcal{L}}}{k}\right)\frac{\mathbf{x} + \frac{1}{k}\langle \mathbf{o}, \mathbf{x} \rangle_{\mathcal{L}}\mathbf{o}}{\|\mathbf{x} + \frac{1}{k}\langle \mathbf{o}, \mathbf{x} \rangle_{\mathcal{L}}\mathbf{o}\|_{\mathcal{L}}} \tag{11}$$

where $\mathbf{x} \in \mathcal{H}^d, \mathbf{o} \in \mathcal{H}^d$ and $\mathbf{x} \neq \mathbf{o}$. The resulting vectors serve as item embeddings, denoted by $\mathbf{M}^E_i$, where the superscript ^{E} indicates that these item embeddings are in a Euclidean space:

$$\mathbf{M}^E_i = \log_\mathbf{o}(\boldsymbol{\theta}_i) \tag{12}$$

We initialize positional embeddings $\mathbf{P}_i$ in the tangent space $\mathcal{T}_\mathbf{o}\mathcal{H}^d$ of the reference point $\mathbf{o}$, where both the item embeddings and positional embeddings are in the same tangent space. Following [8,26], we directly fuse them to get the final input:

$$\widehat{\mathbf{E}} = \begin{bmatrix} \mathbf{M}^E_{s_1} + \mathbf{P}_1 \\ \mathbf{M}^E_{s_2} + \mathbf{P}_2 \\ \cdots \\ \mathbf{M}^E_{s_n} + \mathbf{P}_n \end{bmatrix} \tag{13}$$

3.3 Sequence Learning with Self-attention Mechanism

Recall that there is no notion of vector space structure in hyperbolic space. Euclidean vector operations and common Euclidean deep learning models are undefined in hyperbolic space, where we can't directly apply them. However, the tangent space at any point in the Lorentz manifold is Euclidean, allowing us to perform Euclidean operations undefined in hyperbolic space. Therefore, we perform self-attention mechanism proposed by SASRec [8] mentioned in 2.3 in the tangent space $\mathcal{T}_\mathbf{o}\mathcal{H}^d$ of the origin.

As mentioned above, the final input embeddings $\widehat{\mathbf{E}}$ are in the tangent space $\mathcal{T}_\mathbf{o}\mathcal{H}^d$. Thus we could directly apply self-attention block defined in 2.3 to them.

Self-attention Layer: In our case, the self-attention operation takes the embedding $\widehat{\mathbf{E}}$ as input, first linear transforms it to three matrices and then sends them into an attention layer:

$$\mathbf{S} = \text{SA}(\widehat{\mathbf{E}}) = \text{Attention}(\widehat{\mathbf{E}}\mathbf{W}^Q, \widehat{\mathbf{E}}\mathbf{W}^K, \widehat{\mathbf{E}}\mathbf{W}^V) \tag{14}$$

where the linear transformation matrices $\mathbf{W}^Q, \mathbf{W}^K, \mathbf{W}^V \in \mathbb{R}^{d \times d}$.

Causality: We modify the attention by disconnecting all links between $\mathbf{Q}_i$ and $\mathbf{K}_j (j > i)$ to prevent information leaks.

Point-Wise Feed-Forward Network: We apply a two-layer point-wise feed-forward network to all $\mathbf{S}_i$ identically (sharing parameters):

$$\mathbf{F}_i = \text{FFN}(\mathbf{S}_i) = \text{ReLU}(\mathbf{S}_i \mathbf{W}^{(1)} + \mathbf{b}^{(1)})\mathbf{W}^{(2)} + \mathbf{b}^{(2)} \tag{15}$$

where $\mathbf{W}^{(1)}, \mathbf{W}^{(2)}$ are $d \times d$ matrices and $\mathbf{b}^{(1)}, \mathbf{b}^{(2)}$ are d-dimensional vectors. Note that there is no interaction between $\mathbf{S}_i$ and $\mathbf{S}_j$ $(i \neq j)$, demonstrating that we still prevent information leaks (from later to earlier).

Stacking Self-attention Blocks: We stack self-attention blocks (i.e. a self-attention layer and a feed-forward network), and the b-th $(b > 1)$ block is:

$$\begin{aligned} \mathbf{S}^{(b)} &= \text{SA}(\mathbf{F}^{(b-1)}) \\ \mathbf{F}^{(b)}_i &= \text{FFN}(\mathbf{S}^{(b)}_i), \quad \forall i \in \{1, 2, \ldots, n\} \end{aligned} \tag{16}$$

and the 1-st block is defined as $\mathbf{S}^{(1)} = \mathbf{S}$ and $\mathbf{F}^{(1)} = \mathbf{F}$.

We adopt residual connection, dropout and layer normalization to alleviate problems of overfitting and unstable training process:

$$f'(x) = x + \text{Dropout}\left(f(\text{LayerNorm}(x))\right)$$

where $f(x)$ represents the self-attention layer or the feed-forward network.

3.4   Prediction Layer

After b self-attention blocks which adaptively and hierarchically extract information of previously visited items, given the first t items, we predict the next item based on $\mathbf{F}_t^{(b)}$. We use the opposite number of squared hyperbolic distance to predict user's preference score for item i, where the hyperbolic distance $d_{\mathcal{L}}$ has been defined in Eq. 2:

$$r_{i,t} = -d_{\mathcal{L}}(\mathbf{F}_t^{(b)}, \mathbf{M}_i)^2 \tag{17}$$

where $r_{i,t}$ means the possibility of item i being the next interacted item given the first t items (i.e. $s_1, s_2, \ldots, s_t$), and $\mathbf{M} \in \mathbb{R}^{|\mathcal{I}| \times d}$ is the item embedding matrix mentioned in 3.2. When item i is closer to user in the embedding space (with a small distance), i is more preferred by the user and gets a high score $r_{i,t}$, where a high score means a high relevance, and we can generate recommendations by ranking these scores.

3.5   Model Training

We apply data augmentation to generate more training data. For each position of each user training sequence $(\mathcal{S}_1, \mathcal{S}_2, \ldots, \mathcal{S}_{|\mathcal{S}|-1})$, we process it to a fixed length sequence $s = \{s_1, s_2, \ldots, s_n\}$ via truncation or padding items. Specifically, for a position p, if $p < n$, then we repeatedly add a 'padding' item to the right until the length is n. If $p > n$, we consider the most recent n actions. It means that for each position, we will get a training sequence of length n. The expected output o_p at position p is $\mathcal{S}_{p+1}$ for $1 \leq p < |\mathcal{S}| - 1$. For position $p = |\mathcal{S}| - 1$, $o_p = \mathcal{S}_{|\mathcal{S}|}$. For example, suppose a user interacts with item $1, 2, 3, 4, 5$ in time and $n = 3$, the data augmentation will generate these cases: $s1 = \{1, 0, 0\}, o_1 = 2$; $s2 = \{1, 2, 0\}, o_2 = 3$; $s3 = \{1, 2, 3\}, o_3 = 4$; $s4 = \{2, 3, 4\}, o_4 = 5$

The proposed model takes a sequence s as input, the corresponding sequence o as expected output, and we use the cross entropy loss as the objective function:

$$L = -\sum_{\mathcal{S} \in \mathcal{S}^{all}} \sum_{p \in [1, 2, \ldots, |\mathcal{S}|-1]} \log\left(\text{softmax}(r_{o_p, p})\right) \tag{18}$$

For a single data, we regard the prediction of next item as a multi-class classification task, where each item in the item set represents a class.

Since there are both hyperbolic and Euclidean parameters in our model, specifically, the item embeddings are hyperbolic while all the other parameters (e.g. position embedding and parameters of self-attention blocks) are Euclidean as they are defined in the tangent space of the origin. The Euclidean parameters can be optimized via Euclidean optimization, where we use the Adam [9] optimizer.

The hyperbolic item embeddings are optimized by the Riemannian Adam (RAdam) optimizer [1] based on the package *Geoopt*[1] [10]. RAdam is similar

---

[1] https://github.com/geoopt/geoopt.

Table 1. Statistics of the datasets after preprocessing.

Dataset	# Users	# Items	# Actions	# Sparsity
Beauty	22,363	12,101	198,502	99.93%
Games	24,304	10,673	231,780	99.91%
Steam	25,390	4,090	328,278	99.68%

to Adam while considering the geometry of the hyperbolic manifold. Following [1,19], given an item embedding $\boldsymbol{\theta}_i^{(t)}$ at iteration t, our optimization procedure consists of the following steps:

1. Compute the gradient of the loss in the Euclidean space $\nabla L \in \mathbb{R}^{d+1}$.
2. Compute the Riemannian gradient $\nabla^{\mathcal{H}^d} L$ by first computing $\boldsymbol{h}^{(t)} = g_{\mathcal{L}}^{-1} \nabla L$
   and then projecting $\boldsymbol{h}^{(t)}$ into $T_{\boldsymbol{\theta}_i^{(t)}} \mathcal{H}^d$: $\nabla^{\mathcal{H}^d} L = \boldsymbol{h}^{(t)} + \frac{\langle \boldsymbol{\theta}_i^{(t)}, \boldsymbol{h}^{(t)} \rangle_{\mathcal{L}}}{k} \boldsymbol{\theta}_i^{(t)}$
   $\nabla^{\mathcal{H}^d} L$ is a vector in the tangent space of $\mathcal{H}^d$. Then, compute the direction of descent $\mathrm{dir}(\nabla^{\mathcal{H}^d} L)$ according to the optimizer algorithm (e.g. Adam).
3. Estimate the update step using the exponential map and update the embedding with learning rate η: $\boldsymbol{\theta}_i^{(t+1)} = \exp_{\boldsymbol{\theta}_i^{(t)}}\left(-\eta\, \mathrm{dir}(\nabla^{\mathcal{H}^d} L)\right)$.

4   Experiments

4.1   Experimental Setup

Dataset. We conduct experiments on three datasets from real world applications with varying domains and sparsity. Datasets statistics after preprocessing (see details in below) are shown in Table 1.

- **Beauty and Video Games**: the two datasets are obtained from Amazon review datasets in [17], which include a series of datasets built on product reviews crawled from Amazon.com. Top-level product categories are treated as individual datasets. We select two subcategories: Beauty and Video Games. This dataset is highly sparse.
- **Steam**: a dataset introduced in [8], was crawled from *Steam*, a large online video game distribution platform. The dataset contains users' reviews on games. We remove duplicate interactions.

We followed the same preprocessing procedure as in [5,8,22]. For all the datasets, we treat users' interaction or rating in the implicit feedback setting and sort a user's interactions by the timestamps ascendingly. We filter cold items and inactive users with fewer than five related actions. For data partition, we split the sequence S of each user u into three parts: (1) the last action $S_{|S|}$ for testing, (2) the second last action $S_{|S|-1}$ for validation, and (3) all the remaining actions for training. Note that when testing, the input sequences include both training actions and the validation action.

Baseline Models. To verify the effectiveness of proposed method, following [31], we select a variety of baseline models for extensive comparison. The first group contains general non-sequential recommendation methods that only consider user interaction without utilizing the sequence order of historical actions:

- **PopRec**: ranks items by popularity measured by the number of interactions.
- **Bayesian Personalized Ranking (BPR)** [21]: models the pairwise interactions via matrix factorization and optimizes by a pair-wise Bayesian Personalized Ranking loss.

The second group includes deep-learning based sequential recommendation models of various architectures:

- **GRU4Rec** [7] is a RNN-based method which uses Gated Recurrent Unit (GRU) to model user action sequences for session-based recommendation. Each user's feedback sequence is regarded as a session.
- **Caser** [25] is a CNN-based method applying horizontal and vertical convolutional operations on the embedding matrix of the L most recent items.
- **SASRec** [8] is a unidirectional Transformer-based sequential recommendation model, which adopts a self-attention mechanism to identify relevant items for predicting the next item.
- **HGN** [16] uses hierarchical gating networks to capture both long-term and short-term user interests and applies an item-item product module to explicitly capture the item relations.
- **SRGNN** [27] models session sequences as graph-structured data, uses gated graph neural network to obtain node vector and uses an attention network to combine the global preference and current interests for recommendation.

Implementation Details. For fair comparison, we implement all the baselines based on RecBole [30], a unified, comprehensive and efficient framework for reproducing and developing recommendation algorithms for research purpose. For all the baselines, we set the embedding dimension to 50, use a batch size of 2048 and optimize them with Adam [9]. All hyper-parameters are set according to suggestions from the original papers. For each method, the grid search is applied to find the optimal settings of hyper-parameters using the validation set. We adopt early-stopped training if performance on the validation set doesn't improve for 10 epochs.

For our proposed HSA, we use 2 self-attention blocks and set the curvature $c = -1$. We implement HSA with Pytorch, based on RecBole [30] and the codes of [2]. We use the Riemannian Adam optimizer [1] with a learning rate of 0.01, and the batch size is 2048. The dimension of embedding is also 50, the same as baselines. The dropout rate is set to 0.5 for Amazon Beauty, Video Games and Steam considering their sparsity. The maximum sequence length n is set to 20 for Amazon Beauty, Video Games and 50 for Steam.

Table 2. Recommendation performance. The best performance and the second best performance methods are boldfaced and underlined respectively.

Dataset	Metric	PopRec	BPR	GRU4Rec	Caser	SASRec	HGN	SR-GNN	HSA
Beauty	Hit@10	0.2818	0.4073	0.4929	0.4401	0.5026	<u>0.5106</u>	0.4885	**0.5234**
	NDCG@10	0.1476	0.2452	0.3309	0.2828	<u>0.3485</u>	0.3419	0.3263	**0.3634**
Games	Hit@10	0.3746	0.5926	0.7378	0.6432	<u>0.7555</u>	0.7061	0.7079	**0.7609**
	NDCG@10	0.2134	0.3745	0.5188	0.4246	<u>0.5400</u>	0.4899	0.4901	**0.5480**
Steam	Hit@10	0.6259	0.6399	0.6807	0.6810	<u>0.6852</u>	0.6682	0.6764	**0.6871**
	NDCG@10	0.3897	0.4101	0.4508	0.4460	<u>0.4538</u>	0.4351	0.4434	**0.4575**

Evaluation Metrics. Two common Top-N metrics, Hit Rate@10 and NDCG@10 (Normalized Discounted Cumulative Gain) are adopted to evaluate recommendation model performance [5,6]. Hit@10 calculates the fraction of times that the ground-truth next item is listed in the top 10 items, while NDCG@10 is a position-relative metric where higher positions are assigned with larger weights. Note that since there is only one test item for each user, Hit@10 is equivalent to Recall@10, and is proportional to Precision@10. To save time and speed up evaluation, we avoid heavy computation on all user-item pairs. Instead, following the strategy in [6,11], for each user u, we uniformly sample 100 negative items and rank the ground-truth item with these sampled items. Based on the rankings of the 101 items, Hit@10 and NDCG@10 can be calculated.

4.2 Experimental Results

The experimental results on three datasets are presented in Table 2. We find that non-sequential recommendation approaches (i.e., PopRec and BPR) are obviously inferior to sequential recommendation approaches, which indicates that capturing sequential patterns is important for this task. As for the sequential recommendation approaches, the self-attention based SASRec performs the best among all the baselines in most of the metrics, which is better than RNN-based model GRU4Rec, CNN-based model Caser and GNN-based model SR-GNN. It's probably because self-attention mechanism can adaptively attend items within the range according to their relevance, better capturing sequential patterns. Besides, HGN is competitive with SASRec in part of datasets (e.g., Hit@10 in Beauty). Considering HGN utilizes hierarchical gating networks to model user interests, it indicates that hierarchically modeling is helpful to improve the sequential recommendation performance. However, gated GNN based model SR-GNN does not perform very well, possibly due to it is designed for session-based recommendation scenarios, where the sessions/sequences usually are shorter.

Finally, our proposed model HSA performs consistently better than all the baselines in all the datasets. The improvement on Beauty is notable, though improvement on Steam is marginal. This is because Steam is a challenging dataset since all the baselines with various architectures only marginally boost

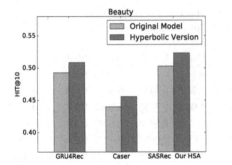 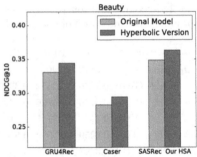

Fig. 2. Performance (Hit@10, NDCG@10) comparison of baseline models enhanced by our hyperbolic embedding learning on Amazon Beauty dataset.

the performance compared to PopRec, a simple baseline, which shows the difficulty of achieving improvement on it. Different from SASRec, we embed items with self-attention mechanism in the hyperbolic space. The hyperbolic space grows exponentially and has more powerful representation ability to learn better embeddings and improve recommendation accuracy than the common Euclidean space which only grows polynomially. Also, hyperbolic geometry can well model the hierarchical structure (e.g., the power-law distribution) in the recommendation datasets, which is showed in Sect. 2.1. The result demonstrates that hyperbolic space is effective for sequential recommendation.

4.3   Performance as a Plugin on Baselines

The key contribution of our HSA is the hyperbolic embedding learning, which is a general technique that can be applied to other sequential recommendation models. Therefore, we examine whether our hyperbolic embedding learning can boost other models. Similar to HSA, the item and/or user embeddings are in hyperbolic space, using the Lorentz model. Then those embeddings are projected to tangent space of the origin, where the same sequence encoding architecture is adopted and we get representation of sequences. Finally, hyperbolic distance is applied to measure the relevance of user behavior sequences and candidate items, making recommendation based on the relevance. We select classic RNN-based GRU4Rec [7], CNN-based Caser [25] and self-attention based SASRec [8] which is the Euclidean counterpart of our proposed HSA as the base models.

The results are shown in Fig. 2. First, equipped with our hyperbolic embedding learning, all the baselines achieve better performance. It shows that hyperbolic embedding learning is generally effective to improve the capacity of model and leverage latent hierarchical structures in datasets, for various architectures. Then, our HSA is still better than all the baselines and their variants. This is perhaps because it adopts self-attention which effectively identify important items and capture long-range dependencies in sequential recommendation task.

5 Conclusion

We have presented a hyperbolic self-attention scheme for sequential recommendation, of which the hyperbolic embedding learning can serve as a general plugin for improving existing sequential recommendation models. The extensive experimental results on public benchmarks show the effectiveness of our approaches. Source code will be made publicly available.

References

1. Becigneul, G., Ganea, O.E.: Riemannian adaptive optimization methods. In: International Conference on Learning Representations (2019)
2. Chami, I., Ying, Z., Ré, C., Leskovec, J.: Hyperbolic graph convolutional neural networks. In: Advances in Neural Information Processing Systems, vol. 32 (2019)
3. Feng, S., Tran, L.V., Cong, G., Chen, L., Li, J., Li, F.: HME: a hyperbolic metric embedding approach for next-poi recommendation. In: SIGIR, pp. 1429–1438 (2020)
4. Guo, N., et al.: HCGR: hyperbolic contrastive graph representation learning for session-based recommendation. arXiv preprint arXiv:2107.05366 (2021)
5. He, R., Kang, W.C., McAuley, J.: Translation-based recommendation. In: Proceedings of the Eleventh ACM Conference on Recommender Systems, pp. 161–169 (2017)
6. He, X., Liao, L., Zhang, H., Nie, L., Hu, X., Chua, T.S.: Neural collaborative filtering. In: Proceedings of the 26th International Conference on World Wide Web, pp. 173–182 (2017)
7. Hidasi, B., Karatzoglou, A., Baltrunas, L., Tikk, D.: Session-based recommendations with recurrent neural networks. In: International Conference on Learning Representations (2016)
8. Kang, W.C., McAuley, J.: Self-attentive sequential recommendation. In: 2018 IEEE International Conference on Data Mining (ICDM), pp. 197–206. IEEE (2018)
9. Kingma, D.P., Ba, J.: Adam: a method for stochastic optimization. arXiv preprint arXiv:1412.6980 (2014)
10. Kochurov, M., Karimov, R., Kozlukov, S.: Geoopt: riemannian optimization in pytorch. arXiv preprint arXiv:2005.02819 (2020)
11. Koren, Y.: Factorization meets the neighborhood: a multifaceted collaborative filtering model. In: Proceedings of the 14th ACM SIGKDD International Conference on Knowledge Discovery and Data Mining, pp. 426–434 (2008)
12. Krioukov, D., Papadopoulos, F., Kitsak, M., Vahdat, A., Boguná, M.: Hyperbolic geometry of complex networks. Phys. Rev. E **82**(3), 036106 (2010)
13. Li, A., Yang, B., Chen, H., Xu, G.: Hyperbolic neural collaborative recommender. arXiv preprint arXiv:2104.07414 (2021)
14. Li, J., Ren, P., Chen, Z., Ren, Z., Lian, T., Ma, J.: Neural attentive session-based recommendation. In: Proceedings of the 2017 ACM on Conference on Information and Knowledge Management, pp. 1419–1428 (2017)
15. Liu, Q., Wu, S., Wang, D., Li, Z., Wang, L.: Context-aware sequential recommendation. In: 2016 IEEE 16th International Conference on Data Mining (ICDM), pp. 1053–1058. IEEE (2016)

16. Ma, C., Kang, P., Liu, X.: Hierarchical gating networks for sequential recommendation. In: Proceedings of the 25th ACM SIGKDD International Conference on Knowledge Discovery & Data Mining, pp. 825–833 (2019)
17. McAuley, J., Targett, C., Shi, Q., Van Den Hengel, A.: Image-based recommendations on styles and substitutes. In: SIGIR, pp. 43–52 (2015)
18. Nickel, M., Kiela, D.: Poincaré embeddings for learning hierarchical representations. In: Advances in Neural Information Processing Systems, vol. 30 (2017)
19. Nickel, M., Kiela, D.: Learning continuous hierarchies in the lorentz model of hyperbolic geometry. In: International Conference on Machine Learning, pp. 3779–3788. PMLR (2018)
20. Ravasz, E., Barabási, A.L.: Hierarchical organization in complex networks. Phys. Rev. E **67**(2), 026112 (2003)
21. Rendle, S., Freudenthaler, C., Gantner, Z., Schmidt-Thieme, L.: BPR: bayesian personalized ranking from implicit feedback. In: Proceedings of the Twenty-Fifth Conference on Uncertainty in Artificial Intelligence, pp. 452–461 (2009)
22. Rendle, S., Freudenthaler, C., Schmidt-Thieme, L.: Factorizing personalized markov chains for next-basket recommendation. In: Proceedings of the 19th International Conference on World Wide Web, pp. 811–820 (2010)
23. Sun, F., Liu, J., Wu, J., Pei, C., Lin, X., Ou, W., Jiang, P.: Bert4rec: sequential recommendation with bidirectional encoder representations from transformer. In: Proceedings of the 28th ACM International Conference on Information and Knowledge Management, pp. 1441–1450 (2019)
24. Sun, J., Cheng, Z., Zuberi, S., Pérez, F., Volkovs, M.: HGCF: hyperbolic graph convolution networks for collaborative filtering. In: Proceedings of the Web Conference 2021, pp. 593–601 (2021)
25. Tang, J., Wang, K.: Personalized top-n sequential recommendation via convolutional sequence embedding. In: Proceedings of the Eleventh ACM International Conference on Web Search and Data Mining, pp. 565–573 (2018)
26. Vaswani, A., Shazeer, N., Parmar, N., Uszkoreit, J., Jones, L., Gomez, A.N., Kaiser, L., Polosukhin, I.: Attention is all you need. Advances in neural information processing systems **30** (2017)
27. Wu, S., Tang, Y., Zhu, Y., Wang, L., Xie, X., Tan, T.: Session-based recommendation with graph neural networks. In: Proceedings of the AAAI Conference on Artificial Intelligence, vol. 33, pp. 346–353 (2019)
28. Yuan, F., Karatzoglou, A., Arapakis, I., Jose, J.M., He, X.: A simple convolutional generative network for next item recommendation. In: Proceedings of the twelfth ACM International Conference on Web Search and Data Mining, pp. 582–590 (2019)
29. Zhang, S., Yao, L., Sun, A., Tay, Y.: Deep learning based recommender system: a survey and new perspectives. ACM Comput. Surv. (CSUR) **52**(1), 1–38 (2019)
30. Zhao, W.X., et al.: Recbole: towards a unified, comprehensive and efficient framework for recommendation algorithms. In: Proceedings of the 30th ACM International Conference on Information & Knowledge Management, pp. 4653–4664 (2021)
31. Zhou, K., Yu, H., Zhao, W.X., Wen, J.R.: Filter-enhanced MLP is all you need for sequential recommendation. In: Proceedings of the ACM Web Conference 2022, pp. 2388–2399 (2022)

CFGCon: A Scheme for Accurately Generating Control Flow Graphs of Smart Contracts

Nengyu Xia[1], Yixin Zhang[1], Wei Ren[1(✉)], and Xianyi Chen[2]

[1] School of Computer Science, China University of Geosciences, Wuhan, China
weirencs@cug.edu.cn
[2] Engineering Research Center of Digital Forensics, Ministry of Education, Nanjing University of Information Science and Technology, Nanjing, China

Abstract. Smart contracts are a significant component that allows decentralized applications (DApps) to automate the exchange of digital assets without third-party surveillance. To build trust, smart contracts are designed to be immutable, resulting in design flaws that may remain unrevealed in deployed contracts. Many analysis tools are developed to identify various vulnerabilities that could be targeted by hackers after deployment and thus cause financial losses. However, these approaches based on graph classification rely much on the quality of control flow graphs (CFGs) generated from the bytecode of smart contracts. In this paper, we propose a novel generator named CFGCon to convert byte-codes of smart contracts to CFGs. After targeting the difficulties for the existing CFG generators, a program counter is designed to deal with the opcodes with loops or instructions that need to read the current counter. Experimental results show that our proposed CFGCon reached a much higher success rate than other state-of-art CFG generators on the dataset containing 579 open source contracts and 10,000 non-open source contracts from Ethereum. At the same time, the analysis speed of CFGCon is similar to that of the current mainstream tools.

Keywords: smart contract · bytecode · Control Flow Graph

1 Introduction

Enabled by smart contracts, Ethereum has gained considerable popularity due to its various decentralized applications (DApps), such as transactions for digital assets [15], industrial Internet of things [12], crowdsourcing [18], etc. Smart contracts serve as a fundamental technology that ensures parties running those DApps faithfully follow the immutable agreements as programmed [20]. However, the immutability of the contract codes may be cut both ways. On one hand, it brings convenience to decentralized parties for managing billions of dollars of assets by simply calling smart contracts. On the other hand, smart contracts may suffer from the vulnerabilities in their code like all computer programs, which may be targeted by hackers and cause unexpected financial losses. The famous TheDAO incident, in which 3.6 million Ether coins were stolen, results in a loss of more than 60 million

© The Author(s), under exclusive license to Springer Nature Singapore Pte Ltd. 2024
X. Song et al. (Eds.): APWeb-WAIM 2023, LNCS 14333, pp. 265–279, 2024.
https://doi.org/10.1007/978-981-97-2387-4_18

dollars [6]. Therefore, identifying the loopholes both in deployed and undeployed contracts becomes an urgent need in recent years.

To deal with this situation, some automated tools have been developed to analyze smart contracts given by the source codes written in some quasi-Turing complete programming language such as Solidity [11]. However, their availability is rather limited considering that most smart contracts published on Ethereum are non-open source [22]. For smart contracts where only the bytecodes are available, graph classification methods becomes a popular solution to detect vulnerabilities. Each smart contract is transformed into a control flow graph (CFG) in which the complete information is supposed to be represented in a structured form. Then the graph classification methods can be carried out based on the generated CFGs. For example, to determine whether a smart contract has a certain type of defect, one may compute the graph similarity between the CFG of the tested contract and other contracts which are confirmed to have such a defect. Therefore, the quality of the generated CFGs is the basis of smart contract analyzing methods based on graph classification. How to generate contract CFGs accurately becomes a vital problem.

Some existing analysis tools provide a CFG generation algorithm as a by-product that follows the procedure described below [19,21]. The basic blocks are first separated according to the meaning of the opcode. Then the jump directed edge between the basic blocks is determined by simulating stack operation. However, we discovered that some of the contracts deployed on Ethereum failed to be resolved in actual tests. It is worth noting that the number of smart contracts that are not successfully resolved could be substantial since there are numerous smart contracts published on Ethereum. Our goal is to parse as many of them as possible to make sure we can conduct the next step of vulnerability detection and other analysis. In this paper, we focus on the success rate of generating smart contract bytecode CFG as the main performance metric. Firstly, the existing methods are manually implemented. Then the model is used to test the test set to get the test results, and the stack analysis and debugging of the failed bytecode set are carried out to find out the cause of the analysis failure and make corrections. After adding module functions, the performance-enhanced model is finally obtained.

The contributions of the paper are as follows:

1. We propose the CFGCon, a parsing tool to generate CFGs for smart contracts, especially for those only provided with bytecodes in Ethereum. Its success rate is higher than other CFG generators to the best of our knowledge.
2. By designing a program counter, CFGCon solves the problem that the mainstream CFG generation tools are difficult to parse the opcodes with loops or the instructions that need to read the current counter value.

The rest of the paper is organized as follows: Related work is reviewed in Sect. 2. Section 3 presents preliminaries. We propose new algorithms for CFG generation in Sect. 4. Section 5 describes the experimental results. Finally, Sect. 6 concludes the paper.

2   Related Work

CFG represents all paths traversed by a program during execution [23]. It shows the possible flow of all basic block execution in a process in the form of a graph and can also reflect the real-time execution process. There is an endless stream of tools for the CFG generation of smart contracts. Many tools, such as *Oyente* [19], *Octopus* [10], *Mythril* [21], and *DefectChecker* [7], generate CFG through symbolic execution and take it as the input of smart contract vulnerability analysis module for further analysis. *SAFEPAY* [17] analysis framework determines whether there is an unfair payment problem in the smart contract by analyzing the contract CFG. *sCompile* [5] analyzes the constructed contract CFG, automatically identifies the critical program path including multiple function calls in the smart contract, and identifies the security vulnerabilities in the contract based on this. *SAFEVM* [1] uses the C program verification engine to extract and analyze the CFG of the input Ethereum smart contract and generate the report of the contract vulnerability analysis in the output.

CFG can be used for more than just smart contract vulnerability detection. CFG is widely used in the analysis of smart contract bytecode because it preserves program semantics. For example, in the detection of gas-Inefficient smart contracts, *Gaschecker* [8], requires the CFG for program analysis. *TEETHER* [14] uses CFG to solve the automatic identification of contract vulnerabilities and the generation of contract vulnerabilities. [3] detects the gas upper limit of the contract with CFG. *Elysium* [24] realizes the automatic repair of smart contracts by inferring context information from bytecode CFG and combining it with the patch. *Ethir* [2] improved the problem that Oyente could not generate all possible directed edges of CFG to apply high-level analysis to infer properties of EVM code. [26] implements smart contract bytecode similarity detection across optimized options and compiled versions through CFG. By analyzing the CFG of the bytecode of the smart contract, *GASPER* [9] locates the gas consumption mode of the contract. It determines whether the problem of high gas consumption caused by insufficient optimization exists in the contract. [13] analyzes the CFG of the contract to achieve super optimization of the smart contract, that is, to find the best translation of the instruction block. *STAN* [16] system generates a description of the smart contract bytecode to help users understand the contents of the contract. *EtherSolve* [10] proposed a new static analysis algorithm based on Ethereum operand stack symbol execution, which improved the generation success rate compared to previous tools. However, it is found that the success rate of parsing is not high in some cases, such as when the CFG with loops or the instruction set of the analyzed bytecode needs to read the current program counter value.

The vast majority of tools only use CFG as intermediate data as the input of the next module, but the success rate of these tools can be further improved, and the algorithm can also be optimized. This paper focuses on improving the success rate of CFG generation and simplifying the algorithm.

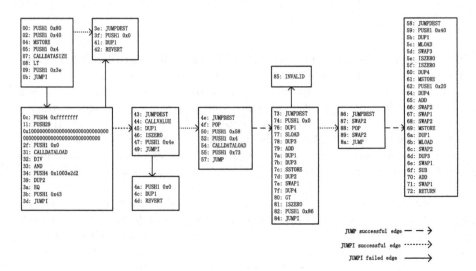

Fig. 1. A Control Flow Graph example.

3   Preliminaries

The general procedure of the smart contract bytecode CFG generation is introduced as follows:

1) Bytecode to opcode

The bytecodes of smart contracts on Ethereum correspond to the opcodes, and the best way to parse the bytecodes of contracts is to disassemble them into opcodes that better explain the semantics. The Ethereum Virtual Machine (EVM) is a stack virtual machine with a 256-bit word length for running smart contracts on the Ethereum blockchain [25]. EVM uses single-byte opcodes, therefore all opcodes are defined in the range from 00 to FF. Because opcodes are limited to one byte, the EVM instruction set can hold a maximum of 256 instructions, of which more than 140 have to define so far.

2) Division of basic blocks

The basic block represents the node of CFG. The basic block is a collection of programs with only one entry and only one exit [4]. Once the basic block is activated, it will be executed to the end without any other branch routes. Jumps between basic blocks are implemented using JUMP and JUMPI operations, and jump positions start with JUMPDEST.

3) Get the next hop position of the basic block

Traversing the resulting basic block set, one notable difference between EVM and other virtual machine languages is the use of a stack to hold jump addresses. In this paper, we simulate the stack operation of the EVM combined with the program counter (PC) to find the next hop position of each basic block. The program counter represents the position of the next instruction in the bytecode. The value of the program counter is increased or assigned after each command is executed.

After the above three operations, a CFG with a basic block as node and jump direction as the directed edge will be obtained. The example figure is shown in Fig. 1.

4 Proposed Scheme

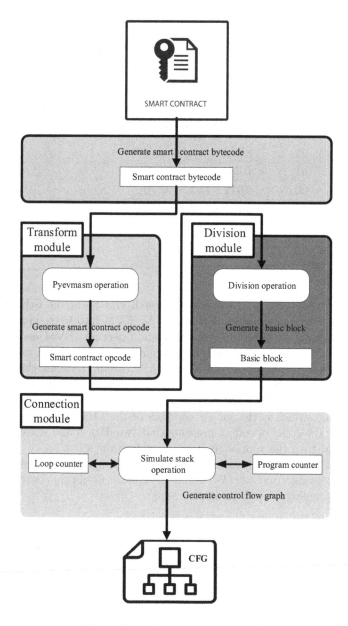

Fig. 2. Overview of system model.

4.1   Overview of the System Model

This model consists of three parts, namely bytecode to opcode, division of basic block, and obtaining the next jump position of the basic block. The overall design diagram is shown in Fig. 2.

Firstly, the mainstream smart contract *Bytecode* on the market is obtained and converted into *Opcode* through disassembly operation (pyevmasm). Then each independent *Basic Block* is obtained through division. Finally, the simulation stack is used to obtain the final result (CFG).

Table 1. Conversion rules for bytecode and opcode

Bytecode	Opcode	Simulate stack ($Top, Bottom$)
00	STOP	–
01	ADD	$a,b \Rightarrow a + b$
02	MUL	$a,b \Rightarrow a * b$
03	SUB	$a,b \Rightarrow a - b$
04	DIV	$a,b \Rightarrow a / b$
05	SDIV	$a,b \Rightarrow a / b$
06	MOD	$a,b \Rightarrow a \% b$
07	SMOD	$a,b \Rightarrow a \% b$

4.2   Transform Module

Contracts are deployed and executed through the EVM on Ethereum. The smart contract source program is compiled by EVM into bytecodes, which are composed of many corresponding opcodes and operands. EVM uses a series of instructions (i.e., opcodes) to perform different tasks. Conversion rules for bytecode and opcode are shown in Table 1. The first column presents the smart contract hex bytecode. The second column indicates the opcode name corresponding to the bytecode. And the third column represents the operation corresponding to each opcode, which is the simulated stack of the Ethereum virtual machine. For example, the ADD opcode represents that two data a and b are taken from the stack, and then the sum of the two is put into the stack.

In the analysis of smart contracts using CFG, the bytecode is first required to be disassembled into opcodes. We adopt pyevmasm to obtain contract opcodes through disassembly operations in this paper.

4.3   Division Module

According to predefined rules, the opcodes are divided to obtain basic blocks which are used as nodes of the CFG. This module iterate over the opcodes obtained in the previous step and divide the opcode set into basic blocks according to the following rules:

1) The operation codes JUMP, JUMPI, STOP, RETURN, INVALID, REVERT, and SELFDESTRUCT represent the end of the basic block and indicate the end of the basic block.
2) The beginning of the target basic block is located when the JUMPDEST opcode is encountered, indicating that the basic block starts execution.
3) The opcodes other than the above opcodes are interpreted as the middle part of the basic block and can be directly added to it. The only special part is the beginning of the entire opcode set, which represents the beginning of the first basic block.

The specific algorithm is shown in Algorithm 1.

Algorithm 1: Basic block

Input: The set of contract opcode, OP_n
Output: The set of partitioned basic block, B_n
while *not at end of OP_n* **do**
    if *Opcode in "JUMP JUMPI STOP RETURN INVALID REVERT SELFDESTRUCT"* **then**
        | this is blockend;
    else if *Opcode == "JUMPDEST" or the beginning of the entire opcode set OP_n* **then**
        | this is blockbegin;
    else
        | this is blockpart;
    end
end
$B_n \leftarrow$ Combine all basic blocks;
Return B_n;

4.4 Connection Module

The next hop position of each basic block is obtained by traversing the basic block. The original basic block and the basic block where the next hop is located form a directed edge, which is regarded as the directed edge of the CFG. Basic block jumps are divided into conditional jumps (JUMPI) and unconditional jumps (JUMP). JUMP reads the address directly from the top of the stack and jumps. JUMPI reads two pieces of data at a time from the stack. The first piece of data serves as the destination address, and the second piece of data serves as a judgment condition. If the judgment condition is true, the algorithm jumps to the destination address; otherwise, it jumps to the block next to the JUMPI address of the basic block. The mock stack operations perform according to the following rules (shown in Algorithm 2):

1) When an opcode starting with PUSH is encountered, the next opcode is pushed onto the stack.

2) When an opcode starting with DUP is encountered, the element corresponding to the numeric position following DUP is copied to the top of the stack. DUP2, for example, copies the second top element to the top of the stack.

3) When encountering an opcode at the beginning of SWAP, SWAP the element from the first top of the stack corresponding to the numeric position after SWAP with the first top of the stack element. SWAP15, for example, swaps the 16th top-stack element with the first top-stack element.

4) When the AND opcode is encountered, two elements in a row are removed from the stack. This situation needs further judgments: if these two elements are not unknown arguments, the bitwise AND operation is performed on them, and the result is pushed onto the stack; If either or both elements are positional arguments, the string "UnknownArgument" is pushed onto the stack.

5) When the GETPC opcode is encountered, the next instruction position needs to be read from the program technician and pushed onto the stack.

6) When the JUMP opcode is encountered, the algorithm takes the top element of the stack as the JUMP address and calculates the block number of the target block. If the first instruction of the target block is JUMPDEST, the algorithm adds a directed edge pointing from the basic block of the JUMP instruction to the basic block of the target block; otherwise, the JUMP fails.

7) When the JUMPI opcode is encountered, the algorithm reads the JUMPI opcode in succession and determines whether the value of the second JUMPI opcode is true. If true, the algorithm reads the JUMPI opcode in succession and determines whether the first instruction of the basic block with the JUMPI opcode as the initial address is JUMPDEST. If so, a directed edge is added from the basic block where the JUMPI instruction resides to the basic block whose initial address is the first element at the top of the stack. If not, JUMPI fails to jump and a directed edge is added from the basic block of the JUMPI instruction to the basic block of the JUMPI instruction's next address.

8) When other opcodes are encountered, stack operation shall be simulated according to the influence of the opcode on the entry and exit of stack elements. The number of stack out and stack elements shall be defined according to the specific meaning of the opcode, and no specific operation shall be performed. For example, the ADD command takes out two elements and adds them to the stack. In comparison, the simulated stack operation symbolically takes out two elements at the top of the stack and pushes the string "UnknownArgument" directly onto the stack.

Algorithm 2: Acquisition of directed edge

Input: The set of partitioned basic block, B_n; The program counter, PC; The simulation of the stack, S;

Output: The set of basic blocks at the tail of directed edge, EB_n; The set of basic blocks at the head of directed edge, EE_n;

while *not at end of B_n* **do**

    if B_i *in "PUSHx"* **then**

        | $PC += x + 1$; S.push(B_{i+1});

    else if B_i *in "DUPx"* **then**

        | $PC += 1$; a = The xth element in S; S.push(a);

    else if B_i *in "SWAPx"* **then**

        | $PC += 1$; a = S.pop(); The xth element in S $<=>$ a; S.push(a);

    else if $B_i ==$ *"AND"* **then**

        $PC += 1$; a = S.pop(); b = S.pop();

        if *a and b* $!=$ *"UnknownArgument"* **then**

            | S.push(a&b);

        else

            | S.push("UnknownArgument");

        end

    else if $B_i ==$ *"GETPC"* **then**

        | S.push(PC); $PC += 1$;

    else if $B_i ==$ *"JUMP"* **then**

        $PC = S$.pop(); TargetBlockNumber = PC;

        EB_n.append("CurrentBlockNumber");

        EE_n.append("TargetBlockNumber");

    else if $B_i ==$ *"JUMPI"* **then**

        a = S.pop(); b = S.pop();

        if *b* **then**

            $PC = $ a; TargetBlockNumber = PC;

            EB_n.append("CurrentBlockNumber");

            EE_n.append("TargetBlockNumber");

        else

            $PC += 1$; CurrentNextBlockNumber = PC;

            EB_n.append("CurrentBlockNumber");

            EE_n.append("CurrentNextBlockNumber");

        end

    else

        $PC += 1$;

        S.pop()* Operation times and S.push("UnknownArgument")* Operation times;

    end

end

Return EB_n and EE_n;

---

5   Experiment and Performance Evaluation

5.1   Dataset

We took the dataset of smart contracts from Ethereum given in *Defectchecker*, which contains the bytecodes of 579 open source smart contracts and 180,000 non-open source smart contracts. In this experiment, 10,000 contract samples were randomly selected from the non-open source contract dataset to test the success rate of our proposed CFGCon. At the same time, in order to find out the possible reasons for the failure of the analysis, this paper adds open source contract data to find problems directly on the basis of the source code. We use the open source contract samples combined with part of the non-open source contract samples to form a dataset with a total of 1000 contracts for further testing. A comparison experiment has been carried out by measuring the success rates of Oyente, Octopus, EtherSolve and our proposed CFGCon using the dataset of 1000 contracts.

5.2   General Test for CFGCon

This experiment is designed to test the ability of the proposed CFGCon in general. We randomly selected 10,000 contracts from 180,000 non-open source smart contracts. Then the proposed CFGCon was applied to resolve their bytecodes. Table 2 shows a summarized result of the experiment.

Table 2. CFGCon test results

	CFGCon
Number of contracts successfully analyzed	9998
Number of failed contracts analyzed	2
The total number of contracts	10000
Analysis success rate	99.9%

The test results showed that 9998 out of 10,000 contracts were successfully resolved, with a success rate of 99.9%. The bytecodes of the two failed contracts are further analyzed. As shown in Fig. 3, it can be found from the opcode snippet that there are programming problems with their bytecodes. We believe that this is the reason causing the operation process to fail. When the opcode flow is executed to the opcode "SWAP14" marked in red in Fig. 3, the algorithm halts since there are no 15 elements in the simulated stack and the operation cannot be performed.

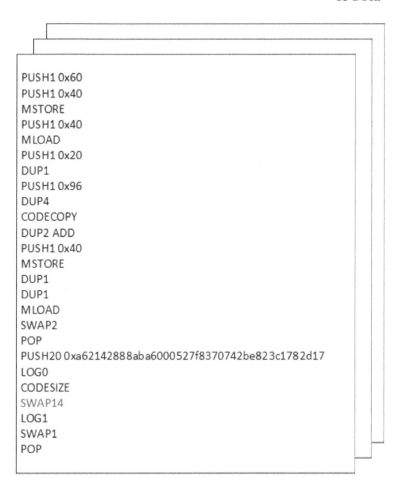

Fig. 3. Opcode snippet.

5.3   Performance Comparision with Existing Approaches

A comparative experiment was conducted using Oyente, Octopus, EtherSolve and the proposed CFGCon with 1000 smart contract bytecodes published on Ethereum. The results are shown in Table 3 and Fig. 4.

Oyente has successfully analyzed 861 out of 1000 smart contract samples, reaching a success rate of 86.1%. Octopus only achieved a low success rate, only 12%. To the best of our knowledge, EtherSolve is the smart contract CFG generation tool with the highest analysis success rate. EtherSolve has successfully analyzed 949 out of 1000 smart contract samples, reaching a success rate of 94.9%. However, our proposed CFGCon can resolve all 1000 smart contract samples, including 51 contracts that failed to resolve by EtherSolve.

In our experiment, although Octopus can parse the CFG of all contracts, 88% of results found through analysis to be incomplete flow graphs. The reason

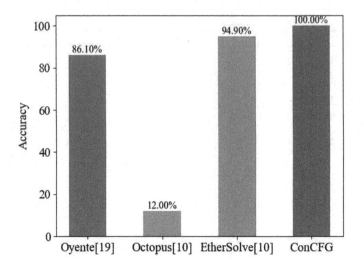

Fig. 4. Comparison diagram of test results.

Table 3. Comparison table of test results

	Oyente [19]	Octopus [10]	EtherSolve [10]	CFGCon
Number of contracts successfully analyzed	861	120	949	1000
Number of failed contracts analyzed	139	880	51	0
The total number of contracts	1000	1000	1000	1000
Analysis success rate	86.1%	12.0%	94.9%	100%

is that Octopus cannot resolve all jump destination addresses, resulting in the loss of many paths in the CFG. We performed further analysis on the part of the contract that EtherSolve failed to analyze. The possible reasons for EtherSolve analysis failure are concluded as follows:

1) The contract for an empty method cannot be resolved, accounting for 10% of the 51 analyzed failed contracts.
2) Flowgraphs with loops are difficult to parse, accounting for 10% of the 51 failed contracts analyzed.
3) Difficult parsing operations that required reading program counters, which account for 40% of the 51 contracts.
4) The remaining contracts that failed to parse are attributed to other unknown reasons, such as the problem with the program mentioned in the previous section. These reasons might be very particular since the proposed CFGCon can resolve these contracts.

Throughout the analysis, it should be noted that the generation of CFG for smart contracts could encounter many problems, which is the reason for improving the CFG generator to achieve a higher success rate. In a conclusion, our proposed CFGCon is promising to parse almost all the smart contract bytecodes except the ones with programming problems. Compared with EtherSolve, the success rate of smart contract bytecode CFG generation has been improved by 5.1%, which means that our proposed CFGCon can parse thousands of smart contracts in Ethereum more than other state-of-art methods.

Additionally, we compared the parsing speed of different tools by testing their average time consumption for each contract. The result is that the average time of CFGCon is 3.9 s, which is similar to the 2.7 s of EtherSolve. At the same time, it is also relatively close to the fastest Oyente's 2.5 s. The above comparison tools can be used in application scenarios, and have a high usage rate. Even though our proposed scheme is slightly inferior to the other three schemes in parsing speed, our scheme is overall better due to higher success rate.

6   Conclusion

In this paper, we design a generation tool parsing smart contracts of bytecodes to CFGs which can be used for further vulnerability detection and smart contract analysis. The main procedure includes a transform module for converting bytecodes to opcodes, a division module for generating basic blocks, and a connection module for simulating stack operations. By witnessing the drawbacks of the existing CFG generators, we design a program counter to parse the opcodes with loops more effectively. It is also helpful to deal with special instructions which need to read the current counter. As a result, our proposed CFGCon has a success rate of 99.9% in the experiment of resolving 10,000 contracts deployed on Ethereum. A comparative experiment is also carried out between the proposed CFGCon and EtherSolve, the tool that achieves the highest success rate so far. CFGCon outperforms EtherSolve by successfully resolving 51 more contracts out of 1000 contract samples. In a conclusion, the CFGCon is capable of parsing more smart contracts from bytecode format to CFGs. Other studies such as vulnerability detection and inefficient gas detection can use CFGCon as a basic tool to enlarge their study samples. Future study of this work lies in analyzing the contracts which CFGCon fails to resolve in detail and further improving the success rate.

Acknowledgement. The research was financially supported by the Provincial Key Research and Development Program of Hubei (No. 2020BAB105), the Knowledge Innovation Program of Wuhan - Basic Research (No. 2022010801010197), the Opening Project of Engineering Research Center of Digital Forensics, Ministry of Education (No. 20220103), and the Opening Project of Nanchang Innovation Institute, Peking University (No. NCII2022A02).

References

1. Albert, E., Correas, J., Gordillo, P., Román-Díez, G., Rubio, A.: SAFEVM: a safety verifier for Ethereum smart contracts. In: Proceedings of the 28th ACM SIGSOFT International Symposium on Software Testing and Analysis (STA 2019), pp. 386–389 (2019)
2. Albert, E., Gordillo, P., Livshits, B., Rubio, A., Sergey, I.: ETHIR: a framework for high-level analysis of ethereum bytecode. In: Lahiri, S.K., Wang, C. (eds.) ATVA 2018. LNCS, vol. 11138, pp. 513–520. Springer, Cham (2018). https://doi.org/10.1007/978-3-030-01090-4_30
3. Albert, E., Gordillo, P., Rubio, A., Sergey, I.: Running on fumes: preventing out-of-gas vulnerabilities in Ethereum smart contracts using static resource analysis. In: Ganty, P., Kaâniche, M. (eds.) VECoS 2019. LNCS, vol. 11847, pp. 63–78. Springer, Cham (2019). https://doi.org/10.1007/978-3-030-35092-5_5
4. Almakhour, M., Sliman, L., Samhat, A.E., Mellouk, A.: Verification of smart contracts: a survey. Perv. Mobile Comput. **67**, 101227 (2020)
5. Chang, J., Gao, B., Xiao, H., Sun, J., Cai, Y., Yang, Z.: sCompile: critical path identification and analysis for smart contracts. In: Ait-Ameur, Y., Qin, S. (eds.) ICFEM 2019. LNCS, vol. 11852, pp. 286–304. Springer, Cham (2019). https://doi.org/10.1007/978-3-030-32409-4_18
6. Chen, J., Xia, X., Lo, D., Grundy, J., Luo, X., Chen, T.: Defining smart contract defects on Ethereum. IEEE Trans. Software Eng. **48**(1), 327–345 (2020)
7. Chen, J., Xia, X., Lo, D., Grundy, J., Luo, X., Chen, T.: DefectChecker: automated smart contract defect detection by analyzing EVM bytecode. IEEE Trans. Software Eng. **48**(7), 2189–2207 (2021)
8. Chen, T., et al.: GasChecker: scalable analysis for discovering gas-inefficient smart contracts. IEEE Trans. Emerg. Top. Comput. **9**(3), 1433–1448 (2020)
9. Chen, T., Li, X., Luo, X., Zhang, X.: Under-optimized smart contracts devour your money. In: Proceedings of the 2017 IEEE 24th International Conference on Software Analysis, Evolution and Reengineering (SANER 2017), pp. 442–446. IEEE (2017)
10. Contro, F., Crosara, M., Ceccato, M., Dalla Preda, M.: EtherSolve: computing an accurate control-flow graph from Ethereum bytecode. In: Proceedings of the 2021 IEEE/ACM 29th International Conference on Program Comprehension (ICPC 2021), pp. 127–137. IEEE (2021)
11. Grieco, G., Song, W., Cygan, A., Feist, J., Groce, A.: Echidna: effective, usable, and fast fuzzing for smart contracts. In: Proceedings of the 29th ACM SIGSOFT International Symposium on Software Testing and Analysis (STA 2020), pp. 557–560 (2020)
12. He, S., Ren, W., Zhu, T., Choo, K.-K.R.: BoSMoS: a blockchain-based status monitoring system for defending against unauthorized software updating in industrial Internet of Things. IEEE Internet Things J. **7**(2), 948–959 (2019)
13. Hernández Cerezo, A.: Integrating the EVM super-optimizer gasol into real-world compilers (2021)
14. Krupp, J., Rossow, C.: TEETHER: gnawing at ethereum to automatically exploit smart contracts. In: Proceedings of the 27th USENIX Security Symposium (USENIX Security 2018), pp. 1317–1333 (2018)
15. Li, T., et al.: FAPS: a fair, autonomous and privacy-preserving scheme for big data exchange based on oblivious transfer, ether cheque and smart contracts. Inf. Sci. **544**, 469–484 (2021)

16. Li, X., Chen, T., Luo, X., Zhang, T., Yu, L., Xu, Z.: STAN: towards describing bytecodes of smart contract. In: Proceedings of the 2020 IEEE 20th International Conference on Software Quality, Reliability and Security (QRS 2020), pp. 273–284. IEEE (2020)

17. Li, Y., Liu, H., Yang, Z., Ren, Q., Wang, L., Chen, B.: SAFEPAY on Ethereum: a framework for detecting unfair payments in smart contracts. In: Proceedings of the 2020 IEEE 40th International Conference on Distributed Computing Systems (ICDCS 2020), pp. 1219–1222. IEEE (2020)

18. Lin, C., He, D., Huang, X., Choo, K.-K.R.: OBFP: optimized blockchain-based fair payment for outsourcing computations in cloud computing. IEEE Trans. Inf. Forensics Secur. **16**, 3241–3253 (2021)

19. Luu, L., Chu, D.-H., Olickel, H., Saxena, P., Hobor, A.: Making smart contracts smarter. In: Proceedings of the 2016 ACM SIGSAC Conference on Computer and Communications Security (CCS 2016), pp. 254–269 (2016)

20. Mohanta, B.K., Panda, S.S., Jena, D.: An overview of smart contract and use cases in blockchain technology. In: Proceedings of the 2018 9th International Conference on Computing, Communication and Networking Technologies (ICCCNT 2018), pp. 1–4. IEEE (2018)

21. Mueller, B.: Smashing Ethereum smart contracts for fun and real profit. HITB SECCONF Amsterdam **9**, 54 (2018)

22. Shi, C., Xiang, Y., Yu, J., Gao, L., Sood, K., Doss, R.R.M.: A bytecode-based approach for smart contract classification. In: Proceedings of the 2022 IEEE International Conference on Software Analysis, Evolution and Reengineering (SANER 2022), pp. 1046–1054. IEEE (2022)

23. Tolmach, P., Li, Y., Lin, S.-W., Liu, Y., Li, Z.: A survey of smart contract formal specification and verification. ACM Comput. Surv. **54**(7), 1–38 (2021)

24. Torres, C.F., Jonker, H., State, R.: Elysium: automagically healing vulnerable smart contracts using context-aware patching. CoRR (2021)

25. Wood, G., et al.: Ethereum: a secure decentralised generalised transaction ledger. Ethereum Project Yellow Paper **151**(2014), 1–32 (2014)

26. Zhu, D., Yue, F., Pang, J., Zhou, X., Han, W., Liu, F.: Bytecode similarity detection of smart contract across optimization options and compiler versions based on triplet network. Electronics **11**(4), 597 (2022)

Hypergraph-Enhanced Self-supervised Heterogeneous Graph Representation Learning

Yuanhao Zhang, Chengxin He, Longhai Li, Bingzhe Zhang, Lei Duan, and Jie Zuo[✉]

School of Computer Science, Sichuan University, Chengdu, China
{zhangyuanhao,hechengxin,lilonghai,zhangbingzhe}@stu.scu.edu.cn,
{leiduan,zuojie}@scu.edu.cn

Abstract. Heterogeneous graphs are widely used to model complex systems in the real world, such as social networks, biomedical networks, and citation networks. Learning heterogeneous graph embeddings (i.e., representations) provides a way to perform deep learning-driven downstream tasks, such as recommendation and prediction. However, existing heterogeneous graph neural networks mainly capture pairwise relations in heterogeneous graphs, while real-world relations are often more complex and not limited to pairs. In this paper, we propose a novel method to capture relations beyond pairwise in heterogeneous graphs, namely HHGR. First, we construct hypergraphs from heterogeneous graphs and preserve semantic information of network schema and meta paths. Second, we design a cross-view contrast module to aggregate information on different aspects. Further, to enhance the performance of HHGR, we propose a semantic positive sampling strategy, which chooses proper positive samples according to structure and attribute semantics. Extensive experiments conducted on various real-world datasets demonstrate the state-of-the-art performance of HHGR.

Keywords: Heterogeneous graph · Hypergraph · Graph neural network · Self-supervised representation learning

1 Introduction

Recently, heterogeneous graph (HG), i.e. heterogeneous information network (HIN), has been an extremely hot topic in data mining as its powerful ability to model many real-world scenarios, ranging from social networks [31], biomedical networks [4], cybersecurity [11] to healthcare systems [2]. Heterogeneous graph neural networks (HGNNs) have drawn attentions due to the superiority in message passing with complex heterogeneity and the ability to capture rich semantics implied in HGs.

This work was supported in part by the National Natural Science Foundation of China (61972268), and the Joint Innovation Foundation of Sichuan University and Nuclear Power Institute of China.

X. Song et al. (Eds.): APWeb-WAIM 2023, LNCS 14333, pp. 280–295, 2024.
https://doi.org/10.1007/978-981-97-2387-4_19

Most HGNN methods rely on heterogeneous relations and meta paths to capture semantic information [8,27], which describe composite relations among nodes and provide an efficient way to capture the semantics in HGs. However, they are designed for pairwise relations. In the real world, relations are complex and not always appear in pairs. For example, the relations "multiple authors co-authoring a paper" and "multiple businesses purchased by a user" are not easily described by meta paths. Moreover, most real-world graphs follow a long-tail distribution in terms of node type and node degree [32]. For nodes with few connections (i.e., tail nodes), the receptive field is limited. And most HGNNs rely on message passing to capture local structural features (low-pass filters), resulting in poor performance of tail nodes.

As an efficient modeling tool, hypergraph convolution provides a new way to break the above limitations. First, hypergraphs have the characteristic that one edge contains any number of nodes, so they have a natural advantage for the expression of data relations beyond pairs. Second, hypergraph convolution provides an extra manner to aggregate intra- and inter- information from hyperedges, which can alleviate the poor performance of tail cases. Specifically, HWNN [24] utilizes simple graph snapshots from HGs to construct the hypergraph and takes the Wavelet basis to perform localized hypergraph convolution. Meta-HGT [15] considers relations contained in meta paths and designs type-aware hypergraph encoder to learn the representation of nodes.

Despite the advantages of these methods, most of them still require labels as supervised signals to train. And in some real-world scenarios, obtaining labels is often challenging and costly since they rely heavily on domain knowledge. Fortunately, self-supervised learning (SSL), which aims to spontaneously explore supervised signals from the data itself, is a promising solution for the setting without explicit labels. Contrastive learning is one kind of SSL, which focuses on maximizing the similarity between positive samples and minimizing the similarity between negative samples. However, there are few works that combine hypergraph convolution with contrastive learning on heterogeneous graphs because it is nontrivial. In practice, we need to carefully consider the characteristics of HGs and address the following challenges:

(1) *How to construct hypergraphs from heterogeneous graphs.* Different types of nodes and edges make it hard to capture the implied semantics relation in HGs. Most methods for constructing hypergraphs, such as attribute-based [7] and cluster-based methods [24], are not elaborately designed for HGs. Therefore, they ignore the rich semantics and characteristics of HGs.

(2) *How to select proper contrastive views.* For HGs, it is easy to extract different views according to the heterogeneity. However, if the contrastive views are similar, a weak supervised signal will be obtained [28]. And inappropriate contrastive views may destroy the ability to filter noises. We assume that contrastive views should contain different semantics, so that information on various aspects can be integrated.

(3) *How to choose high quality positive and negative samples.* Choosing positive and negative samples is an important part of contrastive learning. The neg-

ative samples which are "false negative" and the positive samples which are "false positive" hurt the generalization ability of contrastive frameworks. Further, we assume that proper positive and negative samples should consider both heterogeneous structures and node attributes.

In this paper, we focus on the self-supervised HG representation learning and propose a novel method named HHGR (short for Hypergraph-enhanced self-supervised Heterogeneous Graph Representation learning), which contains different modules to address the challenge mentioned above. First, we propose a hypergraphs construction module to extract hyperedges considering the network schema and the composite semantics represented by meta paths. And we designed a cross-view contrast module that maximizes the agreement between views and aggregated views, as well as the representations between each view. In this way, aggregated representations can learn the weights of different views and filter noises from each view. Further, considering the rich semantics in HGs, we assume to choose positive samples according to the structure and attribute semantics and design a semantic positive sampling module. Finally, HHGR can learn rich semantics and discriminative information in HGs.

The contributions of our work are summarized as follows:

- We propose a self-supervised heterogeneous graph representation learning method, namely HHGR. It considers unpaired relation in the real world and constructs hypergraph views to capture complex semantics in HGs.
- To choose proper positive samples, we propose a semantic positive sampling strategy, which considers the structure and attribute semantics.
- We conduct extensive experiments on three real-world heterogeneous graph datasets and the results demonstrate the state-of-the-art performance of HHGR in terms of effectiveness compared with baselines.

2   Related Work

2.1   Heterogeneous Graph Embedding

Heterogeneous graph embedding focuses on learning a low-dimensional vector of nodes, which contains rich information specially the heterogeneous structure semantics (i.e., meta paths) [27]. The metapath2vec [5] uses random walk guided by meta paths to model the context of nodes with respect to semantic information. RGCN [23] proposes to learn multiple convolution matrices corresponding to each edge type. Instead of aggregating one-hop neighbors, HAN [27] uses meta-path-based graphs to learn the structure of semantic information. MAGNN [8] extends HAN [27] by considering nodes along the meta paths and using an attention mechanism to aggregate attributes of nodes. HGT [12] uses mutual attention to aggregate the information from heterogeneous neighbors.

Although the above methods exploit the semantic information contained in HGs, they do not consider the high-order relation beyond pairwise which will limit the representation of nodes.

2.2   Hypergraph Embedding

A hypergraph is a generalization of a graph in which an edge can join any number of vertices, providing a natural way to capture complex relations. DHNE [25] learns the embedding by modeling the tuple-wise relation and keeps the first-order and the second-order proximities. HGNN [7] encodes high-order correlation into the hypergraph structure. Moreover, Hyper-SAGNN [33] extends the self-attention mechanism to the hypergraph. HyperGCN [30] transforms the hypergraph into several simple graphs and uses the graph convolution network to learn node embeddings. Further, HWNN [24] uses a wavelet basis for heterogeneous hypergraphs and learns the embedding according to different meta paths. HeteHG-VAE [6] constructs hyperedges by certain events but do not consider the high-order relation in meta paths. Meta-HGT [15] extends HGT [12] into hypergraphs and constructs hypedgraph in hierarchical hyperedges. However, most of above methods depend on the supervision signal from labels.

2.3   Self-supervised Learning on Graphs

Self-supervised learning (SSL) through well-designed pre-text tasks to learn representations from unlabeled data, has become a promising learning paradigm for graph data [16]. Especially, DeepWalk [22] introduces a random-walk-based approach to keep the context information of nodes. GraphSage [10] extends it and proposes a novel GNN to learn the embedding of nodes inductively. DGI [26] considers the graph-level information and utilizes the infomax [14] to contrast it with local-level information. GRACE [35] and GCA [36] propose to preserve node-level discrimination. Moreover, DMGI [21] designs a consensus regularization to learn the embedding from different meta-path-based graphs. HeCo [28] uses a view mask mechanism, which combines the characteristics of the network schema and meta-path-based graphs. STENCIL [34] integrates structure information when choosing negative samples which can enhance contrastive learning. HGCML [29] performs intra-metapath and inter-metapath contrasts to model the consistency between meta-path views.

3   Problem Formulation

In this section, we introduce some necessary definitions and then formulate the problem of self-supervised heterogeneous graph embedding.

Definition 1 (Heterogeneous Graph). *A heterogeneous graph is defined as* $\mathcal{G} = (\mathcal{V}, \mathcal{E}, \mathcal{F}, \mathcal{R}, \varphi, \phi)$, *where* $\mathcal{V}$ *denotes the set of nodes,* $\mathcal{E}$ *the set of edges,* $\mathcal{F}$ *the set of node types and* $\mathcal{R}$ *the set of edge types. Each node* $v \in \mathcal{V}$ *associated with a type mapping function* $\varphi : \mathcal{V} \to \mathcal{F}$, *and each edge* $e \in \mathcal{E}$ *with a type mapping function* $\phi : \mathcal{E} \to \mathcal{R}$, *where* $|\mathcal{F}| + |\mathcal{R}| > 2$ *in the heterogeneous graph.*

Definition 2 (Network Schema). *The network schema is the meta template for a heterogeneous graph* $\mathcal{G}$, *denoted as* $T_{\mathcal{G}} = (\mathcal{F}, \mathcal{R})$, *where* $\mathcal{F}$ *and* $\mathcal{R}$ *denote the types of nodes and edges, respectively.*

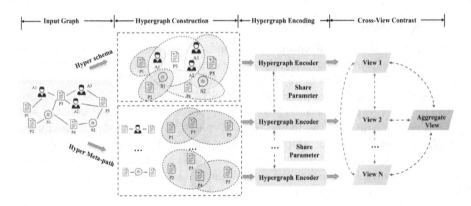

Fig. 1. The illustration of the proposed HHGR framework.

Definition 3 (Meta Path). *A meta path $\mathcal{P}$ is a path defined on the heterogeneous graph and denoted in the form of $F_1 \xrightarrow{R_1} F_2 \xrightarrow{R_2} \cdots \xrightarrow{R_l} F_{l+1}$, where $F \in \mathcal{F}$ and $R \in \mathcal{R}$. And $R = R_1 \circ R_2 \circ \ldots \circ R_l$ is a composite relation between F_1 and F_{l+1}, where $\circ$ denotes the composition operator on relations.*

Definition 4 (Hypergraph). *A hypergraph is defined as $\mathcal{G}_h = (\mathcal{V}_h, \mathcal{E}_h)$, where $\mathcal{V}_h = \{v_1, v_2, ..., v_N\}$ and $\mathcal{E}_h = \{e_1, e_2, ..., e_M\}$ denote the set of nodes and edges, respectively. For any hyperedge $e \in \mathcal{E}_h$ can contain more than two nodes. The hypergraph can be represented by an incidence matrix $H \in \mathbb{R}^{N \times M}$. When the node $v_i \in \mathcal{V}_h$ is connected by $e_j \in \mathcal{E}_h$, $H_{i,j} = 1$, otherwise 0.*

Problem Description (Self-supervised Heterogeneous Graph Embedding). Given a heterogeneous graph $\mathcal{G}$, the problem of *self-supervised heterogeneous graph embedding* is to learn low-dimensional representations $h_v \in \mathbb{R}^d$ for the node $v \in \mathcal{V}$ with $d \ll |\mathcal{V}|$ where the supervision signals are generated from data itself.

4   The Design of HHGR

As shown in Fig. 1, HHGR constructs the hypergraphs according to the semantics of HGs and utilizes hypergraph encoder to achieve information from each view. Second, an attention mechanism is employed to aggregate the information from different views. Finally, HHGR contrasts representation not only between every single view but also between single view and aggregated view.

4.1   Hypergraph Construction

Heterogeneous graphs contain different types of nodes and edges, which implies rich and complex semantics. Although hypergraphs have the superiority in modeling complex relations, how to construct different hypergraphs and preserve rich semantics is nontrivial for HGs. We assume that the manner of constructing hypergraphs needs to preserve different semantics. Further, multiple suitable hypergraph views can complement with each other, and the representations could achieve the semantics from different aspects.

Hyper Schema View. Network schema is the blueprint of HGs, which contains local structure and rich semantics. For node $v \in \mathcal{V}$, which connects with different types of neighbors according to the network schema, we take the node v and multiple different types of connected nodes into one hyperedge. In this way, we construct the hyper schema view to learn the unpaired relation contained in network schema.

Hyper Meta-path View. For HGs, using meta paths is an important method to capture the high-order relation. For example, the meta path "Author - Paper - Author" (A-P-A) indicates the co-author relation between two authors. However, only considering one meta path instance is limited by the pairwise relation. We propose to construct hyper meta paths to capture unpaired relations implied in HGs. Specifically, for the target node P3 with the meta path A-P-A, we take the central node P3 as the anchor and choose all meta path A-P-A instances that center on the node P3 into one hyperedge. Note that we assume the type of start and end nodes in chosen meta paths is the same so we only choose nodes with the same type of the start node. In this way, the hyper meta path A-P-A indicates multiple authors who contribute to the work of the paper P3.

4.2 Hypergraph Encoder

There are different types of nodes in the heterogeneous graph and the attributes of different types of nodes may fall in different feature space [27]. Therefore, we first project them into the same space. For node $v \in \mathcal{V}$ with the type $\varphi(v)$ and feature vector $\mathbf{x}_v \in \mathbb{R}^{d_{\varphi(v)}}$, we design the type-special matrix $\mathbf{W}_{\varphi(v)}$ to transform it into a common space:

$$\mathbf{x}'_v = \mathbf{W}^2_{\varphi(v)} \sigma(\mathbf{W}^1_{\varphi(v)} \cdot \mathbf{x}_v + \mathbf{b}^1_{\varphi(v)}) + \mathbf{b}^2_{\varphi(v)}, \tag{1}$$

where $\sigma(\cdot)$ denotes a non-linear activation function.

For hyper meta-path views, only one type of node is contained, and HHGR employs the hypergraph neural network [1] to encode the information of nodes. Given a hypergraph with N nodes and M hyperedges, HHGR encodes information as follow,

$$\mathbf{h}^{(l+1)} = \sigma(\mathbf{D}^{-1/2} \mathbf{H} \mathbf{W} \mathbf{B}^{-1} \mathbf{H}^T \mathbf{D}^{-1/2} \mathbf{h}^{(l)} \Theta), \tag{2}$$

where $\mathbf{W} \in \mathbb{R}^{M \times M}$ denotes the weights of each hyperedge, $\Theta \in \mathbb{R}^{d \times d}$ denotes the learning parameters, $\mathbf{D}$ and $\mathbf{B}$ are the degree matrices of the nodes and hyperedges, respectively. Note that $\mathbf{h}^{(l+1)}$ denotes the output of l-layer hypergraph neural networks, and we take $\mathbf{x}'$ as the original input.

For the hyper schema view, not all types of nodes make the same contribution to hyperedges. To learn the importance of hyperedges for each node and capture different types of information of the node in the hyper schema view, HHGR designs a type-aware attention mechanism to learn the weights on different hyperedges for target nodes. For node $v \in \mathcal{V}_h$ with the type $\varphi(v)$ and hyper edge $e \in \mathcal{E}_h$, the attention score is as follow,

$$\mathbf{H}_{ve} = \frac{exp(LeakyReLU(sim(\mathbf{h}_v, \mathbf{h}_e)))}{\sum_{k \in \mathcal{E}_{h,v}} exp(LeakyReLU(sim(\mathbf{h}_v, \mathbf{h}_k)))}, \tag{3}$$

where $\mathbf{h}_e$ denotes the features of hyperedges and $\mathcal{E}_{h,v}$ denotes the set of hyperedges which contain node v. Further, the similarity function is formulate as follow,

$$sim(v, e) = \mathbf{a}_{\varphi(v)}^T [\mathbf{h}_v || \mathbf{h}_e], \tag{4}$$

and then the information from hypergraphs is aggregated according to Eq. (2). Note that HHGR employs the feature of the center node in each hyperedge as the features of hyperedges. Further, to avoid overfitting and learn the essential information between different views, we share the parameters on Θ in Eq. (2).

4.3   Cross-View Contrast

Different views describe different aspects of relation and how to effectively integrate these information is nontrivial. Hyper schema and meta-path views represent different kinds of semantics implied in HGs. We assume that they should be considered together and employ a semantic attention to learn the importance of different semantics. Given the semantic views $S = \{S_1, S_2, ..., S_N\}$, we calculated as follow,

$$\mathbf{z} = \sum_{S_i \in S} \beta^{S_i} \cdot \mathbf{h}^{S_i}. \tag{5}$$

where h denotes the final representation. And the coefficients can be calculated as follow,

$$\beta^{S_i} = \frac{exp(\alpha^{S_i})}{\sum_{S_j \in S} exp(\alpha^{S_j})}, \tag{6}$$

$$\alpha^{S_i} = \frac{1}{|\mathcal{V}|} \sum_{v \in \mathcal{V}} \mathbf{a}^T \cdot tanh(\mathbf{W} \cdot \mathbf{h}_v^{S_i} + \mathbf{b}), \tag{7}$$

where $\mathbf{a} \in \mathbb{R}^d$ denotes the attention vector, weight matrix $\mathbf{W} \in \mathbb{R}^{d \times d}$ and bias vector $\mathbf{b} \in \mathbb{R}^d$.

The setting of contrastive objects is a necessary component of contrastive learning. We focus on learning the representation combined with the information from different aspects. In order to guide the attention scores from different views, we first contrast the representation in each view with the final representation,

$$\mathcal{L}(\mathbf{h}_v^{S_i}, \mathbf{z}) = -log \frac{\sum_{u \in Pos(v)} exp(sim(\mathbf{h}_v^{S_i}, \mathbf{z}_u)/\tau)}{\sum_{k \in Pos(v) \cup Neg(v)} exp(sim(\mathbf{h}_v^{S_i}, \mathbf{z}_k)/\tau)}, \tag{8}$$

where τ denotes the temperature coefficient, $Pos(v)$ and $Neg(v)$ denote the set of positive and negative samples for node v, respectively. Despite the final representation can learn suitable weights for different semantics, it is insufficient to filter the noises contained in every single view. If any views contain so many noises, the performance of HHGR will be destroyed. Therefore, we contrast representations between different views, as follow,

$$\mathcal{L}(\mathbf{h}_v^{S_i}, \mathbf{h}^{S_j}) = -log\frac{\sum_{u \in Pos(v)} exp(sim(\mathbf{h}_v^{S_i}, \mathbf{h}_u^{S_j})/\tau)}{\sum_{k \in Pos(v) \cup Neg(v)} exp(sim(\mathbf{h}_v^{S_i}, \mathbf{h}_k^{S_j})/\tau)}, \tag{9}$$

where $sim(\cdot, \cdot)$ denotes the cosine similarity. Finally, the overall loss function can be formulated as follow,

$$\mathcal{L} = \frac{1}{|\mathcal{V}|} \sum_{v \in \mathcal{V}} \sum_{S_i \in S} (\sum_{S_j \in S} \mathcal{L}(\mathbf{h}_v^{S_i}, \mathbf{h}^{S_j}) + \mathcal{L}(\mathbf{h}_v^{S_i}, \mathbf{z}) + \mathcal{L}(\mathbf{z}_v, \mathbf{h}^{S_i})) \tag{10}$$

where $\mathcal{V}$ denotes the set of nodes.

4.4 Semantic Positive Samples

In this paper, we focus on choosing high-quality positive samples based on semantic information from structure and attributes.

Structure Semantic Positive Samples. The structure of the heterogeneous graph contains rich semantic information (i.e., meta paths). Although HGCML [29] and STENCIL [34] employ structure information (i.e., Personalized PageRank [20]) to choose positive or negative samples, they do not consider the different types of nodes in HGs, which causes the lose of semantic integrity. We propose to utilize random-walk-based methods [5] to capture the semantic information. Given a meta-path scheme $\mathcal{P} : F_1 \xrightarrow{R_1} F_2 \xrightarrow{R_2} \cdots F_k \xrightarrow{R_k} F_{k+1} \cdots \xrightarrow{R_{l-1}} F_l$, for node v_k the transition probability at step i is defined as follow,

$$p(v^{i+1}|v_k^i, \mathcal{P}) = \begin{cases} \frac{1}{|N_{k+1}(v_k^i)|} & (v^{i+1}, v_k^i) \in \mathcal{E}, \varphi(v^{i+1}) = F_{k+1}, \\ 0 & (v^{i+1}, v_k^i) \in \mathcal{E}, \varphi(v^{i+1}) \neq F_{k+1}, \end{cases} \tag{11}$$

where $N_{k+1}(v_k^i)$ denotes the F_{k+1} type of neighbors of node v_k^i, $\varphi(v_k^i) = F_k$ and $\mathcal{E}$ denotes the set of edges in HGs. Considering the multiple meta paths in HGs, we concatenate the random walk sequences guided by each meta path, and feed them into the Skip-Gram [19] model to get the structure semantic representation $h_s \in \mathbb{R}^d$ of each node. We first construct a similarity matrix $M \in \mathbb{R}^{N \times N}$ among N nodes, as follow,

$$M_{v,u}^s = \mathbf{1}\{u \in TopK(\{sim(v, k) : k \in \mathcal{V}\})\}, \tag{12}$$

where $u \in \mathcal{V}$ and $sim(\cdot, \cdot)$ denote the cosine similarity. Then we select top-K similar node pairs for each node.

Table 1. Statistics of the datasets.

Datasets	# Nodes	# Edges	# Features	Meta-paths
DBLP	author (A): 4057 paper (P): 14328 term (T): 7723 conference (C): 20	P-A: 19645 P-T: 85810 P-C: 14328	334	APA APCPA APTPA
ACM	paper (P): 4019 author (A): 7167 subject (S): 60	P-P: 9615 P-A: 13407 P-S: 4019	4000	PAP PSP
Yelp	business (B): 2614 user (U): 1286 service (S): 4 rating level (L): 9	B-S: 5228 B-L: 5228 B-U: 61676	82	BUB BLB BSB

Attribute Semantic Positive Samples. Attributes of nodes are important information, which contains rich semantics. We propose to consider the attribute correlations between node pairs. And we use the same method to select suitable positive samples according to Eq. (12), and the similarity matrix denoted as M^a. Finally, the positive samples of target node v can be formulated as follow,

$$Pos(v) = \{u \in \mathcal{V} | M^s_{v,u} > 0 \text{ or } M^a_{v,u} > 0\} \tag{13}$$

In this way, we select positive samples from two different kinds of semantics and treat all retained nodes as negative samples of v, denoted as $Neg(v)$.

5   Experiments

5.1   Experimental Settings

Datasets: we experimented on three real-world HG datasets, where the detailed information of them is shown in Table 1.

- **DBLP** [13]. We use the DBLP dataset which contains 14328 papers (P), 4057 authors (A), 20 conferences (C) and 8789 terms (T). The target nodes in this dataset are authors, which are divided into four classes. And the attributes of these kinds of nodes are bag-of-word representations of keywords extracted from their published papers.
- **ACM** [13]. This is a subset of ACM containing 4019 papers (P), 7167 authors (A) and 60 subjects (S). In this dataset, the target nodes are papers, which are divided into three classes. And the attributes of this kind of nodes are bag-of-word representations of their keywords.
- **Yelp** [17]. This dataset consists of 2614 businesses (B), 1286 users (U), 4 services (S) and 9 rating levels (L). In this dataset, the target nodes are businesses, which are divided into three classes. And the attributes of this kind of nodes are bag-of-word representations of their related keywords.

Table 2. Node classification results (%). The training data is shown in the second column, where **A** denotes the adjacency matrices, **X** denotes node features, and **Y** denotes ground-truth. The implement of STENCIL is not available, and we showed the performance reported in the original paper.

Methods	Training	DBLP		ACM		Yelp	
	Data	Macro-F1	Micro-F1	Macro-F1	Micro-F1	Macro-F1	Micro-F1
node2vec	**A**	90.43	91.05	69.40	74.20	54.01	73.13
metapath2vec	**A**	90.76	91.53	65.09	65.00	54.04	73.00
DGI	**A, X**	90.69	91.30	89.09	89.15	55.21	73.39
GRACE	**A, X**	89.76	90.88	88.72	88.72	53.99	73.10
DMGI	**A, X**	92.09	92.66	91.76	92.53	54.02	73.14
HeCo	**A, X**	91.02	91.56	88.25	88.15	53.99	73.11
STENCIL	**A, X**	92.33	92.81	90.72	90.76	–	–
HGCML	**A, X**	92.51	93.03	90.58	90.41	60.60	74.78
HHGR	**A, X**	**94.55**	**94.91**	**92.94**	**92.84**	**94.13**	**93.54**
RGCN	**A, X, Y**	88.27	89.34	86.67	87.59	63.26	72.13
HGNN	**A, X, Y**	76.05	75.27	78.85	79.69	90.98	89.72
HAN	**A, X, Y**	91.17	92.05	89.40	89.22	90.19	89.24
MAGNN	**A, X, Y**	93.79	94.24	90.78	90.73	92.96	92.30
HWNN	**A, X, Y**	92.77	93.35	91.20	91.92	86.49	87.38

Baselines: we compared HHGR with four categories of baselines, including unsupervised homogeneous methods (i.e., node2vec [9], DGI [26], GRACE [35]), unsupervised heterogeneous methods (i.e., metapath2vec [5], DMGI [21], HeCo [28], STENCIL [34], HGCML [29]), semi-supervised heterogeneous methods (i.e., HAN [27], RGCN [23], MAGNN [8]) and semi-supervised hypergraph methods (i.e., HGNN [7], HWNN [24]).

Implementation Details: for unsupervised methods in the homogeneous graph (i.e., node2vec [9], DGI [26], GRACE [35]), we test them on several meta-path-based graphs and then report the best result. For random-walk-based methods (i.e., node2vec [9], metapath2vec [5]), we set the number of walks per node to 5, the walk length to 50 and the context size to 7. For semi-supervised methods (i.e., RGCN [23], HAN [27], MAGNN [8], HGNN [7], HWNN [24]), we randomly use 20% of training data as the training set. In terms of other parameters, we followed the settings in their original papers.

The proposed HHGR employs a one-layer hypergraph GNN as the encoder. We search to tune the learning rate from 0.0001 to 0.001 and the number of positive samples from 0 to 128. Moreover, we set early stop patience to 5, node dimension to 64 and trained HHGR using Adam as optimizer. The source code and datasets were publicly available on Github[1].

---

[1] https://github.com/scu-kdde/HGA-HHGR-2023.

5.2 Node Classification

Node classification is a classic task to evaluate the quality of embeddings. We conducted experiments compared with both unsupervised and semi-supervised methods. For each evaluation, we ran 10 times and reported the average results. In addition, we employed Macro F1 and Micro F1 as metrics.

As the results shown in Table 2, the proposed HHGR achieved the state-of-the-art performance in Macro F1 and Micro F1 on all three datasets. Compared with the random-walk-based method, the proposed GNN-based HHGR gets large improvements on the ACM dataset, which demonstrates the superiority of GNN. For self-supervised methods, although most of the self-supervised heterogeneous graph methods consider the semantic information contained in meta paths, they regard them as one kind of pairwise relations. HHGR considers comprehensive information more than pairwise relations and constructs different hypergraph views. Compared with SOTA method HGCML, HHGR improved by 2.04 on Macro F1 and 1.88 on Micro F1 for the DBLP dataset. HHGR even gets competitive performance with several supervised methods, which demonstrates the effectiveness of contrastive views and semantic positive samples.

5.3 Node Clustering

To further evaluate the effectiveness of HHGR, we employed k-means on the obtained embedding of all target nodes and adopted normalized mutual information (NMI) and adjusted rand index (ARI) to evaluate the quality of the clustering results. Note that we did not compare with semi-supervised methods (i.e., RGCN [23], HGNN [7], HAN [27], MAGNN [8], HWNN [24]) because they have known the labels in the training set and are guided by the validation set.

As the results shown in Table 3, we can find that HHGR outperforms most methods. For homogeneous methods, lacking information on different types of nodes, they may fall into sub-optimum and not get good performance. For heterogeneous methods, they only consider the pairwise relation contained in the HG but ignore the unpaired relation, such as the community structures contained in citation network DBLP and ACM. However, HHGR constructs hypergraphs and learns comprehensive information from different hypergraph views. Further, compared with STENCIL [34], HHGR improved by 2.82 on NMI and 2.32 on ARI for the DBLP dataset. Besides, we can observe that most of the methods did not get good performance on the Yelp dataset. We assume that simply aggregating the information from meta-path-based neighbors may contain a different degree of noise. In general, HHGR considers not only the semantics in the meta paths but also the information in the network schema. And by sharing the parameters in hypergraph GNN and combining with two different kinds of semantic positive samples, HHGR achieves discriminative representations.

5.4 Visualization

For more intuitive evaluation, we projected the learned embeddings into two-dimensional space on the ACM dataset and used t-SNE [18] to visualize the

Table 3. Evaluation results (%) on node clustering.

Methods	Trainng	DBLP		ACM		Yelp	
	Data	NMI	ARI	NMI	ARI	NMI	ARI
node2vec	A	67.19	72.96	15.02	18.33	38.71	42.45
metapath2vec	A	74.30	78.50	21.22	21.00	38.66	42.33
DGI	A, X	60.62	60.42	58.13	57.18	39.12	34.60
GRACE	A, X	62.06	64.13	53.38	54.39	38.66	42.33
DMGI	A, X	70.96	76.96	65.11	**74.96**	40.92	44.69
HeCo	A, X	70.99	76.67	59.53	57.59	39.02	42.53
STENCIL	A, X	76.60	81.58	67.93	72.65	–	–
HGCML	A, X	76.17	81.66	63.26	65.53	37.87	33.10
HHGR	A, X	**79.42**	**83.90**	**69.82**	73.14	**67.82**	**72.11**

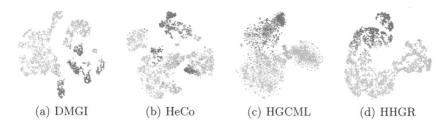

(a) DMGI   (b) HeCo   (c) HGCML   (d) HHGR

Fig. 2. The visualizations of the embedding on ACM. The Silhouette scores for (a)(b)(c)(d) are 0.33, 0.17, 0.40 and 0.42, respectively.

embeddings of nodes. We compared our method with three heterogeneous methods (i.e., DMGI [21], HeCo [28], HGCML [29]). Besides, principal components analysis (PCA) was employed to initialize t-sne [18] for all methods. The results are shown in Fig. 2, where different colors represent different classes.

As shown in Fig. 2, there are no clear bounds between different types of nodes in DMGI [21] and HeCo [28]. We can see that some nodes belonging to the same class did not get together and mix with other types of nodes. And for HGCML [29], despite some types of nodes getting together, there still exists a large proportion of overlapping. However, HHGR has a clear bound with a different type of nodes and outperforms other methods on silhouette scores.

5.5 Ablation Studies

We evaluated the effectiveness of different parts of the proposed HHGR by comparing it with three variants of the HHGR. (1) To evaluate the effectiveness of the schema-based hypergraph view, we eliminated it and only used the meta-path-based hypergraph view, denoted as HHGR-s. (2) For the view aggregating in the cross-view contrast module, which fuses the information from different

Table 4. The effectiveness of the design for each modules in HHGR.

Variants	DBLP		ACM		Yelp	
	Macro-F1	Micro-F1	Macro-F1	Micro-F1	Macro-F1	Micro-F1
HHGR-s	94.25	94.63	92.58	92.53	79.98	82.53
HHGR-avg	93.84	94.27	92.50	92.40	89.78	89.73
HHGR-p	93.50	93.90	92.59	92.50	89.96	90.01
HHGR	**94.55**	**94.91**	**92.94**	**92.84**	**94.13**	**93.54**

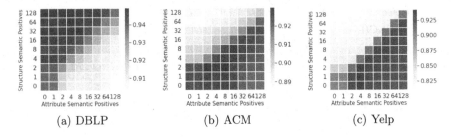

(a) DBLP   (b) ACM   (c) Yelp

Fig. 3. The parameter sensitivity results of positive samples.

views with the attention mechanism, we remove it and just average the embedding from different views, denoted as HHGR-avg. (3) Further, we replace the semantic positive samples and only use the node itself as the positive sample, denoted as HHGR-p.

As shown in Table 4, we can observe that HHGR outperforms all of the variants, indicating the effectiveness of the modules designed in HHGR. We assume that the neighbor-based hypergraph view contains different semantic information from the meta-path-based hypergraph view. And the performance of HHGR will degrade if remove one of them. Further, different hypergraph views play different roles, and simply employing mean operation is not suitable. However, HHGR employs an adaptive weight for different views based on the attention mechanism. In this way, HHGR can learn complementary information from different views. Moreover, a good strategy for taking positive and negative samples is important for contrastive learning. Semantic positive sampling strategy considers not only the semantics from attributes but also the semantics from structure, which improves the performance of HHGR on all three datasets.

5.6   Analysis of Hyper-parameters

To evaluate the stability of HHGR, we systematically investigated the sensitivity of the thresholds of semantic positive sampling. We conduct node classification experiments on all three datasets and use Micro F1 as metrics. As shown in Fig. 3, attribute and structure semantic samples can improve the performance of HHGR. Especially, a large number of structure semantic samples brought obvious improvements on the DBLP dataset. The structure semantic samples contain the position information of nodes which can recognize the author sharing similar research interests [3]. For ACM and Yelp datasets, HHGR encounters a large drop with the increasing of structure semantic samples. We assume that random-walk-based features may bias the node with a large degree and density connections. Therefore, HHGR learns the representation of nodes with unsuitable directions. However, we can see that HHGR remains stable with an equal number of attribute and structure semantic samples.

6 Conclusion

In this paper, we consider the complex relations in the real world and propose HHGR to extract relations and construct hypergraphs to learn the rich semantics in heterogeneous graphs. Further, HHGR employs contrastive learning to integrate the representations from different views. To enhance the performance of HHGR, we propose a semantic positive sampling strategy, which chooses the positive samples according to structure and attribute semantics. Extensive experiments on real-world heterogeneous graph datasets demonstrate the effectiveness of proposed method.

References

1. Bai, S., Zhang, F., Torr, P.H.: Hypergraph convolution and hypergraph attention. Pattern Recognit. **110**, 107637 (2021)
2. Cao, Y., Peng, H., Yu, P.S.: Multi-information source HIN for medical concept embedding. In: Lauw, H.W., Wong, R.C.-W., Ntoulas, A., Lim, E.-P., Ng, S.-K., Pan, S.J. (eds.) PAKDD 2020. LNCS (LNAI), vol. 12085, pp. 396–408. Springer, Cham (2020). https://doi.org/10.1007/978-3-030-47436-2_30
3. Cui, H., Lu, Z., Li, P., Yang, C.: On positional and structural node features for graph neural networks on non-attributed graphs. In: CIKM, pp. 3898–3902 (2022)
4. Davis, A.P., et al.: The comparative toxicogenomics database: update 2017. Nucleic Acids Res. **45**(D1), D972–D978 (2017)
5. Dong, Y., Chawla, N.V., Swami, A.: metapath2vec: scalable representation learning for heterogeneous networks. In: SIGKDD, pp. 135–144 (2017)
6. Fan, H., et al.: Heterogeneous hypergraph variational autoencoder for link prediction. TPAMI **44**(8), 4125–4138 (2022)
7. Feng, Y., You, H., Zhang, Z., Ji, R., Gao, Y.: Hypergraph neural networks. In: AAAI, pp. 3558–3565 (2019)
8. Fu, X., Zhang, J., Meng, Z., King, I.: MAGNN: metapath aggregated graph neural network for heterogeneous graph embedding. In: WWW, pp. 2331–2341 (2020)
9. Grover, A., Leskovec, J.: node2vec: scalable feature learning for networks. In: SIGKDD, pp. 855–864 (2016)
10. Hamilton, W.L., Ying, Z., Leskovec, J.: Inductive representation learning on large graphs, In: NIPS, pp. 1024–1034 (2017)
11. Hu, B., Zhang, Z., Shi, C., Zhou, J., Li, X., Qi, Y.: Cash-out user detection based on attributed heterogeneous information network with a hierarchical attention mechanism. In: AAAI, pp. 946–953 (2019)

12. Hu, Z., Dong, Y., Wang, K., Sun, Y.: Heterogeneous graph transformer. In: WWW, pp. 2704–2710 (2020)
13. Jin, D., Huo, C., Liang, C., Yang, L.: Heterogeneous graph neural network via attribute completion. In: WWW, pp. 391–400 (2021)
14. Linsker, R.: Self-organization in a perceptual network. Computer **21**(3), 105–117 (1988). https://doi.org/10.1109/2.36
15. Liu, J., Song, L., Wang, G., Shang, X.: Meta-HGT: metapath-aware hypergraph transformer for heterogeneous information network embedding. Neural Netw. **157**, 65–76 (2023)
16. Liu, Y., Pan, S., Jin, M., Zhou, C., Xia, F., Yu, P.S.: Graph self-supervised learning: a survey. TKDE (Early Access) (2022). https://doi.org/10.1109/TKDE.2022.3172903
17. Lu, Y., Shi, C., Hu, L., Liu, Z.: Relation structure-aware heterogeneous information network embedding. In: AAAI, pp. 4456–4463 (2019)
18. Van der Maaten, L., Hinton, G.: Visualizing data using t-SNE. J. Mach. Learn. Res. **9**(11), 2579–2605 (2008)
19. Mikolov, T., Sutskever, I., Chen, K., Corrado, G.S., Dean, J.: Distributed representations of words and phrases and their compositionality. In: NIPS, pp. 3111–3119 (2013)
20. Page, L., Brin, S., Motwani, R., Winograd, T.: The pagerank citation ranking: Bringing order to the web. Tech. rep., Stanford InfoLab (1999)
21. Park, C., Kim, D., Han, J., Yu, H.: Unsupervised attributed multiplex network embedding. In: AAAI, pp. 5371–5378 (2020)
22. Perozzi, B., Al-Rfou, R., Skiena, S.: DeepWalk: online learning of social representations. In: SIGKDD, pp. 701–710 (2014)
23. Schlichtkrull, M., Kipf, T.N., Bloem, P., van den Berg, R., Titov, I., Welling, M.: Modeling relational data with graph convolutional networks. In: Gangemi, A., et al. (eds.) ESWC 2018. LNCS, vol. 10843, pp. 593–607. Springer, Cham (2018). https://doi.org/10.1007/978-3-319-93417-4_38
24. Sun, X., et al.: Heterogeneous hypergraph embedding for graph classification. In: WSDM, pp. 725–733 (2021)
25. Tu, K., Cui, P., Wang, X., Wang, F., Zhu, W.: Structural deep embedding for hyper-networks. In: AAAI, vol. 32 (2018)
26. Velickovic, P., Fedus, W., Hamilton, W.L., Liò, P., Bengio, Y., Hjelm, R.D.: Deep graph infomax. In: ICLR. p. Poster (2019)
27. Wang, X., et al.: Heterogeneous graph attention network. In: The world Wide Web Conference, pp. 2022–2032 (2019)
28. Wang, X., Liu, N., Han, H., Shi, C.: Self-supervised heterogeneous graph neural network with co-contrastive learning. In: SIGKDD, pp. 1726–1736 (2021)
29. Wang, Z., Li, Q., Yu, D., Han, X., Gao, X., Shen, S.: Heterogeneous graph contrastive multi-view learning. ArXiv **abs/2210.00248** (2022)
30. Yadati, N., Nimishakavi, M., Yadav, P., Nitin, V., Louis, A., Talukdar, P.P.: Hyper-GCN: a new method for training graph convolutional networks on hypergraphs. In: NeurIPS, pp. 1509–1520 (2019)
31. Yang, J., Leskovec, J.: Defining and evaluating network communities based on ground-truth. Knowl. Inf. Syst. **42**(1), 181–213 (2015)
32. Yun, S., Kim, K., Yoon, K., Park, C.: LTE4G: long-tail experts for graph neural networks. In: CIKM, pp. 2434–2443 (2022)
33. Zhang, R., Zou, Y., Ma, J.: Hyper-SAGNN: a self-attention based graph neural network for hypergraphs. In: ICLR (2020)

34. Zhu, Y., Xu, Y., Cui, H., Yang, C., Liu, Q., Wu, S.: Structure-enhanced heterogeneous graph contrastive learning. In: SDM, pp. 82–90 (2022)
35. Zhu, Y., Xu, Y., Yu, F., Liu, Q., Wu, S., Wang, L.: Deep graph contrastive representation learning. ArXiv **abs/2006.04131** (2020)
36. Zhu, Y., Xu, Y., Yu, F., Liu, Q., Wu, S., Wang, L.: Graph contrastive learning with adaptive augmentation. In: Proceedings of the Web Conference 2021, pp. 2069–2080 (2021)

LAF: A Local Depth Autoregressive Framework for Cardinality Estimation of Multi-attribute Queries

Qianwen Cheng, Hao Li, Dawei Wang, Yue Zhang, and Zhaohui Peng[✉]

School of Computer Science and Technology, Shandong University,
Qingdao 266237, China
{chengqw,lihao_bd,202000130125,202235205}@mail.sdu.edu.cn, pzh@sdu.edu.cn

Abstract. Cardinality estimation is significant for database query optimization, which affects the query efficiency. Most existing methods often use a uniform approach to model strongly and weakly correlated attributes and seldom make comprehensively use of data information and query information. Some methods have poor accuracy due to simple structure, while others suffer from low efficiency due to complex structure. The problem of cardinality estimation that strong and weak association coexist among attributes can not be well solved by these methods or their simple combinations. Therefore we propose LAF, a new Local deep Autoregressive Framework, which performs fine-grained modeling for attributes with strong and weak correlation. LAF utilizes mutual information to identify the strong and weak association between attributes, applying the local strategy to construct deep autoregressive models to learn the joint distribution for strongly correlated attributes and outputting corresponding local estimations, using lightweight regression model to capture the complex mapping between local estimations with weak correlation and cardinality, and LAF combines information entropy to sort attributes in descending order. Not only do we enable local deep autoregressive models to learn from data information, but also make lightweight regression model to learn from query information. Extensive experimental evaluations on real datasets show that accurate result is achieved while estimation time is significantly shortened, and model size is controlled within a reasonable range.

Keywords: Query optimization · Cardinality estimation · Multi-attribute Queries

1 Introduction

Cardinality estimation has a major role in query optimization of relational databases, whose main task is to estimate the number of tuples that satisfy the query before executing it [19]. Only based on high-quality estimation results can the optimizer select the excellent execution plan and ensure good query efficiency. Due to the increasing amount of data and the complex relationship among attributes, cardinality estimation of multi-attribute queries remains one

© The Author(s), under exclusive license to Springer Nature Singapore Pte Ltd. 2024
X. Song et al. (Eds.): APWeb-WAIM 2023, LNCS 14333, pp. 296–311, 2024.
https://doi.org/10.1007/978-981-97-2387-4_20

of the most challenging issues. The difficulty of it mainly comes from constructing accurate joint distribution in finite space and responding quickly within a controlled time [14]. Existing methods have made plenty of attempts for this problem but it is still not completely solved.

Traditional techniques such as histogram and sketch directly store lossy compressed information based on AVI [21], resulting in unsatisfactory results due to poor consideration of dependency among attributes [1,9]. For multi-dimensional queries, the methods based on sampling or KDE may be obviously inaccurate due to the lack of rich samples or significantly inefficient due to the large sample size [15]. The query-driven methods [5,8] that only utilize query information lack robustness and generalization when dealing with new queries.

Recent data-driven learning methods have made some progress in cardinality estimation [27]. The outstanding processing capability of autoregressive model for multi-dimensional data provide the guarantee for constructing condensed data information containing complex relationships among attributes [21]. Available techniques using autoregressive model [22,24] for cardinality estimation are basically based on the global strategy by creating a single global network over the entire relation, which means that both strongly and weakly associated attributes are decomposed lossless in the same manner. Although the global autoregressive models can well model strong-correlation among attributes, they omit the processing of weak-correlation among attributes, making the modeling of weak-correlation more complex. That not only increases the complexity of the model, but also slows down the estimation speed.

Our Solution. To systematically model attributes with different degrees of association, we propose a Local deep Autoregressive Framework (**LAF**), which utilizes mutual information to divide the universal set of attributes into multiple subsets according to the strength of association among attributes, so that the attributes within subset are strongly associated and the attributes between subsets are weakly associated. LAF establishes local deep autoregressive models for subsets, learning the joint distribution of strongly correlated attributes and outputting corresponding local estimations. A lightweight regression model is constructed to capture the complex mapping between all weakly correlated local estimations and true cardinality. Attributes are sorted in ascending order according to their information entropy, allowing autoregressive models to learn with increasing difficulty. During the training of LAF, local deep autoregressive models directly learn from data information while the lightweight regression model effectively draws knowledge from query information.

Contributions. The contributions of our work are the following:

1) We propose a fine-grained modeling method for modeling strongly and weakly correlated attributes differently, which effectively balances the accuracy and efficiency of cardinality estimation for multi-attribute queries. Moreover, we propose to utilize information entropy to sort attributes in descending order, which further improves the estimation accuracy.

2) We propose a comprehensive training scheme that effectively utilizes data information and query information, balancing the accuracy and robustness of cardinality estimation model for multi-attribute queries.
3) Extensive experimental evaluations show that accurate result is achieved while estimation time is significantly shortened, and model size is controlled within a reasonable range.

Outline. The remainder of the paper is organized as follows. We first briefly introduce related work in Sect. 2 and then describe problem in Sect. 3. Our model is discussed in detail in Sect. 4. We show extensive experiments on multiple datasets in Sect. 5 before concluding in Sect. 6.

2   Related Work

We briefly review prior cardinality estimation work, including traditional methods, query-driven learning methods, and data-driven learning methods.

The traditional methods based on AVI ignore the dependencies among attributes in the table, resulting in low-quality estimation. Multi-dimensional histogram [18] appears for capturing the correlation among attributes, but the space consumption of it rapidly increases with the number of dimension grows. Sampling-based methods suffer from poor accuracy for highly selective predicates and expensive estimation cost because calculating result requires a full scan over the sample [9]. For multi-dimensional queries, the estimation accuracy of KDE [7] is poor due to the difficulty in adjusting the bandwidth parameter.

Query-driven learning methods apply supervised learning strategy. Representative work includes self-tuning histogram [1] and some models. Various models have been applied to fit the mapping between query and cardinality [11], such as neural network [8], tree model [5], and convolutional network [12]. However, they suffer from the requiring of executed queries to train models, and may introduce terrible errors due to not learning the knowledge outside of the training set [13].

Data-driven learning methods solve the cardinality estimation problem as unsupervised learning problem and learn data representation from relational tables. Bayesian network [20] models the dependencies among attributes as directed acyclic graph, but it suffer from inefficiency. SPN [10] partially uses AVI to simulate the global joint distribution with partial joint distribution, and its performance heavily depends on whether the assumptions are valid. Recently proposed neural density estimator has shown advantages in estimation accuracy. Naru [24] and UAE [22] leverage deep autoregressive model to learn conditional probability distribution. They use global strategy for coarse-grained modeling, creating a single global autoregressive network for the entire relationship table and modeling strong and weak correlation among attributes in a unified way. However, weak-correlation among attributes do not need to be complexly modeled, which will lead to over complex learning and inferior efficiency. To obtain the accurate estimation result and improve the estimation speed, we propose LAF, which performs fine-grained modeling for attributes with different correlation degrees.

3   Problem Description

3.1   Supported Queries

A relation T with n attributes $A = \{a_1, a_2, ..., a_n\}$, and $Dom\,(A_i)$ contains all the values of attribute A_i. The cardinality of the query q is represented by $N_{card}\,(q)$, and $|T|$ denotes the total amount of tuples in relation T.

We consider point query and range query on single relation or multi-relation. The form of point query is $A_i = a_i$ AND $A_j = a_j$ AND ... for attributes $\{A_i, A_j\} \subseteq A$, $a_i \in Dom\,(A_i)$, $a_j \in Dom\,(A_j)$. Range query is conjunction of predicates, each of which contains a range constraint ($\neq, <, \leq, >, \geq$), equality constraint ($=$) or IN clause on a numeric or categorical attribute [7,17]. If the query does not contain predicate on attribute A_k, it will be expressed as $\min\,(Dom\,(A_k)) \leq A_k \leq \max\,(Dom\,(A_k))$.

3.2   Density Estimation

Cardinality estimation for query q is to estimate the number of the tuples satisfy q before executing it. If the tuple with the value $A_1 = a_1, A_2 = a_2, ..., A_n = a_n$ satisfies q, the problem becomes to estimate the number of tuples with that specifies value.

$$N_{card}\,(q) = P\,(q) \cdot |T| \qquad (1)$$

As shown in Eq. (1), the number of all tuples $|T|$ is known, and the estimation task can be completed only by getting the data distribution of tuples. Therefore, in local deep autoregressive models, we solve the problem of getting local estimations as density estimation problem, modeling to learn the joint distribution of strong-correlated attributes, $\widehat{P}\,(A_m, A_n...)$.

4   Local Deep Autoregressive Framework

4.1   Overview

Motivation. The proposed neural density estimator builds global network for all attributes, modeling attributes with different degrees of correlation in a unified approach [22–24]. In reality, strongly and weakly correlated attributes are often appearing simultaneously. For example, in the health status register, there is a weak or no correlation between "registrant's serial number" and "registrant's exercise duration," but a strong correlation between "registrant's exercise duration" and "registrant's health status," with a very high probability that people who exercise regularly will be in good health. It is common for weak and strong correlation attributes to be in the same relational table. However, modeling weak correlation among attributes in the same way as dealing with strong correlation among attributes will increase the complexity of the model and decrease the estimation efficiency. Establishing a global network for all attributes complicates the simple problem of learning the distribution of weakly correlated attributes.

Therefore strong and weak correlation among attributes need to be modeled differently.

Challenges. It is nontrivial to perform fine-grained modeling for attributes with different degrees of correlation. There are two challenges we need to deal with. One is that the existing autoregressive estimators do not model strong and weak correlation among attributes differently, and the other is that the existing cardinality estimation methods rarely use both data information and query information to train different structures in the model.

Overview of High-Level Idea. Both challenges call for designing new model framework. We propose LAF, as shown in Fig. 1, which uses local strategy for fine-grained modeling of strong and weak correlation among attributes, and utilizes data information and query information to comprehensively train the model.

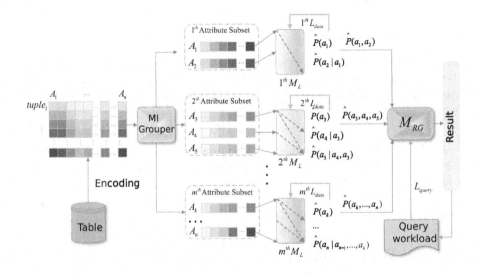

Fig. 1. Overview of LAF.

4.2   Encoding Tuples and Dividing Subsets

Encoding Tuples. Local deep autoregressive models treat relational table as multi-dimensional distributions of data. We first scan all attributes to obtain their domains and then encode them. Binary encoding [8] is our choice because of its spatial efficiency. For example, $Dom(A_k) = \{a_{k_1}, a_{k_2}, a_{k_3}, a_{k_4}\}$, the encoding results of attribute A_k are 00, 01, 10, 11, and the minimum storage required in binary encoding is $\sum_{i=1}^{|Dom(A_k)|} |Dom(A_k)|$. In addition, binary encoding method is beneficial to reduce the model size.

Dividing Subsets. We distinguish weak or strong correlation by computing mutual information (MI) between attributes. In information theory, MI [3] is a metric of interdependence between random variables. It is used to measure how much the degree of uncertainty about the random variable Y is reduced when the random variable X is known. As shown in Eq. (2), we calculate the MI between attributes, $MI_{set} = \{MI_{1,2}, MI_{1,3}, ..., MI_{1,n}, ..., MI_{n-1,n}\}$, where $MI_{i,j}$ represents the MI between A_i and A_j and $MI_{i,j} = MI_{j,i}$. If $MI_{i,j}$ is greater than the threshold, that is, when the value of A_i has been determined, the uncertainty of the value of A_j is reduced. That means when the value of A_i is known in a tuple, there is a high probability that A_j is a specific value corresponding to A_i, therefore we think A_i and A_j are strong-correlation attributes.

LAF deals with weak correlation and strong correlation in different ways adaptively. Specifically, strong-correlation attributes with MI greater than multiple thresholds are recursively aggregated in the same attribute subset, while weak-correlation attributes are divided into different subsets. We summarize the division strategy in Algorithm 1.

$$MI\left(X;Y\right) = \sum_{x \in X} \sum_{y \in Y} P\left(x,y\right) \log\left(\frac{P\left(x,y\right)}{P\left(x\right)P\left(y\right)}\right) \tag{2}$$

4.3 Model Architectures

Constructing Model. LAF models the strong-correlated attributes as local deep autoregressive model M_{L} and uses lightweight regression model M_{RG} to deal with weak correlation among attributes. We leverage deep autoregressive model to approximate the joint distribution of strong-correlation attributes, since it is expressive enough to model complex distribution [6]. Autoregressive decomposition mechanism uses chain rule to decompose the likelihood of the data sample P. For a set containing m strongly related attributes, deep autoregressive model decomposes the joint distribution into m conditional probabilities and gets the local estimation $P(x)$ from the product of conditional probability distributions, as shown in Eq. (3).

$$P(A_1 = a_1, ..., A_m = a_m) = \prod_{i=1}^{m} P(A_i = a_i | A_1 = a_1, ..., A_{i-1} = a_{i-1}) \tag{3}$$

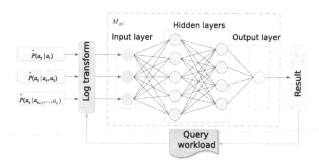

Fig. 2. Regression Model M_{RG}.

For attributes with strong-correlation, constructing a M_L for each attribute subset obtained by the Algorithm 1. Here we use ResMADE [16], a multi-layer perceptron with information masking technique, which has excellent performance. The problem of attribute order in M_L will be discussed later. Each trained M_L outputs a local estimation related to attributes in the corresponding subset. For attributes with weak-correlation, they are modeled in different M_L, and the lightweight regression model M_{RG} is used to capture the mapping between mathematical processed local estimations and real cardinality through supervised learning, as shown in Fig. 2. The weak association among attributes is learned by M_{RG}, and AVI is not used throughout the process.

The workflow of LAF simplifies the representation of weak-correlation while allowing M_L to focus on learning the joint distribution of attributes with strong-correlation and applying M_{RG} to learn from query information, as shown in Fig. 1. Compared with the global autoregressive model, the number of attributes of each local deep autoregressive model is greatly reduced, the complexity of the model is vastly reduced, and the estimation delay is significantly reduced.

Algorithm 1. Divide Attribute Subsets

Input: parameter β_1, β_2, attributes set $A = \{A_1, A_2, ..., A_n\}$, MI_{set}
Output: attribute subset division results D_{set}

1: $D_{set} = \emptyset$ ▷ initialize D
2: $\alpha_1 = \beta_1 * \max(MI_{set})$, $\alpha_2 = \beta_2 * (\max(MI_{set}) + \min(MI_{set}))$ ▷ two thresholds
3: $G = \{A_p, A_q\}$ ▷ A_p and A_q correspond to $\max(MI_{set})$
4: $GA = \{A_i | MI(A_p, A_i) > \alpha_1 \lor MI(A_q, A_i) > \alpha_1\}$
5: $s = |GA|$ ▷ records the number of elements in initial GA
6: **for** $i = 1 \to s$ **do**
7:     $MI_{i_{mean}} = \dfrac{\sum_{j=1}^{|G|} MI\left(A_{GA_i}, A_{G_j}\right)}{|G|}$ ▷ A_{GA_i} is element in GA, A_{G_j} is in G,
       $MI(X, Y)$ is MI between X and Y.
8:     **if** $MI_{i_{mean}} > \alpha_1$ **then**
9:         $G = G \cup A_{GA_i}$
10:    **end if**
11: **end for**
12: $D_{set} = D_{set} \cup G$, $A = A - G$
13: update MI_{set} ▷ remove elements in MI_{set} that related to attributes in G
14: $G = \emptyset$, $GA = \emptyset$
15: **if** $\max(MI) \geq \alpha_2$ **then**
16:    goto line 3
17: **else**
18:    **for** $j = 1 \to |A|$ **do**
19:        $G = A_j$, $D_{set} = D_{set} \cup G$ ▷ remaining attributes become separate subset
20:    **end for**
21: **end if**
22: **return** D_{set}

Sorting Columns Using Information Entropy. The M_L obtained by different attribute input sequence is various, and it requires a reasonable attribute order to better learn joint distribution. In information theory, information entropy(IE) is used to describe the uncertainty of each possible event. It can be seen from Eq. (4) that the greater the number of values of random variables, the greater the degree of chaos of data distribution, and the greater the IE. When the distribution is uniform, IE is maximum.

We sort attributes in ascending order according to its IE, that is, the attribute with the lowest IE enters the M_L first. Let the model learn the data distribution of attributes with low IE (low IE means centralized data distribution and less learning difficulty) before learn it with high IE (high IE means scattered data distribution and greater learning difficulty), so that model can learn from easy to difficult, which is also in line with our human learning process. Use Eq. (4) to calculate IE, where a_{k_i} is the element in $Dom(A_k)$, and $H(A_k)$ is the IE of A_k.

$$H(A_k) = - \sum_{i=1}^{|Dom(A_k)|} P(a_{k_i})\log_2(a_{k_i}) \tag{4}$$

4.4   Model Training with Data and Query

Training with Data. Every M_L is trained with the data-driven approach, whose input is the tuple values of the corresponding attributes and output is the predicted density estimation. The data-driven loss L_{data} is the cross-entropy between real distribution $P(x)$ and estimated distribution $\widehat{P}(x)$, and the weights of M_L are learned from the data by minimizing its L_{data}.

$$L_{data} = - \sum_{x \in T} P(x) \log \widehat{P}(x) \tag{5}$$

Training with Queries. M_{RG} is trained by query-driven approach. Each trained M_L outputs a local estimation, and the final result is obtained after all the local estimations pass through M_{RG}. Obviously, the query-driven loss L_{query} should be defined as the error between real selectivity $sel(q)$ of the query q and estimated selectivity $\widehat{sel}(q)$, and we choose MSELoss.

$$L_{query} = \frac{1}{|Q|} \sum_{i=1}^{|Q|} \left(\widehat{sel}(q_i) - sel(q_i) \right)^2 \tag{6}$$

4.5   Estimating Cardinality

Point Query. It has been introduced that the solve the cardinality estimation problem as a density estimation problem, so the trained LAF can answer the cardinality estimation problem of point query directly. Specifically, all values of attributes that are not involved in the query are set to satisfy it.

Range Query. Density estimator can not answer cardinality estimation of range query directly, and it is not reasonable to use enumeration when the number of satisfied tuples is relatively large. So we use progressive sampling to solve it, which can effectively extract high-quality data samples from relational tables. Progressive sampling is a Monte Carlo method and has been proved by predecessors that it is unbiased [24].

Join Query. For multi-table join queries, it is common practice to collect join results using a join sampler and then train the estimator on this result so we deal with it as [23] did, adding *virtual indicators* and *fanout columns* to the LAF process. LAF clusters the sample tuples collected according to the weight [26] into several groups and builds a M_L for each group. Results are obtained through *fanout scaling* calculation and M_{RG}.

5   Experiment

We conduct extensive experiments on real datasets, proving that LAF performs well in answering cardinality estimation problem involving multi-attribute queries. We first describe the experimental settings in Sect. 5.1. Then the cardinality estimation results of LAF are reported in Sect. 5.2, and we report the estimation accuracy of different attribute orders in Sect. 5.3.

5.1   Experimental Settings

Datasets. We choose three challenging real single table datasets: Census, DMV and IOV. The statistical data amounts of them are shown in Table 1. Attributes in each datasets are distributed on both numerical and categorical attributes, and have significantly wide range of values. For the multi-table experiments, we evaluate LAF on the IMDB dataset and IOV-large dataset.

Table 1. Dataset characteristics. 'Cols' refers to the number of columns; 'Domain' means per-column domain size; 'Join' is the product of all domain sizes.

Dataset	Rows	Cols	Domain	Join
Census	45K	14	2-26K	10^{20}
DMV	12.4M	13	2-30.9K	10^{24}
IOV	1M	9	2-46K	10^{26}

Census [4]. The dataset was extracted from the 1994 Census database. We use the following 14 columns with large differences in data types and range sizes (numbers in parentheses): Age (74), Workclass (7), Fnlwgt (26741), Education (16), Education-num (16), Marital-status (7), Occupation (14), Relationship (6), Race (5), Sex (2), Capital-gain (121), Capital-loss (97), Hours-per-week (96), Native-country (41).

DMV [25]. The dataset contains vehicle registrations in New York. We use the following 13 columns: Record Type (3), Registration Class (73), City (30959), State (86), Zip (11180), County (63), Model Year (125), Make (23658), Body Type (59), Fuel Type (8), Scofflaw Indicator (2), Suspension Indicator (2), Revocation Indicator (2). This dataset is used to test the performance of various methods when the NDVs is large.

IOV. The dataset comes from the internet of vehicles platform of a large automobile enterprises, containing data of 560 commercial vehicles in April 2022. It consists of the following attributes: Longitude (4019), Latitude (556), StartDistance (6852), EndDistance (7035), OilConsumption (46585), Unit (13), Time(2), Accelerator(5), State(4). The data distribution in IOV is the most complex.

IMDB. The dataset contains movie-related data. We use the provided JOB-light with 70 queries and generate a more complex workload with 2000 queries.

IOV-Large. It is from the same platform as IOV and contains three relational tables, involving configuration information and driving data of vehicles.

Performance Metric. We use q-error defined in Eq. (7) to evaluate the cardinality estimation accuracy of the model and report the entire q-error distribution (50%, 90%, 95%, 100% quantile) of the model for each workload.

$$q - error = \max(1, \frac{sel(q)}{\widehat{sel(q)}}, \frac{\widehat{sel(q)}}{sel(q)}) \tag{7}$$

Query Workload. For every dataset, construct multi-dimensional queries containing range predicates and equivalence predicates. Generate queries with dimensions ranging from $D/2$ to D and cover all possible attribute subsets of the dataset which contains D attributes. In each subset, range predicates and equivalence predicates are randomly placed in columns. We construct a workload containing 20K queries for each dataset, which has great changes in capacity and actual selectivity. Figure 3 plots the selectivity distribution of queries in these workloads.

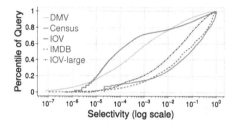

Fig. 3. Distribution of Selectivity.

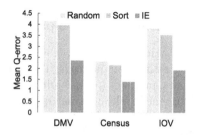

Fig. 4. Mean Q-error.

Baseline. We compared the LAF with others, including traditional methods (1–4) and data-driven learning models (5–7) and query-driven learning models (8–11).

1. AVI. Based on AVI, multiply the selectivity of predicate on each attribute directly, for example $sel\,(A_i = a_i \wedge A_j = a_j) = sel\,(A_i = a_i) \times sel\,(A_j = a_j)$.
2. EBO. The correlation between some columns is considered. It sorts the selectivity of predicate on each attribute and selects the first $i\,(i \leq 4)$ selectivity with the lowest value, then the result is calculated by Eq. (8).

$$sel\,(q) = sel_0 \times sel_1^{\frac{1}{2}} \times sel_2^{\left(\frac{1}{2}\right)^2} \times sel_3^{\left(\frac{1}{2}\right)^3} \; (sel_0 < sel_1 < sel_2 < sel_3) \quad (8)$$

3. Sample. It uses uniform random sampling, no more than 5 percent of the tuples are selected from dataset as the sample set during the experiment.
4. MHIST [18]. This method is to build a multidimensional histogram on entire dataset. In experiments, we choose Maxdiff as the partition constraint.
5. MSCN [12]. A state-of-the-art method uses one-hot vector to describe attributes and operators, joining average results on predicate set as query coding. In single table experiment, we delete the connection module referring to [22].
6. LW-NN [5]. It solves the cardinality estimation problem as a regression problem, using a single network to learn mapping between query and its cardinality.
7. LW-XGB [5]. Compared with LW-NN, this method uses the tree-based model XGBoost to solve estimation problem.
8. Bayes [2]. A probability graph model method. We adopt the same implementation method as [24].
9. DeepDB [10]. This method is based on the structure of sum-product network, and uses relational sum-product network to model joint data distribution.
10. Naru [24]. Using the global autoregressive model to learn the distribution of underlying data, and NeuroCard [23] is an extension of it to join queries.
11. UAE [22]. A method uses Gumbel-Softmax technique to develop differentiable progressive sampling, and uses data information and query information to learn joint data distribution in a single model.

5.2 Performance Comparison

Estimation Accuracy. Table 2 reports q-error distribution for different cardinality estimation methods. Their accuracy can be roughly ranked as data-driven learning methods > query-driven learning methods > traditional methods.

In general, the accuracy of LAF is outstanding. On three datasets, the median q-error (50%) of LAF is close to 1, which is the optimal value. LAF has the best tail error performance on IOV, and Naru has it on DMV and Census. The accuracy of Naru and UAE is equivalent to LAF. Their high accuracy stems from using of autoregressive models to effectively learn the joint distribution. The

Table 2. Estimation errors on single table.

Estimator	Census				DMV				IOV			
	50th	90th	95th	Max	50th	90th	95th	Max	50th	90th	95th	Max
Traditional Methods												
AVI	1.33	4.90	7.51	364	8.00	59.52	198	$1 \cdot 10^5$	3.01	40.00	99.68	$2 \cdot 10^5$
EBO	1.91	5.10	17.70	819	2.11	42.26	90.56	$8 \cdot 10^4$	2.54	26.00	48.24	$1 \cdot 10^6$
Sample	1.47	4.00	68.00	$1 \cdot 10^4$	1.45	29.13	140	$6 \cdot 10^4$	5.07	20.90	135	$2 \cdot 10^5$
MHIST	1.20	4.82	152.45	$3 \cdot 10^4$	1.78	24.23	336.50	$1 \cdot 10^5$	9.05	271.89	$3 \cdot 10^4$	$3 \cdot 10^5$
Learning Methods												
MSCN	1.83	5.60	18.74	558	2.72	64.03	435.31	$3 \cdot 10^4$	3.70	49.00	98.50	538
LW-NN	2.57	8.00	75.39	854	3.04	155	487.75	$4 \cdot 10^4$	10.00	198.91	544	$5 \cdot 10^4$
LW-XGB	1.70	5.30	8.00	125.50	1.61	27.00	56.00	$2 \cdot 10^4$	6.25	156	365.50	$1 \cdot 10^4$
Bayes	1.28	3.71	7.50	191	1.37	9.00	18.06	$1 \cdot 10^4$	5.55	129.76	281	571.75
DeepDB	2.00	4.66	6.60	127.59	1.36	55.71	242.45	$2 \cdot 10^4$	3.00	101.49	599	$1 \cdot 10^4$
Naru	1.17	2.24	**2.49**	**10.63**	1.18	2.22	3.38	**249**	1.66	6.00	15.00	61.25
UAE	1.20	2.55	2.97	46.00	1.11	2.10	**3.36**	257	1.63	4.90	13.25	90.00
LAF (Ours)	**1.15**	**2.03**	2.61	31.75	**1.10**	**2.07**	3.39	265	**1.42**	**3.50**	**11.00**	**32.24**

other two data-driven learning models, DeepDB and Bayes, perform worse than LAF. The error of Bayes mainly comes from its approximate structure. The performance of DeepDB decreases rapidly on DMV with strong attribute correlation, especially in tail error, that is because of the independence assumption.

The accuracy of Query-driven learning methods is unstable. MSCN, LW-NN and LW-XGB perform well on the Census, but poorly on DMV and IOV. The precision of the query-driven method depends on the similarity between the testing set and the training set.

LAF largely outperforms Traditional methods which are limited by AVI or the problem that the sample is not enough representative. Although they perform well in the middle error, tail errors of them are terrible.

In addition, LAF also achieves optimal or equivalent overall performance for join queries, as shown in Table 3. LAF produces the lowest median error and exhibits the accuracy advantage in all error quantiles on IMDB and IOV-large.

Table 3. Estimation errors on multi-table.

Estimator	IMDB				IOV-large			
	50th	90th	95th	Max	50th	90th	95th	Max
MSCN	1.73	10.96	72.02	$1 \cdot 10^3$	4.73	35.60	108.74	$2 \cdot 10^4$
DeepDB	2.54	11.06	62.43	$1 \cdot 10^3$	3.80	24.66	76.26	$1 \cdot 10^4$
NeuroCard	1.62	8.89	25.30	255	2.17	**6.24**	49.00	363.80
UAE	1.43	7.63	**17.36**	**161**	2.20	6.85	29.70	465
LAF (Ours)	**1.31**	**6.01**	21.62	172.34	**1.85**	6.53	**20.61**	**176.75**

Estimation Latency. Figure 5 reports the average latency of data-driven learning methods and their speed can be ranked as LAF>DeepDB>Naru> UAE>Bayes.

LAF and DeepDB run the fastest, but LAF performs much better than DeepDB in accuracy, especially at the tail error on DMV. LAF is much faster than Naru and UAE, and the estimation accuracy of them is similar. On DMV, it outperforms Naru by 3.19× and UAE by 3.21×. On Census, it outperforms Naru and UAE by 2.11×, 2.19×, respectively. And it outperforms Naru by 2.22× and UAE by 2.25× on IOV. Compared with other methods, Bayes is several orders of magnitude slower.

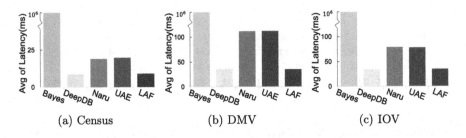

(a) Census   (b) DMV   (c) IOV

Fig. 5. Estimator Latency [log scale]

Model Size. We compared model size and average accuracy of these methods, and the results are shown in Fig. 6. MHIST consumes a large memory but estimation accuracy is not ideal. Query driven learning methods save space but have poor accuracy. Data driven learning methods perform outstanding in accuracy, especially Naru, UAE and LAF, because these three methods use autoregressive models. The model size of LAF is smaller than that of Naru and UAE on all datasets, and it performs better in controlling model size effectively.

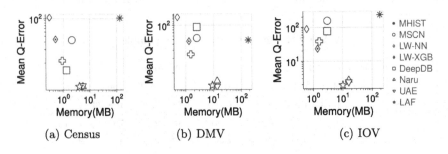

(a) Census   (b) DMV   (c) IOV

Fig. 6. Model Size [log scale]

5.3 Attribute Ordering

We examine the effect of attribute order on estimation accuracy in the local model of LAF, and the results are shown in Fig. 4. Random means attributes are entered into the local model in random order. Original refers to sort attributes by original order, and IE means sorting attributes in ascending order by their information entropy.

The results show that sorting attributes according to their IE from low to high can significantly improve estimation accuracy, because the distribution of attributes with low IE is more concentrated than attributes with high IE, and it is easier to learn. The number of attributes in DMV is large, and the average q-error of sorting attributes by IE is nearly twice lower than that of sorting attributes randomly. The number of attributes in the IOV is small, and using IE to sort attribute has little impact on estimation accuracy. The more attributes contained in dataset, the more obvious the improvement of estimation accuracy obtained by sorting attributes with IE.

6 Conclusion

We propose a novel local deep autoregressive framework LAF, performing fine-grained modeling for strong and weak correlation among attributes, and fully using data information and query information to comprehensively train the model. LAF utilizes the local strategy to build local deep autoregressive models to learn joint distribution for strongly associated attributes and constructs a lightweight regression model to get the complex mapping between weakly associated local estimations and cardinality, drawing valid information from both the data and the query resource. Extensive experiments show that LAF significantly reduces the latency while maintaining accurate estimation, and the model size is also reasonably controlled.

Acknowledgement. This work is supported by National Natural Science Foundation of China (No. 62072282), Industrial Internet Innovation and Development Project in 2019 of China, Shandong Provincial Key Research and Development Program (No. 2019JZZY010105).

References

1. Bruno, N., Chaudhuri, S., Gravano, L.: STHoles: a multidimensional workload-aware histogram. In: Proceedings of the 2001 ACM SIGMOD International Conference on Management of Data, pp. 211–222 (2001)
2. Chow, C., Liu, C.: Approximating discrete probability distributions with dependence trees. IEEE Trans. Inf. Theory **14**(3), 462–467 (2006)
3. Cover, T.M., Thomas, J.A.: Elements of Information Theory. Wiley, Hoboken (2012)
4. Dua, D., Graff, C.: UCI machine learning repository (2017). https://archive.ics.uci.edu/ml/index.php

5. Dutt, A., Wang, C., Nazi, A., Kandula, S., Narasayya, V., Chaudhuri, S.: Selectivity estimation for range predicates using lightweight models. Proc. VLDB Endow. **12**(9), 1044–1057 (2019)
6. Goodfellow, I., Bengio, Y., Courville, A.: Deep Learning. MIT Press, Cambridge (2016)
7. Gunopulos, D., Kollios, G., Tsotras, V.J., Domeniconi, C.: Selectivity estimators for multidimensional range queries over real attributes. Proc. VLDB Endow. **14**(2), 137–154 (2005)
8. Hasan, S., Thirumuruganathan, S., Augustine, J., Koudas, N., Das, G.: Deep learning models for selectivity estimation of multi-attribute queries. In: Proceedings of the 2020 ACM SIGMOD International Conference on Management of Data, pp. 1035–1050 (2020)
9. Heimel, M., Kiefer, M., Markl, V.: Self-tuning, GPU-accelerated kernel density models for multidimensional selectivity estimation. In: Proceedings of the 2015 ACM SIGMOD International Conference on Management of Data, pp. 1477–1492 (2015)
10. Hilprecht, B., Schmidt, A., Kulessa, M., Molina, A., Kersting, K., Binnig, C.: DeepDB: learn from data, not from queries! Proc. VLDB Endow. **13**(7), 992–1005 (2020)
11. Kim, K., Jung, J., Seo, I., Han, W.S., Choi, K., Chong, J.: Learned cardinality estimation: An in-depth study. In: Proceedings of the 2022 ACM SIGMOD International Conference on Management of Data, pp. 1214–1227 (2022)
12. Kipf, A., Kipf, T., Radke, B., Leis, V., Boncz, P., Kemper, A.: Learned cardinalities: estimating correlated joins with deep learning. arXiv preprint arXiv:1809.00677 (2018)
13. Kwon, S., Jung, W., Shim, K.: Cardinality estimation of approximate substring queries using deep learning. Proc. VLDB Endow. **15**(11), 3145–3157 (2022)
14. Leis, V., Gubichev, A., Mirchev, A., Boncz, P., Kemper, A., Neumann, T.: How good are query optimizers, really? Proc. VLDB Endow. **9**(3), 204–215 (2015)
15. Leis, V., Radke, B., Gubichev, A., Kemper, A., Neumann, T.: Cardinality estimation done right: Index-based join sampling. In: CIDR (2017)
16. Nash, C., Durkan, C.: Autoregressive energy machines. In: International Conference on Machine Learning, pp. 1735–1744. PMLR (2019)
17. Park, Y., Zhong, S., Mozafari, B.: QuickSel: quick selectivity learning with mixture models. In: Proceedings of the 2020 ACM SIGMOD International Conference on Management of Data, pp. 1017–1033 (2020)
18. Poosala, V., Ioannidis, Y.E.: Selectivity estimation without the attribute value independence assumption. In: VLDB, vol. 97, pp. 486–495 (1997)
19. Sun, J., Li, G., Tang, N.: Learned cardinality estimation for similarity queries. In: Proceedings of the 2021 ACM SIGMOD International Conference on Management of Data, pp. 1745–1757 (2021)
20. Tzoumas, K., Deshpande, A., Jensen, C.S.: Lightweight graphical models for selectivity estimation without independence assumptions. Proc. VLDB Endow. **4**(11), 852–863 (2011)
21. Wang, X., Qu, C., Wu, W., Wang, J., Zhou, Q.: Are we ready for learned cardinality estimation? Proc. VLDB Endow. **14**(9), 1640–1654 (2021)
22. Wu, P., Cong, G.: A unified deep model of learning from both data and queries for cardinality estimation. In: Proceedings of the 2021 ACM SIGMOD International Conference on Management of Data, pp. 2009–2022 (2021)
23. Yang, Z., et al.: NeuroCard: one cardinality estimator for all tables. Proc. VLDB Endow. **14**(1), 61–73 (2020)

24. Yang, Z., et al.: Deep unsupervised cardinality estimation. Proc. VLDB Endow. **13**(3), 279–292 (2019)
25. Zanettin, F.: State of New York. Vehicle, snowmobile, and boat registrations (2019). https://catalog.data.gov/dataset/vehicle-snowmobile-and-boat-registrations
26. Zhao, Z., Christensen, R., Li, F., Hu, X., Yi, K.: Random sampling over joins revisited. In: Proceedings of the 2018 ACM SIGMOD International Conference on Management of Data, pp. 1525–1539 (2018)
27. Zhu, R.: Flat: fast, lightweight and accurate method for cardinality estimation. Proc. VLDB Endow. **14**(9), 1489–1502 (2021)

MGCN-CT: Multi-type Vehicle Fuel Consumption Prediction Based on Module-GCN and Config-Transfer

Hao Li, Qianwen Cheng, Zhaohui Peng$^{(\boxtimes)}$, Yashu Tan, and Zengzhe Chen

School of Computer Science and Technology, Shandong University, Qingdao, China
{lihao_bd,chengqw,tys}@mail.sdu.edu.cn, pzh@sdu.edu.cn,
chenzengzhe28@gmail.com

Abstract. Accurate vehicle fuel consumption prediction is crucial to reduce pollutant emissions and save commercial vehicle operating costs. With the support of Internet of Vehicles data, data-driven multivariate time series forecasting methods have been adopted for fuel consumption prediction. Different types of vehicles are composed of modules with different configurations and contain different domain knowledge. However, existing methods rarely consider these differences, and cannot be adjusted according to the vehicle configuration when facing multiple types of vehicles. Moreover, the number of vehicle samples for some personalized configurations is not enough to support the training of the model. To solve the above problems, we propose the multi-type vehicle fuel consumption prediction model based on Module Graph Convolution Network and Configuration Transfer(MGCN-CT). First, in order to express the vehicle module domain knowledge and driving data uniformly, a module graph embedded with domain knowledge is proposed. Then a module graph convolutional network is proposed to model the spatio-temporal dependence of the module graph and realize fuel consumption prediction. Finally, a configuration transfer module based on a configuration classifier is proposed to realize the fuel consumption prediction of a few-sample personalized configuration vehicles. The effectiveness of the model is verified through extensive experiments on real datasets. Compared with the baseline methods, our method achieves superior accuracy for fuel consumption prediction.

Keywords: Fuel Consumption Prediction · Module Graph Convolutional Network · Configuration Transfer Learning

1 Introduction

The wide application of vehicles provides impetus for social development, but the huge amount of vehicles also leads to the rapid consumption of energy [12]. Accurate fuel consumption prediction can help drivers plan routes, which is of great significance in saving fuel consumption. With the rapid development

X. Song et al. (Eds.): APWeb-WAIM 2023, LNCS 14333, pp. 312–327, 2024.
https://doi.org/10.1007/978-981-97-2387-4_21

of Internet of Vehicles technology, vehicle sensors can collect multiple vehicle operating states and provide a wealth of vehicle driving data [31]. Since the multivariate time series prediction method has achieved good results in many tasks, it is used in the research of vehicle fuel consumption prediction.

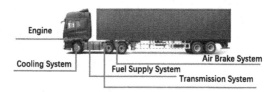

Fig. 1. Commercial vehicle structure

Taking the IVD dataset used in this paper as an example, commercial tractors can be divided into five modules as shown in Fig. 1, and each module contains its own domain knowledge. For example, the domain knowledge of the engine system can be expressed as the output power of the vehicle, and the domain knowledge of the cooling system can be expressed as the heat dissipation. Multiple modules such as engine and cooling system need fuel to provide energy, and different types of vehicles will lead to great differences in the working principle of the same module. Therefore, it is necessary to consider domain knowledge when predicting fuel consumption of multiple types of vehicles. Most of the existing methods are to model the temporal correlation and spatial correlation of vehicle driving data to predict the fuel consumption in the future. Due to the lack of domain knowledge in these methods, the accuracy of fuel consumption prediction for multiple types of vehicles is not high. Some personalized configuration vehicles cannot meet the model training requirements due to the small number of vehicles. The transfer learning model uses methods such as data transfer [13] and model fine-tuning [8] to improve the learning ability of the few-sample model, but these methods cannot cope with complex and changeable vehicle data, and may even produce negative transfer phenomena.

The challenges of the accurate fuel consumption prediction is as follows. First, it is difficult to directly apply multi-module domain knowledge to deep learning. Due to the different functions of each module of the vehicle, the expression of its domain knowledge is also different, and the components of the same module with different configurations will also change the domain knowledge. Therefore, it is necessary to design a unified way to extract domain knowledge and apply it to the modeling process. Moreover, vehicle driving data is not suitable for applying to the transfer learning. Vehicle driving data cannot represent the difference between vehicle types, and vehicle configuration is the root cause of the fuel consumption gap. Therefore, it is necessary to apply vehicle configuration information to the transfer process.

In order to solve the above challenges, we propose the domain knowledge vector formula to extract the domain knowledge of each module in a unified manner, and construct a module graph to uniformly represent the vehicle domain knowledge and driving data. Then a module graph convolutional network is proposed to model the spatio-temporal dependence of the module graph, and apply domain knowledge to fuel consumption prediction. What's more, we build a vehicle configuration classifier to learn configuration differences between vehicle types, and use it to construct transfer constraints for few-shot fuel consumption prediction for personalized configuration vehicles.

In summary, we propose a multi-type vehicle fuel consumption prediction model based on Module Graph Convolution Network and Configuration Transfer (MGCN-CT). The main contributions are as follows:

- A domain knowledge embedded module graph convolutional network is proposed to jointly express module domain knowledge and vehicle driving data, and obtain accurate fuel consumption prediction results under the guidance of domain knowledge through spatiotemporal dependency modeling.
- Using vehicle configuration information as an auxiliary domain, a transfer framework based on configuration classifiers is used to improve the fuel consumption prediction effect of few-sample personalized configuration vehicles and provide interpretable transfer results.
- Experiments on real vehicle datasets demonstrate that our method outperforms existing baseline methods. In addition, ablation experiments also demonstrate the effectiveness of each module in the model.

2   Related Work

2.1   Fuel Consumption Prediction

Accurate fuel consumption prediction is of great significance to the development of the transportation industry [16]. The current methods of fuel consumption prediction can be divided into two categories:

Physical Principles Based Models: The dynamic heat transfer model of the engine cooling system [19] and the Willan internal combustion engine model [3] are used to estimate fuel consumption. Powertrain-based models evaluate vehicle emissions and fuel consumption using airflow as an indicator of instantaneous fuel consumption [26]. The single-cylinder air-fuel ratio estimation algorithm based on the Kalman filter [29] can accurately estimate the single-cylinder air-fuel ratio of the asymmetric exhaust runner gas engine.

Machine Learning Based Models: ANN [1] and GRNN [30] are used to simplify the engine process and simulate the driver's driving behavior, and predict fuel consumption. Intelligent time series modeling methods [24] can be used to predict NOx emissions and fuel consumption. The key factors affecting vehicle fuel consumption are analyzed through three views of environment, driver and vehicle [15].

The current fuel consumption prediction methods do not combine domain knowledge and machine learning, resulting in inaccurate fuel consumption prediction.

2.2   Time Series Prediction

Vehicle driving status information is multivariate time series data, so multivariate time prediction methods provide support for the study of fuel consumption prediction problems.

MFDMA [6] achieves good results on backward analysis. VARMA [9] can provide a parsimonious representation of linear data generation processes. LS-SVM has achieved good results in short-term forecasting in meteorological time series [21]. GPCM [27] is used to model stationary signals as continuous-time white noise processes. Music sentiment can be predicted using a deep RNN attention mechanism [2] composed of LSTMs. RNN-GNN [4] encodes geographic knowledge for crop yield prediction.

However, the above methods are all data-driven and predict a specific scenario. When faced with multiple types of vehicle fuel consumption predictions, if the model cannot be adjusted in time based on domain knowledge, it will lead to problems such as a decrease in model accuracy.

2.3   Transfer Learning

Transfer learning can apply the knowledge of the source domain to the learning of the target domain, effectively improving the model learning ability in the case of few samples [25].

Joint distribution adaptation methods [11] can reduce the distribution difference between source and target domains. AudiBERT integrates multiple models involved with training audio and text into a deep learning framework [28]. ST-GFSL [18] reconstructes the graph structure during meta-learning, avoiding structural bias between different datasets. Weighted cost learning and importance concept sampling methods [23] can avoid model loss due to negative transfer effects. The Siamese network data structure [5] enables fine-tuning of training parameters and optimizes with relative distance loss between datasets.

How to use vehicle configuration information to transfer vehicle fuel consumption information remains to be studied.

3   Preliminary

In this section, we give the relevant symbolic representations and formal definitions for the vehicle fuel consumption prediction task and the transfer learning task.

Notations: When the vehicle is running, the on-board sensors will collect the operating status information of each component. We can obtain multivariate time series data(MTS) describing the operating state of the vehicle, including various attributes such as vehicle speed, water temperature, and engine oil pressure. We use $MTS = \{x_1, x_2, ..., x_i, ..., x_T\}$ to express the multivariate time series data, where T is the length of the time series, and x_i represents the running status of the i-th time node. $x_i = \{x_i^1, x_i^2, ..., x_i^{N_s}\}$, where N_s represents

the dimension of the collected data features. We use $Y = \{y_1, y_2, ..., y_i, ..., y_T\}$
to represent the collected fuel consumption sequence data, and y_i represents the
fuel consumption from $i-1$ to i moment. Vehicle configuration information such
as engine type, driving form, etc. is represented by $C = \{c_1, c_2, ..., c_{N_c}\}$, where
N_c represents the number of recorded vehicle configurations.

Prediction Task Definition: Given Δt historical MTS data of vehicles
$X = \{x_{t-\Delta t+1}, x_{t-\Delta t-2}, ..., x_t\}$, our goal is to build a suitable model f, to
predict the sum of fuel consumption in the future Δh time period $\widetilde{y} = sum(y_{t+1}, y_{t+2}, ..., y_{\Delta h})$, which is recorded as:

$$\widetilde{y} = f(X) \tag{1}$$

Transfer Learning Task Definition: Divide the general-purpose vehicle data
into source domains D^s, which contains vehicle MTS data and configuration
data. Divide the personalized configuration vehicle data into target domain D^t.
Use D^s and D^t to predict the sum of fuel consumption $\widetilde{y}_t$ of vehicles in the target
domain in the future Δh time period, recorded as:

$$\widetilde{y}_t = \Gamma(D^s, D^t) \tag{2}$$

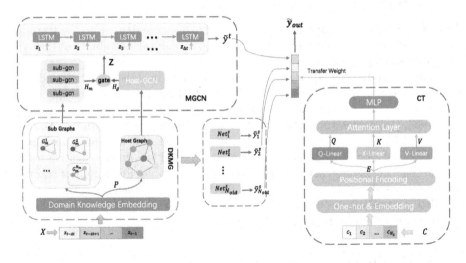

Fig. 2. Framework of MGCN-CT

4   Method

In this section, we will introduce the model MGCN-CT as shown in Fig. 2. The
model framework consists of three parts: (1)Module graph with domain knowl-
edge(DKMG). This module will extract the domain knowledge of different modules to form a domain knowledge vector and generate a module graph. (2)Mod-
ule graph convolution network(MGCN). This module models spatio-temporal
dependencies on module graph to predict fuel consumption. (3)Configuration
transfer module based on classifier(CT). This module uses a classifier to obtain
transfer weights to achieve knowledge transfer between different vehicle types.

4.1   Module Graph with Domain Knowledge

A vehicle is a modular device with rich domain knowledge, containing multiple modules such as engine, cooling system, gearbox, etc. For example, the cooling system contains multiple parts, and there is a high degree of coupling between the parts.

Guided by the domain knowledge of vehicles, we construct a module graph embedded with domain knowledge for MTS modeling. The engine is an important component module of the vehicle, and its theoretical output power can be calculated by the following formula:

$$W = \frac{2\pi \times T \times n}{60} \tag{3}$$

where T represents the engine torque, and n represents the engine speed. However, since generators and other conditions will occupy part of the output power, the above formula cannot be directly applied to actual scenarios. Guided by domain knowledge, we obtain the following formula:

$$P_w = W_w V_w^\top + b_w \tag{4}$$

where P_w represents the output power of the engine module, called the domain knowledge vector of the engine module, V_w represents the set of sub-vectors collected in the engine module. Other modules can also use Formula(4) to obtain the corresponding domain knowledge vector.

The N_s features in MTS are divided into corresponding sets according to the modules they belong to, so as to express the running status of the corresponding modules, which is represented as $V = \{V_1, V_2, ..., V_{N_m}\}$, where N_m represents the number of modules. This article uses $G_m^i = (V_m^i, E_m^i, A_m^i)$ to represent the sub-graph of module i, where V_m^i is the module feature set V_i, E_m^i is the edge set, $A_m^i \in R^{K_i \times K_i}$ is the adjacency matrix, and K_i is the feature dimension of V_i. When calculating the domain knowledge vector P_i of the module i, we can get the constraint weight $W_i = [w_1, w_2, ..., w_{K_i}]$. Constraint weights represent the contribution of different features to the domain knowledge vector, and the change of contribution over time reflects the mutual influence between module sub-variables. We use the weight constraints to calculate the adjacency matrix in the module sub-graph as follows:

$$A_{ij}^k = |\frac{cov(w_i, w_j)}{\sigma w_i \sigma w_j}| \ for \ i, j \in W_k \tag{5}$$

where cov represents the covariance and σ represents the standard deviation. The module sub-graph contains the domain knowledge of the module, and represents the mutual influence of the sub-variables within the module.

Each module of the vehicle is not completely independent, but is related to each other. We use $G = (V, E, A)$ to represent the host-graph, where V is the set of module sub-graphs, $|V| = N_m$, E is the edge set, $A \in R^{N_m \times N_m}$ is the adjacency matrix. Since the domain knowledge vector $P = [P_1, P_2, ..., P_{N_m}]$

is the embodiment of the domain knowledge of the module, we use P_i as the representative value of the node it belongs to, and bring it into Formula(5) to calculate the adjacency matrix. The host-graph is constructed through P, which contains the overall domain knowledge of the vehicle and expresses the correlation between modules.

4.2 Module Graph Convolution Network

Existing graph convolutional networks can only handle graph convolutions of single graphs, and are not suitable for the module graph structure containing sub-graphs in this paper. Therefore, we propose the Module Graph Convolutional Network(MGCN) to extract the spatio-temporal dependencies of module graphs. **Spatial Dependence:** When extracting spatial dependencies, MGCN is divided into two-step convolutions. The first step is to use sub-gcn to convolve the module sub-graph G_m^i, treat the module sub-graph as an ordinary graph structure, and use Formula 6 to perform graph convolution on module i:

$$H_i^{l+1} = \sigma(\widetilde{P}_i H_i^l W_i^l + b_i^l) \tag{6}$$

where $\widetilde{P}_i = \widetilde{D}^{-\frac{1}{2}} \widetilde{A}_m^i \widetilde{D}^{-\frac{1}{2}}$ is the normalized adjacency matrix, H_i^l is the hidden state of layer l, when $l = 0$, $H_i^0 = V_m^i$, W_i^l and b_i^l are the weight and bias items of layer l respectively.

The first step of graph convolution obtains the mutual influence relationship H_i between module sub-variables, and the second step of graph convolution needs to use host-gcn to extract the mutual influence relationship between modules. We use Formula 6 to perform the second convolution on the host-graph G to obtain the spatial correlation vector H_g between modules.

In order to retain the spatial correlation between modules and within modules, we introduce the concept of gate, and use the fusion gate to combine the two parts of correlation

$$g = \sigma(W_f \cdot H_g) \tag{7}$$

$$Z = g \cdot H_g + (1 - g) \cdot \sum_{i=1H_g}^{m} H_i \tag{8}$$

Time Dependence: The vehicle operating state in the previous period has very important information for fuel consumption prediction. Some vehicle states do not change abruptly, but gradually change, such as vehicle speed, water temperature, etc., which all have a process of rising or falling. Knowing the changing trend of the vehicle over a period of time is of great significance for accurate fuel consumption prediction. RNN and its variants have achieved good results in capturing time dependence, so we use LSTM to capture the time dependence of the vehicle's operating state, and predict the fuel consumption of the vehicle for a period of time in the future. Input the spatial dependence information extracted in the Δt time period into LSTM to get the prediction result.

$$\widetilde{y} = lstm(Z_{t-\Delta t+1}, ..., Z_{t-1}, Z_t) \tag{9}$$

4.3 Configuration Transfer Module Based on Classifier

Vehicle manufacturers will record vehicle configuration data $C = \{c_1, c_2, ..., c_{N_c}\}$, such as engine type, rear axle type, etc. General-purpose vehicles can be subdivided into N_{old} categories, and the configuration differences between vehicles in each subcategory are much smaller than those between other subcategories and personalized configuration vehicles. Therefore, we construct an attention-based configuration classifier to learn the differences between vehicles and use it for transfer learning.

Use the encoder to encode the configuration data C, expressed as $V \in R^{N_e \times N_c}$, where N_e is the encoded vector dimension.

$$v_i = f_{encoder}(c_i) \quad for \quad v_i \in C_i \tag{10}$$

In order to make better use of configuration data, we use location codes PE to reflect the sequence of configurations, and combine configuration information with location codes to generate new configuration information representation E.

$$E = PE \oplus V \tag{11}$$

In the process of building the classifier, we use the self-attention mechanism to extract features from the configuration data E. Use three different linear transformation layers to convert E into Q, K, V as the dependency vector calculated by the attention mechanism, and then use the following formula to calculate the feature results of the feature under the attention mechanism.

$$AE = softmax(\frac{QK^\top}{\sqrt{d_k}})V \tag{12}$$

Use a multi-layer perceptron for classification tasks, and set the number of neural units in the output layer network to $N_{old} + 1$, corresponding to N_{old} general-purpose vehicles and personalized configuration vehicles.

After sufficient training of the classifier model, the output of the classifier after $softmax$ is used as the transfer weight between the source domain and the target domain. Because vehicle fuel consumption is related to the configuration to a certain extent, and the softmax output of the classifier represents the similarity between the personalized configuration vehicle configuration and the general vehicle configuration, using this similarity can be used as a transfer dependency.

4.4 Pre-train and Prediction

Pre-train: Divide the vehicle MTS of general-purpose vehicles into source domains $M^s = \{M_1^s, M_2^s, ..., M_{N_{old}}^s\}$. For each type in M^s, using its sufficient data to train the MGCN as the source domain network, we can obtain the source domain network set $Net^s = \{Net_1^s, Net_2^s, ..., Net_{N_{old}}^s\}$. Then use the personalized configuration vehicles data as the target domain data M^t, and use it to train a new MGCN as the source domain network Net^t.

Using the configuration data, we train the configuration classifier. Then the classifier is used as a connection module between the source domain network and the target domain network, and all model parameters in the source domain network and some model parameters in the classifier model are fixed to make it untrainable.

The data of the target domain is sent to the MGCN-CT model, and some parameters of the target domain network and classifier are fine-tuned to obtain the trained model.

Prediction: When using the MGCN-CT for prediction, we input the collected driving data of personalized configuration vehicles into Net^s and Net^t to obtain the calculated fuel consumption prediction results $[\widetilde{y}_1^s, \widetilde{y}_2^s, ..., \widetilde{y}_{N_{old}}^s]$ and $\widetilde{y}^t$. Then we input the vehicle configuration information of the personalized configuration vehicles into the configuration transfer module, and the transfer weight $W = [w_1^s, w_2^s, ..., W_{N_{old}}^s, w^t]$ for each network can be obtained. We can use W to calculate the final fuel consumption prediction result.

$$\widetilde{y}_{out} = w^t \widetilde{y}^t + \sum_{i=1}^{N_{old}} w_i^t \widetilde{y}_i^s \tag{13}$$

5 Experiments

5.1 Experiments Setup

Datasets: We use two datasets to verify the effect of the model.

- The Internet of Vehicles Dataset (IVD)[1] is the data collected from a commercial vehicle manufacturer in April 2022. It includes a total of 500 vehicles of general classification T1-T5, 100 of each type, and a total of 60 vehicles of P1-P3 of personalized configuration, 20 of each type. The sensor collects data every 30 s, including 33 dynamic attributes such as vehicle speed, water temperature, and engine speed. In addition, the dataset also contains vehicle 41 static attributes such as engine type and clutch type.
- Vehicle Energy Data-set (VED) [22][2] is a public dataset that mainly collects the driving data in Michigan, USA from November 2017 to November 2018. According to the research content of this paper, 264 ICE vehicles and 90 HEVs are selected to verify the effect of the model. The dataset contains 22 dynamic attributes, and 6 static attributes.

We regard P1-P3 as target domain data, T1-T5 as source domain data in IVD to participate in the experiments. And we take 24 ICEs, 10 HEVs as target domain data, and other vehicles as source domain data in VED.

Baseline Methods: In order to evaluate the effect of the MGCN-CT on the fuel consumption prediction of multi-type vehicles, we select 10 methods as comparison methods, including two statistical methods (AR, ARIMA), and two machine

---

[1] https://www.fawjiefang.com.cn.
[2] https://github.com/gsoh/VED.

learning methods (SVR, LR-REF [10]), four deep learning methods (LSTM, LSTNet [14], T-GCN [32], AST-GCN [7]), two transfer learning methods (TL-DCRNN [20], RELF -ATGNN [17]).

Since the comparison methods 1–8 are non-transfer learning models, this experiment first uses the source domain data to train the basic model, and then uses the target domain data to continue training the model to ensure the effect of the model.

Evaluation Metrics: We use Mean Absolute Error (MAE), Root Mean Square Error (RMSE) and Mean Absolute Percentage Error (MAPE) as evaluation criteria for experimental models.

Experimental Implementation Details: The method proposed in this paper is implemented based on python 3.7, and the deep learning tool TensorFlow is used to complete the construction of the deep learning network. The comparison methods are implemented according to the environment required by the source code. If the source code is not provided, we reproduce the model according to the description of the paper. The experiments are completed on a Linux server, equipped with 2 16-core CPUs, 8 RTX 3080 graphics cards, 384G memory, and the system version is Ubuntu 20.04 LTS. All the experimental results are repeated ten times and the average value is taken.

5.2   Fuel Consumption Prediction Performance

Table 1 and Table 2 show the performance of MGCN-CT on IVD and VED datasets. As shown in the table, the MGCN-CT model achieves the best performance in all evaluation metrics.

It can be seen from the experimental results that due to the strong nonlinear fitting ability of the deep learning methods, the performance of these models is significantly better than the statistical methods and the traditional machine

Table 1. The performances of MGCN-CT and baseline models on the IVD dataset

Models	P1			P2			P3		
	MAE	RMSE	MAPE	MAE	RMSE	MAPE	MAE	RMSE	MAPE
AR	0.152	0.519	15.9	0.164	0.578	17.2	0.147	0.499	15.5
ARIMA	0.144	0.495	16.1	0.159	0.621	18.3	0.149	0.462	15.6
SVR	0.135	0.453	15.6	0.132	0.354	15.5	0.132	0.437	15.1
LR-RFE	0.131	0.414	15.2	0.139	0.465	16.6	0.126	0.425	14.2
LSTM	0.127	0.326	14.1	0.133	0.441	15.2	0.121	0.355	14.9
LSTNet	0.125	0.235	13.7	0.128	0.326	14.4	0.116	0.276	13.1
T-GCN	0.102	0.177	14.5	0.117	0.232	13.7	0.101	0.157	13.1
AST-GCN	0.114	0.163	13.3	0.121	0.305	14.6	0.110	0.133	13.6
TL-DCRNN	0.095	0.147	12.6	0.110	0.238	13.7	0.094	0.137	12.4
RELF-ATGNN	0.092	0.141	13.2	0.101	0.245	13.9	0.100	0.136	12.3
MGCN-CT	**0.083**	**0.134**	**11.7**	**0.099**	**0.221**	**12.8**	**0.082**	**0.130**	**11.5**

Table 2. The performances of MGCN-CT and baseline models on the VED dataset

Models	ICE			HEV		
	MAE	RMSE	MAPE	MAE	RMSE	MAPE
AR	0.177	0.623	16.6	0.162	0.552	15.4
ARIMA	0.178	0.578	16.4	0.145	0.541	16.3
SVR	0.169	0.551	15.5	0.154	0.523	15.1
LR-RFE	0.152	0.523	14.5	0.152	0.532	15.4
LSTM	0.156	0.511	14.8	0.132	0.462	14.1
LSTNet	0.155	0.526	14.1	0.131	0.455	14.5
T-GCN	0.132	0.324	13.2	0.119	0.332	13.6
AST-GCN	0.131	0.375	13.9	0.107	0.303	12.9
TL-DCRNN	0.120	0.268	13.4	0.109	0.221	11.8
RELF-ATGNN	0.116	0.274	12.9	0.104	0.236	11.5
MGCN-CT	**0.104**	**0.223**	**11.4**	**0.101**	**0.197**	**10.1**

learning methods. LSTM and LSTNet do not consider the spatial dependence of MTS, and the performance is slightly worse than that of the other four deep learning methods. Both T-GCN and AST-GCN are models proposed with sufficient training data, and their model effects are not as good as TL-DRCNN and RELF-ATGNN proposed for the case of few samples. The MGCN-CT model takes vehicle domain knowledge into account when modeling MTS spatio-temporal dependencies, and uses the vehicle configuration to learn the transfer knowledge between models, making the transfer learning process more accurate in the research scenario of this problem, thus achieving the best performance.

5.3   Ablation Studies

A total of 5 different variant models have been designed, namely M1: remove the module graph, M2: remove the graph convolution of the sub-graph, M3: remove the graph convolution of the host-graph, M4: remove the CT module and use RELF-ATGNN for transfer learning, M5: remove the feature extraction method based on the attention mechanism in CT. When conducting the experiment, we select T1-5 in the IVD dataset as the source domain dataset, and P1 as the target domain dataset. Except for the above descriptions related to the relevant models, other experimental conditions and parameter settings are exactly the same. The results of the ablation experiment are shown in Fig. 3.

First of all, M1 has the worst effect, indicating that the model cannot cope with the complex operating state of the vehicle without the guidance of domain knowledge. The model effects of M2 and M3 have improved, indicating that the guidance of domain knowledge can effectively improve the model accuracy. When performing module graph convolution, M3 retains more detailed information than M2, and the model works better. Secondly, the performance of M4 is worse than that of the MGCN-CT model, indicating that the transfer learning

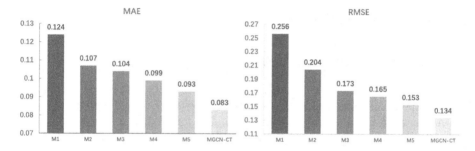

Fig. 3. Ablation experiment results

method of RELF-ATGNN is difficult to play a role in complex vehicle operating conditions. The effect of M5 is better than that of M4 but worse than that of MGCN-CT, indicating that it is feasible to apply configuration data as an auxiliary domain to transfer knowledge between different vehicle types, and the configuration data feature extraction method based on the attention mechanism can improve the transfer effect.

5.4 MGCN Module Performance

In order to evaluate the performance of the MGCN, this section selects four deep learning methods (LSTM, LSTNet, T-GCN, AST-GCN) and T1 in the IVD dataset and ICE in the VED dataset to participate in the experiments.

Table 3. The performances of MGCN and baseline models

Models	T1			ICE		
	MAE	RMSE	MAPE	MAE	RMSE	MAPE
LSTM	0.114	0.152	13.1	0.132	0.163	15.4
LSTNet	0.105	0.121	12.0	0.109	0.131	14.3
T-GCN	0.086	0.095	11.3	0.112	0.142	15.9
AST-GCN	0.095	0.112	12.1	0.102	0.115	12.5
MGCN	**0.081**	**0.091**	**10.6**	**0.094**	**0.109**	**10.9**

The experimental results are shown in Table 3, and the MGCN model has achieved the best performance. From the experimental results, except for the MGCN model, the performance of the other four methods has been significantly improved when there is sufficient training data. The T-GCN and AST-GCN model the spatiotemporal dependence of MTS, and their performance is better than the LSTM and LSTNet that only model time dependence. The MGCN obtained the best results, indicating that using domain knowledge to express the correlations within and between modules, and modeling the temporal and spatial dependencies of the module graph embedded with domain knowledge can effectively improve the accuracy of fuel consumption prediction.

5.5   CT Module Performance

Applicability analysis of CT: Using the pre-training and prediction methods of the MGCN-CT model, we choose LSTM, LSTNet, T-GCN, and AST-GCN models to replace the MGCN model. In this experiment, we select T1-5 data and P1 data in the IVD data set, and ICE data in the VED data set to participate in the experiment, and use MAE and RMSE as evaluation indicators.

As shown in Fig. 4, after using the CT module for prediction based on transfer learning, the accuracy of all methods has improved, which shows that the CT module proposed in this paper is versatile and suitable for transfer learning of various models.

Visual Display of CT Module: We extract the two transfer weights applied to P1 and P2 data in the IOV dataset, and visualized them in the form of a heat map.

The CT module migration weight visualization diagram is shown in Fig. 5. The darker the color in the figure, the greater the impact weight of the source domain model on the target domain model. When transferring the fuel consumption knowledge of the same target domain vehicle, the weight distribution among the source domain vehicles is relatively concentrated. For example, the P1 vehicle has a larger proportion than the T1 and T2 vehicles except for its own weight. Through further analysis of the data set, we find that the engine used in P1 is widely used in T2 models, and the chassis architecture and clutch model of P1 are the same as most of the vehicles in T2 models. This can explain why the migration weight of P1 is biased towards T1 and T2.

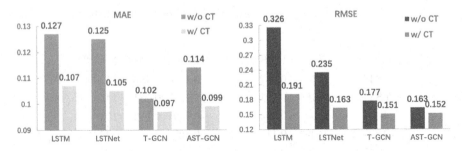

Fig. 4. CT module performance

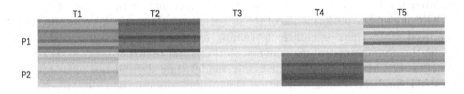

Fig. 5. Visualization of transfer weights

6   Conclusion

In this paper, we propose a multi-type commercial vehicle fuel consumption prediction method MGCN-CT based on module graph convolution and configuration transfer. First, the domain vector formula is used to uniformly express the vehicle module domain knowledge and construct a module graph. Then, the module graph convolutional network is used to model the spatio-temporal dependence of the module graph embedded with domain knowledge to obtain fuel consumption prediction results. Finally, using the transfer learning framework based on the configuration classifier, the fuel consumption prediction of the few-sample personalized configuration vehicles is realized. A large number of experiments prove that MGCN-CT outperforms the existing baseline methods in solving the problem of fuel consumption prediction, and the effectiveness of each module in the model is proved by ablation experiments. MGCN can achieve excellent fuel consumption prediction results with sufficient data. The CT module is suitable for transfer learning of various models. In the future, we will further study how to optimize the complexity of the model, reduce the training time of the model, and improve the efficiency of the model.

Acknowledgements. This work is supported by National Natural Science Foundation of China (No. 62072282, No. 62172443), Industrial Internet Innovation and Development Project in 2019 of China, Shandong Provincial Key Research and Development Program (Major Scientific and Technological Innovation Project) (No.2019JZZY010105) and CAAI Huawei MindSpore Open Fund.

References

1. Aliev, K., Narejo, S., Pasero, E., Inoyatkhodjaev, J.: A predictive model of artificial neural network for fuel consumption in engine control system. In: Esposito, A., Faudez-Zanuy, M., Morabito, F.C., Pasero, E. (eds.) Multidisciplinary Approaches to Neural Computing. SIST, vol. 69, pp. 213–222. Springer, Cham (2018). https://doi.org/10.1007/978-3-319-56904-8_21
2. Chaki, S., Doshi, P., Patnaik, P., Bhattacharya, S.: Attentive RNNs for continuous-time emotion prediction in music clips. In: AAAI, pp. 36–46 (2020)
3. Dhaou, I.B.: Fuel estimation model for eco-driving and eco-routing. In: 2011 IEEE Intelligent Vehicles Symposium (IV), pp. 37–42. IEEE (2011)
4. Fan, J., Bai, J., Li, Z., Ortiz-Bobea, A., Gomes, C.P.: A GNN-RNN approach for harnessing geospatial and temporal information: application to crop yield prediction. In: AAAI, vol. 36, pp. 11873–11881 (2022)
5. Feng, K., Chaspari, T.: A siamese neural network with modified distance loss for transfer learning in speech emotion recognition. arXiv preprint arXiv:2006.03001 (2020)
6. Gu, G.F., Zhou, W.X., et al.: Detrending moving average algorithm for multifractals. Phys. Rev. E **82**(1), 011136 (2010)
7. Guo, S., Lin, Y., Feng, N., Song, C., Wan, H.: Attention based spatial-temporal graph convolutional networks for traffic flow forecasting. In: AAAI, vol. 33, pp. 922–929 (2019)

8. Guo, Y., Shi, H., Kumar, A., Grauman, K., Rosing, T., Feris, R.: Spottune: transfer learning through adaptive fine-tuning. In: Proceedings of the IEEE/CVF Conference on Computer Vision and Pattern Recognition, pp. 4805–4814 (2019)

9. Helmut, L.: Chapter 6 forecasting with varma models. Handbook of Economic Forecasting **1**, 287–325 (2006)

10. Hoffman, A.J., van der Westhuizen, M.: Empirical model for truck route fuel economy. In: 2019 IEEE Intelligent Transportation Systems Conference (ITSC), pp. 914–919. IEEE (2019)

11. Jia, S., Deng, Y., Lv, J., Du, S., Xie, Z.: Joint distribution adaptation with diverse feature aggregation: a new transfer learning framework for bearing diagnosis across different machines. Measurement **187**, 110332 (2022)

12. Kim, C., Ostovar, M., Butt, A., Harvey, J.: Fuel consumption and greenhouse gas emissions from on-road vehicles on highway construction work zones. In: International Conference on Transportation and Development 2018, pp. 288–298. American Society of Civil Engineers Reston, VA (2018)

13. Kumar, S., Shukla, A.K., Muhuri, P.K.: Anomaly based novel multi-source unsupervised transfer learning approach for carbon emission centric GDP prediction. Comput. Ind. **126**, 103396 (2021)

14. Lai, G., Chang, W.C., Yang, Y., Liu, H.: Modeling long-and short-term temporal patterns with deep neural networks. In: The 41st International ACM SIGIR Conference on Research & Development in Information Retrieval, pp. 95–104 (2018)

15. Li, Y., Zeng, I.Y., Niu, Z., Shi, J., Wang, Z., Guan, Z.: Predicting vehicle fuel consumption based on multi-view deep neural network. Neurocomputing **502**, 140–147 (2022)

16. Li, Z., Chen, M., Yao, H.: "Trucks Trailer Plus" fuel consumption model and energy-saving measures. In: Zhu, R., Zhang, Y., Liu, B., Liu, C. (eds.) ICICA 2010. CCIS, vol. 106, pp. 204–211. Springer, Heidelberg (2010). https://doi.org/10.1007/978-3-642-16339-5_27

17. Lin, W., 0044 DW, D.W.: Residential electric load forecasting via attentive transfer of graph neural networks. In: IJCAI, pp. 2716–2722 (2021)

18. Lu, B., Gan, X., Zhang, W., Yao, H., Fu, L., Wang, X.: Spatio-temporal graph few-shot learning with cross-city knowledge transfer. In: Proceedings of the 28th ACM SIGKDD Conference on Knowledge Discovery and Data Mining, pp. 1162–1172 (2022)

19. Lu, L., Chen, H., Hu, Y., Gong, X., Zhao, Z.: Modeling and optimization control for an engine electrified cooling system to minimize fuel consumption. IEEE Access **7**, 72914–72927 (2019)

20. Mallick, T., Balaprakash, P., Rask, E., Macfarlane, J.: Transfer learning with graph neural networks for short-term highway traffic forecasting. In: 2020 ICPR, pp. 10367–10374. IEEE (2021)

21. Mellit, A., Pavan, A.M., Benghanem, M.: Least squares support vector machine for short-term prediction of meteorological time series. Theoret. Appl. Climatol. **111**, 297–307 (2013)

22. Oh, G., Leblanc, D.J., Peng, H.: Vehicle energy dataset (VED), a large-scale dataset for vehicle energy consumption research. IEEE Trans. Intell. Transp. Syst. **23**(4), 3302–3312 (2020)

23. Omondiagbe, O.P., Licorish, S.A., MacDonell, S.G.: Preventing negative transfer on sentiment analysis in deep transfer learning. In: Proceedings of the Workshop on Deep Learning for Search and Recommendation (2022)

24. Ozmen, M.I., Yilmaz, A., Baykara, C., Ozsoysal, O.A.: Modelling fuel consumption and NOX emission of a medium duty truck diesel engine with comparative time-series methods. IEEE Access **9**, 81202–81209 (2021)
25. Shaker, A., Yu, S., Onoro-Rubio, D.: Learning to transfer with von neumann conditional divergence. In: AAAI, vol. 36, pp. 8231–8239 (2022)
26. Shaw, S., Hou, Y., Zhong, W., Sun, Q., Guan, T., Su, L.: Instantaneous fuel consumption estimation using smartphones. In: VTC2019-Fall, pp. 1–6. IEEE (2019)
27. Tobar, F., Bui, T.D., Turner, R.E.: Learning stationary time series using gaussian processes with nonparametric kernels. In: Advances in Neural Information Processing Systems, vol. 28 (2015)
28. Toto, E., Tlachac, M., Rundensteiner, E.A.: AudiBERT: a deep transfer learning multimodal classification framework for depression screening. In: Proceedings of the 30th ACM International Conference on Information & Knowledge Management, pp. 4145–4154 (2021)
29. Wang, T., Chang, S., Liu, L., Zhu, J., Xu, Y.: Individual cylinder air-fuel ratio estimation and control for a large-bore gas fuel engine. Int. J. Distrib. Sens. Netw. **15**(2), 1550147719833629 (2019)
30. Xu, Z., Wei, T., Easa, S., Zhao, X., Qu, X.: Modeling relationship between truck fuel consumption and driving behavior using data from internet of vehicles. Comput.-Aided Civil Infrastruct. Eng. **33**(3), 209–219 (2018)
31. Yang, F., Wang, S., Li, J., Liu, Z., Sun, Q.: An overview of internet of vehicles. China Commun. **11**(10), 1–15 (2014)
32. Zhao, L., et al.: T-GCN: a temporal graph convolutional network for traffic prediction. IEEE Trans. Intell. Transp. Syst. **21**(9), 3848–3858 (2019)

Hardware and Software Co-optimization of Convolutional and Self-attention Combined Model Based on FPGA

Wei Hu[1,2], Heyuan Li[1,2(✉)], Fang Liu[3,4], and Zhiyv Zhong[1,2]

[1] College of Computer Science, Wuhan University of Science and Technology, Wuhan, China
huwei@wust.edu.cn, liheyuanwust@qq.com
[2] Hubei Province Key Laboratory of Intelligent Information Processing and Real-time Industrial System, Wuhan, China
[3] College of Computer Science, Wuhan University, Wuhan, China
liufangfang@whu.edu.cn
[4] Department of Information Engineering, Wuhan Institute of City, Wuhan, China

Abstract. Since Transformer was proposed, the self-attention mechanism has been widely used. Some studies have tried to apply the self-attention mechanism to the field of computer vision CV. However, since self-attention lacks some inductive biases inherent to CNNs, it cannot achieve good generalization in the case of insufficient data. To solve this problem, researchers have proposed to combine the convolution module with the self-attention mechanism module to complement the inductive bias lacking by the self-attention mechanism. Many models based on this idea have been generated with good results. However, traditional central processor architectures cannot take good advantage of the parallel nature of these models. Among various computing platforms, FPGA becomes a suitable solution for algorithm acceleration with its high parallelism. At the same time, we note that the combined modules of convolution and self-attention have not received enough attention in terms of acceleration. Therefore, customizing computational units using FPGAs to improve model parallelism is a feasible solution. In this paper, we optimize the parallelism of the combined model of convolution and self-attention, and design algorithm optimization for two of the most complex generic nonlinear functions from the perspective of hardware-software co-optimization to further reduce the hardware complexity and the latency of the whole system, and design the corresponding hardware modules. The design is coded in HDL, a hardware description language, and simulated on a Xilinx FPGA. The experimental results show that the hardware resource consumption of the ZCU216 FPGA-based design is greatly reduced compared to the conventional design, while the throughput is increased by 8.82× and 1.23× compared to the CPU and GPU, respectively.

Keywords: Attention Mechanism · Conv · Accelerators · FPGA · Co-optimization

© The Author(s), under exclusive license to Springer Nature Singapore Pte Ltd. 2024
X. Song et al. (Eds.): APWeb-WAIM 2023, LNCS 14333, pp. 328–342, 2024.
https://doi.org/10.1007/978-981-97-2387-4_22

1   Introduction

Transformer [1] has rapidly become one of the focal points in the field of machine learning research since it was proposed in 2017. When Transformer was first proposed, it was first used for machine translation tasks. In recent years, researchers have also tried to apply Transformer to the field of computer vision. Dosovitskiy et al. first segmented images into PATCH blocks and transformed them into linear embedding sequences that were directly fed into the encoder of the standard Transformer, achieving results close to the best CNN model in the JFT-300 dataset [2]. Many subsequent works have been devoted to improve ViT, such as: Crossvit [3], CCT [4], Conformer [5], etc.

The main difficulty in applying the Transformer to computer vision is that the Transformer lacks some inductive biases inherent to CNNs, such as translational invariance and localization, and therefore does not generalize well during training. Combining the Transformer with a convolutional module to complement the lack of inductive bias of the Transformer is a straightforward solution, while the ability of the attention mechanism to capture remote dependencies also compensates for the lack of global feature extraction ability of the CNN, and the two complement each other. Many models have been developed based on this idea and show good results, such as DS-Net [6], Lite Transformer [7], AAConv [8], etc. These models integrate well the advantages of self-attention mechanisms with convolutional modules.

When choosing a computing platform, it is crucial issue to make full use of computing resources to improve the efficiency of the model. The combined model of convolution and self-attention has its own structure of parallelism, which can greatly improve the operation efficiency of the model by parallel computing. However, when using CPUs for inference, the parallelism of the model is not fully utilized due to the limitation of the device's own hardware structure, and the computational efficiency is not fully utilized. Besides, the self-attention module and the convolutional module itself have the characteristics of matrix computation intensive and complex data flow, but to our knowledge, there are few studies on hardware design for these characteristics. Therefore, we believe that designing an efficient hardware architecture for the combined model of convolutional and self-attention modules is an important issue.

FPGAs are a suitable solution for algorithm acceleration in various computing platforms due to their high parallelism, low latency, and programmability. Compared to generalized GPUs, FPGAs are reconfigurable and can be custom designed as needed to maximize the parallelism benefits of the model. In addition, the high parallelism and low-power design of FPGAs make them a better choice for specific scenarios. Therefore, we choose FPGAs as our acceleration platform. Some prior works have proposed various FPGA-based accelerators for Attention and Conv modules, respectively, which do reduce the resource consumption of a single specific module, such as [9–12], etc. However, these accelerators are not fully applicable to modules with parallel execution of convolution and self-attention mechanisms. According to the available studies, few studies have been found to target modules with parallel execution of convolution and self-attention mechanisms. In summary, deploying computational modules on customized FPGAs is a feasible solution.

In this paper, we use Attention Augmented Convolutional Networks [8] as the base model, address the problem that the combined convolutional and self-attention models fail to be sufficiently parallel when inference training on CPUs, and design a framework using the idea of hardware-software collaboration to fully exploit the parallelism of the models on FPGAs to realize the convolutional module and the self-attention module in parallel. At the same time, the convolution and self-attention modules are designed separately in hardware to further optimize their latency and resource utilization. In addition, this paper optimizes the computational modules from the software point of view. A software and hardware co-optimization algorithm is designed for the two most complex non-linear functions in the model and is highly optimized to further reduce the hardware complexity and the latency of the whole system, fully embodying the idea of combining software and hardware. The main contributions of this paper can be summarized as follows:

1. The resource utilization and throughput of the convolution and self-attention modules are improved by a hardware-software collaborative approach, and the two module cells are computed in parallel on the FPGA device. To ensure high hardware utilization in this design, we have designed the computational flow well.
2. Using hardware-software co-design, we optimize the two most complex non-linear function modules, Softmax and LayerNormer, to make them more hardware-friendly and greatly reduce their time consumption and required hardware resources.
3. We propose another FPGA-based architecture design to optimize the Attention Augmented Convolutional module, and through experimental verification, we achieve 8.82× and 1.23× speedups compared with CPU and GPU, respectively.

2   Background

2.1   Conv+Transformer

Compared with the self-attention mechanism, convolutional neural networks (CNNs) possess stronger local modeling ability and translational invariance. To combine the advantages of both, researchers have started to explore methods that combine convolution and self-attention.

Dual-stream Network for Visual Recognition [6] generates two feature maps with different resolutions by introducing the Dual-stream Blocks (ds - block) architecture Dual-stream Network, which preserves both local and global information of the image through two parallel branches. S. Han et al. in 2020 proposed an efficient mobile NLP architecture Lite Transformer [7] whose key is that one set of heads focuses on local context modeling (via convolution) while the other set focuses on remote relationship modeling (via attention). AAConv [8] uses the self-attention mechanism to augment convolution as an alternative to the convolution-only task, which is the most directly relevant to this paperwork, so in this paper, we will present this model in more detail, as described in Subsect. 2.2.

These hybrid models have shown excellent results. It has been shown that combining CNN and self-attention mechanisms can achieve better performance in computer vision tasks. Therefore, the exploration of hybrid CNN and self-attention mechanisms has become one of the hot issues in computer vision today.

2.2 Attention Augmented Convolutional

Convolutional networks have become the paradigm for many computer vision applications, but their apparent weakness is that they operate only in local neighborhoods, losing global information. In contrast, Transformer introduces a self-attention technique that captures remote interactions well, and the two techniques can complement each other. In experiments, it is found that the best results can be obtained when convolution and self-attention are combined. Based on this idea, Attention Augmented Convolutional Networks use the self-attention mechanism to enhance convolution as an alternative to the convolution-only task. Its model is divided into two branches, which perform the Conv module and the improved version of the self-attention module, and finally splice the two in the channel dimension.

Figure 1 shows a module of AAConv.

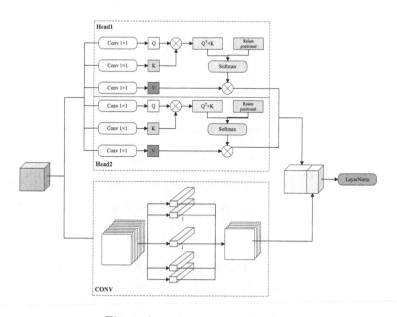

Fig. 1. Attention augmented conv.

The Attention Augmented Convolutional Networks model itself has an obvious parallel structure. However, the parallelism of the model cannot be fully exploited due to the limitation of CPU hardware architecture, resulting in underutilization of the model's parallelism. In contrast, the hardware structure of FPGAs consists of a large number of programmable logic gates and programmable memory cells, which can be combined and connected on demand to perform different operations simultaneously and achieve highly parallel computation. Therefore, improving the parallelism of the model and optimizing the model by combining hardware and software in FPGAs is a feasible approach.

2.3 Hardware Optimization

In recent years, neural networks have been widely used in computer vision, natural language processing, speech recognition, and other fields. However, the high computational complexity, high energy consumption, and high memory requirements of neural networks have led to many challenges in their application on mobile and embedded devices. In order to solve these problems, researchers have proposed various approaches, among which software and hardware co-optimization is a common approach.

Existing studies mainly focus on the optimal design of the self-attention mechanism module and the convolutional module, respectively. FTRANS [13] proposed an effective acceleration framework using with Transformer-based large scale language representations (for transformer-based large scale language representations) for the problem of accuracy degradation caused by traditional BCM compression, and proposed an improved BCM-based method for compressing the Transformer to reduce the weight occupation in the Transformer; A3 [9] proposed a greedy algorithm-based approximation mechanism for the attention mechanism to avoid exhaustive search; and designed dedicated hardware pipeline to achieve significant speedup on state-of-the-art conventional hardware; Co-Design [10] proposed a new hardware-software collaborative structural weight pruning scheme to control the size variation among compressed weights by imposing stricter restrictions on weight shapes and sizes, while proposing a unified computational model in the hardware design phase to handle Transformer inference FFConv [14] proposes an efficient pipelined high-throughput convolution engine based on the Winograd minimum filtering algorithm, which improves the throughput, resources and efficiency compared to previous design. These optimization methods can reduce the computational, energy, and storage costs of neural networks on mobile and embedded devices and promote the widespread diffusion of neural networks in practical applications.

However, there are few studies on hardware optimization for modules that execute in parallel with convolution and self-attention mechanisms. As more and more models adopt the idea based on the combination of convolution and self-attention mechanisms and show good results, their optimization through hardware and software co-design will become an important topic.

3 Algorithm and Hardware Co-optimization

3.1 Quantitative

For deep learning models, using fewer bits helps reduce the bandwidth and storage requirements of deep neural networks, and also reduces the hardware cost per operation, while the loss of model accuracy caused by reasonable quantization is negligible; for FPGA chips, the available computational resources are limited, and a smaller computational law cell design usually also means higher peak performance. To sum up, reasonable quantization is necessary to achieve a balance between model accuracy and hardware performance.

Common quantization methods can be divided into linear quantization and nonlinear quantization. Linear quantization finds the nearest fixed point representation of each weight and activation, as shown in Eq. 1, where S is the scale, which represents the proportional relationship between real numbers and integers, and Z is the zero point, which represents the integer corresponding to the 0 in real numbers after quantization. In contrast to linear quantization, nonlinear quantization independently assigns values to different binary codes.

$$q = round(\frac{r}{S} + Z) \tag{1}$$

The 8-bit quantization is a common quantization strategy in existing studies. A Survey of FPGA-based Neural Network Inference Accelerate [15] compares several typical quantization methods in several papers. For linear quantization, 8bit is a clear cutoff that ensures negligible loss of accuracy. Lite Transformer [7], and MobileNets [16] both use an 8bit quantization strategy. In the case that the requirement for accuracy is not so high, an 8-bit linear quantization design is a good quantization scheme that can ensure the relative balance between model accuracy and hardware performance.

In this paper, it is found that by using 8-bit quantization, the model accuracy is only lost by 0.1. Therefore, without introducing other models, all experiments in this paper will use an 8bit linear quantization scheme for data representation.

3.2 Overall Architecture

Currently, major deep learning frameworks rely heavily on CPUs and GPUs when training models; however, these hardware structures themselves have pipeline depth limitations that prevent deep parallelism, resulting in wasted parallelism inherent in the models. In contrast, FPGAs consist of a large number of programmable logic gates and programmable memory units, which are highly customizable and scalable. On FPGAs, programmable logic gates can be combined and connected on demand to support many different operations at the same time, which makes FPGAs highly parallel and computationally capable. In addition, FPGAs offer the advantages of low power consumption, fast data transfer, and easy integration. In summary, FPGAs, as a new hardware architecture, have rich advantages, so in this study, we choose to perform software and hardware co-optimization on FPGAs to fully utilize their parallel computing features to improve the training efficiency and performance of deep learning models.

To maximize the parallel advantage of FPGA, the structure of the Attention Augmented Convolutional Networks model is analyzed and designed in this paper. The model consists of three parts: Conv, Attention, and Combine. To further optimize the model performance, we further analyze these three parts and decompose them into smaller modules, such as the matrix multiplication module, Softmax module, LayerNorm module, and convolution module. These modules can be reused between different parts. By this design, we generate resource reuse and reduce the required hardware resources while fully utilizing the parallel computing capability of FPGAs. At the same time, we have carefully orchestrated the data flow, including the optimization of data flow and control flow between modules. Specifically, the first step was to analyze the module computation and data transfer volumes to determine how the data flow between modules should be transferred. Then, we optimized the control flow between modules to achieve efficient computation. Finally, our proposed design solution can fully utilize the parallelism advantage of FPGA and improve the model performance. Figure 2 shows the overall architecture.

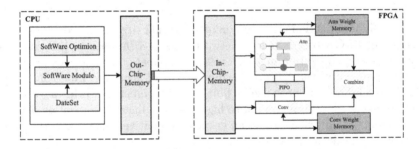

Fig. 2. Overall architecture.

In the following, we will introduce the fully parallelized module and the hardware-software co-design part in detail. The hardware-software co-design is mainly for the Softmax module and the LayerNorm module.

3.3   Hardware Layer Design

Based on the analysis in the previous section, to take full advantage of FPGA device parallelism, this study uses sequential function sequences to create a concurrent task-level pipeline architecture while preserving the behavior of sequential execution of fetches in the original code as much as possible. Specifically, PIPO-type communication channels are established based on the variables in the Attention function and Conv function, and each channel contains additional signals to indicate whether the channel is full or empty. By using separate PIPO buffers, Attention and Conv can be executed independently, and the throughput is limited only by the availability of the input and output buffers. Therefore, this study achieves full parallelism between the Attention module and the Convolution module, which works better than the partial parallelism achieved by the pipeline.

However, in most cases, the Attention module involves a large number of intensive computations, and its execution time is usually longer than that of the convolution module. To improve the efficiency of parallel operations and make the execution time of both modules as equal as possible, this study employs two steps to solve this problem. First, we allocate more hardware resources to the Attention module, which has a longer execution time. Second, we adjust the ratio of the Attention channel to the Convolution channel to achieve a balance between the two channels. In this study, we use a 1:4 channel ratio between the Attention channel and the convolution channel. It is experimentally demonstrated that the 1:4 channel ratio can achieve a balance between accuracy and performance.

To improve the efficiency of parallel operations, this study adopts a full parallelization strategy, but full parallelization is inevitably accompanied by additional hardware overhead. To offset these additional hardware overheads, along with parallel operations, this study also performs a reasonable arrangement of data streams to ensure that the resource utilization of the module is maximized. We adopted a data flow optimization strategy to multiplex hardware resources by rationally orchestrating data flows to connect a set of logic circuits to multiple data flows. With data flow optimization, each logic circuit can be shared in multiple data flows, and one hardware resource is used for different processing stages of multiple data flows, which greatly reduces the number of logic circuits needed, thus achieving the goal of saving hardware resources.

3.4   Softmax Module

The Softmax function is used to map the input values onto a probability distribution. As one of the most commonly used classification algorithms, there has been a lot of work thinking about how to speed up the Softmax module. Our approach is improved based on the Online normalizer calculation for softmax [17], which optimizes Softmax in an algorithmic perspective and designs a Softmax optimization method, which is a good representation of the software and hardware co-optimization mindset. In this subsection, we will introduce it in detail.

The traditional SoftMax function is defined in the following form Eq. 2.

$$y_i = \frac{e^{x_i}}{\sum_{j=1}^{V} e^{x_j}} \tag{2}$$

where e^{x_i} can be considered as the score, V represents the dimension of input x, $\sum_{j=1}^{V} e^{x_j}$ can be considered as the normalization factor. However, when the model is applied to the actual hardware since the range of data that the hardware can represent is limited, there is a risk of overflow in the above formula when performing exponential operations, and the maximum value of the input vector needs to be subtracted before calculation to avoid overflow, and the Softmax formula can be expressed as Eq. 3.

$$y_i = \frac{e^{x_i - \max_{k=1}^{v} x_k}}{\sum_{j=1}^{v} e^{x_j - \max_{k=1}^{v} x_k}} \tag{3}$$

The implementation of Softmax requires three loops, one to calculate the maximum value of the input vector, one to perform the normalization calculation, and finally the exponential calculation to calculate the output value, which contains up to four memory accesses. At the same time, exponential operation, as the main operation in Softmax function, requires a lot of computational resources to perform high-precision exponential operation on the hardware platform. Therefore, it is feasible to optimize Softmax from two perspectives of reducing memory access and optimizing exponential operations by means of software and hardware co-optimization.

On the software side, to reduce memory access, in this paper the maximum value and the denominator summation are calculated consecutively. Instead of subtracting the maximum value of the vector before applying the exponent, the maximum value of the vector so far is subtracted, and when a new maximum value is found, the cumulative value is adjusted. This reduces the memory accesses for the Softmax function computation from four to three. To reduce the resource consumption of exponential operations, we use an approximate polynomial with an 8-level Taylor expansion instead of the exp function. Also, to further save computational resources, the Softmax module will filter the input data and compute only positive values, and directly set the result to 0 if the input is negative. the pseudo-code of the Online-Taylor Softmax function is show in Algorithm 1.

Algorithm 1: Online-Taylor Softmax

    Input : x[DIM]
    Output: y[DIM]
1   $m_0 \leftarrow 0$;
2   $sum_0 \leftarrow 0$;
3   **for** $i \leftarrow 1$ **to** DIM **do**
4   **if** $x_i \leq 0$ **then**
5         $continue$;
6   **else**
7         $m_i \leftarrow max(m_{i-1}, x_i)$;
8         $sum_i \leftarrow sum_{i-1} \times Taylor(m_{i-1} - m_i) + Taylor(x_i - m_i)$;
9   **end**
10 **end**
11 **for** $j \leftarrow 1$ **to** DIM **do**
12      $y_j \leftarrow \frac{Taylor(x_j - m)}{sum}$;
13 **end**

On the hardware side, this study performs pipeline operations on the innermost layer of the loop to increase parallelism and reduce the overall module time consumption. Matrix partitioning is performed on the input matrix to eliminate data dependencies. Ultimately, the Softmax module is shown in Fig. 3.

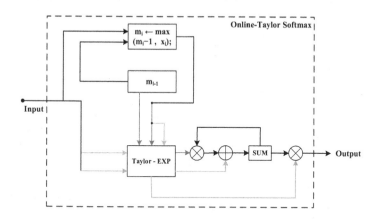

Fig. 3. The architecture of Softmax module.

3.5   LayerNorm Module

LayerNorm is a common normalization technique used in deep learning models, which avoids the effect of traditional batch normalization on the effectiveness of the model by normalizing the activation values on each layer. However, the variance computation in LayerNorm involves large and intensive matrix multiplication and data access, which consumes a large number of hardware resources. In addition, in the traditional Transformer-based model, the output results of each multi-headed attention module need to be normalized by LayerNorm, which also means that the LayerNorm module is on top of the critical path of the system and becomes a bottleneck to reduce the system latency [18]. Therefore, to reduce the consumption of hardware resources and lower the overall system latency, it is necessary to optimize the LayerNorm module to make it more hardware-friendly.

For the conventional LayerNorm module, the layer normalization function is:

$$y = \frac{x - E\left[x\right]}{\sqrt{Var[x] + \epsilon}} * \gamma + \beta \tag{4}$$

where ϵ is a very small number, in order to ensure that the denominator is not zero. E represents the mean of the input elements and var represents the variance of the input elements. For the variance, the traditional method of calculating the variance is shown in Eq. 5.

$$Var[x] = \frac{1}{d_{model}} \sum_{k=1}^{d_{model}} \left[(x_k - E[x])^2 \right] \tag{5}$$

For this way of calculating the variance, each element needs to be accessed twice: once to calculate the mean and once to calculate its deviation from the mean, which will lead to unnecessary waste of resources as the hardware reads the same data multiple times. In this paper, we use an iterative numerical stable variance algorithm for variance calculation, as shown in the following equation:

$$S[x_n] = \sum_{k=1}^{n} \left[(x_k - E[x])^2 \right] \tag{6}$$

$$Var[xn] = \frac{1}{d_{\text{model}}} \left[S[x_{n-1}] + (\frac{n-1}{n})(x_n - E[x_{n-1}])^2 \right] \tag{7}$$

The d_{model} represents the dimension of the input and uses the average of each value and all previous values to calculate the deviation so that only two for loops are needed to complete all calculations of the LayerNorm module, each input element is used only once, no additional hardware overhead is needed to store it, and no important numbers are lost at any stage.

To avoid II (Initiation Interval) conflicts, we chunk the input matrix, thus extending the read and write ports and further increasing the read and write speed of the system; at the same time, we also pipeline the innermost level of each for loop in the module and set the task interval to 1 to make the loop structure as streamlined as possible and increase the system parallelism to reduce the minimum latency of the overall system. The data flow of the entire module is also rearranged to make it more suitable for computation on hardware. Ultimately, the LayerNorm module requires very few hardware resources to complete. the overall architecture of the LayerNorm module is shown in Fig. 4.

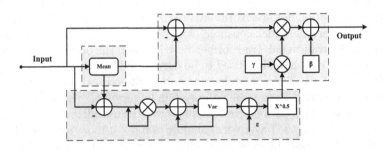

Fig. 4. The architecture of LayerNorm module.

4   Experiments

4.1   Experimental Setup

In this paper, our proposed module is evaluated using the Xilinx Zynq Ultra-Scale+ ZCU216 Evaluation Platform development board with Vits HLS 2022.1, and the CPU and GPU comparison experiments are conducted. In this experiment, 12th Gen Intel(R) Core(TM) i5-12490F is selected as CPU and NVIDIA GeForce RTX3090 as GPU. the code structure under C++14 framework is used for CPU and FPGA, and the code structure under Pytorch framework is used for GPU (Table 1).

Table 1. Experimental hardware setup

Platform	CPU	GPU	FPGA
Type of Device	12th GEN Intel Core i5-12490	NVIDIA GeForce RTX 3090	Xilinx Zynq UltraScale+ ZCU216 Evaluation Platform
Code heading	C++14	Pytorch	C++14
Clock heading	–	–	10ns

4.2 Channel Ratio Comparison Experiment

In this section, we investigate the effect of different attention channel to convolution channel ratios on the accuracy and hardware implementation performance. We compare the top-5 accuracy of the Attention augmented conv model on the CIFAR10 dataset for different ratios (1:4, 2:3, 1:1, 3:2, 4:1) of attention channel to convolution channel ratios, and experimentally test the hardware performance. The experimental results are shown in Table 2.

Table 2. varying ratios of channels Comparison

Ratios of Channels	1:4	2:3	1:1	3:2	4:1
Top-5 Acc	92.44	90.32	89.56	88.52	87.07
Latency	525420	699945	786113	878156	1051554
BRAM	313	362	402	437	445
DSP	41	41	41	41	41
FF	7880	7920	7946	7987	8016
LUT	17705	18543	19354	19732	21457

The results show that the hardware running speed increases slightly with the increase of the attention channel ratio. This is because the computation of the attention module is larger than that of the convolution module, and the parallelism of the model is reduced when the attention channel ratio is too large. When the ratio of the attention channel to the convolution channel is 1:4, the accuracy and hardware performance of the model reaches a balance.

4.3 Hardware Implementation Results

For the two complex nonlinear functions, after the software and hardware co-optimization, we compared the Softmax module and LayerNorm module separately with the pre-optimization, and the experimental results are shown in Table 2, where the Baseline represents the consumption of the two nonlinear functions without optimization. The results show that after the optimization, the DSP resource consumption of the Softmax module is reduced by 36%, the FF resource consumption is reduced by 73%, and the LUT resource consumption is reduced by 58%. For the LayerNorm module, after optimization, the DSP resource consumption is reduced by 78%, the FF resource consumption is reduced by 66%, and the LUT resource consumption is reduced by 75% (Table 3).

Table 3. Softmax and LayerNorm resource consumption

Instance	Module	Latency	DSP	FF	LUT
Baseline	Softmax	1140	22	6258	5570
	LayerNorm	890	74	6955	7716
Ours	Softmax	362	14	1650	2315
	LayerNorm	663	16	2346	1861

After optimization, our module achieves a high degree of parallelism with reasonable resource allocation and data flow. At the same time, we have a reasonable arrangement of data flow and resources, and we get the experimental results shown in Table 4, where the Baseline represents the data before optimization. The experimental results show that the runtime of the optimized model is improved by 4.99×, the BRAM resource consumption is reduced by 65%, the DSP resource consumption is reduced by 72%, the FF resource consumption is reduced by 73%, and the LUT resource consumption is reduced by 75%. Also, we compared the changes in resource usage as the batch size increased, and the experimental results are shown in Fig. 5 As the batch size increases, the running time of the whole module tends to increase linearly, but the usage of hardware resources such as DSP, FF, and LUT remains the same.

Table 4. Total resource consumption

Instance	Baseline	+Co-design	+Co-design +DataFlow	+Co-design +DataFlow +Quant(Ours)
Latency	2622545	630170	645680	525420
BRAM	891	997	857	313
DSP	147	157	67	41
FF	30109	80473	52188	7880
LUT	72141	89557	63117	17705

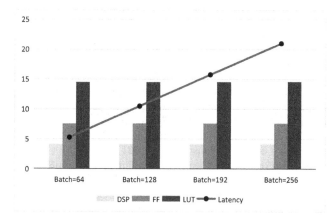

Fig. 5. Comparison of batch resources.

4.4 Multi-platform Comparison

We compared the throughput and operation speed on three different platforms, GPU, CPU, and FPGA, using the same model and hyperparameters, and the results are shown in Table 5 The final results show that our FPGA design achieves an 8.82x throughput improvement compared to the CPU; our FPGA design also improves the throughput by 1.23x compared to the GPU.

Table 5. Multi-platform Comparison

	Baseline	CPU	GPU	Ours
Execution time (ms)	26.22	46.378	6.425	5.254
Relative Speedup	1×	0.55×	4.08×	4.99×
Throughput(Gops)	14.33	7.89	58.48	71.52

5 Conclusion

In this paper, using Attention Augmented Convolutional Networks as the base model, we try to solve the problem that the model based on the combination of Conv and Attention cannot fully exploit the inherent parallelism of the model when inference computation is performed on a conventional central processor. We use software and hardware co-optimization design to rearrange the data stream and redesign the algorithm flow for the two generic nonlinear functions with high complexity in the model, which reduces resource consumption and speeds up the operation. We tested our ideas on the Xilinx Zynq UltraScale+ ZCU216 Evaluation Platform development board, achieving a high degree of parallelism and a significant reduction in hardware resource consumption.

References

1. Vaswani, A., et al.: Attention is all you need. arXiv (2017)
2. Dosovitskiy, A., Beyer, L., Kolesnikov, A., Weissenborn, D., Houlsby, N.: An image is worth 16×16 words: transformers for image recognition at scale (2020)
3. Chen, C.F.R., Fan, Q., Panda, R.: Crossvit: cross-attention multi-scale vision transformer for image classification. In: Proceedings of the IEEE/CVF International Conference on Computer Vision, pp. 357–366 (2021)
4. Hassani, A., Walton, S., Shah, N., Abuduweili, A., Li, J., Shi, H.: Escaping the big data paradigm with compact transformers (2021)
5. Peng, Z., et al.: Conformer: local features coupling global representations for visual recognition (2021)
6. Mao, M., et al.: Dual-stream network for visual recognition (2021)
7. Lin, J., Han, S., Lin, Y., Wu, Z., Liu, Z.: Lite transformer with long-short range attention (2020)
8. Bello, I., Zoph, B., Le, Q., Vaswani, A., Shlens, J.: Attention augmented convolutional networks. In: 2019 IEEE/CVF International Conference on Computer Vision (ICCV) (2020)
9. Ham, T.J., et al.: A^3: accelerating attention mechanisms in neural networks with approximation. In: IEEE (2020)
10. Zhang, X., Wu, Y., Zhou, P., Tang, X., Hu, J.: Algorithm-hardware co-design of attention mechanism on FPGA devices. ACM Trans. Embedded Comput. Syst. (TECS) **20**(5s), 1–24 (2021)
11. Chen, Y., Zhang, N., Yan, J., Zhu, G., Min, G.: Optimization of maintenance personnel dispatching strategy in smart grid. World Wide Web **26**(1), 139–162 (2023)
12. Xu, D., Chen, Y., Cui, N., Li, J.: Towards multi-dimensional knowledge-aware approach for effective community detection in LBSN. In: World Wide Web, pp. 1–24 (2022)
13. Li, B., Pandey, S., Fang, H., Lyv, Y., Ding, C.: FTRANS: energy-efficient acceleration of transformers using FPGA. In: ACM (2020)
14. Ahmad, A., Pasha, M.A.: FFConv: an FPGA-based accelerator for fast convolution layers in convolutional neural networks. ACM Trans. Embedded Comput. Syst. **19**(2), 1–24 (2020)
15. Guo, K., Zeng, S., Yu, J., Wang, Y., Yang, H.: [DL] a survey of FPGA-based neural network inference accelerators. ACM Trans. Reconfigurable Technol. Syst. (TRETS) **12**(1), 1–26 (2019)
16. Howard, A.G., et al.: MobileNets: efficient convolutional neural networks for mobile vision applications (2017)
17. Milakov, M., Gimelshein, N.: Online normalizer calculation for softmax (2018)
18. Lu, S., Wang, M., Liang, S., Lin, J., Wang, Z.: Hardware accelerator for multi-head attention and position-wise feed-forward in the transformer. In: System-on-Chip Conference (2020)

FBCA: FPGA-Based Balanced Convolutional Attention Module

Wei Hu[1,2], Zhiyv Zhong[1,2(✉)], Fang Liu[1,3], and Heyuan Li[1,2]

[1] College of Computer Science, Wuhan University of Science and Technology,
Wuhan, China
`huwei@wust.edu.cn`, `liufangfang@whu.edu.cn`, `zhongzhiyv@qq.com`
[2] Hubei Province Key Laboratory of Intelligent Information Processing and
Real-time Industrial System, Wuhan, China
[3] College of Computer Science, Wuhan University, Wuhan, China

Abstract. Large-scale computation and data processing are common tasks in machine learning. While traditional central processors are capable of performing these tasks, their computational speed is often inadequate when dealing with large-scale data sets and deep neural networks. As a result, many accelerators have emerged, such as graphics processors, field-programmable gate arrays, etc. FPGA have become a widely used type of accelerator compared to other accelerators due to their high flexibility, high performance, low power consumption, and low latency. However, most of the existing FPGA accelerators only accelerate single modules of CNN, RNN, and attention modules, and few cases of joint acceleration for different types of network combinations are mentioned. Therefore, this work is based on the hardware design of a model with a combination of convolutional and attention modules, and the way they combine to process the data is a perfect fit for the core of hardware acceleration. On the hardware device, the data in this model can flow into the computation at the same time to obtain parallel processing speed. We use a cut that is more suitable for hardware parallelism to process the data coming into both modules, thus making the best use of resources and keeping the time of both modules close to each other. In the same way, for the most computationally heavy loop structure, we have adapted the array structure for faster computation. We also parallelize the design of the serial linear layer in the attention module after the efforts in this paper, the model is further streamlined and accelerated, and finally, our model achieves a speedup of 12.5 times with only a 0.25 decrease in BLEU.

Keywords: FPGA · Hardware and Software co-optimization · CNN · Attention

1 Introduction

The Transformer model [1] has been successfully applied to natural language processing problems with unprecedented results due to its ability to capture a wide

X. Song et al. (Eds.): APWeb-WAIM 2023, LNCS 14333, pp. 343–357, 2024.
https://doi.org/10.1007/978-981-97-2387-4_23

range of global information in its middle attention module. However, along with the extensive research on the Transformer model, The performance improvement of the model has reached a bottleneck, and the performance gain from increasing the model size is gradually diminishing, while the training time and computational resources are increasing in cost. Many researchers believe that this is due to the inability of the attention module to extract local information and therefore use the convolution module to supplement it. such as CCT [2], ConVit [3], MOBILEVIT [4], etc. In language modeling tasks, models such as ConVit [3] and MOBILEVIT [3] have achieved better performance than traditional Transformer models, making language modeling tasks more efficient and accurate, while CCT [2] has made a breakthrough in machine translation tasks. These models have made great progress in the field of natural language processing by combining the attention module with the convolution module, allowing the models to extract features more efficiently when dealing with localized information, thus improving the performance of the models.

However, in the inference phase, the parallelism of the model is not exploited due to the limitations of the CPU architecture itself, and the computational efficiency is low. FPGAs as specialized integrated circuits, can reprogram their internal circuitry at runtime, thus enabling efficient parallel computation. Therefore, the acceleration of such models using FPGA is tailor-made. In the accelerator designed for the model, data can flow into both modules and be computed simultaneously in the modules. However, when we want to process the convolution and attention modules in parallel on FPGA, the huge difference in computation between the two results in poor parallelism. Lite Transformer [5] solves this problem perfectly, Its allocation of the number of channels for the input of the two modules can balance the processing speed of the two modules. So our work will be carried out on Lite Transformer [5] with co-optimization of software and hardware for accelerator design.

In order to solve the time wastage caused when the modules are serialized, we hardware both and arrange them into the same module. At the same time, we modified the software model accordingly in order to reduce the huge difference in running time between the two sides. These measures further accelerated the overall model. After our processing, the speed of the model is significantly improved. The contributions of this paper are as follows.

1. We have hardwareized and improved the dynamic convolution and Attention modules, deployed them on FPGAs, and optimized them to further reduce the latency.
2. We redistributed the number of channels for the input of the two modules so that their computational speed in the two parts no longer differs too much, while the BULE value only decreases by 0.26.
3. We design a balanced attention convolution module with FPGA-based architecture. And experimentally verify that our hardware and software co-optimization improves the throughput by a factor of 12.5 and 1.81 compared to CPU and GPU, respectively.

2   Related Works

Since the introduction of the Transformer model, the model has taken the lead in the field of natural language processing. Among them, the attention module's grasp of global information is its advantage in language processing. However, the grasp of global information requires a large amount of computation, which leads to the problem of too much computation. To solve this problem, a model consisting of the convolutional module and attention module together is proposed, reducing huge computation with the convolution module.

Convolutional modules are ideal for the extraction of local information, but it is more difficult to grasp the distal information. At the same time, simple convolutional modules cannot dynamically adapt to the input, which is essential in the processing of language problems.

In Lite Transformer [5], on the other hand, the advantages of the convolution and attention modules are combined, while discarding the parts that are difficult to take care of in both modules. In this model, the number of input channels is cut in half for the data stream and passed into the two modules separately to enable the processing of different dimensions of the same input. This design allows the convolution module to focus on the extraction of local information, while the attention module processes global information. It can achieve better results than the general Transformer model.

Lite Transformer [5] uses a dynamic convolution module [6], and its specific flow is shown in Fig. 1. The traditional convolution operation involves convolving a fixed convolution kernel with the input feature map. In contrast, dynamic convolution introduces a learnable control module into the convolution operation, by which the weight values are changed, and the new, input-dependent convolution weight values are used to perform the convolution operation with the input values. The dynamic convolution module enables the model to adaptively adjust the weights of the convolution kernel, thus improving the model's ability to capture the input information. In the translation tasks of this paper, due to the strong variability of each input value, it is difficult to adjust the adaptation by ordinary convolution, while the dynamic convolution module is the best choice to handle these tasks.

There are already many accelerator designs for convolution and Transformer [7] on the market, for convolutional modules, FTDL [8] proposes a new overlay architecture for convolution and full linkage layers, specifically designed for modern FPGA, making the FPGA more relevant to convolution design and thus achieving better throughput rates. Synetgy [9] made changes to the software model to make ShuffleNetV2 [10] a more hardware-accelerated architecture by changing the residual join to a channel split and then combining and replacing the 3*3 convolution with a 1*1 convolution and shift operation to make the convolution operation more suitable for the hardware architecture, the method, although not directly applicable to dynamic convolution, gave us the idea to change the model structure to adapt to the hardware. For attention modules, OPTIMUS [11] is believed that the computational complexity of the decoding layer increases with time due to the feedback structure of the network, so its computational model is optimized to greatly reduce the attention module execution time. However, these accelerators only accelerate a single model, while the area of joint acceleration of the convolution and attention modules has rarely been addressed, which is what our work is trying to do.

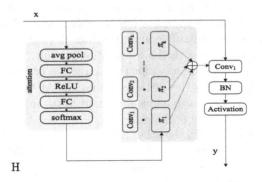

Fig. 1. Dynamic convolution module

In the field of neural networks, there are many methods used to optimize model performance and reduce the consumption of computational and storage resources. Among them, pruning for unimportant weights is a common technique. For example, the Deep Compression [12] uses pruning techniques to set the unimportant weights to zero to obtain a sparse matrix, followed by compressed storage of the sparse matrix using compression algorithms such as CSR or CSC. This method not only reduces the storage space but also can improve the computational speed. In addition, weight quantization [13] is also a common optimization method. This method converts the weights from 32-bit floating-point numbers to integers of lower bits, such as 8 or 4 bits. Weight quantization can improve the speed of model inference because integer operations are faster to compute and less resource intensive. It is important to note that these methods do not conflict with our work, but are orthogonal.

3 Software Design

3.1 Lite Transformer

In this study, Lite Transformer is chosen as the benchmark model, and the acceleration is designed for its convolution module and attention module. As a representative of the streamlined Transformer, the BLEU index of Lite Transformer on the translation task is improved by 3 points compared to the Transformer of the same size.

The hierarchical structure of the model is shown in Fig. 2. First, the model divides the input data into two parts according to the channel hierarchy, which is equally divided in the original model. One part is sent to the attention module for processing, and the other part is sent to the convolution layer after processing by GLU (gated linear unit). In the convolutional layer, the input array is linearly transformed into a weight matrix, which is then reorganized so that the convolution operation becomes a matrix multiplication operation and is finally output. Finally, the two parts of the channel are joined together and the information from both ends is mixed and output.

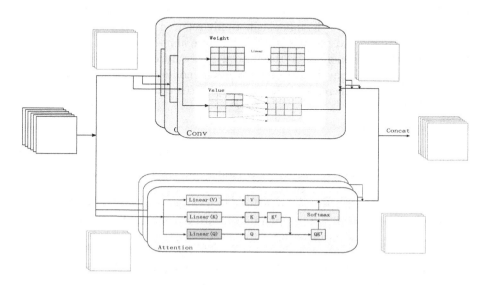

Fig. 2. Lite Transformer module structure

The selection of Lite Transformer as the research object in this paper is not only due to its excellent model size and structure but also because its overall concept is well-suited for hardware model construction. Through weight visualization experiments of the attention and convolution modules of Lite Transformer, we discovered that the convolution module is crucial for local information extraction, whereas the attention module is more appropriate for extracting global information, which partially contradicts the previously proposed channel cutting scheme. This paper posits that the key to addressing language problems lies in the extraction of local information, as the vast majority of important information typically appears in its proximity. Furthermore, we assert that the convolutional module plays a more significant role in the model's accuracy, and hence, the number of channels of the convolutional module is increased to further enhance its efficacy in task processing.

This idea corresponds to the design of the hardware part of this paper, where the hardware acceleration is based on the parallelism of the attention module and the convolution module. On top of the parallelism of the two modules, the problem of computational imbalance arises. Even with the use of dynamic convolution, which is more comprehensive in terms of information collection and leads to a large increase in computation, there is still a large gap between the computation of the convolution module and the attention module. As a result, parallelism is not satisfied, the overall time is basically equal to the time of the attention module. However, with our idea to improve the Lite Transformer, the number of channels is mostly given to the convolution module and a small portion to the attention module, which equalizes the computation of the convolution module and the attention module. With this improvement, the computation of the attention module is significantly reduced. After the improvement of the cut channel, the total time of the model is not only reduced but also the time of the convolution module and the attention module are almost balanced.

The Lite Transformer, after the special hardware design and the channel reallocation for the convolution and attention modules, is significantly more efficient without any decrease in the accuracy of the computational task.

3.2   Quantization

Quantization technology has now become an important technology for processing neural networks. Quantization of the model can further reduce the size of the model. In our model, the 32-bit floating-point number is reduced to 8 bits, and the overall size of the model can be reduced to four times above; FPGA hardware resources are fixed, so processing high-digit floating-point numbers requires more storage space and computing resources, while low-digit integers can be calculated more efficiently on FPGA. Therefore, for FPGA, the model after quantizing the accuracy is more in line with the hardware conditions of FPGA, which can significantly accelerate the calculation speed and reduce power after quantization; and for the model itself, after quantization, the risk of over-fitting of the model can be reduced and the model can be improved.

Most of the quantization methods can be roughly divided into two forms: symmetric quantization and asymmetric quantization. Symmetric quantization is to limit the representation range of floating-point numbers to a symmetrical interval, then divide the entire interval into equal parts, and quantizes the model parameters and activation values into one of N integers, while asymmetric quantization restricts the representation range to one in an asymmetric range. Compared with the two, the realization of symmetric quantization is simple, but the precision is lost. Asymmetric quantization can quantify data to a wider range, so that the accuracy is higher and representation is more flexible, but the implementation is more complicated.

To further speed up the model, this paper will use symmetric quantization to further compress the model. The quantification results are shown in Fig. 3, we find that the model accuracy does not change significantly when we use 16-bit quantization, and the size of our model is reduced by a factor of 4 when we use 8-bit quantization with only 0.1 large loss in accuracy. So we finally adopt 8-bit quantization to adapt our model.

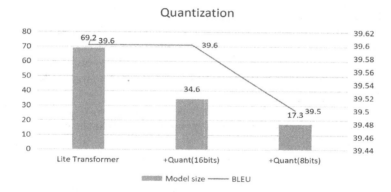

Fig. 3. Model size and BLEU score

4 Hardware Design

4.1 The Accelerator Architecture

Figure 4 shows the overall architecture design of our accelerator, and we design the hardware of the dynamic convolution and attention modules separately and combine them to form a balanced attention convolution module. After the data are cut, they enter the two modules separately for parallel computation, and we will then describe the design of parallelization in each module in detail.

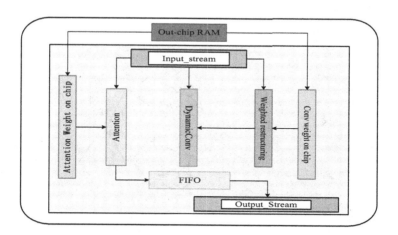

Fig. 4. The accelerator architecture

4.2 Matrix Multiplication Module

The purpose of this study is to investigate the computational process of matrix multiplication in the attention module, which can be considered as a series of cyclic iterations of vector multiplication. Specifically, for the input matrix A and the weight matrix W, the dimension of the A matrix is MN and the dimension of the W matrix is NQ. In the matrix multiplication process, it is necessary to perform a dot product operation for each vector in the second dimension of the W matrix with the vector in the first dimension of the A matrix to finally obtain the matrix of M*Q.

To achieve the best parallel effect, this paper proposes a vector expansion and parallel dot product method. Specifically, we vector unfold the second dimension of the weight matrix W and the first dimension of the input matrix A and perform parallel dot product operations on the unfolded vectors. By this method, we achieve the unification of the matrix multiplication computation model and expand the loop as much as possible to obtain better parallelism.

The matrix multiplication calculation model proposed in this paper is shown in Fig. 5. First, the input matrix A and the weight matrix W are vector expanded and stored in BRAM to reduce the time of off-chip access. Then, a parallel dot product calculation is performed for the unfolded vectors.

A pipelined design is used for the dot product computation. Specifically, the PE1 array performs the parallel computation and passes the result of the parallel computation through the tree structure to the PE2 array below for the accumulation operation. Such a round-trip loop gets the final result, which is stored in a register until the matrix multiplication operation ends when the register size is M*Q. By this design, the whole cycle of matrix multiplication is fully expanded, the sequential nature of the multiplication-addition operation is reorganized into a common operation at both ends, and the time complexity is changed from N to logN.

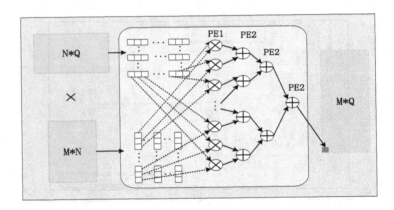

Fig. 5. Streamline multiplication structure

4.3 Convolutional Computing Module

The original convolution module algorithm is not suitable for deployment on FPGA. Convolution modules deployed on FPGA are faced with the choice of loop tiling, and the choice of loop tiling in loop nesting is diverse. The effect of using tiling operations at different loop layers is different, and incorrect tiling not only does not reduce the loop time but also produces an increase in the number of resources used.

We use the method in Roofline [14] to analyze the convolutional structure and first solve the data dependency problem. Data dependency refers to the relationship between different array accesses in a loop and the loop variables. Generally, we can classify the data dependency in loops into three cases.

1. Irrelevant: If there is no relationship between the accesses of a certain array and the loop variables, the array is considered to be irrelevant to the loop.
2. Independent: If the access to an array is related to the loop variable, but the access to the array is independent when the loop variable takes different values, then the array is considered independent of the loop.
3. Related: If the access to the array is related to the loop variable, and the access to the array is not independent when the loop variable takes different values, then the array is considered related to the loop.

For the three relations, the different relations will have different effects on the parallel operation of the convolutional layer. Irrelevant data dependencies do not affect the hardware parallelism, independent data dependencies can be expanded for loops, while relevant data dependencies cannot be expanded and need to be reduced to independent or irrelevant. Therefore, in nested loops, the data-dependent loops need to be reordered and placed at the innermost level to reduce the data dependencies for unfolding.

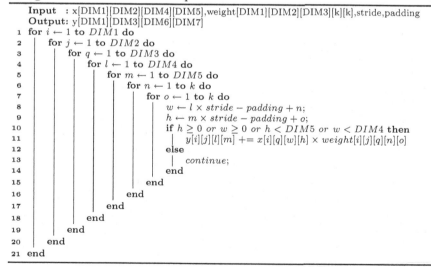

Algorithm 1: Convolution operations

    Input   : x[DIM1][DIM2][DIM4][DIM5],weight[DIM1][DIM2][DIM3][k][k],stride,padding
    Output: y[DIM1][DIM3][DIM6][DIM7]

1 **for** $i \leftarrow 1$ **to** $DIM1$ **do**
2 | **for** $j \leftarrow 1$ **to** $DIM2$ **do**
3 | | **for** $q \leftarrow 1$ **to** $DIM3$ **do**
4 | | | **for** $l \leftarrow 1$ **to** $DIM4$ **do**
5 | | | | **for** $m \leftarrow 1$ **to** $DIM5$ **do**
6 | | | | | **for** $n \leftarrow 1$ **to** k **do**
7 | | | | | | **for** $o \leftarrow 1$ **to** k **do**
8 | | | | | | | $w \leftarrow l \times stride - padding + n;$
9 | | | | | | | $h \leftarrow m \times stride - padding + o;$
10 | | | | | | | **if** $h \geq 0$ or $w \geq 0$ or $h < DIM5$ or $w < DIM4$ **then**
11 | | | | | | | | $y[i][j][l][m] += x[i][q][w][h] \times weight[i][j][q][n][o]$
12 | | | | | | | **else**
13 | | | | | | | | $continue;$
14 | | | | | | | **end**
15 | | | | | | **end**
16 | | | | | **end**
17 | | | | **end**
18 | | | **end**
19 | | **end**
20 | **end**
21 **end**

Algorithm 1 shows the pseudo-code of our convolution module, and we will analyze the relationship between the input array, the weight array and the output array with the loop variables as shown in Table 1.

Table 1. Variable cycle relationship

Heading level	X	Weights	Output
i	*Independent*	Independent	Independent
j	*Irrelevant*	Independent	Independent
q	*Independent*	Independent	Irrelevant
l	*Related*	Irrelevant	Independent
m	*Related*	Irrelevant	Independent
n	*Related*	Independent	Irrelevant
o	*Related*	Independent	Irrelevant

Loop variables have no dependencies, so we can move the i, j, and q loops to the innermost level for expansion. After such expansion, the loops can create their hardware structures to execute operations, thus processing the whole operation process in parallel, and after such processing, the loop structure is simplified and the latency is greatly reduced.

4.4   Layer Parallel Optimization

In the attention module, the input data is first dimensionally reduced by the linear layer, because the input dimensionality of the adapted tasks such as image and audio/video is too large, which leads to too much computation and affects the model effect. However, for the language problem in this paper, the input dimensionality is not very large due to the nature of the language itself. In order not to lose information, we use a flat linear layer and do not change the input and output dimensions. After this treatment, serial linear layers take up too much of the computation in the attention module. So in this paper, the process of qkv linear transformation is optimized so as to reduce the linear layer time significantly and further reduce the overall module running time.

The attention module of the Transformer model includes self-attention and general attention modules. Since q, k, and v in the self-attention module are the same input, their operations are performed for the same address. However, read and write operations to the same address can generate conflicts, so the execution of the linear transformations of q, k, and v must be performed sequentially. Due to the large amount of computation in the linear layer, sequential execution increases the delay time of the module. Therefore, we propose the idea of parallelizing the q, k, and v linear transforms. In the q, k, v linear transform phase of the self-attention module, we make two copies of the values in q and store them in separate hardware addresses. With this copy operation, q, k, and v of the same size and shape can be operated together in the same clock cycle, thus achieving a parallelization of the three linear transformations. This optimization scheme significantly reduces the software latency and improves the operation efficiency of the model. The process diagram is shown in Fig. 6.

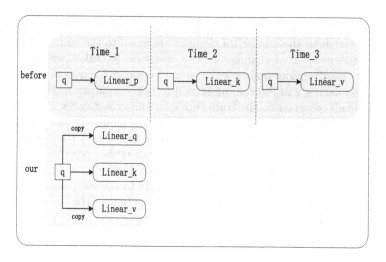

Fig. 6. Linear layer parallel optimization

5   Experiment

5.1   Experimental Setup

As shown in Table 2, the hardware part of this paper was developed and designed using Vivado's high-level programming language HLS and validated using the WMT14. en-fr dataset and the hardware and software co-design units were developed and designed on the Xilinx ZCU102 development board, which has an ARM core running on the software layer and an FPGA programmable array running on the hardware layer. The CPU model is the 12th GEN Intel Core i5-12490F and the GPU model is the NVIDIA GeForce RTX 3090. The CPU and FPGA use the code structure in the C++14 framework and the GPU uses the coding structure in the Pytorch framework.

Table 2. Experimental hardware setup

Platform	CPU	GPU	FPGA
Device	12th GEN Intel Core i5-12490	NVIDIA GeForce RTX 3090	Xilinx ZCU102
Code heading	C++14	Pytorch	C++14
Clock heading	-	-	10ns

5.2   Experimental Results

We make different cuts for the number of channels of the model and compare their computational latency and resource consumption the experiments are shown in Table 3. In this study, we redistribute the number of channels of the input convolutional and attentional modules and experimentally analyze the effect of different channel assignments on the model performance. The results show that as the number of channels assigned to the convolution module increases, the number of channels in the attention module decreases, and the decrease in the attention module leads to a decrease in the overall latency, along with a linear decrease in the computational resource utilization. When the number of channels assigned to the convolution module reaches 88, the time of the convolution module and the attention module is the same, the latency is reduced by a factor of 7 and the BRAM is reduced by a factor of 3 compared with the original 1:1 cut channel model, and modules can realize parallel computation. However, when the number of channels assigned to the convolution module is less than 88, the delay and resource usage starts to increase again because the computation time of dynamic convolution exceeds that of the attention module.

Table 3. Channel cutting resource comparison

Channel scale	Latency			BRAM	DSP	FF	LUT	BLUE
	Litetransfomer module	Attention	Dynamic Conv					
1:1	7402747	7397705	178843	435	375	43338	60425	39.6
200:298	4818946	4818904	772862	263	399	42091	57837	39.1
152:344	2788498	2783456	898142	231	379	38332	55063	38.7
100:398	1212445	1207403	1033862	153	377	35676	53529	39.04
Ours	1070224	1035951	1065182	145	367	34375	52543	39.34
60:436	1149265	959000	1138262	189	377	33900	56034	39.13

We test the effect of linear layer parallel optimization on the forward propagation delay of the attention module, and the results are represented in Fig. 7. After our parallel optimization for linear layers, the serial three linear layers are completed in parallel at the time of one linear layer processing. The experimental results show that the models formed by different cutting ratios are optimized to reduce the latency of the attention module by more than three times, which proves the effectiveness of the method. After linear layer parallel optimization the latency of the model after the 88:398 channel count cut is only 2.9ms which is 30 times less than the latency without optimization and cut channels.

We perform a generalization analysis of the convolutional layer optimization, and apply the optimization method to a model formed by cutting different numbers of channels, the results of which are shown in Fig. 8. The baseline model

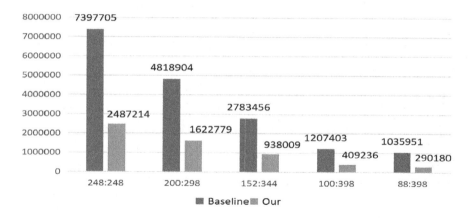

Fig. 7. Comparison of parallel optimization in linear layer

is a convolutional module without convolutional layer optimization. The experimental results show that our optimization method can reduce the delay of the convolutional module by more than 3 times, which proves the effectiveness and feasibility of the method.

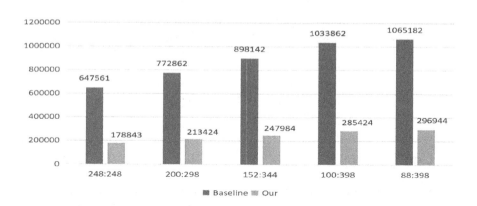

Fig. 8. Comparison of Convolutional layer optimization

Finally, we combine the attention module, the convolution module, and the quantization optimization on the model after different numbers of channel cuts the experiments are shown in Table 4. After our specialized modular design and assigning more channels to dynamic convolution, the latencies of both our modules are significantly reduced, and the latencies of both modules converge to the same, achieving better parallelism. Our latency is 12 times faster than before optimization. 86.8% reduction in BRAM, 85.06% reduction in DSP, 70.82% reduction in FF, and 74.95% reduction in LUT.

Table 4. Final optimization results

Channel scale	Latency			BRAM	DSP	FF	LUT	BLUE
	Litetransfomer module	Attention	Dynamic Conv					
1:1	2487214	2482172	178843	451	375	43727	60971	39.6
200:298	1622779	1617737	213424	271	399	42091	57837	39.1
152:344	943051	938009	247984	239	380	38836	55763	38.7
100:398	414278	409236	285424	157	367	36169	54221	39.04
Ours	301986	290180	296944	153	364	34694	53054	39.34
Ours+Quant	160266	152629	116325	57	56	12645	15134	39.23

5.3 Multi-platform Comparison

Finally, we adopt the optimized attention and convolution modules and use the 88:398 channel allocation to customize the whole module design. In the model of this paper, the convolution module and the attention module are executed in parallel on the FPGA, and the final experimental results are shown in Table 5. The execution latency of our design on GPU and CPU is shorter than other design solutions.

Table 5. Comparison with CPU and GPU

Device	CPU	GPU	FPGA
Execution time (ms)	20.8	2.98	1.60
Relative Speedup	1×	6.97×	12.5×

6 Conclusion

In this paper, we have designed the model with the combination of convolution and attention to hardware. Also, due to the huge difference in computing time between the two modules, we redistribute the number of channels passed into the two modules and perform parallel operations for the linear layers in the convolution and attention modules, and finally, our model is faster than both CPU and GPU.In the future, we will further explore the speed-matching relationship between convolution and attention and hope to propose a generalized way to solve the problem of the large computational speed gap between the two, to generalize the hardware design for such models.

References

1. Vaswani, A., et al.: Attention is all you need. arXiv (2017)
2. Hassani, A., Walton, S., Shah, N., Abuduweili, A., Li, J., Shi, H.: Escaping the big data paradigm with compact transformers. arXiv preprint arXiv:2104.05704 (2021)
3. d'Ascoli, S., Touvron, H., Leavitt, M.L., Morcos, A.S., Biroli, G., Sagun, L.: ConViT: improving vision transformers with soft convolutional inductive biases. In: International Conference on Machine Learning, pp. 2286–2296. PMLR (2021)
4. Mehta, S., Rastegari, M.: MobilEViT: light-weight, general-purpose, and mobile-friendly vision transformer. arXiv preprint arXiv:2110.02178, 2021
5. Wu, Z., Liu, Z., Lin, J., Lin, Y., Han, S.: Lite transformer with long-short range attention. arXiv preprint arXiv:2004.11886 (2020)
6. Chen, Y., Dai, X., Liu, M., Chen, D., Yuan, L., Liu, Z.: Dynamic convolution: attention over convolution kernels. In: Proceedings of the IEEE/CVF Conference on Computer Vision and Pattern Recognition, pp. 11030–11039 (2020)
7. Xu, D., Chen, Y., Cui, N., Li, J.: Towards multi-dimensional knowledge-aware approach for effective community detection in LBSN. In: World Wide Web, pp. 1–24 (2022)
8. Shi, R., et al.: FTDL: a tailored FPGA-overlay for deep learning with high scalability. In: 2020 57th ACM/IEEE Design Automation Conference (DAC), pp. 1–6. IEEE (2020)
9. Yang, Y., et al.: Synetgy: algorithm-hardware co-design for convnet accelerators on embedded FPGAs. In: Proceedings of the 2019 ACM/SIGDA International Symposium on Field-Programmable Gate Arrays, pp. 23–32 (2019)
10. Ma, N., Zhang, X., Zheng, H.T., Sun, J.: ShuffleNet V2: practical guidelines for efficient CNN architecture design. In: Proceedings of the European Conference on Computer Vision (ECCV), pp. 116–131 (2018)
11. Park, J., Yoon, H., Ahn, D., Choi, J., Kim, J.-J.: Optimus: Optimized matrix multiplication structure for transformer neural network accelerator. Proc. Mach. Learn. Syst. **2**, 363–378 (2020)
12. Song, H., Mao, H., Dally, W.J.: Deep compression: compressing deep neural networks with pruning, trained quantization and huffman coding. In: ICLR (2016)
13. Chen, Y., Zhang, N., Yan, J., Zhu, G., Min, G.: Optimization of maintenance personnel dispatching strategy in smart grid. World Wide Web **26**(1), 139–162 (2023)
14. Zhang, C., Li, P., Sun, G., Guan, Y., Cong, J.: Optimizing FPGA-based accelerator design for deep convolutional neural networks. In: The 2015 ACM/SIGDA International Symposium (2015)

Multi-level Matching of Natural Language-Based Vehicle Retrieval

Ying Liu[1](✉), Zhongshuai Zhang[1], and Xiaochun Yang[2]

[1] Beijing Institute of Technology, Beijing, China
{3120201051,zszhang}@bit.edu.cn
[2] Northeastern University, Liaoning, China
yangxc@mail.neu.edu.cn

Abstract. Utilizing natural language to retrieve vehicles of specific types and motion states in videos holds great significance for analyzing traffic conditions. But natural language and vehicle video contain rich semantics, including static and dynamic information about vehicles. Additionally, the flexibility of natural language allows for multiple expressions of sentences with identical semantics. To make full use of the information in it, we divide the natural language and video data into different levels and divide them into the representation of overall and local information. We propose information enhancement methods for different data levels, followed by generating embedded representations for layered data using representation learning networks. Finally, the overall cross-modal similarity is calculated by applying weighted measures. Experimental results demonstrate the method's capability to enhance the accuracy of retrieving vehicles in specific states from videos using natural language.

Keywords: cross-modal · vehicle retrieval · data enhancement

1 Introduction

Retrieving specific vehicles in video data has many applications in urban planning, traffic engineering, and law enforcement [12,13]. For example, traffic police can use natural language to search for a criminal vehicle with a specific color and type. The natural language contains rich semantic information, including not only static information about the vehicle (such as type, and color), and dynamic information (vehicle movement) but also global information about the vehicle (following a white car, at the cross). Making full use of this information to query can get more accurate retrieval results, so developing a natural language vehicle retrieval system is necessary. Figure 1 shows the retrieval method through natural language. It can be seen that the natural language not only contains the static and dynamic information of the vehicle but also indicates the relationship between the vehicle and the environment. Our primary objective is to fully leverage the information embedded in both natural language and video data.

There are mainly two traditional retrieval methods. One is image-to-image retrieval, which needs to provide vehicle pictures in advance, and then perform matching based

The work is partially supported by the National Natural Science Foundation of China (Nos. U22A2025, 62072088, 62232007), Ten Thousand Talent Program (No. ZX20200035), Liaoning Distinguished Professor (No. XLYC1902057), and 111 Project (B16009).

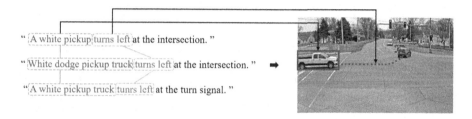

Fig. 1. Natural language describes the appearance and motion of the vehicle.

on vehicle pictures to find vehicles with specific trajectories [5], which limits the use in actual situations. The other method involves predefining tags in the video and performing matching based on these tags, but it requires significant manual preprocessing time. Natural language-based video retrieval differs from the aforementioned methods by matching query intent and finding relevant video segments. The whole process can be regarded as a cross-modal retrieval problem. Currently, a mainstream solution for cross-modal retrieval is to find a common space and map the representations of different modalities to the same space for measurement [16]. Another way of thinking is to convert data of different modalities into the same modal [10], such as converting video data into text data and unifying them in the text data space for measurement. However, this measurement approach involves two separate modeling steps. The first step involves converting video data into text data, while the second step focuses on modeling the matching between two sets of text data. In our paper, we use the first method, which is the common subspace mapping, to train a multi-level end-to-end representation learning network, enabling the generation of representations for both natural language and vehicle videos.

Because of the flexibility of text expression, a sentence with the same semantics can have different expressions, such as "A white jeep turns right at the intersection" and "At the intersection, a jeep turns right". Words in sentences have multiple expressions, and their positions can also be different, with many uncertainties. If we only learn the whole sentence representation as in [12], this uncertainty will have an impact on the results. The video of the vehicle is also serialized data, which contains not only the spatial information of the vehicle but also the change of the vehicle's position with time. In [1], only the position of the vehicle movement is traced, without considering the change of time dimension. This position-tracing method ignores the direction of the vehicle. For example, the vehicle turns left to the upper side of the current frame from the left side of the current frame. If only position markers are considered without considering time markers, such position changes can also be seen as turning right to the left side of the current frame from the upper side of the current frame. This affects the accuracy of the final matching results. Moreover, due to the sparsity of video data, there is less paired data, which is not good for sequences. To sum up the above reasons, we hope to improve the sequencing capability by using data enhancement methods.

The enhancement mainly includes several aspects. For text, we first standardize the text to avoid ambiguity caused by the flexible expression of natural language. Due to the diversity of text expression, we extract different sentence components from the text to judge synchronously. We use existing analysis tools to perform part-of-speech tagging and extract vehicle instance and vehicle motion from the sentence, that is, the static information and dynamic information of the vehicle, corresponding to the information at the same level in the video data. At the same time, this paper also constructs a word frequency dictionary, which extracts the high-frequency words in the text according to the dictionary to strengthen the important information. In the above example, the vehicle instance extracted after semantic analysis is "a white jeep" and the vehicle motion is "turn right". We use Glove [11] as a text encoder to get the initial vector representation of information at each layer.

For vehicle video, several enhancement methods are considered. The first is the track marker. Unlike in [1], we not only consider the position marker but also consider the time marker. Extract the vehicle background, paste the crop of the vehicle into the background, eliminating the impact caused by other vehicle movements in the environment, and form a frame sequence that only contains the motion of the queried vehicle as the global motion of the vehicle. At the same time, the vehicle will have different deflection angles with different motion states, and the change in the vehicle state also reflects the motion characteristics of the vehicle. Therefore, this paper makes full use of the re-ID model to represent the details of the vehicle, taking the crop sequence of the vehicle as a part of the global motion, and synchronizing it with the newly constructed video frame as the global motion representation. In the video data of the current vehicle, we extract the information of the same level corresponding to the text data. So the second enhancement method is to consider the change of position and the static characteristics of the vehicle separately and extract the bounding box of the vehicle as the static characteristics. In the above example, the corresponding is "a white jeep". For the dynamic characteristics of vehicles, we focus on the angle change of vehicle motion, and use the difference between position coordinate vectors to form a sequence of angle changes as the sequence of vehicle motion, which corresponds to the text as "turn right". We use the pre-trained ResNet101-ibn [6] and re-ID model [19] to get the vector representation of global frames and crops.

It can be seen from the above that we need to divide the data of different modes into different levels, and then adopt some data enhancement strategies to effectively use the data information of each level to enable them to obtain the data we need for training. Then the two modal data are encoded, and the representations of the same level in different modalities are mapped to the joint embedding space for measurement, and finally weighted to calculate the final loss. Throughout this process, we encounter the following challenges: (1) Thoroughly and accurately analyzing data features to effectively leverage the information in natural language and vehicle videos. (2) Implementing efficient data augmentation techniques to maximize the utilization of information for matching purposes. (3) Selecting appropriate model components to construct an end-to-end network that can generate informative representations and accurately measure the matching process.

The following are the main contributions of this paper:

- In order to make full use of the information in the two modalities, we have divided the natural language and vehicle video into different levels. The sentence is deconstructed based on its components, extracting the event, event subject, and action. A similar approach is applied to partition the video.
- For different levels of data in natural language and vehicle videos, we have implemented a range of data enhancement techniques. These measures enable the models to emphasize relevant information, disregard irrelevant details, and contribute positively to the matching results.
- Build a layered end-to-end network to realize the matching of different layers and weighted the losses of each layer to get the final matching result.

2   Related Work

2.1   Vehicle Retrieval

Vehicle retrieval has gone through a relatively long period of development, from image retrieval to video retrieval. Stanford University proposed a classic proprietary model NoScope [8] system, which mainly searches through keywords and performs binary classification tasks. Only image frames can be retrieved. The TASTI [9] system proposed later mainly retrieves vehicles through SQL language, and establishes a cluster index for vehicle type and other characteristics, so as to realize fast vehicle frame retrieval. Later, the MARIS [2] system based on predicate query was proposed, and the CNN network was used to realize vehicle tracking, eliminate the uncertainty of the trajectory, and thus realize the retrieval of vehicle trajectory segments.

2.2   Vehicle Re-identification

Vehicle re-identification mainly uses vehicle images to retrieve vehicles with the same characteristics in the video, which can be based on target detection algorithms such as Yolo [3], and track tracking can be realized on this basis through vehicle recognition. DeepSort [14] is a widely used target tracking algorithm based on the re-id algorithm. It can measure and match the results based on the cosine distance of feature similarity and the Mahalanobis distance based on the target state, and can successfully perform re-id to achieve target tracking. When performing retrieval, we can use DeepSort to extract the trajectory of individual vehicles from the full video for matching with natural language.

The above MIRIS system is also a system for retrieving the video clips of the vehicle. It mainly solves the problem of finding the movement of the same object on different video frames for a specific predicate, then, the uncertain trajectory is further judged by adding frames.

2.3   Video Retrieval Based on Natural Language

Retrieving videos through natural language is to match data of different modalities to solve the problem of cross-modal retrieval. In the beginning, most of them used text to retrieve pictures. With the development of video processing technology, video text retrieval gradually appeared. The video text retrieval model is mostly a combination of existing models. In order to improve the accuracy, the loss function will be designed more strictly. Most of the solutions to the cross-modal retrieval problem are to map the data of different modalities to the same subspace for measurement and calculate the distance between the data of different modalities in this space, so as to judge the correlation between the data.

Many existing cross-modal detection ideas have achieved relatively good results. Chen et al. proposed a graph reasoning model [?] to embedding representations between different components of text data, and to represent the correlation between different components of sentences. Hui et al. proposed a collaborative temporal model [7], modeling from two dimensions of time and space, to achieve the matching of different sentences to videos. These models have relatively high learning significance.

3   Method

In this section, we will detail the data enhancement strategy and network design. We divide the text data and video data into three levels: global, object, and motion. First, we enhanced the data of natural language and vehicle videos. We perform semantic analysis on the text, extract hierarchical data, and adopt some data enhancement strategies. For video data, we conducted vehicle motion modeling and instance extraction to match the corresponding level of text data and optimized the frame serialization modeling in time. Finally, we will introduce the modeling of the whole network in detail, and optimize the loss calculation.

3.1   Data Enhancement

To make full use of different levels of information in natural language and video, we have divided natural language and video into different levels: global, object, and motion, and enhance text data and video data. The enhanced data strengthen the characteristics of events and subjects, making them pay more attention to the characteristics of the corresponding levels, and the difference is more significant, which is conducive to the matching of different modal data. Next, we will give a detailed description of the data layering and enhancement methods of the two modal data.

Natural Language Extraction. To reduce the impact of various expressions of natural languages on representation generation, we make full use of the three description statements corresponding to each video. We found that the subject in natural language and the motion of vehicles usually use relatively fixed words, so we use 'Spacy'[1] to label the sentence components, and analyze the text based on 'verb'. Based on the position of the verb, the subject of the event is extracted forward and the action of the event is extracted backward, as shown in Fig. 2. There are three natural language descriptions for each training data, and the subjects and actions of the three natural languages are extracted. When executing the parsing algorithm, it was found that the sentence components of some description sentences were not obvious after semantic analysis. For

---

[1] https://spacy.io/.

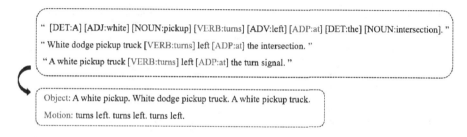

Fig. 2. Use 'Spacy' for part of speech tagging.

such sentences, we used the reverse translation technology [1] to retranslate the sentences, as shown in Fig. 3. We first translate it into another language, then translate it into English to make its sentence components obvious, and then implement the text parsing method. We divide the four types of vehicle movement status into "straight", "stop", "turn left" and "turn right". We use regular expressions to divide the sentences containing "turn left", "turn right" and "stop" into these three categories, and the rest are "straight". Then, we spliced the subject and action of the three sentences respectively as the enhanced object and motion to generate the embedded representation of these two levels.

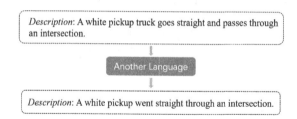

Fig. 3. Reverse translation makes sentence components obvious.

In the text description, some words often appear, such as the color, type, and movement state of the vehicle. For these keywords, we also need to pay attention to them in the matching process. We have built a word frequency dictionary, as shown in Fig. 4, to extract the words that appear more frequently in the sentence, and then extract the keywords in the sentence to form a new description, which is used as the global motion together with the original natural language description.

Vehicle Video Extraction. To match the text partition at the same level, the video needs to be divided similarly. For the hierarchical division of video, because the camera position is fixed, the environment in the background does not change, only the movement of different vehicles. In the original frame sequence, when the queried vehicle moves, other vehicles also move. If the vehicle's global motion representation is generated from the original frame sequence, there will be many uncertainties. To eliminate

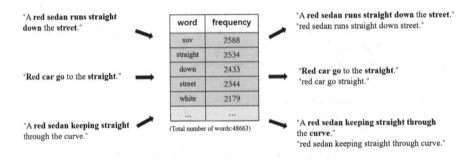

"A red sedan runs straight down the street."

"Red car go to the straight."

"A red sedan keeping straight through the curve."

word	frequency
suv	2588
straight	2534
down	2433
street	2344
white	2179
...	...

(Total number of words:48663)

"A red sedan runs straight down the street."
"red sedan runs straight down street."

"Red car go to the straight."
"red car go straight."

"A red sedan keeping straight through the curve."
"red sedan keeping straight through curve."

Fig. 4. Extract sentence focus according to word frequency dictionary.

the impact of other moving vehicles, we only focus on the motion state of the queried vehicle, so we use background modeling [1] to focus on the vehicle driving environment. We process all frames of the same video with the following mean value to get the background information of vehicle motion. Let $f_1, f_2, \ldots, f_n$ be a sequence of frames:

$$Background = \frac{f_1 + f_2 + \cdots + f_n}{n} = \frac{1}{n} \sum_{i=1}^{n} f_i \tag{1}$$

where f_i $(1 \leq i \leq n)$ is the i-th frame. $Background$ is the generated background image.

Then according to the bounding box of the vehicle, the vehicle in each frame is intercepted and placed in the background picture, thus generating the motion sequence of the vehicle in the environment, as shown in Fig. 5.

Fig. 5. Motion modeling: Extract the background and place the vehicle in the background.

Because the position of the camera is relatively fixed, the vehicle presents different angles in front of the camera with the change of movement, so the crop also reflects the movement of the vehicle, as shown in Fig. 6. Therefore, the crop sequence and the frame sequence are used as the global motion auxiliary result judgment.

For the object, randomly select one crop as the static feature. The motion is mainly aimed at the change of vehicle position coordinates. The vector change is more representative of the type of motion (such as a left turn) than the coordinate change, so we

Fig. 6. Vehicle changes in straight and turning states

can get the sequence of vector changes according to the vehicle position coordinates:

$$a_j = (x_j, y_j) - (x_i, y_i)(1 <= i, j <= n) \tag{2}$$

where (x_j, y_j) is the vehicle's position in the j-th frame, and a_j is the vector of vehicle position change from the i-th frame to the j-th frame. Now we get the sequence of vehicle position changes is $(a_1, a_2, \ldots, a_n)$.

3.2 Representation Learning

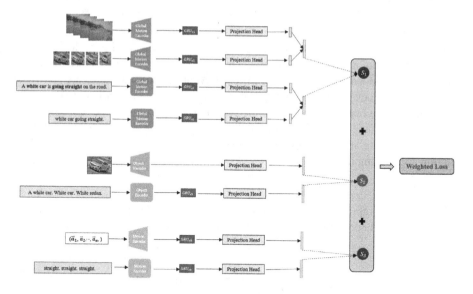

Fig. 7. The overall design of the system. The encoders of each modality are used to obtain the embedding representations of the two modalities with different levels, and then mapped to the same subspace for similarity measurement, and weighted to calculate the loss.

According to the video and text data at different levels obtained in Sect. 3.1, specific modeling is carried out, as shown in Fig. 7. The visual feature coder and text feature coder are respectively referenced for video and text to obtain their respective representations. Then map the representation of the two modes to the common subspace for measurement, compare the cosine similarity of the two modes, and calculate the loss. According to the loss results, the backpropagation and training model can obtain better matching accuracy.

Text Feature Encoding. For text data, we use Glove to get the embedded representation of events (global motion), event focus, object, and motion. Then use Bi-GRU to serialize and represent each text, and take the last layer as the final feature representation. Finally, the representation of the three levels is mapped using the full connection layer and normalized using LN (Layer Normalization). Take global motion as an example:

$$h^t_{s_{global}} = BiGRU\{[g_t, h^{s_{global}}_{t-1}]\} \tag{3}$$

where g_t is the output of the text feature encoder. Take the output of the last layer and send it to the full connection layer and LN.

$$h^s_{global} = W_\beta h_{s_{global}} + b_\beta \tag{4}$$

$$f^s_{global} = LN(W_1 h^s_{global}) \tag{5}$$

where f^s_{global} is the final global motion representation. Then the two features of global motion and event focus are contacted to form a new global motion representation.

Visual Feature Encoding. We use Resnet101-ibn pre-trained on ImageNet to get the frame representation. The re-ID [19] model is retrained as the vehicle feature encoder to obtain the embedded representation of the crop. After the image features are generated for the crop sequence and frame sequence, the final representation is obtained using Bi-GRU, and the feature is mapped to the common subspace using the full connection layer and BN (Batch Normalization), and then the two features are contacted. The other two levels of data are also mapped in the same way, as shown in Fig. 7. Take global motion as an example:

$$h^t_{v_{global}} = BiGRU\{[g_t, h^{v_{global}}_{t-1}]\} \tag{6}$$

where g_t is the output of the visual feature encoder. Take the output of the last layer and send it to the full connection layer and BN.

$$h^v_{global} = W_\alpha h_{v_{global}} + b_\alpha \tag{7}$$

$$f^v_{global} = BN(W_2 h^v_{global}) \tag{8}$$

where f^v_{global} is the final global motion representation.

3.3 Loss Calculation

After mapping the representations of the two modalities to the same subspace, we measure the cosine similarity of the embedded representations of the two modalities. Taking the global motion as an example, the specific calculation method is:

$$D = cos(\boldsymbol{q}, \boldsymbol{v}) = \frac{\boldsymbol{q} \cdot \boldsymbol{v}}{|\boldsymbol{q}| \cdot |\boldsymbol{v}|} \tag{9}$$

where q and v are representations of query and video in the joint embedding space.

Then we calculate InfoNCE loss. Given that the batch size is K. $f_i^{sentence}$ and f_i^{video} have the same semantic meaning, so they are considered positive pairs. Other $f_i^{sentence}$ and $f_{j \neq i}^{video}$ have different semantics, treating them as negative pairs. So the positive pair is 1, and the negative pair is $K - 1$. The loss from text to video can be calculated as follows:

$$L_{s2v} = -\frac{1}{K} \sum_{i=1}^{K} \log \frac{exp(D_{(i_{sentence}, i_{video})}/\tau)}{\sum_{j=1}^{K} exp(D_{(i_{sentence}, j_{video})}/\tau)} \tag{10}$$

Similarly, the loss calculation from video to text can be expressed as:

$$L_{v2s} = -\frac{1}{K} \sum_{i=1}^{K} \log \frac{exp(D_{(i_{video}, i_{sentence})}/\tau)}{\sum_{j=1}^{K} exp(D_{(i_{video}, j_{sentence})}/\tau)} \tag{11}$$

$D_{(i,i)}$ is the cosine distance between the i-th text and the j-th video, and τ is the temperature coefficient initialized to 0.2. Finally, we get the loss of this level (take 'global' as an example):

$$S_{global} = \alpha_1 L_{s2v} + \alpha_2 L_{v2s} \tag{12}$$

where $\alpha_1 = \alpha_2 = 1$. To get the loss of three levels, we use the weighted calculation method to get the final loss:

$$S_{info} = \lambda_1 * S_{global} + \lambda_2 * S_{object} + \lambda_3 * S_{motion} \tag{13}$$

where λ is the weight of each granularity, we set $\lambda_1 = 4$, $\lambda_2 = 1$, $\lambda_3 = 1$. S_{global}, S_{object} and S_{motion} are the losses of the three levels in Fig. 7.

4 Experiment

4.1 Settings

Environment. The operating system is Ubuntu18.04.5 LTS, the CPU is Intel(R) Xeon(R) Silver 4216 CPU @ 2.10 GHz, 64 processors, 256G memory. The GPU is Quadro RTX 6000, and 24G GPU memory. The Python version used is 3.7, and the version of Pytorch is 1.7.1+cu92. The PyTorch-lightning used is 1.2.4.

Dataset. The dataset used in the experiment is from AI City Challenge2022 track2. The dataset contains 2155 pieces of training data, each of which contains three query descriptions, a sequence of frames where vehicles appear, and the bounding box sequence of vehicles in the frame. The dataset contains 184 test data, which is divided into two parts: query and tracks. The query is described in three different ways.

4.2 Metrics

Vehicle video retrieval through natural language description is mainly evaluated by MRR (Mean Reciprocal Rank), a common evaluation index for retrieval tasks. The formula is as follows:

$$MRR = \frac{1}{|Q|} \sum_{i=1}^{|Q|} \frac{1}{rank_i} \tag{14}$$

where $|Q|$ is the total number of queries, and $rank_i$ is the rank of the correct answer corresponding to a query in all results. Recall@5 and Recall@10 results are also used as evaluation indicators. The formula of Recall@5 is as follows:

$$Recall@5 = \frac{1}{|Q|} \sum_{i=1}^{|Q|} \alpha \tag{15}$$

α indicates whether the best match answer to the i-th question is in the top 5 of the returned results. If so, then $\alpha = 1$; otherwise, $\alpha = 0$. Recall@10 is calculated in the same way.

4.3 Comparison Results

We show the performance comparison between our proposed method and the other methods. The comparison model is described as follows:

- Baseline [4]: It is the baseline proposed by AI City Challenge 2021 track5. A relatively simple matching method is adopted, and only ResNet50 is used to identify the vehicle bounding box, and 'Bert' is used to generate the representation of the entire sentence for the text description, and then compared in the joint space.
- SSM [17]: It is the method to get second place in AI City Challenge2022 track2.
- MGRS [15]: It is the method to get first place in AI City Challenge2022 track2.
- MLM: It is the method proposed by our paper. It mainly divides the level of natural language and vehicles optimizes the way of loss calculation, and combines the losses of three levels to get the final matching result. And optimize the sorting results through re-ranking.

We compared the results of the three methods and obtained the MRR of the three methods, as shown in Table 1.

Table 1. Comparison of different systems.

Method	MRR
Baseline	0.0254
SSM	0.4392
MGRS	0.6606
MLM (our)	**0.8918**

Our model achieves better results under the same experimental dataset than other methods. The comparison results prove that the multi-level matching idea proposed in this paper can fully utilize the overall and detailed information in natural language and vehicle videos, and improve the accuracy of retrieval.

The performance of the model is affected by the number of iterations. In order to find the best iteration number, we tested five different iterations and compared the MRR under different epochs to avoid excessive training and wasting training time. The MRR under various iterations is shown in Fig. 8. MRR tends to be stable in 80 to 100 epochs.

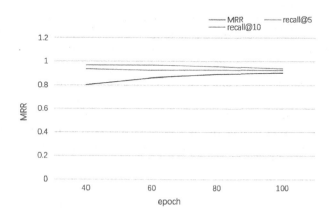

Fig. 8. Influence of training epochs on accuracy.

4.4   Ablation Experiment

In order to analyze the influence of different components of the model on the experimental results, we conducted ablation experiments. We analyze the impact of different levels of the model on the results. The specific model composition is as follows:

- MLM_{global}: In terms of partition level, the model only recognizes the global motion, but lacks the similarity comparison between vehicle instance and vehicle motion.
- MLM_{object}: Based on MLM_{global}, the similarity comparison of the vehicle instance is added, which is the second level in Fig. 7.
- MLM_{motion}: Based on MLM_{global}, the similarity comparison of the vehicle motion is added, which is the third level in Fig. 7.
- MLM: Methods proposed in this paper. Integrate the matching results of three levels.

Experiments show the MRR, Recall@5, and Recall@10 of different models in Table 2. It can be seen from the experimental results that the vehicle instance and vehicle motion have similar effects on the experimental results, and both improve the accuracy of the system. According to the improvement ratio of the experimental results, we assigned weights to the losses of different levels when calculating the losses at the end, and summed them up according to the weights to obtain the final total loss.

Table 2. Comparison of model scores for different module combinations.

Method	MRR	Recall@5	Recall@10
MLM_{global}	0.8587	0.9348	0.9620
MLM_{object}	0.8666	0.9239	0.9674
MLM_{motion}	0.8721	**0.9457**	**0.9783**
MLM	**0.8897**	0.9402	0.9620

Re-ranking. In the experiment, we used k-reciprocal coding [18] to reorder the retrieval results. Reordering is often used in retrieval tasks and is often used as a post-processing method to improve retrieval performance. The main idea is that the closer the two features are, the more likely they are to be correct results, so as to improve the ranking of similar vectors and achieve the purpose of improving accuracy. In this experiment, after obtaining the vectors of two modes, re-ranking is used to rearrange them. Each video has three text descriptions, and three distance matrices are calculated respectively. After adding, the final distance matrix is used. The retrieval results are sorted according to the distance matrix. Through reordering, the retrieval effect has been slightly improved, mainly as shown in Table 3:

Table 3. The effect of re-ranking on matching results.

Method	MRR	Recall@5	Recall@10
MLM	0.8897	**0.9402**	**0.9620**
MLM (re-ranking)	**0.8918**	0.9239	0.9348

4.5   Result

We show the retrieval results returned by different natural language descriptions. The results are labeled with specific retrieved vehicles and video frames of vehicle movement. In Fig. 9, we randomly selected several described retrieval results. From the retrieval results, it can be seen that this model meets the requirements of natural language vehicle video retrieval. However, there may also be retrieval matching errors, such as the fourth vehicle in the picture, where the color of the vehicle is affected by the things on the vehicle and is recognized incorrectly. And the three natural language descriptions of a vehicle may also have different descriptions due to cognitive differences, such as different views on vehicle type and color. These will all have an impact on the retrieval results.

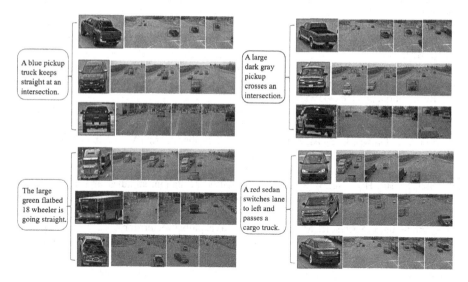

Fig. 9. Example of retrieval results.

5   Conclusion

In this paper, we propose a query system for vehicle video retrieval based on natural language. The paper introduces a division of video and natural language into distinct levels and discusses the process of dividing modal data into these levels to obtain vehicle-related information at various levels. And some data enhancement strategies are adopted to enhance the global motion, static characteristics, and position movement of the vehicle, highlighting the data characteristics of concern. The system integrates global, static, and dynamic information from the vehicle, computes InfoNCE loss for different levels, and applies weighting to calculate the total loss. Finally, the system proposed in this paper achieves 89.18% accuracy on the test set. In the follow-up research, we will continue to focus on and try more effective data enhancement methods, pay attention to the loss and efficiency of model training, and expect to achieve better results.

References

1. Bai, S., et al.: Connecting language and vision for natural language-based vehicle retrieval. In: Proceedings of the IEEE/CVF Conference on Computer Vision and Pattern Recognition, pp. 4034–4043 (2021)
2. Bastani, F., et al.: MIRIS: fast object track queries in video. In: Proceedings of the 2020 ACM SIGMOD International Conference on Management of Data, pp. 1907–1921 (2020)
3. Bochkovskiy, A., Wang, C.Y., Liao, H.Y.M.: YOLOv4: optimal speed and accuracy of object detection. arXiv preprint arXiv:2004.10934 (2020)
4. Feng, Q., Ablavsky, V., Sclaroff, S.: CityFlow-NL: tracking and retrieval of vehicles at city scale by natural language descriptions. arXiv preprint arXiv:2101.04741 (2021)
5. Gao, G., Shao, H., Wu, F., Yang, M., Yu, Y.: Leaning compact and representative features for cross-modality person re-identification. World Wide Web **25**(4), 1649–1666 (2022)

6. He, K., Zhang, X., Ren, S., Sun, J.: Deep residual learning for image recognition. In: Proceedings of the IEEE Conference on Computer Vision and Pattern Recognition, pp. 770–778 (2016)

7. Hui, T., et al.: Collaborative spatial-temporal modeling for language-queried video actor segmentation. In: Proceedings of the IEEE/CVF Conference on Computer Vision and Pattern Recognition, pp. 4187–4196 (2021)

8. Kang, D., Emmons, J., Abuzaid, F., Bailis, P., Zaharia, M.: NoScope: optimizing neural network queries over video at scale. arXiv preprint arXiv:1703.02529 (2017)

9. Kang, D., Guibas, J., Bailis, P.D., Hashimoto, T., Zaharia, M.: TASTI: semantic indexes for machine learning-based queries over unstructured data. In: Proceedings of the 2022 International Conference on Management of Data, pp. 1934–1947 (2022)

10. Mai, S., Hu, H., Xing, S.: Modality to modality translation: an adversarial representation learning and graph fusion network for multimodal fusion. In: Proceedings of the AAAI Conference on Artificial Intelligence, vol. 34, pp. 164–172 (2020)

11. Pennington, J., Socher, R., Manning, C.D.: GloVe: global vectors for word representation. In: Proceedings of the 2014 Conference on Empirical Methods in Natural Language Processing (EMNLP), pp. 1532–1543 (2014)

12. Sun, Z., Liu, X., Bi, X., Nie, X., Yin, Y.: DUN: dual-path temporal matching network for natural language-based vehicle retrieval. In: Proceedings of the IEEE/CVF Conference on Computer Vision and Pattern Recognition, pp. 4061–4067 (2021)

13. Wang, F., Xu, J., Liu, C., Zhou, R., Zhao, P.: On prediction of traffic flows in smart cities: a multitask deep learning based approach. World Wide Web 24, 805–823 (2021)

14. Wojke, N., Bewley, A., Paulus, D.: Simple online and realtime tracking with a deep association metric. In: 2017 IEEE International Conference on Image Processing (ICIP), pp. 3645–3649. IEEE (2017)

15. Zhang, J., et al.: A multi-granularity retrieval system for natural language-based vehicle retrieval. In: Proceedings of the IEEE/CVF Conference on Computer Vision and Pattern Recognition, pp. 3216–3225 (2022)

16. Zhang, P.F., Luo, Y., Huang, Z., Xu, X.S., Song, J.: High-order nonlocal hashing for unsupervised cross-modal retrieval. World Wide Web 24, 563–583 (2021)

17. Zhao, C., et al.: Symmetric network with spatial relationship modeling for natural language-based vehicle retrieval. In: Proceedings of the IEEE/CVF Conference on Computer Vision and Pattern Recognition, pp. 3226–3233 (2022)

18. Zhong, Z., Zheng, L., Cao, D., Li, S.: Re-ranking person re-identification with k-reciprocal encoding. In: Proceedings of the IEEE Conference on Computer Vision and Pattern Recognition, pp. 1318–1327 (2017)

19. Zhu, X., Luo, Z., Fu, P., Ji, X.: VOC-ReID: vehicle re-identification based on vehicle-orientation-camera. In: Proceedings of the IEEE/CVF Conference on Computer Vision and Pattern Recognition Workshops, pp. 602–603 (2020)

Improving the Consistency of Semantic Parsing in KBQA Through Knowledge Distillation

Jun Zou[1], Shulin Cao[2], Jing Wan[1(✉)], Lei Hou[2], and Jianjun Xu[3]

[1] Beijing University of Chemical Technology, Beijing 100029, China
wanj@mail.buct.edu.cn
[2] Tsinghua University, Beijing 100084, China
[3] Beijing Caizhi Technology Co., Ltd., Beijing 100081, China

Abstract. Knowledge base question answering (KBQA) is an important task that involves analyzing natural language questions and retrieving relevant answers from a knowledge base. To achieve this, Semantic Parsing (SP) is used to parse the question into a structured logical form, which is then executed to obtain the answer. Although different logical forms have unique advantages, existing methods only focus on a single logical form and do not consider the semantic consistency between different logical forms. In this paper, we address the issue of consistency in semantic parsing, which has not been explored before. We show that improving the semantic consistency between multiple logical forms can help increase the parsing performance. To address the consistency problem, we present a dynamic knowledge distillation framework for semantic parsing (DKD-SP). Our framework enables one logical form to learn some useful hidden knowledge from another, which improves the semantic consistency of different logical forms. Additionally, it dynamically adjusts the supervised weight of the hidden knowledge as the student model's ability changes. We evaluate our approach on the KQA Pro dataset, and our experimental results confirm its effectiveness. Our method improves the overall accuracy of the seven types of questions by 0.57%, with notable improvements in the accuracy of Qualifier, Compare, and Count questions. Furthermore, in the compositional generalization scenario, the overall accuracy improved by 4.02%. Our codes are publicly available on https://github.com/zjtfo/SP_Consistency_By_KD.

Keywords: knowledge distillation · semantic parsing · consistency · KoPL · KBQA

1 Introduction

Knowledge base question answering (KBQA) is a technique that utilizes the rich semantic information in knowledge bases to comprehend questions and retrieve answers [1]. In recent years, KBQA has become a crucial component of various intelligent applications, and as a result, it has gained significant attention from researchers. There are two primary approaches to KBQA: Information Retrieval

X. Song et al. (Eds.): APWeb-WAIM 2023, LNCS 14333, pp. 373–388, 2024.
https://doi.org/10.1007/978-981-97-2387-4_25

(IR) and Semantic Parsing (SP). IR-based methods retrieve and rank answers from the knowledge base by constructing a question-specific graph that provides comprehensive information related to the question. In contrast, SP-based methods transform unstructured questions into structured logical forms and execute them on the knowledge base to retrieve the final answers. SP-based methods have gained increasing attention due to their strong reasoning ability and better interpretability. Various logical forms of semantic parsing exist, including KoPL [2], SPARQL [3], λ-DCS [4], SQL [5], and others, each with unique advantages.

The concept of semantic consistency has been previously applied in information extraction and question answering [6,7]. In this study, we expand this concept to consider the semantic consistency between different logical forms. For example, as illustrated in Fig. 1, the SPARQL parser mistakenly interprets *"start_time"* as *"end_time"* for the question *"What is the basketball team where LeBron James played in 2010?"*, while the KoPL parser correctly identifies the temporal relation. Therefore, there is a lack of semantic consistency between the corresponding KoPL program and the SPARQL query.

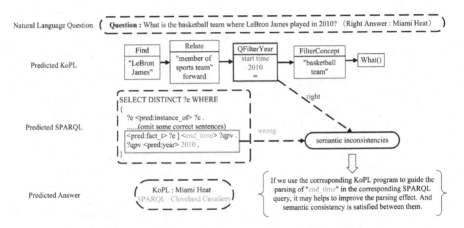

Fig. 1. A practical example of semantic inconsistencies between different logical forms.

Previous research on consistency of expression forms has primarily focused on tasks such as summary generation [8,9], dialogue [7,10], visual Q&A [11,12], and machine translation [13]. These works have proposed various ideas including (1) automatic evaluation protocols or metrics for measuring consistency, (2) novel learning algorithms such as gradient-based interpretability methods and contrasting gradient learning-based methods to improve consistency, (3) new datasets for consistency identification to facilitate research, and (4)auxiliary loss functions for the constraint enhancement of the model's ability to understand the context. However, no previous studies have explored semantic consistency in different logical forms for KBQA. To address this gap, we propose to enhance the parsing effectiveness of models by improving the semantic consistency of different logical forms in KBQA.

Knowledge distillation (KD) is a widely-used approach for model compression and model enhancement in natural language processing [14]. It allows for the transfer of knowledge from a large, complex teacher model to a smaller, simpler student model. KD can also be used to transfer hidden knowledge from one logical form to another, improving semantic consistency between different forms. In this paper, we propose to use KD to address the problem of semantic consistency between different logical forms of semantic parsing in KBQA. However, we face two main challenges: (1) **KD guidance method** - it is difficult for one logical form to guide another logical form without conveying incorrect knowledge; (2) **KD objective adaptation** - it is challenging to adjust the supervised contributions of different alignment goals to improve students' abilities.

To address these challenges, we propose a dynamic knowledge distillation framework for semantic parsing (DKD-SP) that aligns hidden layer representations with the output distribution of the teacher model. We also introduce a novel method to adjust the weights of different alignment goals dynamically during training, based on the semantic consistency between different logical forms.

In this paper, we present three primary contributions, which are described as follows:

- To the best of our knowledge, our work is the first to explore semantic consistency between different logical forms of semantic parsing in KBQA, from the perspective of knowledge distillation.
- We propose a comprehensive and effective framework for KD-based semantic consistency learning.
- We validate the effectiveness and soundness of our approach by conducting extensive experiments on the KQA Pro dataset.

2   Method

In this subsection, we present our framework DKD-SP, consisting of three essential modules: a *teacher guidance model*, a *student dynamic distillation learning model* and a *dynamic weight assignment model*. Figure 2 illustrates an overview of our framework's implementation.

2.1   Problem Definition

When given an input question x, our framework operates as follows: the logical form A-generated teacher model generates a logical form $\hat{t_A}$, while the student dynamic distillation learning model produces a logical form $\hat{s_B}$. The seq2seq model produces a logical form $\hat{t_B}$ based on $\hat{t_A}$. Subsequently, the dynamic weight assignment model assesses the semantic consistency between $\hat{t_A}$ and $\hat{s_B}$ based on their semantic information. Here, t_A and s_B represent the golden logical form A and golden logical form B from the original dataset, respectively. Our objective is to improve the generation of $\hat{s_B}$ using $\hat{t_B}$.

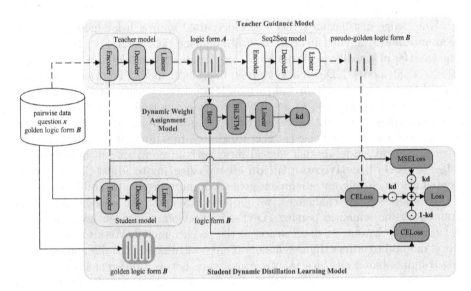

Fig. 2. Overall implementation of DKD-SP. Dotted lines represent teacher inference, and solid lines represent student training.

2.2   Teacher Guidance Model

The goal of our proposed framework is to transfer knowledge from logical form A to logical form B, such that the student model is capable of generating the latter. To achieve this goal, the teacher guidance model facilitates knowledge transfer. Specifically, the logical form A-generated teacher model is employed to convert natural language questions into logical form A. However, since logical form A and B exist in different vector spaces, direct knowledge transfer is not feasible. To address this challenge, we introduce a seq2seq model that leverages logical form A to generate logical form B. The resulting logical form B acts as another pseudo-golden label for the student model.

Logical Form A-Generated Teacher Model. The teacher model responsible for generating logical form A utilizes a transformer-based encoder-decoder architecture to extract features from natural language questions. The model's output is then passed through a linear classifier with softmax activation to predict the probabilities of different tokens in logical form A. In our DKD-SP framework, we utilize a generative pre-trained language model like Bart or T5 as the encoder-decoder structure. Equations (1) and (2) define the network architecture of the logical form A-generated teacher model, outlining the input and output of the model, respectively.

$$h = transformer(x) \tag{1}$$

$$\hat{t_A} = softmax(Wh_i + b) \tag{2}$$

where x represents the input question. $h = \{h_i\}_{i=1}^{L}$ and h_i represents the decoder output at position i, L denotes the length of the output tokens. $\hat{t_A}$ represents the predicted logical form A, while W and b are trainable parameters.

Seq2seq Model. To improve the performance of the model in generating logical form B using the knowledge contained in logical form A, we design the model network to resemble the logical form A-generated teacher model, which is described by Eqs. (3) and Eqs. (4),

$$a = transformer(\hat{t_A}) \tag{3}$$

$$\hat{t_B} = softmax(W_a a_i + b_a) \tag{4}$$

where $a = \{a_i\}_{i=1}^{L_a}$ and a_i represents the decoder output at position i, L_a denotes the length of the output tokens. $\hat{t_B}$ represents the predicted logical form B, while W_a and b_a are trainable parameters.

During the process of knowledge distillation learning, the seq2seq model enables the transfer of knowledge from logical form A to logical form B with the assistance of the logical form A-generated teacher model. This allows the student model to be trained with the guidance of the transformed logical form B.

2.3 Student Dynamic Distillation Learning Model

In this subsection, we aim to address the first challenge of performing knowledge distillation between different logical forms. To achieve this, we propose three guidance methods: Logical form-based guidance, Encoder-based guidance, and Hybrid guidance. The logical form-based guidance method aligns the output distribution, while the encoder-based guidance aligns hidden layer representations. The hybrid guidance method combines the two alignments. To ensure that the model learns correct knowledge, we utilize labels from the dataset as supervision during the knowledge distillation learning process. In this process, we use logical form A as the teacher and logical form B as the student. Our student dynamic distillation learning model adopts a transformer-based encoder-decoder structure for feature extraction and utilizes a linear classifier with softmax activation on the model output.

Logical Form-Based Guidance. The primary goal of this approach is to mimic the final predictions of the teacher model. Given an input x, the student model generates logical form B denoted by $\hat{s_B}$, while the predicted logical form A is represented by $\hat{t_A}$ using the logical form A-generated teacher model. However, it is not possible to directly align the logical forms $\hat{t_A}$ and $\hat{s_B}$ due to their different vector spaces. To address this challenge, we propose a seq2seq model that generates the logical form $\hat{t_B}$ based on $\hat{t_A}$. We train $\hat{t_B}$ as another pseudo-golden label for the student model. To simulate the student's prediction output, we combine the soft predictions of the teacher model and the hard predictions corresponding to the golden logical form. We also introduce different losses to enhance the training process. The objective function is presented in Equations (5),

$$L_{lf} = loss_{distillation} * kd + loss_{golden} * (1 - kd)$$
$$= L_{CE}(\hat{s_B}, \hat{t_B}) * kd + L_{CE}(\hat{s_B}, s_B) * (1 - kd) \tag{5}$$

where kd is the dynamically changing weight coefficient, it is discussed in detail in the following subsection.

Encoder-Based Guidance. Logical form-based guidance typically relies on the output of the last layer and may not fully utilize the intermediate feature supervision of teacher models, which can be crucial for learning structural knowledge. To overcome this limitation, we propose an encoder-based guidance method that utilizes the alignment between the implicit vectors of the student model's encoder and those of the teacher model's intermediate layers. We formulate the knowledge transfer as an objective function, as demonstrated in Equations (6). The intermediate layer vectors of the teacher model guide the student model's encoder to generate representations that are comparable to the teacher model.

$$L_{encoder} = loss_{mse} * kd + loss_{golden} * (1 - kd)$$
$$= \sum_{i=0}^{n} L_{MSE}(L_2(s_{ev\_i}), L_2(t_{ev\_i}))) * kd + L_{CE}(\hat{s_B}, s_B) * (1 - kd) \tag{6}$$

where MSE is the mean square error loss function, and L_2 is the L_2 operation for normalization. The number of layers in the model encoding layer is denoted by n. s_{ev_i} represents the i-th layer feature vector of the student model encoding layer, while t_{ev_i} denotes the i-th layer feature vector of the teacher model encoding layer.

Hybrid Guidance. Each of the two aforementioned guidance methods has its own advantages, and it is reasonable to combine them in order to jointly guide the student learning from the teacher model. The resulting training loss is shown in Equations (7).

$$L_{hybrid} = loss_{mse} * kd + loss_{distillation} * kd + loss_{golden} * (1 - kd) \tag{7}$$

By utilizing the three types of guidance, the student model can dynamically absorb knowledge and effectively learn valuable hidden knowledge during the training process.

2.4 Dynamic Weight Assignment Model

In this subsection, we address the second challenge of adjusting the supervised contributions of different objective functions. To overcome this challenge, we propose a dynamic weight assignment model that adjusts the weights based on the semantic consistency between different logical forms. The dynamic weight assignment model comprises Bert, BiLSTM, and a linear layer. Bert and BiLSTM are employed to extract the features of the input sample, i.e., logical forms A and B, and then pass them through a linear layer for classification. The network structure of the model can be formulated as Eq. (8) and Eq. (9),

$$m = BiLSTM(transformer(\hat{t_A}, \hat{s_B}))$$ (8)

$$\hat{k} = softmax(W_k m_i + b_k)$$ (9)

where $m = \{m_i\}_{i=1}^{L_m}$ and m_i denotes the decoder output at position i. L_m denotes the length of the output tokens. $\hat{k}$ denotes the predicted result. The $\hat{k} = 1$ indicates the semantic is consistent between $\hat{t_A}$ and $\hat{s_B}$, and 0 otherwise. W_k and b_k are trainable parameters.

For a given input x, $\hat{s_B}$ represents the logical form B generated by the student model, and $\hat{t_A}$ denotes the predicted logical form A generated by the logical form A-generated teacher model. The calculation of the dynamic weight kd is presented in Eqs. (10),

$$kd = \frac{1}{bs} \sum_{i=1}^{bs} \hat{k}$$ (10)

where bs is the batch size.

3   Experiments and Result Analysis

3.1   Experimental Settings

Datasets. We conduct experiments on the KQA Pro dataset, which is currently the most extensive dataset for NLQ-to-SPARQL and NLQ-to-KoPL, consisting of 117,970 question-SPARQL and question-KoPL pairs based on a Wikidata subset. The dataset includes seven types of natural language question: Multi-hop, Qualifier, Compare, Logical, Count, Verify, and Zero-shot. We split the dataset into training, validation, and test sets with an 8:1:1 ratio. For training the dynamic weight assignment model, we used positive and negative examples in a 1:5 ratio and split them into training, validation, and test sets with an 8:1:1 ratio.

Baselines. IR-based and SP-based models are the two major complex KBQA models. In order to verify the effectiveness of our proposed method, we compare it with several baseline models, including IR-based models such as KVMemNet [15], EmbedKGQA [16], RGCN [17], and Blind GRU [18], as well as SP-based models such as RNN [19] and Bart [20]. The results of the above baseline models on the KQA Pro dataset are all from the work of Cao et al. [2].

Evaluation Metrics. Following prior research [2], we evaluate the performance of our model using accuracy as the metric. Additionally, we introduce a semantic consistency metric, LFC, to demonstrate the effectiveness of our proposed method in improving semantic consistency. LFC is defined as the proportion of the correct answers predicted by both logical form A and logical form B. The calculation formula for LFC is presented in Equations (11),

$$LFC = \frac{1}{t_{num}} \sum_{i=1}^{c_{num}} (1 \mid ans_i \in M \cap N) \tag{11}$$

where t_{num} is the number of questions in the test set and c_{num} is the number of correct answers predicted by the model. ans_i represents the correct answer predicted by the model. M and N denote the sets of answers predicted by the model using logical form A and B, respectively. The numerator of the equation increments by 1 when ans_i is a member of the intersection of M and N.

Implementation Details. We use the optimizer Adam [21] for both our proposed model and the baseline models. The initial learning rate is set to 3e−5 and gradually decreased to 1e−5 during training. The models are trained for 25 epochs and their performance is reported on the test set. The batch size is 16 for the student model and 8 for the teacher model. For static knowledge distillation, we set kd to 0.8. The encoder-decoder-based model in our proposed framework can be any generative pre-trained language model, such as Bart and T5. Following prior research [2], we select the Bart pre-trained language model for our experiments. In our proposed framework, logical form A is either KoPL or SPARQL, while logical form B is SPARQL or KoPL, respectively.

Model Training. In our proposed framework, the logical form A-generated teacher model, the seq2seq model, and the dynamic weight assignment model are all trained before conducting knowledge distillation training on the student model. For **the logical form A-generated teacher model**, we use the fine-tuned BART model provided by Cao et al. [2] directly. **The seq2seq model** is trained directly on the training corpus $D_{train}^{A-B} = (t_A, s_B)$. For samples $(t_A, s_B) \in D_{train}^{A-B}$, we use the loss function shown in Equations (12). After three rounds of training, the average translation accuracy reached 99.63%.

To construct corresponding negative examples based on (t_A, s_B), we denote the labeled training data as $D_{train}^{classifier} = (t'_A, s'_B, k)$, where $k = 1$ indicates semantic consistency between t'_A and s'_B, and 0 otherwise. For samples $(t'_A, s'_B, k) \in D_{train}^{classifier}$, we use the loss function shown in Equations (13). We directly train **the dynamic weight assignment model** on $D_{train}^{classifier}$ and obtain an average accuracy of 78.5% after three rounds of training.

$$L_a(t_A, s_B) = L_{CE}(\hat{t_B}, s_B) \tag{12}$$

$$L_k(t'_A, s'_B, k) = L_{CE}(\hat{k}, k) \tag{13}$$

During the process of knowledge distillation training on **the student model**, both the logical form A-generated teacher model and the student model generate

$\hat{t_A}$ and $\hat{s_B}$, respectively, based on the same input x. The dynamic weight assignment model then uses these outputs to obtain the dynamic weight value kd. The encoding layer vectors of the logical form A-generated teacher model and the pseudo-golden logical forms generated by the seq2seq model guide the training of the student model. Moreover, the student model learns from the labeled data in the dataset as a supervision signal. Finally, the weight value kd is utilized to combine the different loss values and obtain the final training loss.

3.2   Experimental Results

Overall Results. Table 1 displays the accuracy comparison on KQA Pro, where the reported values represent the average accuracy over three runs. Our proposed method surpasses all baselines, with an average improvement of 0.57% over the Bart KoPL baseline and 0.23% over the Bart SPARQL baseline. Particularly noteworthy is that our approach yields a 1.64% absolute accuracy improvement in the Compare category and a 1.58% absolute accuracy improvement in the Count category, demonstrating the advantages of knowledge distillation in broadening knowledge diversity.

Table 1. The results on the test set, and the evaluation metric uses accuracy. Mul, Qua, Com, Log, Cou, Ver, and Zer represent the seven problem categories of Multihop, Qualifier, Compare, Logical, Count, Verify, and Zero-shot, respectively.

Models		Overall	Mul	Qua	Com	Log	Cou	Ver	Zer
KVMemNet		16.61	16.50	18.47	1.17	14.99	27.31	54.70	0.06
EmbedKGQA		28.36	26.41	25.20	11.93	23.95	32.88	61.05	0.06
Blind GRU		34.36	33.25	28.82	25.77	34.17	39.43	62.15	0.06
RGCN		35.07	34.00	27.61	30.03	35.85	41.91	65.88	0.00
RNN SPARQL		41.98	36.01	19.04	66.98	37.74	50.26	58.84	26.08
RNN KoPL		43.85	37.71	22.19	65.90	47.45	50.04	42.13	34.96
Bart SPARQL		89.68	88.49	83.09	96.12	88.67	85.78	92.33	87.88
Bart KoPL		90.55	89.46	84.76	95.51	89.30	86.68	93.30	89.59
Ours	Bart SPARQL	**89.91**	**88.62**	82.63	96.07	**89.42**	**87.36**	**92.54**	**88.20**
	Bart KoPL	**91.12**	**90.06**	**85.79**	**97.15**	**89.63**	86.76	**93.51**	**90.04**

To further assess the efficacy of our method in improving semantic consistency, we evaluate the LFC metric for semantic consistency. As shown in Table 2, the baseline LFC is 86.93%. Our proposed method, using both static and dynamic knowledge distillation methods, results in LFC improvements of 0.76% and 0.82%, respectively. The only difference between dynamic and static knowledge distillation is that the three weight coefficients in Equations (7) are set as hyperparameters in the former and fixed during training. The experimental results of LFC provide further evidence of the effectiveness of our proposed approach.

Table 2. The results on the test set, and the evaluation metric uses LFC.

Models		Overall	Mul	Qua	Com	Log	Cou	Ver	Zer
Bart SPARQL/KoPL		86.93	63.24	18.74	17.25	24.20	9.21	11.16	11.23
Static	Bart SPARQL/KoPL	**87.69**	**63.80**	**19.02**	**17.32**	**24.43**	**9.34**	**11.20**	**11.34**
Dynamic	Bart SPARQL/KoPL	**87.75**	**63.91**	**19.00**	**17.33**	**24.51**	**9.42**	**11.21**	**11.38**

Ablation Study. In this section, we investigate the effectiveness of the dynamic weight assignment model through an ablation study. Specifically, we disabled the dynamic weight assignment model and refer to this variant as w/o *kd*.

The results presented in Table 3 reveal several key observations. Firstly, we observe that dynamic knowledge distillation leads to a higher improvement in performance compared to static knowledge distillation. Secondly, the weight setting of static knowledge distillation may result in suboptimal learning on unnecessary alignments and incur significant time and resource costs to achieve optimal model performance. These findings further validate the effectiveness of our proposed dynamic knowledge distillation framework and reinforce its superiority over the static counterpart.

Table 3. The results of ablation experiments, and the evaluation metric uses the accuracy.

Models		Overall	Mul	Qua	Com	Log	Cou	Ver	Zer
Ours	Bart SPARQL	**89.91**	**88.62**	82.63	96.07	**89.42**	**87.36**	**92.54**	**88.20**
	Bart KoPL	**91.12**	**90.06**	**85.79**	**97.15**	**89.63**	86.76	**93.51**	**90.04**
w/o *kd*	Bart SPARQL	89.88	88.58	**82.95**	**96.16**	89.12	86.91	92.20	87.94
	Bart KoPL	90.74	89.50	84.72	96.68	89.15	86.00	93.37	89.40

Comparison Between Different Guidance. Previous experiments have demonstrated the contribution of the teacher model to the improvement of the proposed method. In this section, we conduct a comparative analysis of the effects of three guidance modes provided by the teacher model. The reported accuracy values are averaged over three runs to counter the effect of randomness on the training process. The experimental results are presented in Table 4.

The following observations are derived from the results presented in Table 4:

(1) Logical form-based guidance: The results demonstrate that the logical form-based guidance is effective for KoPL, resulting in an average improvement of 0.54% and 0.34% in accuracy, respectively. This finding confirms the benefit of imitating the output distribution of the teacher model.

(2) Encoder-based guidance: The proposed method's accuracy still improves even when we switch to encoder-based guidance, with an average improvement of 0.19% and 0.57%, respectively. This result suggests that the intermediate layers contain useful information for distillation.

(3) Hybrid guidance: As a result of the differences between logical form-based and encoder-based guidance, the proposed hybrid guidance can achieve better but not optimal results, with an average improvement of 0.38% and 0.40%, respectively.

Table 4. The results of comparison experiments(Take Bart KoPL as an example), and the evaluation metric uses the accuracy.

Guidance		Overall	Mul	Qua	Com	Log	Cou	Ver	Zer
Bart KoPL		90.55	89.46	84.76	95.51	89.30	86.68	93.30	89.59
Static	Logical form-based	**91.09**	**89.94**	**85.71**	96.68	**89.72**	**86.91**	93.37	**89.78**
	Encoder-based	90.74	89.50	84.72	96.68	89.15	86.00	93.37	89.40
	Hybrid	90.93	89.76	85.11	**97.15**	89.33	86.46	**93.51**	89.66
Dynamic	Logical form-based	90.89	89.82	**85.82**	96.96	**89.72**	86.91	93.16	88.77
	Encoder-based	**91.12**	**90.06**	85.79	**97.15**	89.63	86.76	**93.51**	90.04
	Hybrid	90.95	89.92	85.32	96.96	89.24	**87.06**	93.09	89.78

Compositional Generalization. To investigate pre-trained language models' ability in compositional generalization, we adopt the dataset partitioning method proposed by Cao et al. [2] in KQA Pro. We use this dataset as the training, validation, and test sets, containing 106,182, 5,899, and 5,899 examples, respectively. In their work, the performance of BART KoPL drops from 90.55% to 77.86%. However, our proposed method achieves 81.88% accuracy on this dataset, resulting in a 4.02% absolute improvement over their baseline. This result shows that our proposed method is capable of leveraging valuable information hidden in other logical forms, resulting in further improvements in the model's compositional generalization ability.

Case Study. In this section, we present a case study to verify the effectiveness of our method. Specifically, we examine two cases, as illustrated in Table 5:

(1) For the question *"Who is shorter: Thomas Jefferson, signer of the U.S. Declaration of Independence, or Salma Hayek?"*, the KoPL parser correctly predicts the answer, while the SPARQL parser predicts incorrectly. By applying knowledge distillation, the KoPL program can provide additional supervisory signals to rectify the mispredicted parameters of the SPARQL parser.

(2) For the question *"What is the Giphy handle of the prime subject of Steve Jobs, featuring Michael Fassbender as a cast member?"*, the SPARQL parser gives correct output while the KoPL parser predicts incorrectly. Our findings reveal that the KoPL parser wrongly predict human concepts as activity concepts. Under the appropriate guidance of the SPARQL query, the KoPL parser correct the error.

These two cases demonstrate that our dynamic knowledge distillation framework provides constructive supervision signals to enhance the student model at intermediate stages.

4   Related Work

In this subsection, we present related work.

Table 5. Case studies for two different scenarios. Only the semantically inconsistent parts are shown, with green color indicating correct predictions, red color indicating incorrect predictions, and blue color indicating corrected predictions.

Natural Language Question		Before KD		After KD	
		KoPL	*SPARQL*	*KoPL*	*SPARQL*
		...	...		...
Case One	Who is shorter: Thomas Jefferson, signer of the U.S. Declaration of Independence, or Salma Hayek?	*Find* United States Declaration of Independence	?e_1<famous_people>?e.	*Find* United States Declaration of Independence	?e_1<signat_ory>?e.
		...	...	...	...
		...	...	...	...
Case Two	What is the Giphy handle of the prime subject of Steve Jobs, featuring Michael Fassbender as a cast member?	*FilterConcept* activity	?c<pred:name> "human".	*FilterConcept* human	?c<pred:name> "human".
		...	...	...	...

Semantic Parsing. SP-based methods convert natural language questions into logical forms, which are then processed by an execution engine to obtain answers. Several approaches have been proposed to improve the accuracy and efficiency of this process. For instance, [19] uses an LSTM-based seq2seq model to handle this task and proposes a seq2tree model to consider its structural features better. Similarly, [22] demonstrates that general sequence encoders ignore useful syntactic information and argues that using syntactic graphs and a graph2seq model allows the model to learn more syntactic information. [23] proposes a novel framework for constructing a unified multi-domain semantic parser with a multi-strategy distillation mechanism to assist weakly supervised training. [24] proposes a two-stage semantic parsing framework to alleviate the shortage of dataset annotations. [25] proposes an approach based on graph isomorphism to answer questions in small and medium-sized knowledge graphs. Additionally, other works, such as [26,27], investigate combinatorial generalization in semantic parsing. However, these methods do not consider the impact of different logical forms.

Knowledge Distillation. Knowledge distillation is not only the key to making large pre-trained models practical but has also shown vast promise in solving downstream tasks using end-to-end models. Recently, researchers have proposed several new techniques to improve the performance of knowledge distillation. For instance, [28] presents a novel text-based adversarial training algorithm, MATE-KD, that enhances the performance of it. [29] proposes a unified multilingual sequence tagging model. Several works have emphasized the importance of student models imitating intermediate features of teacher models [30,31]. [32] introduces a novel cross-modal contrastive distillation (CCD) scheme to address the gap between visual and text modalities in the distillation process. In addition to these static knowledge distillation methods, researchers have also investigated dynamic knowledge distillation. [33] discusses the influence of dynamic knowledge distillation on students' learning ability, while [34] introduces a new framework named Iterative Self Distillation for QA (ISD-QA), which leverages the "dark knowledge" embedded in large pre-trained models to provide supervision for commonsense question answering. Our work is inspired by these studies.

Consistency. *In abstract generation tasks*, the issue of fact consistency in summaries has been addressed by previous studies [8,9]. *In dialogue tasks*, the challenge of consistency is to solve the dialogue dependency problem. [7] proposes a novel framework ExCorD, and [10] introduces a new dataset CI-ToD for consistency identification. *In VQA tasks*, models capable of answering seemingly difficult questions but failing in simpler related subquestions highlight the issue of inconsistency. To address this, [11] proposes a contrasting gradient learning-based approach, while [12] proposes region representation scaling to improve model consistency. *In MT tasks*, [13] encourages linguistic translation consistency in document-level machine translation. Furthermore, the problem of consistency extends to other domains, such as semantic image segmentation [35] and natural language annotations [36]. In this work, we aim to address the issue of semantic consistency between different logical forms in semantic parsing using a well-designed dynamic teacher-student framework.

5 Conclusion and Future Work

In this paper, we propose DKD-SP, a dynamic knowledge distillation framework designed to enhance the consistency of various logical forms in the semantic parsing of KBQA. In our proposed framework, the student model is responsible for the semantic parsing task, while the teacher model provides supplementary supervision signals to improve the student model's performance. To facilitate knowledge transfer, the student model adopts three guidance methods. Furthermore, we introduce a dynamic weight assignment model that enables the student model to adjust its learning process according to its abilities.

We evaluate our proposed method using the KQA Pro dataset, and the results indicate that our framework outperforms the baseline methods regarding its effectiveness for the KBQA semantic parsing task. Currently, we employ the Bart model as the student model, but our framework is flexible enough to support other neural architectures. In the future, we aim to extend our method by integrating other logical forms, such as λ-DCS, to gain additional supervision signals. And we will explore the potential benefits of utilizing multiple teacher models and investigate better methods for dynamic knowledge distillation.

Acknowledgement. This work is supported by the National Key R&D Program of China (2020AAA0105203).

References

1. Luo, Y., Yang, B., Xu, D., et al.: A survey: complex knowledge base question answering. In: Proceedings of ICICSE, pp. 46–52. IEEE (2022)
2. Cao, S., Shi, J., Pan, L., et al.: KQA Pro: a dataset with explicit compositional programs for complex question answering over knowledge base. In: Proceedings of ACL, pp. 6101–6119. ACL (2022)
3. Sun, Y., Zhang, L., Cheng, G., et al.: SPARQA: skeleton-based semantic parsing for complex questions over knowledge bases. In: Proceedings of AAAI, New York, USA, pp. 8952–8959. AAAI (2020)
4. Liang, P., Jordan, M.I., Klein, D.: Lambda dependency-based compositional semantics. In: Proceedings of CCL, Suzhou, China, pp. 389–446. ACL (2013)
5. Zhong, V., Xiong, C., Socher, R.: Seq2sql: generating structured queries from natural language using reinforcement learning. arXiv preprint arXiv:1709.00103 (2017)
6. Veyseh, A., Dernoncourt, F., Dou, D., et al.: A joint model for definition extraction with syntactic connection and semantic consistency. In: Proceedings of AAAI, New York, USA, pp. 9098–9105. AAAI (2020)
7. Gangwoo, K., Hyunjae, K., Jungsoo, P., Jaewoo, K.: Learn to resolve conversational dependency: a consistency training framework for conversational question answering. In: Proceedings of IJCNLP, Bangkok, Thailand, pp. 6130–6141. ACL (2021)
8. Wang, A., Cho, K., Lewis, M.: Asking and answering questions to evaluate the factual consistency of summaries. In: Proceedings of ACL, Seattle, Washington, United States, pp. 5008–5020. ACL (2020)
9. Nan, F., et al.: Improving factual consistency of abstractive summarization via question answering. In: Proceedings of IJCNLP, Bangkok, Thailand, pp. 6881–6894. ACL (2021)
10. Qin, L., Xie, T., Huang, S., Chen, Q., Xu, X., Che, W.: Don't be contradicted with anything! CI-ToD: towards benchmarking consistency for task-oriented dialogue system. In: Proceedings of EMNLP, Punta Cana, Dominican Republic, pp. 2357–2367. ACL (2021)
11. Dharur, S., Tendulkar, P., Batra, D., Parikh, D., Selvaraju, R.: SOrT-ing VQA models: contrastive gradient learning for improved consistency. In: Proceedings of NAACL, Mexico City, pp. 3103–3111. ACL (2021)
12. Yang, S., Zhou, Q., Feng, D., et al.: Diversity and consistency: exploring visual question-answer pair generation. In: Proceedings of EMNLP, Punta Cana, Dominican Republic, pp. 1053–1066. ACL (2021)
13. Lyu, X., Li, J., Gong, Z., et al.: Encouraging lexical translation consistency for document-level neural machine translation. In: Proceedings of EMNLP, Punta Cana, Dominican Republic, pp. 3265–3277 (2021)
14. Gou, J., Yu, B., Maybank, S.J., et al.: Knowledge distillation: a survey. IJCV **129**(6), 1789–1819 (2021)

15. Miller, A.H., Fisch, A., Dodge, J., Karimi, A.H., Bordes, A., Weston, J.: Key-value memory networks for directly reading documents. In: Proceedings of EMNLP, Austin, Texas, pp. 1400–1409. ACL (2016)
16. Saxena, A., Tripathi, A., Talukdar, P.: Improving multi-hop question answering over knowledge graphs using knowledge base embeddings. In: Proceedings of ACL, Seattle, Washington, United States, pp. 4498–4507. ACL (2020)
17. Schlichtkrull, M., Kipf, T.N., Bloem, P., van den Berg, R., Titov, I., Welling, M.: Modeling relational data with graph convolutional networks. In: Gangemi, A., et al. (eds.) ESWC 2018. LNCS, vol. 10843, pp. 593–607. Springer, Cham (2018). https://doi.org/10.1007/978-3-319-93417-4_38
18. Dey R., Salem F.M.: Gate-variants of gated recurrent unit (GRU) neural networks. In: Proceedings of MWSCAS, Boston, USA, pp. 1597-1600. IEEE (2017)
19. Dong, L., Lapata, M.: Language to logical form with neural attention. In: Proceedings of ACL, Berlin, Germany, pp. 33–43. ACL (2016)
20. Lewis, M., et al.: Bart: denoising sequence-to-sequence pre-training for natural language generation, translation, and comprehension. In: Proceedings of ACL, Seattle, Washington, United States, pp. 7871–7880. ACL (2020)
21. Kingma, D.P., Ba, J.: Adam: a method for stochastic optimization. In: Proceedings of ICLR, San Diego, USA (2015)
22. Xu, K., Wu, L., Wang, Z., Yu, M., Chen, L., Sheinin, V.: Exploiting rich syntactic information for semantic parsing with graphtosequence model. In: Proceedings of EMNLP, Brussels, Belgium, pp. 918–924. ACL (2018)
23. Agrawal, P., Dalmia, A., Jain, P., Bansal, A., Mittal, A., Sankaranarayanan, K.: Unified semantic parsing with weak supervision. In: Proceedings of ACL, Florence, Italy, pp. 4801–4810. ACL (2019)
24. Cao, R., et al.: Unsupervised dual paraphrasing for two-stage semantic parsing. In: Proceedings of ACL, Seattle, Washington, United States, pp. 6806–6817. ACL (2020)
25. Aghaei, S., Raad, E., Fensel, A.: Question answering over knowledge graphs: a case study in tourism. IEEE Access 10, 69788–69801 (2022)
26. Oren, I., Herzig, J., Gupta, N., Gardner, M., Berant, J.: Improving compositional generalization in semantic parsing. In: Proceedings of EMNLP, Punta Cana, Dominican Republic, pp. 2482–2495. ACL (2020)
27. Lukovnikov, D., Daubener, S., Fischer, A.: Detecting compositionally out-of-distribution examples in semantic parsing. In: Proceedings of EMNLP, Punta Cana, Dominican Republic, pp. 591–598. ACL (2021)
28. Rashid, A., Lioutas, V., Rezagholizadeh, M.: Mate-kd: masked adversarial text, a companion to knowledge distillation. In: Proceedings of IJCNLP, Bangkok, Thailand, pp. 1062–1071. ACL (2021)
29. Wang, X., Jiang, Y., Bach, N., Wang, T., Huang, F., Tu, K.: Structure-level knowledge distillation for multilingual sequence labeling. In: Proceedings of ACL, Seattle, Washington, United States, pp. 3317–3330. ACL (2020)
30. Aguilar, G., Ling, Y., Zhang, Y., et al.: Knowledge distillation from internal representations. In: Proceedings of AAAI, New York, USA, pp. 7350–7357. AAAI (2020)
31. Mirzadeh, S.I., Farajtabar, M., Li, A., Levine, N., Matsukawa, A., Ghasemzadeh, H.: Improved knowledge distillation via teacher assistant. In: Proceedings of AAAI, New York, USA, pp. 5191–5198. AAAI (2020)
32. Yang, Z., Liu, J., Huang, J., et al.: Cross-modal contrastive distillation for instructional activity anticipation. In: Proceedings of ICPR, Montreal, QC, Canada, pp. 5002–5009. IEEE (2022)

33. Li, L., Lin, Y., Ren, S., Li, P., Zhou, J., Sun, X.: Dynamic knowledge distillation for pre-trained language models. In: Proceedings of EMNLP, Punta Cana, Dominican Republic, pp. 379–389. ACL (2021)
34. Ramamurthy, P., Aakur, S.N.: ISD-QA: iterative distillation of commonsense knowledge from general language models for unsupervised question answering. In: Proceedings of ICPR, Montreal, QC, Canada, pp. 1229–1235. IEEE (2022)
35. Luo, X., Chen, J., Song, T., et al.: Semi-supervised medical image segmentation through dual-task consistency. In: Proceedings of AAAI, pp. 8801–8809. AAAI (2021)
36. Panthaplackel, S., Li, J.J., Gligoric, M., et al.: Deep just-in-time inconsistency detection between comments and source code. In: Proceedings of AAAI, pp. 427–435. AAAI (2021)

DYGL: A Unified Benchmark and Library for Dynamic Graph

Teng Ma[1,2], Bin Shi[1,2(✉)], Yiming Xu[1,2], Zihan Zhao[1,2], Siqi Liang[4], and Bo Dong[2,3]

[1] School of Computer Science and Technology, Xi'an Jiaotong University, Xi'an, China
{mateng0920,xym0924,zzh819}@stu.xjtu.edu.cn
[2] Shaanxi Provincial Key Laboratory of Big Data Knowledge Engineering, Xi'an Jiaotong University, Xi'an, China
{shibin,dong.bo}@xjtu.edu.cn
[3] School of Distance Education, Xi'an Jiaotong University, Xi'an, China
[4] PricewaterhouseCoopers, Singapore, Singapore
1873186749@qq.com

Abstract. Difficulty in reproducing the code and inconsistent experimental methods hinder the development of the dynamic network field. We present DYGL, a unified, comprehensive, and extensible library for dynamic graph representation learning. The main goal of the library is to make dynamic graph representation learning available for researchers in a unified easy-to-use framework. To accelerate the development of new models, we design unified model interfaces based on unified data formats, which effectively encapsulate the details of the implementation. Experiments demonstrate the predictive performance of the models implemented in the library on node classification and link prediction. Our library will contribute to the standardization and reproducibility in the field of the dynamic graph. The project is released at the link: https://github.com/half-salve/DYGL-lib

Keywords: dynamic graph · Reproducibility · deep learning · Library

1 Introduction

The bulk of network science literature focuses on static networks, yet every network existing in the real world changes over time. Dynamic networks add new dimensions to network modeling and edge prediction. This new dimension fundamentally affects network properties, enabling more powerful representations of network data and in turn improving the predictive power of methods using such data.

In general, dynamic graphs can be broadly classified into two categories: Continuous-Time Dynamic Graph (CTDG) and Discrete-Time Dynamic Graph (DTDG). Given an initial graph G_0, a CTDG describes a continuous graph evolving process which can be formulated as sequential events $\{Event_1, \cdots, Event_k\}$, where k is the number of graph events. Specifically, $Event_i = (type_i, e_i)$ denotes

X. Song et al. (Eds.): APWeb-WAIM 2023, LNCS 14333, pp. 389–401, 2024.
https://doi.org/10.1007/978-981-97-2387-4_26

an edge operation of $type_i$ conducting on edge $e_i \in E_i$ at time i. The types of edge operations are restricted as *InsertEdge* and *DeleteEdge* [13]. A DTDG is a sequence of static graph snapshots taken at the time intervals, represented as $S = \{G_0, \cdots, G_{|S|}\}$. A graph snapshot G_i in DTDG can be viewed as a graph captured over a period of time in CTDG, while the detailed evolution process (e.g. $\{Event_1, \cdots Event_k\}$) is omitted.

When the existing dynamic graph model for CTDG [10] is compared with the model for DTDG [3,6], it is often necessary to convert the CTDG data set into the form of DTDG data set for comparison. However, the model for CTDG is only compared with the performance of the model that supports CTDG. In TGN [7], the author selected 11 models as the baseline for the edge prediction task. Among them, there are four models for CTDG, and the remaining models are static graph models. In TGAT [11], the author selected 9 models as the baseline for the edge prediction task. Among them, there is one model for CTDG, and the remaining models are static graph models.

Coincidentally, similar things also happened in the experiments conducted by the DTDG model. The existing DTDG-based models [1,2,6] are more inclined to compare with snapshot-based models when comparing performance. The contrast between the CTDG model and the DTDG model in the experiment is insufficient, so that we cannot see the advantages and disadvantages of different models in different application scenarios. As a result, we cannot better absorb the advantages of the two types of models to develop new models with stronger applicability and better generalization, which hinders the development of this field.

Based on the above questions, we ask the question: (Q.1) Can we create a unified data structure so that it is compatible with CTDG and DTDG so that we can easily convert between the two data sets?

There has been a lot of research on dynamic graphs in node classification, link prediction, and anomaly detection, but we have investigated the literature and found that there is no open source library to unify the entire pipeline including data preparation, model design and implementation, and performance evaluation. Furthermore, as more and more dynamic graph representation learning models are proposed, we find that they are often implemented in different frameworks and contexts, making it difficult to reproduce the results of these methods in a unified way. In particular, The variety and format of the data sets are not uniform. It is increasingly difficult to guarantee the effectiveness of new dynamic graph representation learning and imitation, and it is also increasingly difficult to conduct fair evaluation. So we propose a second challenge: (Q.2) Can we use a unified process, unified indicators and standard data sets to evaluate model performance?

To this end, we propose a unified, scalable and comprehensive dynamic graph library DYGL. The algorithm library is implemented based on PyTorch, and we consider node classification and link prediction. We provide various datasets, mechanisms, models, and utilities to support data preprocessing, model instantiation, and performance evaluation for tasks.

Our contributions. The main contributions of our work can be summarized as:

- We publicly release DYGL, a deep learning library for dynamic graph representation learning models.
- We introduced a unified data structure that is compatible with CTDG and DTDG so that we can easily convert between the two data sets.
- We provide dynamic graph datasets commonly used in finance, paper citations, forum comments, and encyclopedic knowledge fields, and provide related data loaders and iterators based on DYGL.
- We formulate unified experimental and evaluation metrics to evaluate the performance of representation learning models included in the algorithm library on the dataset.

The rest of this article has the following structure. In Sect. 2, we outline the structure of the algorithm library, including the data structure, the main process and the display of various modules. In Sect. 3, we introduce the dataset used and design experiments to evaluate the models covered by the algorithm library. In Sect. 4, we summarize the full text and put forward our own thoughts on some doubts in the original paper.

2   The Framework Design

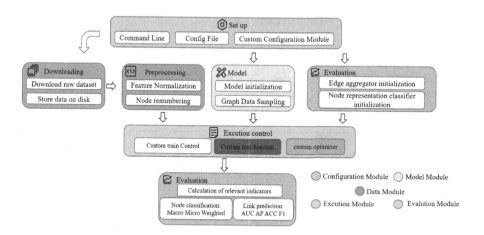

Fig. 1. Overview of the Library

The overall framework of Library is presented in Fig. 1, consisting of five major modules. Our primary goal is to give a general theoretical overview of the framework, discuss the framework design choices, give a detailed practical example and highlight our strategy for the long-term viability and maintenance of the project.

- Configuration Module: Responsible for managing all parameters involved in the framework.
- Data Module: Responsible for downloading datasets, preprocessing datasets, saving and loading datasets.
- Model Module: Responsible for initializing the baseline model or custom model.
- Evaluation Module: Provides a unified downstream task evaluation model, and evaluates algorithm performance through multiple indicators.
- Execution Module: The user defines the loss function, selects the optimizer and the specific process of training.

2.1 Dataset Structures

The library's design required the introduction of custom data structures that efficiently store datasets and provide time-ordered data to models. Not only supports the conversion of CTDG and DTDG, but also can be effectively applied to static graph models.

Algorithm 1. Basic data processing flow.

Input: the name of the dataset $data\_name$, dataset download path url, raw data storage folder raw_dir, the folder where the processed data set is stored $save\_dir$, check if dataset already exists $hash\_key()$, whether to reprocess the data $force\_reload$

Output:

1: $load\_flag \leftarrow !force\_reload$ and $has\_cashe()$
2: **if** $load\_flag$ **then**
3:     LOAD($save\_dir$)
4: **else**
5:     DOWNLOAD($raw\_dir, url, data\_name$)
6:     PROCESS($raw\_dir, url, data\_name$)
7:     SAVE($save\_dir, data\_name$)
8: **end if**

Data Download and Preprocessing. The following will introduce the process of processing data at the basic level of the data set in Algorithm 1. If $load\_flag$ is true or the return value of $has\_cashe()$ is false then we judge that the data needs to be downloaded and processed again(lines 1). Divide the entire data processing process into three steps: downloading the original data(lines 5), processing the original data(lines 6) and saving the processed data set(lines 7). DYGL only plans the specific implementation of the process and we reserve it for user customization.

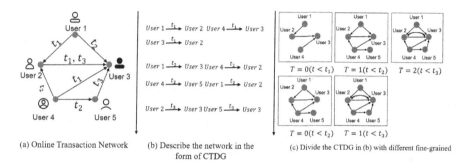

(a) Online Transaction Network   (b) Describe the network in the form of CTDG   (c) Divide the CTDG in (b) with different fine-grained

Fig. 2. An example of temporal interaction networks. (a) An online transaction network with five users. (b) Describe the network in the form of CTDG. (c) Divide the CTDG in (b) with different fine-grained

Dataset Compatibility. Our data set adopts the storage form of *dgl.DGLGraph*, and the DGL library provides algorithms related to static graph. Using the storage structure of *dgl.DGLGraph* makes DYGL suitable for both dynamic graph and static graph. We also provide a **converter**, as long as the user processes it into the format of *dgl.DGLGraph* according to the method in Sect. 2.1, the converter can easily convert data between CTDG and DTDG.

When we convert CTDG to DTDG, because fine-grained selection makes slices have different results, For example, in Fig. 2, if the resolution is too fine, a large number of snapshots will bring intractable computational cost when training deep generative models; if the resolution is too coarse, fine-grained temporal context information (e.g., addition/deletion of nodes and edges) may be lost during temporal aggregation [14]. So we introduced a **manual setting parameter** in the converter, which is convenient for users to adjust the size of slices according to their own needs.

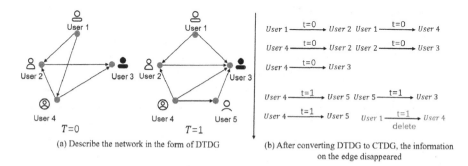

(a) Describe the network in the form of DTDG   (b) After converting DTDG to CTDG, the information on the edge disappeared

Fig. 3. An example of converting DTDG to CTDG. (a) Describe the network in the form of DTDG. (b) After converting DTDG to CTDG, the information on the edge disappeared

2.2 Data Normalization Initiative

When we convert DTDG data into CTDG data, as shown in Fig. 3, the time information on the edge in the dynamic graph disappears. We only know that this edge comes from a certain time slice but not the specific time point when it appears. When performing performance comparisons, models that need to use time-specific information for training cannot perform well. Therefore, we call for the data sets to be processed in the following format, so that more information about the original data can be retained. Adapt the dataset to more models. We give some attribute content descriptions and calling methods of some datasets in Table 1.

Table 1. Dataset struction: It's structure is a key-value data structure (dict) based on python. where the keys are the feature names and the values are the corresponding feature tensors in the data

member variable	Function description
edata["time"]	Timestamps on edges sorted by time
edata["train_edge_mask"]	positive training edge mask for Transductive task
edata["valid_edge_mask"]	positive validation edge mask for Transductive task
edata["test_edge_mask"]	positive testing edge mask for Transductive task
edata["train_edge_observed_mask"]	positive training edge mask for Inductive task
edata["valid_edge_observed_mask"]	positive validation edge mask for Inductive task
edata["test_edge_observed_mask"]	positive testing edge mask for Inductive task
ndata["feat"]	Node features
edata["edge_feat"]	edge features

2.3 Comprehensive Models

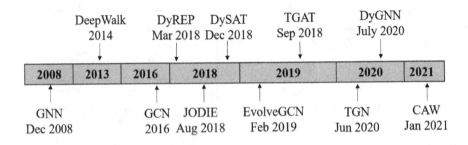

Fig. 4. GNN development time

At present, many models have been proposed for dynamic graphs, but the codes provided by existing papers have the problems of arbitrary writing and complex calling logic. We hope that in the process of using graph representation learning, the representation learning model, loss function, and downstream tasks can be independent of each other. These are available in experiments, better evaluating the model itself free from other disturbances. We have also written a corresponding initialization example for the representation model, which is convenient for users to quickly initialize the Model. We conduct a comprehensive review of papers published in the recent five years (2016–2021) on top-tier conferences and journals, such as AAAI, WSDM, KDD, SIGIR, ICML, ICLR. Through Fig. 4, we show the launch time of these representation learning models, which is convenient for users to use better.

2.4 Fast Initialization and Unified Process

The overall framework of DYGL has been given in Fig. 1, we need to initialize the Configuration Module, Data Module, Model Module, and Evaluation Module through the functions and classes provided in DYGL first, and then it is particularly important to point out that we separate the entire graph representation learning module from the downstream tasks Come out, users can initialize the corresponding content from Evaluation Module according to the tasks they want to perform, and then conduct experiments.

Algorithm 2. The flow of the entire algorithm library for link prediction task.

Input: the name of the dataset *dataset*, the name of the model *model*, the name of the task *task*

Output:

1: $config \leftarrow$ LIB.CONFIGPARSER($[task, dataset, model]$)
2: $data \leftarrow$ LIB.GET_DATA($config['dataset']$)
3: $model \leftarrow$ LIB.GET_MODEL($config, data$)
4: $feat\_dim \leftarrow model.feat\_dim$
5: $eval \leftarrow$ LIB.EVALUATOR.MERGELAYER $(feat\_dim, 1)$
6: $node\_embeddings \leftarrow$ EXECUTOR($config, data, model$)
7: $prob \leftarrow$ EVAL($node\_embeddings$)
8: **function** EXECUTOR($config, data, model$)
9:      The user defines how to train and we just give the specification on the process
10:     **return** $node\_embeddings$
11: **end function**

In order to use DYGL to start training, an example is given for link prediction task, see Algorithm 2 After specifying the *task* to be performed, the *model* and *dataset* to be used in the input, use *lib.ConfigParser* to initialize the parameter manager Configuration Module(line 1), and then initialize each module in turn(lines 2–5). We have not defined the specific training process *Executor*. Users can choose the training method or use the **example** given in the library. This part only needs to give the node embedding required by the edge aggregation module. Finally, use the unified edge aggregation module to calculate various indicators to evaluate the performance of the model(line 7).

3   Experiments with Library

We formulate unified experimental and evaluation metrics, selecting the models and datasets included in DYGL, to evaluate the performance of the included representation learning models.

3.1   Compared Methods

In the following we briefly introduce the selected representation learning models and give their summary in Table 2.

Table 2. A brief description of the models included in the algorithm library.

Model	conferences	Layer	Use dataset type
VGRNN [3]	WSDM	Attention	DTDG
EvolveGCN [6]	AAAI	GCN+LSTM	DTDG
Jodie [5]	KDD	RNN	CTDG
TGAT [11]	ICLR	Attention	CTDG
DyRep [8]	ICLR	RNN	CTDG
CAW [10]	ICLR	Attention	CTDG
TGN [7]	ICML	GRU+Attention	CTDG

- **EvolveGCN** [6] uses a RNN to estimate the GCN parameters for the future snapshots.
- **Jodie** [5] applies RNNs to estimate the future embedding of nodes. The model was proposed for bipartite graphs while we properly modify it for standard graphs if the input graphs are non-bipartite.
- **TGAT** [11] adapts the self-attention mechanism to handle the continuous time by proposing a time encoding, capturing temporal-feature signals in terms of both node and topological features on temporal graphs.
- **DyRep** [8] uses RNNs to learn node embedding while its loss function is built upon temporal point process.
- **CAW** [10] implicitly extracts network motifs via temporal random walks and adopts set-based anonymization to establish the correlation between network motifs.
- **TGN** uses RNNs to learn node embedding while its loss function is built upon temporal point process.
- **VGRNN** [3] generalizes the variational GAE [4] to temporal graphs, which makes the prior depend on the historical dynamics and captures those dynamics with RNNs.

3.2   Datasets

We release new dynamic graph datasets which can be used to test models on link prediction.

- **Reddit** [5]: A dataset tracking active users posting in subreddits. Data is represented as a bipartite graph with nodes being the users or subreddit communities. An edge represents a user posting on a subreddit. Each user's post is mapped to an embedding vector which is used as an edge feature. A Dynamic label indicates whether a user u is banned from posting after an interaction (post) at time t.
- **Wikipedia** [5]: A dataset tracking user edits on Wikipedia pages. The data is also represented as a bipartite graph involving interactions (edits) between users and Wikipedia pages. Each user edit is mapped to an embedding vector which is treated as an edge feature. A dynamic label indicates whether a user u is banned from posting after an interaction (edit) at time t.
- **LastFM** [5]: This dataset tracks songs that users listen to throughout one month. Nodes represent users or songs. Dynamic labels are not present.
- **MOOC** [5]: This dataset tracks actions performed by students on the MOOC online course platform. Nodes represent students or items (i.e., videos, questions, etc.). A dynamic label indicates whether a student u drops out after performing an action at time t.
- **UCI** [10]: This dataset recording online posts made by university students on a forum, but is non-attributed.
- **Brain** [12] the nodes represent the tidy cubes of brain tissue and the edges indicate the connectivity. We apply PCA to the functional magnetic resonance imaging data to generate node attributes. Two nodes are connected if they show similar degree of activation.

Table 3. Data statistics for the datasets

Dataset	# Nodes	# Edges	# Feature dimension	# Time Steps	# Classes
Reddit	11,000	672,447	172(Edges)	30 days/174	2
Wikipedia	9,227	157,474	172(Edges)	30 days/20	2
MOOC	7,144	411,749	–	–	–
LastFM	1,980	1,293,103	–	–	–
UCI	1,899	59.835	–	194 days	–
Brain	5000	1,955,488	100(Nodes)	12	10

Due to missing features in the dataset, we reported the existing feature conditions in Table 3. For cases where the model requires certain features but they are missing, we filled the missing values with positional encoding [9] or all zeros. The filling dimension was set to 128.

3.3 Experimental Setting

The experiments focus on the evaluation of the dynamic graph neural networks implemented in our framework. We compare performance under two downstream tasks.

Evaluation Tasks: Two types of tasks are for evaluation: transductive and inductive link prediction.

Transductive link prediction task allows temporal links between all nodes to be observed up to a time point during the training phase, and uses all the remaining links after that time point for testing. In our implementation, we split the total time range $[0, T]$ into three intervals: $[0, T_{train})$, $[T_{train}, T_{val})$, $[T_{val}, T]$. Links occurring within each interval are dedicated to the training, validation and test sets respectively. For all datasets, we fix $T_{train}/T = 0.7$, and $T_{val}/T = 0.85$.

Inductive link prediction task predicts links associated with nodes that are not observed in the training set. In practice, we follow two steps to split the data: 1) we use the same setting of the transductive task to first split the links chronologically into training/validation/testing sets; 2) we randomly select 10% nodes, remove any links associated with them from the training set, and remove any links not associated with them in the validation and testing sets.

Following most baselines, we randomly sample an equal amount of negative links and consider link prediction as a binary classification problem.

3.4 Results and Discussion

We divided the process of comparing the model performance into two steps. In the first step, we compared the performance of the DTDG model on a unified dataset. In the second step, we compared the performance of the CTDG model on the same dataset. The results evaluating the performance of the DTDG model using the AUC and AP metrics are reported in Table 5, while the results evaluating the performance of the CTDG model using the AUC metric are reported in Table 6.

Table 4. Snapshot split for evaluating snapshot-based baselines

	Reddit	Wikipedia	MOOC	LastFM	UCI	Brain
total snapshots	174	20	20	20	88	12
exact split	122/26/26	14/3/3	14/3/3	14/3/3	62/13/13	8/2/2
referenced baseline	EvolveGCN [6]	CAW [10]	CAW	-	EvolveGCN	STAR [12]

How CTDG Performe in DTDG Model. We make the following decision to evaluate snapshot-based baselines in a fair manner so that their performance is comparable to that derived from flow-based evaluation procedures. The first step

Table 5. Transductive task and Inductive task results for predicting future edges of nodes that have been observed in training data. All results are converted to percentage by multiplying by 100.

Task	Metric	Methods	Reddit	Wikipedia	MOOC	LastFM	UCI	Brain
Inductive	AUC	EvolveGCN	65.61	56.29	50.20	60.23	70.78	65.78
		VGRNN	54.11	62.93	60.10	61.23	62.39	64.52
	AP	EvolveGCN	66.29	53.82	51.53	52.62	76.30	74.12
		VGRNN	52.84	60.99	62.95	67.83	67.50	76.30
Transductive	AUC	EvolveGCN	58.42	60.48	50.36	61.53	78.30	60.36
		VGRNN	51.89	71.20	90.03	75.62	89.43	78.28
	AP	EvolveGCN	54.49	55.84	51.80	70.23	81.63	69.72
		VGRNN	50.87	67.66	83.70	80.24	82.23	78.66

we do is to evenly divide the entire dataset into many snapshots in chronological order. We determine the exact number of snapshots by referring to how many papers are split. For the Wikipedia, LastFM and MOOC datasets not used by any snapshot-based baselines, we split them into a total of 20 snapshots. Next, we need to determine the proportion of these snapshots to allocate to the training, validation, and test sets. In doing so, our guideline is that the ratio of these three sets should be as close as possible to 70:15:15, as this ratio is the one we use to evaluate flow-based baselines and our proposed method. These decisions led to our final split proposal summarized in the Table 4.

How DTDG Performe in CTDG Model. We expand Brain from graphs to interation, and describe their details in the Table 3. From Table 3, we can see that Brain has 12 slices. After converting it to CTDG, if an edge originates from G_i, then the timestamp on that edge is set to i.

3.5 Results

- Continuous network models outperform discrete network models on link prediction tasks.
- TGN is a model proposed in 2020, but its performance is inferior to earlier proposed models such as TGAT and Jodie in some indicators. When the dataset and task are unified, the latter model does not necessarily perform better than the first proposed model on the task.
- From the comparison of Table 6 and Table 5, it can be seen that the continuous graph model is generally better than the discrete graph model. TGN, CAW, and TGAT are continuous network models that outperform Jodie and DyRep, which are also continuous models, on more datasets. The same point is that TGN, CAW and TGAT are all based on graph self-attention mechanism. We speculate that self-attention mechanisms may be the way forward.

Table 6. Performance in Area Under Curve(AUC). All results are converted to percentage by multiplying by 100

Task	Methods	Reddit	wikipedia	MOOC	LastFM	UCI	Brain
Inductive	JODIE	97.85	93.87	87.88	88.79	67.12	75.24
	DyRep	97.42	93.65	84.35	99.71	62.91	77.63
	TGN	98.21	93.52	76.98	99.98	66.01	76.54
	TGAT	98.39	91.29	87.63	91.95	72.14	80.36
	CAW	99.45	99.57	73.91	99.99	79.25	82.41
Transductive	JODIE	97.89	94.35	82.44	68.24	72.18	79.37
	DyRep	96.64	93.30	74.32	58.11	68.96	79.24
	TGN	98.19	95.45	76.45	63.38	71.56	81.50
	TGAT	98.39	93.83	82.56	80.24	91.60	82.45
	CAW	99.20	99.31	77.98	85.42	92.21	89.40

4 Conclusion

This paper proposes a unified, comprehensive, and extensible dynamic library called DYGL. It collects commonly used datasets and state-of-the-art dynamic graph representation learning models, and designs two downstream tasks. We introduced its overall framework, the base class of the constructed dataset, and called on everyone to process the original dataset in the same way. We conducted experiments in a unified way, and set and selected an appropriate baseline. In future work, we will continue to extend the library in multiple ways: to cover more downstream tasks, more efficient GPU-accelerated execution, and continuous replication of state-of-the-art models.

Acknowledgements. This research was partially supported by the National Key Research and Development Project of China No. 2021ZD0110700, the Key Research and Development Project in Shaanxi Province No. 2022GXLH-01-03, the National Science Foundation of China under Grant Nos. 62002282, 62037001, 62250009 and 61721002, the Major Technological Innovation Project of Hangzhou No. 2022AIZD0113, the "Pioneer" and "Leading Goose" R&D Program of Zhejiang No. 2022C01107, the China Postdoctoral Science Foundation No. 2020M683492, the MOE Innovation Research Team No. IRT_17R86, and Project of XJTU-SERVYOU Joint Tax-AI Lab.

References

1. Goyal, P., Chhetri, S.R., Canedo, A.: dyngraph2vec: capturing network dynamics using dynamic graph representation learning. CoRR abs/1809.02657 (2018). http://arxiv.org/abs/1809.02657
2. Goyal, P., Kamra, N., He, X., Liu, Y.: DynGEM: deep embedding method for dynamic graphs. CoRR abs/1805.11273 (2018). http://arxiv.org/abs/1805.11273

3. Hajiramezanali, E., Hasanzadeh, A., Narayanan, K., Duffield, N., Zhou, M., Qian, X.: Variational graph recurrent neural networks. In: Advances in Neural Information Processing Systems, pp. 10700–10710 (2019)
4. Kipf, T.N., Welling, M.: Variational graph auto-encoders (2016)
5. Kumar, S., Zhang, X., Leskovec, J.: Predicting dynamic embedding trajectory in temporal interaction networks. In: Association for Computing Machinery, pp. 1269–1278 (2019). https://doi.org/10.1145/3292500.3330895
6. Pareja, A., et al.: EvolveGCN: evolving graph convolutional networks for dynamic graphs. In: Proceedings of the AAAI Conference on Artificial Intelligence, vol. 34, no. 04, pp. 5363–5370 (2020). https://doi.org/10.1609/aaai.v34i04.5984, https://ojs.aaai.org/index.php/AAAI/article/view/5984
7. Rossi, E., Chamberlain, B., Frasca, F., Eynard, D., Monti, F., Bronstein, M.M.: Temporal graph networks for deep learning on dynamic graphs. CoRR abs/2006.10637 (2020). https://arxiv.org/abs/2006.10637
8. Trivedi, R., Farajtabar, M., Biswal, P., Zha, H.: Representation learning over dynamic graphs. CoRR abs/1803.04051 (2018). http://arxiv.org/abs/1803.04051
9. Vaswani, A., et al.: Attention is all you need. CoRR abs/1706.03762 (2017). http://arxiv.org/abs/1706.03762
10. Wang, Y., Chang, Y., Liu, Y., Leskovec, J., Li, P.: Inductive representation learning in temporal networks via causal anonymous walks. CoRR abs/2101.05974 (2021)
11. Xu, D., Ruan, C., Korpeoglu, E., Kumar, S., Achan, K.: Inductive representation learning on temporal graphs. arXiv preprint arXiv:2002.07962 (2020)
12. Xu, D., Cheng, W., Luo, D., Liu, X., Zhang, X.: Spatio-temporal attentive RNN for node classification in temporal attributed graphs. In: IJCAI, pp. 3947–3953 (2019)
13. Zheng, Y., Wang, H., Wei, Z., Liu, J., Wang, S.: Instant graph neural networks for dynamic graphs (2022). https://doi.org/10.48550/ARXIV.2206.01379, https://arxiv.org/abs/2206.01379
14. Zhou, D., Zheng, L., Han, J., He, J.: A data-driven graph generative model for temporal interaction networks. In: Proceedings of the 26th ACM SIGKDD International Conference on Knowledge Discovery & Data Mining. KDD 2020, New York, NY, USA, pp. 401–411. Association for Computing Machinery (2020). https://doi.org/10.1145/3394486.3403082

TrieKV: Managing Values After KV Separation to Optimize Scan Performance in LSM-Tree

Zekun Yao[1,2], Yang Song[1,2], Yinliang Yue[3(✉)], Jinzhou Liu[1,2], and Zhixin Fan[1,2]

[1] Institute of Information Engineering, Chinese Academy of Sciences, Beijing, People's Republic of China
{yaozekun,liujinzhou,fanzhixin}@iie.ac.cn
[2] School of Cyber Security, University of Chinese Academy of Sciences, Beijing, People's Republic of China
[3] Zhongguancun Laboratory, Beijing 100080, People's Republic of China
yueyl@zgclab.edu.cn

Abstract. Persistent key-value(KV) stores are mainly designed based on the Log-Structured Merge-tree(LSM-tree) for high write performance, yet the LSM-tree suffers from the inherently high I/O amplification which influences the read and write performance when KV stores grow in size. KV separation mitigates I/O amplification by storing only keys in the LSM-tree while values are in separated storage. However, the KV separation breaks the key sequence of values, which influences their range query performance. We propose TrieKV make the most of the hard-disk drives(HDD)'s sequential read performance advantages to improve range query performance. TrieKV uses a dynamic prefix index and a collaborative KV data merging and sorting mechanism to manage values after KV separation. Compared with the typical KV separation storage system WiscKey, TrieKV achieves 2.35× range query performance under HDD. Meanwhile, TrieKV also performs better than WiscKey in all six YCSB workloads.

Keywords: KV Storage · LSM-Tree · Dynamic Prefix Index · Performance Optimization

1 Introduction

The data analysis scenario needs high range query performance. Meanwhile, persistent key-value (KV) storage is a critical part of modern large-scale storage for massive data. The Log-Structured Merge-tree (LSM-tree) [20] is a popular index design in modern persistent KV stores. It uses an out-of-place mechanism to update the data, which means the storage system does not immediately update

Z. Yao and Y. Song—These authors contributed equally to this work.

or delete the old data, but first writes the new version to the storage system, then recycles the outdated by compaction. With this mechanism, LSM-tree converts random writes to sequential and then undergoes compaction to merge and sort all KV pairs. Through compaction, the KV pairs will become ordered in the sequence of keys, and the point query operation can be performed faster through binary searching. At the same time, the ordered data is also conducive to the range query operation.

The LSM-tree also has some problems, which are space amplification [19] and I/O amplification in both writes and reads [9]. Because all delete and update operations are transformed to write operations in a log-structured way, multiple versions of data will exist in the storage system at the same time. The phenomenon that expired data has not been cleaned up immediately and occupies much extra space is space amplification. Read amplification means that the read operation of the LSM-tree needs to look up data layer by layer from top to bottom, which will occupy multiple disk I/Os. During compaction, the data will be continuously read, merged, and sorted. These operations cause write amplification, which means the size of the data actually written to the disk is often larger than the size required by the program. As the height of the LSM-tree grows, the I/O amplification problem is more serious.

In order to reduce I/O amplification in writes and reads, researchers propose some new data organization methods, one of them is WiscKey [16] which proposes the idea of KV separation. Based on the KV separation mechanism, the keys are stored in the LSM-tree while the values are separately stored in an append-only circular log file. Because the LSM-tree only needs to store keys and metadata, the height of the LSM-tree is efficiently reduced, and the I/O amplification is mitigated. However, WiscKey appends values in a log file to achieve a high write performance, which breaks the key sequence of values. Thereby, WiscKey has more random reads during the range query process, which weakens the scan performance.

To overcome the out-of-order state of values after the KV separation and improve the range query performance, we carefully design two techniques. First, we propose a dynamic prefix index method that ensures the order of values within the global key range. Second, a collaborative KV data merging and sorting mechanism is proposed to solve the disorder in partial areas. Based on these two mechanisms, the KV separation storage system TrieKV is designed and implemented. The dynamic prefix index aims at the global area and the collaborative KV data merging and sorting mechanism solves the problems in the partial area, the combination of them forms a KV pair organization scheme that makes values ordered within the global area and gradually becomes ordered at the fine-grained level. This scheme can make better use of the performance advantages of hard-disk drives(HDD) sequential read to improve range query performance.

We implement our TrieKV prototype atop RocksDB [7] and compare its performance via YCSB [4] with RocksDB and WiscKey. Compared with the typical KV separation storage system WiscKey, TrieKV gains a 28.86% point query performance improvement and a 134.88% range query performance improvement under the HDD. Compared with the typical KV aggregation storage system

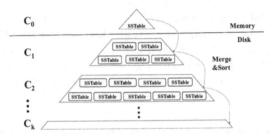

Fig. 1. Structure diagram of LSM-tree

RocksDB, TrieKV gains an 89.42% point query performance improvement and a 162.32% write performance improvement under the HDD. Besides, TrieKV also has an encouraging performance on SSD.

2   Background and Motivation

Modern KV stores use the Log-Structured Merge-tree(LSM-tree) [20] to optimize the performance of writes. The LSM-tree is a hierarchical structure, and the capacity increases layer by layer. Its structure is shown in Fig. 1, the first layer C_0 is in memory but the others are in disks. To get high write performance, LSM-tree converts a large number of random writes into batch writes. When inserting data, a record that is used to ensure crash consistency is written to a log file at first, and then the data will be inserted into the C_0. Meanwhile, the update and delete operations are converted to write operations. When the amount of data in a certain layer reaches the capacity limit, compaction will start. Compaction includes two actions: merge and sort, which integrate the data of the previous layer into the next layer, and ensure that the data of each layer keeps sequence. Because of compaction, LSM-tree can support a fast query operation.

The classical KV stores based on LSM-tree (e.g. LevelDB [8] and RocksDB [7]) store KV pairs as a whole, append them to a memory table and compact them in the background to get a key sequence layout. But these approaches make a serious I/O amplification in both writes and reads, which limits the development of LSM-tree. To mitigate the I/O amplification, WiscKey is presented and its main idea is key-value separation. WiscKey still stores keys and metadata in the LSM-tree, but the values are separately appended to a circular log. To distinguish from the KV separation storage method adopted by WiscKey, the traditional LSM-tree used to store KV pairs is called as KV aggregation storage method.

Existing research could not preserve the order of values after separating KV pairs, which is mainly manifested as storage in the form of additional logs. WiscKey organizes values in a value-log file, named vLog. When a KV pair is inserted, WiscKey separates the key and the value, then appends the value to the vLog. Because of this, WiscKey implements the range query operations by parallel searching the target KV pairs through random I/Os rather than sequential, which makes WiscKey perform undesirable in HDD. HashKV [3] performs hash operations on keys and maps values to different segment groups. Due to the feature of hash, the storage location of values with similar keys is not very close. So there will also be multiple random reads during the range query process. The disordering of values makes range queries mostly implemented in random data sets, resulting in poor performance.

In 2018, Wang proposed KT-store [26] which used a key-and-write hybrid data layout to balance write and range query throughput. KT-Store implements a hybrid layout by taking a fixed-length prefix for keys to map and append their values to the corresponding storage files. The values in KT-Store are globally ordered but partially disordered. When performing a range query operation, KT-Store groups the targets into different storage files determines the start-offset and the end-offset in each storage file, then sequentially reads all data in the range to the memory, and finally filters the target data and returns it. The essence of this approach is to aggregate data through fixed-length prefixes. It has two major shortcomings: (i) it does not solve the order of values in a partial area. In the scenarios where data is randomly distributed, the target data will be very little in the data area read by the range query, which wastes the benefit of disks' sequential reads. Therefore the improvement of range query performance is limited. (ii) the design of the fixed depth prefix needs to estimate the total amount of data in advance to set a reasonable size, so this method is difficult to adapt to the scene of practical applications' dynamic changes in data volume.

After a comparative analysis of the existing KV separation storage system, it is determined what we need to do is to design a new data storage system that can keep the internal values of the stored files as ordered as possible. We propose a variant structure from the following aspects to solve the problems. First, aiming at the problem that fixed prefix index design is difficult to adapt to the change of data volume, the idea of a dynamic prefix index has been put forward. Second, aiming at the problem that the values are not ordered in storage files because of KV separation, we propose a collaborative KV data merging and sorting mechanism named pbCompaction. Finally, we design and implement a KV storage system TrieKV based on these two methods.

3   TrieKV Design

3.1   Architectural Overview

Figure 2 depicts the overall architecture of **TrieKV**. TrieKV is a selective KV separation storage system. When the size of a value is small, the KV separation will not reduce the read amplification but increase the storage space occupation and access delay. So the benefit of KV separation is diminished for small KV pairs. The selective KV separation strategy sets a threshold and only separates the KV pairs when the value size is larger than the threshold. It can take better advantage of the key-value separation.

To make values ordered in the whole range, we promote a trie-index to manage values. The path from the root to each node in the trie index represents its own prefix. Each node points to a data file for storing values called a TrieTable.

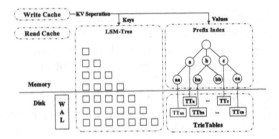

Fig. 2. System architecture diagram of TrieKV

One TrieTable file stores data with the same prefix. To improve write performance, the data will be written to TrieTables in an append-only way. When a TrieTable is full, the system will split this node. The Prefix Index in memory and the TrieTables in disk combined to **DyTrieIndex**. It makes the values globally ordered. Meanwhile, it will carry out node-splitting operations and adjust the prefix depth when the data volume of a TrieTable reaches one threshold.

Although the append-only way written to TrieTables speeds up the write performance, it breaks the key sequence in TrieTables. To make values more ordered in fine grain area, we proposed a **pbCompaction** mechanism based on the node-splitting operation. This mechanism sorts values and carries out garbage collection (GC) in a "piggyback" way during node-splitting. This data organization scheme combined with DyTrieIndex and pbCompaction can make better use of the performance advantages of sequential reads in HDD to improve range query performance.

To prevent the data from losing and damaging, TrieKV adopts an on-disk write-ahead log(WAL). When a user writes a piece of data to write buffer, a piece of log data representing the operation will be appended to the log file immediately. Only when the data is successfully persisted to the disk, the corresponding part of the log will be deleted.

3.2 DyTrieIndex

TrieKV applies **DyTrieIndex**, a dynamic prefix index with its correcting TrieTables, to organize values. We set a prefix index tree as a TrieIndex. The prefix is the path from the index root to the node. Each prefix points to a TrieTable that stores values in HDD. The values with the same prefix are stored in one TrieTable. The size of the TrieTable is fixed and can be set. If traverse the prefix index in the middle order, we can find that the different TrieTables are arranged in a lexicographic order, which shows that the values are ordered within the global key range. In order to improve write performance, TrieTable will write data in an append-only way. The append operation can ensure that values of the same key are stored in the same TrieTable, and the latest data is at the tail. The TrieIndex will be persisted to the disk when the system is shut down, and restored to memory when it is started.

When the data volume of TrieTable reaches the capacity threshold, this TrieTable will split. Node-splitting is the core of dynamic prefix indexing. The purpose of this operation is to avoid too much data in a single TrieTable and adapt to the data volume's changes, which can make the system more flexible. Meanwhile, pbCompaction is based on this operation. The process is shown in Fig. 3: In the case, TrieTable TT_a is full and is splitting. And there are three types of K_1's value stored in it. To split this node, TT_a will be read in memory first. And at the second step, the $< K_1, V_3 >$ pair will be saved, $< K_1, V_1 >$ & $< K_1, V_2 >$ will be deleted as the garbage. Meanwhile, there may be some overlaps between the areas which are divided by the three K_1 pairs. So the third step, the data in the overlapped area will be sorted. Then calculate the new prefix, and create the next-level prefix node with its corresponding TrieTable in the index. Finally, append the data into the new corresponding TrieTable and clear the TT_a. During this process, the second and the third steps operate as a compact operation, namely pbCompaction.

3.3 pbCompaction

The write operation by appending method and the use of the write buffer can mitigate write amplification, but the data in a TrieTable will be divided into several data areas, which may have overlapped key ranges in one TrieTable. The values in one area are written at the same time. Therefore, the data is unique and ordered in one area, but overlapped and disordered between different areas. When a TrieTable is read into memory from the disk during the node-splitting, the system will carry out garbage collection(GC) and sort values, just like the compaction in KV aggregation LSM-tree storage systems. Because this operation compacts the data in a "piggyback" way, we call it **pbCompaction**.

Compared with WiscKey, the pbCompaction of TrieKV carries out GC in an easy way. Since keys are mapped to the storage file according to the longest prefix matching principle, the same key's values must be stored in the same TrieTable. And the append-only method ensures that newer value is stored at the further back end of this TrieTable. Therefore, the value of the same key can be overwritten with the further back version, instead of looking up the LSM-tree to determine the validity of the value during GC.

The overhead of compaction can be hidden in the node-splitting operation. Meanwhile, the time complexity and space complexity of this mechanism is much lower than that of deduplication and then reordering, which saves CPU and memory source. Since the data in the data area is ordered and the new data is located at the latest position in a TrieTable, the pbCompaction can achieve GC more simply and efficiently than WiscKey, which should sort the values with querying the LSM-tree multiple times to ensure the validity of the data. After introducing pbCompaction, the system can sort values in a fine-grained way which can make values tend to be fully-sorted ordering. The pbCompaction further solves the problem of values disordering after the KV separation. The system can take advantage of the HDDs' sequential read to enhance the range query performance.

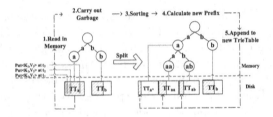

Fig. 3. Schematic diagram of node splitting based on pbCompaction

3.4   Main Procedure

Range Query Procedure need to first check the LSM-tree to obtain the key range. Because there may be some needed data in write buffer, TrieKV also should check it. Then determine the reading range of each TrieTable, and go to the disk for sequential reading. Finally, filter out the data. The operation flow is shown in Fig. 4.

Point Query Procedure includes three stages: search cache, LSM-tree, and TrieTable step-by-step. When querying TrieKV for the value of K_1, the data will be searched in the write buffer first, if hit, data will be returned, or will retrieve from read cache. Similarly, if the data is hit, it will be returned and the data will be adjusted to the head of read cache. If no data is hit in both read cache and write buffer, it will search the LSM-tree continuously. If not found, means K_1 is not in the system. If K_1 is found, it will be judged whether the corresponding value is real value or metadata. If it is real value, it will be returned directly. If it is metadata, TrieKV will parse its TrieTable file information, then the data will be read from the corresponding TrieTable file. If the data is stored in the disk, it will also be added to the head of read cache. If the read cache is full, a piece of data is removed from the end of read cache according to the principle of least recently used. All process has shown in Fig. 5.

Write Procedure should think about the log and cache these two aspects. Figure 5 presents the write procedure. First, generate a log of this operation and insert the log into the WAL. Then append the KV pair to write buffer. If write buffer has been full, switch it to immutable cache and persist it by the background thread. Finally, create a new buffer and a corresponding WAL file again. Until now, the write procedure has been finished for users, and the other operations like persist cache to disk will be done in the background.

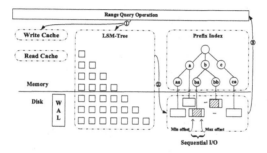

Fig. 4. Range query Procedure flow chart of TrieKV.

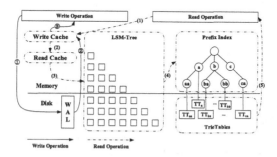

Fig. 5. Get&Put Procedure flow chart of TrieKV.

4   Evaluation

4.1   Experimental Setup

We present evaluation results on TrieKV. We compare the TrieKV with the typical KV aggregation storage system RocksDB and the typical KV separation storage system WiscKey. All experiments are run on a testing machine with ten Intel(R) Xeon(R) Silver 4114 CPU @ 2.20GHz processors, and 128GB memory. The storage devices are a 1.8 TB TOSHIBA MG04ACA200N and a 960 GB SAMSUNG MZ7LH960HAJR-00005. The operating system is CentOS Linux release7.5.1804 (Core) with Linux3.10.0-862.el7.x8664. We use YCSB to generate workload traces which are replayed in a lightweight generator. YCSB generates synthetic workloads with various degrees of read/write ratio, statistical distribution, and value size. We configure YCSB to produce different datasets that are described in the following subsection.

We use RocksDB 6.2.2 with its default parameters. The WiscKey is achieved based on the code that Abhishek Sharma submit to GitHub. For TrieKV, we set one read cache and one write buffer, and their size is 64 MB.

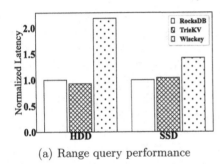
(a) Range query performance

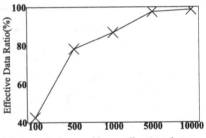
(b) Trend chart of how effective data ratio varies with query length

Fig. 6. Range query Evaluation

4.2 Range Query Performance

The range query performance experiments of the system measure its performance by comparing the range query latency of TrieKV, RocksDB, and WiscKey. At the same time, the paper also counts the effective data percentage of TrieKV's range query, which refers to the ratio of the required target data in each range query, to analyze the effect on the program.

In the experiments, 100 GB data have been written to each KV storage system under HDD and SSD respectively. The size of a single value is 8 KB, and the size of TrieTable is set to 10 MB. The data are randomly distributed uniformly. Then for each experiment, randomly select the initial key, perform a range query operation with a length of 1000, and repeat it 10,000 times to evaluate the average delay. Figure 6(a) shows the latency results, normalizing the latency of each system to RocksDB for ease of comparison. It can be seen from the figure that the latency of TrieKV is significantly lower than WiscKey whether under the HDD(reduced by 134.88%) or SSD(reduced by 36.66%). The TrieKV's performance is a little better than RocksDB under HDD, but a little worse under SSD. This is because RocksDB adopts the form of key-value aggregation storage, and the value can be obtained by the key. The range query only needs to traverse the lexicographic order of the keys in the LSM-tree through the iterator until a suffcient number of KV pairs are obtained. The KV separation storage method reduces the I/O amplification of the LSM-tree, but when reading the value, it needs to query the log structure merge tree first, and then read the disk file after parsing the size and offset of the value, so the delay may be a bit higher.

The percentage of valid data in TrieKV's range query has been counted. From Fig. 6(b) we can find that when the query length is small, the proportion of valid data in the query is not high, which is affected by the data distribution in the TrieTable. As the length of the query increases, the percentage of effective data in the query continues to increase, reaching a maximum of 98.64%. This is because the value of TrieKV is ordered within the overall key range (the level of TrieTables), and data with the same prefix must be stored in the same TrieTable, so when the query length is larger, it will be more concentrated.

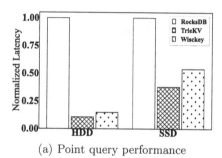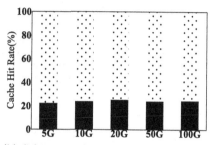

(a) Point query performance

(b) Schematic diagram of read cache hit ratio under Zipfian request distribution

Fig. 7. Point query Evaluation

4.3 Point Query Performance

The point query performance is reflected by the latency. The evaluation method is to first write 100 GB data to the three system randomly, then perform the point query operation. In order to get closer to the scenario where there is a distinction between hot and cold data in practical applications, the data distribution is Zipfian. The number of experiment queries is 10,000 and the average point query latency is normalized to RocksDB's as the measurement. At the same time, TrieKV also further counts the hit ratio of its read cache. The size of a single value is 8KB and repeat point query 10,000 times to evaluate the average delay. The results are shown in Fig. 7(a) and Fig. 7(b).

In terms of read latency under HDD and SSD, TrieKV is reduced by 89.42% and 60.05% compared to RocksDB; compared to WiscKey, it is reduced by 28.86% and 26.01%. Because of the KV separation strategy, the WiscKey point query operation requires one more query process than RocksDB, which will affect its performance when the amount of data is small. But as the amount of data increases, the advantage of KV separation which can effectively reduce read amplification is reflected. On the basis of WiscKey, TrieKV also considers the degree of hot and cold data and uses a read cache to store hot data, so that repeated read of hot data directly hit read cache. It can be seen from Fig. 7(b) that in the point query performance experiment under different data volumes, the probability of hitting the read cache is relatively stable and maintained between 22.36% and 25.49%. It is proved that in the scene of hot and cold data, the design of read cache can effectively reduce the number of accesses to the disk and improve the overall point query performance.

4.4 Write Performance

The systems participating in the comparison experiment are TrieKV, RocksDB, and WiscKey. The measurement index of write performance is the throughput of write operations, and the data is randomly distributed uniformly. In the experiment, the size of a single value is 8 KB, and 100 GB data are written to each

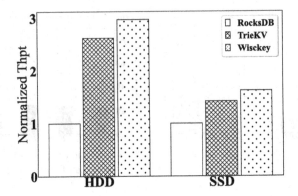

Fig. 8. Write Evaluation

KV storage system to evaluate the write throughput. The result is shown in Fig. 8, normalizing the throughput of each system to RocksDB for ease of comparison. Overall, TrieKV compared to RocksDB's write throughput increased by 162.32% under the HDD and 41.90% under the SSD; but compared to WiscKey decreased by 11.79% under the HDD and 12.58% under the SSD. The reason why RocksDB's write throughput is lowest is that the two adopt a KV separation strategy, which reduces the write amplification of the LSM-tree. Another reason for the higher write throughput of WiscKey is that compared to TrieKV, it writes all values to the same log file, so there is no overhead of node-splitting. In addition, WiscKey does not have write buffer design, and it also eliminates the need for write ahead logs to ensure crash consistency.

4.5 Evaluate TrieKV Under YCSB Default Workloads

Our final set of experiments compares the performance with YCSB standard workloads, which are treated as a basic benchmark for storage systems. Each of the six standard workloads combines one or two operation types and can make us understand the performance of the system. All the standard workloads are based on the Zipf distribution of KV pairs. Workload A is an update heavy workload. B is a read-mostly workload. C is a read-only workload. D is a read latest workload which has 95% reads of the most recently inserted KV pairs. E is a short ranges workload which does the short-range query operations. In E, the max range query length is 100 and the range query length is under the uniform distribution. F is a read-modify-write workload. In F, the KV pairs will be read modified, and written back to the storage system. We perform the six workloads on TrieKV, RocksDB, and WiscKey with the 8KB value size on HDDs. We load 100GB data with Zipfian distribution, then perform each workload and evaluate the throughput.

Figure 9 presents the operations throughput of RocksDB, TrieKV, and WiscKey and normalizing the result to RocskDB. In workload B, C, and D, which are read more frequently, the TrieKV has the highest throughput. In the range

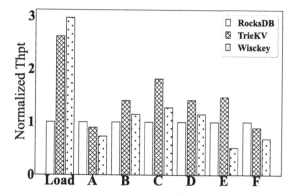

Fig. 9. Diagram of YCSB default workload evaluation results

query scenario workload E, TrieKV also has the highest throughput. In read-write balance scenario workload A and read-modify-write scenario workload F, the throughput of TrieKV is lower than RocksDB but higher than WiscKey. Because these two scenarios are composite operation scenarios that include writing operations rather than a large number of single write scenarios, they generally do not cause write pauses. Therefore, the acceleration effect of RocksDB and TrieKV write buffers is brought into play. Through the YCSB default load test, it can be seen that TrieKV can perform very well under different load scenarios, which further proves the effectiveness of the solution in this article in actual application scenarios.

5   Related Work

To improve the performance and reduce the inherent problem of LSM-tree, many researchers have proved different optimization methods. During these years, researchers focus on some new ways to optimize the KV store based on the LSM-tree. In addition to the automatic tuning [5,15,28] or using new hardware, like FPGA [24], KV-SSD [11] or non-volatile memory [12,25,31], accelerating the point query operation is another focal point. Bloom Filter is one important optimize choice. Monkey [6] and Elastic Bloom Filter [13] are presented in succession to improve the point query performance. Besides HotKey-LSM [27] apply the hot awareness to accelerate the point query operation.

Researchers also focus on how to improve the range query operation. A lot of range filters have been proposed. Rosetta [18] is a probabilistic range filter to optimize the range query operation of the LSM-tree. SuRF [32] is a succinct range filter which can both optimize the point and range query operation. To reduce the number of useless data which may be read to the cache, a partition pruning strategy [10] has been proposed. It used a statistics cache to determine whether a partition contains the desired data and prune the useless partitions.

Compaction overhead is a huge cost for LSM-tree and causes the write amplification. So extensive researchers focus on it. Decreasing the frequency of compaction to reduce the compaction cost is one study line. Yue et al. propose a multi-component Skip-Tree [30]. By adopting a skip strategy during the compaction process, KV pairs will not be compacted level by level before they arrived at the target level, which can reduce the compact operations. To reduce the number of compaction, dCompaction [21] presents virtual compaction, which means only compacting the metadata and delaying the real compaction when the compaction is triggered. In this way, frequent I/O can be avoided. Raju et al. proposed PebblesDB [1] which introduces guards at each level of the LSM-tree and uses skiplist [22] to organize the guard. The idea that relaxes fully-sorted ordering of the KV pairs is widely used by Lightweight Compaction tree [29], SlimDB [23] and SifrDB [17] to reduce the cost of compaction operation.

To mitigate the write amplification problem, WiscKey [16] first proposed KV separation. Based on KV separation, Chan et al. implemented the HashKV [3] which maps values to fix-size segments by the hash way. HashKV also uses hotness awareness [2], and utilize different garbage collection (GC) strategy to reduce the write amplification during the compaction. UniKV [33] is also designed atop KV separation. It leverages data locality to use different strategies to manage the different KV pairs index. To optimize WiscKey, DiffKV [14] manages keys with fully-sorted but values with partially-sorted to improve the performance of both write, read, and scan.

We present the TrieKV which also differentiates to manage the keys and values after KV separation. Different from the DiffKV, TrieKV not only partially sort values but use the pbCompaction to perfect the order of the values, which can manage the values to tend to fully sorted. In this way, TrieKV can make the best use of the feature of HDD and achieve better write, read, and scan performance.

6   Conclusion

This paper presents TrieKV, which makes a balance between range query, point query, and write performance in KV stores. Its novelty lies in leveraging DyTrieIndex and pbCompaction to make the values after separation to keep the state overall in order, while also gradually becoming ordered at a fine-grained level. Meanwhile, we enhance TrieKV with hotness awareness and selective KV separation. Experiments show that TrieKV traded a limited price for a significant performance improvement, verifying the effectiveness of the thesis proposal.

References

1. Pebblesdb: Building key-value stores using fragmented log-structured merge trees. In: The 26th Symposium, pp. 497–514 (2017)
2. Triad: Creating synergies between memory, disk and log in log structured key-value stores. In: Proceedings of the 2017 USENIX Conference on Usenix Annual Technical Conference (2017)

3. Chan, H.H.W., Li, Y., Lee, P.P.C., Xu, Y.: HashKV: enabling efficient updates in KV storage via hashing. In: Proceedings of the 2018 USENIX Conference on Usenix Annual Technical Conference, p. 14 (2018)

4. Cooper, B.F., Silberstein, A., Tam, E., Ramakrishnan, R., Sears, R.: Benchmarking cloud serving systems with YCSB. In: Proceedings of the 1st ACM symposium on Cloud computing - SoCC 2010, Indianapolis, Indiana, USA, p. 143. ACM Press (2010)

5. Dai, Y., et al.: USENIX Assoc: From WiscKey to Bourbon: A Learned Index for Log-Structured Merge Trees, pp. 155–171 (2020)

6. Dayan, N., Athanassoulis, M., Idreos, S.: Optimal bloom filters and adaptive merging for LSM-trees. ACM Trans. Database Syst. **43**(4) (2018). https://doi.org/10.1145/3276980

7. Facebook: Rocksdb documentation. https://rocksdb.org.cn/doc.html

8. Google: Leveldb documentation. https://github.com/google/leveldb/blob/master/doc/index.md

9. Hu, X.Y., Eleftheriou, E., Haas, R., Iliadis, I., Pletka, R.: Write amplification analysis in flash-based solid state drives. In: Proceedings of SYSTOR 2009: The Israeli Experimental Systems Conference on - SYSTOR 2009, Haifa, Israel, p. 1. ACM Press (2009)

10. Huang, C., Hu, H., Wei, X., Qian, W., Zhou, A.: Partition pruning for range query on distributed log-structured merge-tree. Front. Comput. Sci. **14**(3) (2020). https://doi.org/10.1007/s11704-019-8234-x

11. Im, J., Bae, J., Chung, C., Arvind, Lee, S.: Design of LSM-tree-based Key-value SSDs with Bounded Tails. ACM Trans. Storage **17**(2) (2021). https://doi.org/10.1145/3452846

12. Li, C., Chen, H., Ruan, C., Ma, X., Xu, Y.: Leveraging NVME SSDS for building a fast, cost-effective, LSM-tree-based KV store. ACM Trans. Storage **17**(4) (2021). https://doi.org/10.1145/3480963

13. Li, Y., Tian, C., Guo, F., Li, C., Xu, Y., USENIX Assoc: ElasticBF: elastic bloom filter with hotness awareness for boosting read performance in large key-value stores, pp. 739–752 (2019)

14. Li, Y., et al.: Differentiated Key-Value storage management for balanced I/O performance. In: 2021 USENIX Annual Technical Conference (USENIX ATC 21), pp. 673–687. USENIX Association (2021). https://www.usenix.org/conference/atc21/presentation/li-yongkun

15. Lu, K., Zhao, N., Wan, J., Fei, C., Zhao, W., Deng, T.: TridentKV: a read-optimized LSM-tree based KV Store via adaptive indexing and space-efficient partitioning. IEEE Trans. Parallel Distrib. Syst. **33**(8), 1953–1966 (2022). https://doi.org/10.1109/TPDS.2021.3118599, https://ieeexplore.ieee.org/document/9563237/

16. Lu, L., Pillai, T.S., Gopalakrishnan, H., Arpaci-Dusseau, A.C., Arpaci-Dusseau, R.H.: Wisckey: separating keys from values in SSD-conscious storage. ACM Trans. Storage (TOS) **13**(1), 5 (2017)

17. Lu, Z., Cao, Q., Mei, F., Jiang, H., Li, J.: A novel multi-stage forest-based key-value store for holistic performance improvement. IEEE Trans. Parallel Distrib. Syst. **31**(4), 856–870 (2020). https://doi.org/10.1109/TPDS.2019.2950248

18. Luo, S., et al.: A robust space-time optimized range filter for key-value stores, pp. 2071–2086 (2020).https://doi.org/10.1145/3318464.3389731

19. Ouaknine, K., Agra, O., Guz, Z.: Optimization of RocksDB for Redis on Flash. In: Proceedings of the International Conference on Compute and Data Analysis - ICCDA 2017, Lakeland, FL, USA, pp. 155–161. ACM Press (2017)

20. O'Neil, P., Cheng, E., Gawlick, D., O'Neil, E.: The log-structured merge-tree (LSM-tree). Acta Informatica **33**(4), 351–385 (1996)
21. Pan, F.F., Yue, Y.L., Xiong, J.: dcompaction: speeding up compaction of the LSM-tree via delayed compaction. J. Comput. Sci. Technol. **32**(1), 41–54 (2017)
22. Pugh, W.: Skip lists: a probabilistic alternative to balanced trees. In: Workshop on Algorithms & Data Structures (1990)
23. Ren, K., Zheng, Q., Arulraj, J., Gibson, G.: Slimdb: a space-efficient key-value storage engine for semi-sorted data. Proc. VLDB Endow. **10**(13), 2037–2048 (2017). https://doi.org/10.14778/3151106.3151108
24. Sun, X., Yu, J., Zhou, Z., Xue, C.J.: FPGA-based compaction engine for accelerating LSM-tree key-value stores. In: 2020 IEEE 36th International Conference on Data Engineering (ICDE), pp. 1261–1272 (2020). https://doi.org/10.1109/ICDE48307.2020.00113
25. Yao, T., et al.: Matrixkv: reducing write stalls and write amplification in LSM-tree based KV stores with matrix container in NVM. In: The 2020 USENIX Annual Technical Conference, pp. 17–31 (2020)
26. Wang, H., Yue, Y., He, S., Wang, W.: KT-store: a key-order and write-order hybrid key-value store with high write and range-query performance. In: Zhang, F., Zhai, J., Snir, M., Jin, H., Kasahara, H., Valero, M. (eds.) NPC 2018. LNCS, vol. 11276, pp. 64–76. Springer, Cham (2018). https://doi.org/10.1007/978-3-030-05677-3_6
27. Wang, Y., Jin, P., Wan, S.: HotKey-LSM: a hotness-aware LSM-tree for big data storage, pp. 5849–5851 (2020). https://doi.org/10.1109/BigData50022.1010.9377736
28. Wu, F., Yang, M., Zhang, B., Du, D., USENIX Assoc: AC-key: adaptive caching for LSM-based key-value stores, pp. 603–615 (2020)
29. Yao, T., Wan, J., Huang, P., He, X., Wu, F., Xie, C.: Building efficient key-value stores via a lightweight compaction tree. ACM Trans. Storage **13**(4) (2017). https://doi.org/10.1145/3139922
30. Yue, Y., Wang, W., Li, Y., He, B.: Building an efficient put-intensive key-value store with skip-tree. IEEE Trans. Parallel Distrib. Syst. **23**, 961–973 (2017)
31. Zhang, B., Du, D.H.C.: NVLSM: a persistent memory key-value store using log-structured merge tree with accumulative compaction. ACM Trans. Storage **17**(3) (2021). https://doi.org/10.1145/3453300
32. Zhang, H., et al.: Succinct range filters. ACM Trans. Database Syst. **45**(2) (2020). https://doi.org/10.1145/3375660
33. Zhang, Q., Li, Y., Lee, P.P.C., Xu, Y., Cui, Q., Tang, L.: UniKV: toward high-performance and scalable KV storage in mixed workloads via unified indexing. In: 2020 IEEE 36th International Conference on Data Engineering (ICDE), Dallas, TX, USA, pp. 313–324. IEEE (2020)

Bit Splicing Frequent Itemset Mining Algorithm Based on Dynamic Grouping

Wenhe Xu[1] and Jun Lu[1,2(✉)]

[1] College of Computer Science and Technology, Heilongjiang University, Harbin, China
lujun116_lily@sina.com
[2] Jiaxiang Industrial Technology Research Institute of HLJU, Jining, Shandong, China

Abstract. Frequent itemset mining has always been one of the most classic tasks in data mining. It provides effective decision-making and judgment for many problems. A novel MPL (multi-partition list) structure is proposed in this paper combining bit combination and linear table structure. The MPL is composed of arrays where each unit stores a combination of items rather than a single item, which addresses the limitations of maintaining many pointers in the traditional tree structure. In addition, the MPL stores the least valid information required in the mining process. This paper further proposes a bit splicing frequent itemset mining algorithm based on dynamic grouping (BSFIM-DG) for the MPL. The algorithm dynamically calculates the number of grouping by using coverage according to the dataset's characteristics. The candidate itemset is obtained by the bit-splicing method. The length of the MPL to be traversed is determined by the low-bit feature of the candidate itemset. The search space is reduced with the corresponding pruning strategy. Experiments on various open datasets demonstrate that the algorithm has excellent running speed, especially since the support is low. The proposed algorithm has a similar running speed to the BCLT-O and the FP-growth on some datasets. In terms of memory usage, the algorithm is better than the FP-growth and comparable to the BCLT-O, but there is still a particular gap with the Bit-combination algorithm. Nevertheless, as the pace of technology updates and iteration is getting faster and faster, it is very feasible to exchange space for speed.

Keywords: multi-partition list · dynamic grouping · coverage · pruning

1 Introduction

Frequent itemset mining [1] is an important research topic in data mining. It can provide effective decision-making for association rules, correlation analysis, and periodic judgment. Apriori [2] algorithm uses an iterative method to generate candidate k-itemsets from frequent $(k-1)$-itemsets and judges whether the k-itemsets are frequent itemsets by traversing the original dataset, which leads to poor performance. The FP-growth [3] algorithm adopts a highly compressed FP-tree structure to store the original dataset. The FP-growth algorithm uses a divide and conquers strategy. And the conditional FP-trees need to be created recursively during mining, which consumes a lot of memory.

© The Author(s), under exclusive license to Springer Nature Singapore Pte Ltd. 2024
X. Song et al. (Eds.): APWeb-WAIM 2023, LNCS 14333, pp. 417–432, 2024.
https://doi.org/10.1007/978-981-97-2387-4_28

The negFIN algorithm [4] is a frequent itemset mining method that adopts a prefix tree structure based on the bitmap representation of sets. The algorithm adopts the set enumeration tree and the promotion method for mining. The DPT algorithm [5] only uses a prefix tree structure to store all the information and mines frequent itemsets by continuously and dynamically adjusting the prefix tree structure. The PrePost+ algorithm [6] uses the N-list data structure [7] to represent itemsets and directly discovers frequent itemsets on the set enumeration search tree. In particular, an effective child-parent-equivalent pruning strategy is adopted to decrease the search space. The paper [8] proposes a new data structure called frequent pattern lists. The database is divided into several sub-databases in a divide-and-conquer manner, and mining tasks are performed separately.

In Reference [9], by scanning the original database once, an ISSP-tree structure is constructed, maintaining a header table structure named "sameItemLast" to assist tree construction and incremental processing. The FCFPIM algorithm [10] introduces the FCFP-Tree data structure, which also retains the complete information of the original dataset. There is no need to scan and recalculate previous datasets when new datasets are added. Reference [11] proposes a pruning method called Lengthsort. The dataset is partitioned according to the length of transactions, and items and transactions are removed simultaneously before constructing a frequent pattern tree structure.

2   Related Work

The Bit-combination-based algorithm [12] uses binary to represent combinations of items. The candidate itemsets are generated by bit traversal operation that the binary is gradually added 1 from all 0 to all 1. Then the original dataset is scanned to judge whether the binary is frequent. When candidate C is not frequent, the pruning strategy $C = C + C \& (-C)$ is used to reduce the search space. However, the algorithm needs to continuously scan the dataset, which generates a lot of IO overhead.

The BCLT-O [13] uses a linear table structure to store the information in the dataset, and the linear table can be directly scanned during the mining process. Although the algorithm statically divides the linear table into three groups, static grouping for different datasets cannot achieve the best effect.

However, most algorithms will face two main problems: running time and memory consumption. Thus, a bit splicing frequent itemset mining algorithm based on dynamic grouping (BSFIM-DG) is proposed after comparing various mining methods. The main contributions of the algorithm are as follows.

- The BSFIM-DG algorithm is proposed and the MPL is constructed by scanning the original dataset twice.
- The coverage formula is used to group the dataset dynamically. It makes the constructed MPL have a specific order, which is profitable to improve the mining speed.
- A bit splicing operation is proposed to reduce the number of candidate itemset considerably.
- The search space is reduced by continuous superset pruning and group pruning.

3 Proposed Method

3.1 Preprocessing

In this section, a dataset in Table 1 will describe the entire design of the BSFIM-DG algorithm.

Table 1. Original dataset

TID	Original Itemset
001	c, m, j, e
002	b, d, h, f, j
003	a, d, b
004	h
005	e, g, c, d
006	d, m, j, b, k
007	f, m
008	i, h
009	j, h, k, g
010	g, i, m, b, k
011	b, a
012	i, b, j, d, g

Each TID in Table 1 corresponds to a transaction. The support is replaced by the support count. Let the minimum support count minSup = 3.

Table 2. Filtered and sorted

TID	Filtered and Sorted ItemSet	Binary
001	j, m	00100100
002	b, d, j, h	00010111
003	b, d	00000011
004	h	00010000
005	d, g	00001010
006	b, d, j, m, k	10100111
007	m	00100000

(continued)

Table 2. (*continued*)

TID	Filtered and Sorted ItemSet	Binary
008	h, i	01010000
009	j, g, h, k	10011100
010	b, g, m, i, k	11101001
011	b	00000001
012	b, d, j, g, i	01001111

First, the dataset in Table 1 is scanned once, and the occurrence number of each item is counted. Filter according to the preset minSup, and the retained items are sorted in descending order according to their support count. The final result is (b, 6) $\gg$ (d, 5) $\gg$ (j, 5) $\gg$ (g, 4) $\gg$ (h, 4) $\gg$ (m, 4) $\gg$ (i, 3) $\gg$ (k, 3). Each transaction in the dataset is filtered according to the above rules and then sorted in descending order and expressed in binary form, as shown in Table 2.

It can be seen from the Binary column in Table 2 that the Items with a greater support count are at lower index bits. If an item exists, the content of the index bit corresponding to the item is 1. Otherwise, it is 0. For example, there is a transaction jm whose corresponding binary is 00100100.

Next, the TID corresponding to the first item of each transaction after filtering and sorting is recorded in the groupArray. The item column of groupArray is sorted in descending order, and different transactions in each row share at least one item. Figure 1 shows the groupArray after processing Table 2. For example, the TIDs of item j that appears as the first item in Table 2 are T001 and T009, and it is evident that they share at least item j as the root node.

id	item	TID	TID.size
0	b	002, 003, 006, 010, 011, 012	6
1	d	005	1
2	j	001, 009	2
3	g	null	0
4	h	004, 008	2
5	m	007	1
6	i	null	0
7	k	null	0

Fig. 1. Logical diagram of groupArray.

3.2 Dynamic Grouping

After preprocessing, assume that the sorted frequent 1-itemset is $I = \{item_0, item_1, item_2, ..., item_m\}$, $groupArray[item_n]$ represents the total number of transactions with $item_n$ as the first item, where $0 \leq n \leq m$. The *Trans* represent the total number of transactions in the dataset. Further analysis shows that the dataset can be dynamically grouped according to the proportion of transactions corresponding to the top few items in the frequent 1-itemset to the total transactions. The coverage τ is used to represent this proportion in this paper. Equation (1) is the calculation of coverage τ, p can be computed based on Eq. (1).

$$\tau = \frac{\sum_{0}^{p-1} groupArray[item_n]}{Trans} \tag{1}$$

The τ is specified according to the sparseness of the dataset, and subsequent experiments show that τ fluctuates typically between 70% and 90%.

Continue to use the dataset shown in Table 2 as an example and set $\tau = 75\%$ here. It can be seen from Fig. 1 that the number of transactions containing the top 2 items in the groupArray is precisely 75% of all transactions. It is not lower than τ. So the dataset can be divided into four groups. That is, transactions with items b, d, and j as the first item correspond to one group, respectively. And all the remaining transactions are divided into another group. Figure 2 illustrates the logical structure after the dynamic grouping.

Partition	item	TID	TID.size
partition 0	b	002, 003, 006, 010, 011, 012	6
partition 1	d	005	1
partition 2	j	001, 009	2
	g	null	
	h	004, 008	
partition 3	m	007	3
	i	null	
	k	null	

Fig. 2. Dynamic grouping.

3.3 Build MPL

In this section, the linear table and header table are used to assist in constructing the MPL. The linear table consists of a series of nodes T, and each node T stores the following seven pieces of information. $T.name$ represents the item. $T.child$ points to the item's child node index. $T.sibling$ points to the item's sibling node index. $T.father$ points to the item's father node index. $T.point$ indicates the index of the last occurrence of the item. $T.frequency$ stores the support count of the item. $T.bit$ stores the binary of the item combination between the current node and the root node. In addition, the 0th column in the header table stores the sorted frequent 1-itemsets, the 1st column stores the support count of each item, and the 2nd column represents the total number of occurrences of each item in the linear table. For column 3, $First\_node\_link$ points to the item's first occurrence index in the linear table, and $Last\_node\_link$ points to the item's last occurrence index. Based on the above example, the transactions in each partition in Fig. 2 are sequentially mapped into the linear table, and the information in the header table is updated simultaneously. The value at index 0 in the linear table is initialized to Φ. Firstly, each transaction in partition 0 is inserted into the linear table. The first inserted transaction is T002(b, d, j, h).

The first time item *b* is inserted into index 1 of the linear table. A new node *b* be created. That is, *b.name* is item *b*. It can be determined that *b.child*, *b.sibling*, *b.father*, and *b.point* are all 0. And the *b.frequency* is 1, *b.bit* is 00000001. At the same time, the *Count* column of item *b* in the header table increases by 1, and here First_node_link and Last_node_link both point to index 1 of the linear table.

Then, item *d* is inserted into index 2 of the linear table, and a new node *d* also be created. That is, *d.name* is *d*. It can be determined that *d.child*, *d.sibling*, and *d.point* are all 0, and *d.frequency* is 1. The parent node of *d* is *b*, and the index number of node *b* in the linear table is 1. So *d.father* is 1, and *b.child* is updated to 2. *d.bit* stores the binary of the combination b and d. The *d.bit* is 00000011. The *Count* column of item *d* in the header table is increased by 1, and its *First_node_link* and *Last_node_link* both point to index 2 of the linear table.

The remaining items *j* and *h* in T002 are inserted into the linear table in the same way and updated to the header table simultaneously. The result after inserting T002 is shown in Fig. 3.

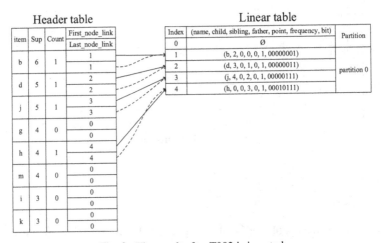

Fig. 3. The result after T002 is inserted.

Next, the second transaction T003(b, d) in partition 0 is processed. Since nodes *b* and *d* already exist in the linear table and are the relationship of parent and child nodes, nodes *b* and *d* in the linear table will be shared. Simply increase *b.frequency* and *d.frequency* by 1. And, the header table also does not need to be updated.

Then, the third transaction T006(b, d, j, m, k) in partition 0 is processed. Inserting items *b*, *d*, and *j* is similar to the processing of transaction T003 and only needs to increase the frequency of the corresponding node by 1. Next, when inserting the item *m*, the item corresponding to the node at index 4 in the linear table is *h* ($h \neq m$). And sibling at index 4 is 0. It has no sibling node, so a new node *m* needs to be created and inserted at index 5. It can be calculated that *m.child*, *m.sibling*, and *m.point* are 0, and *m.frequency* is 1. The parent of node *m* is *j*, so *m.father* is the index 3 of node *j* in the linear table. The *m.bit* is 00100111, which stores the binary form of the combination *b*, *d*, *j*, and *m*. The *h.sibling* is updated to 5 because node *h* and node *m* have the same parent node *j*. At this time, the *Count* column corresponding to item *m* in the header table increases by 1, and here *First_node_link* and *Last_node_link* both point to index 5 of the linear table. Next, to insert the item *k*, a new node *k* needs to be created and inserted into index 6. It can be determined that *k.child*, *k.sibling*, and *k.point* are all 0, *k.frequency* is 1, and *k.bit* is 10100111. Since the parent of node *k* is *m*, then *k.father* is 5 and *m.child* is 6. Similarly, the information at item *k* in the header table needs to be updated.

Insert the remaining transactions in partition 0 into the linear table and update the header table. Finally, the transactions in partition 1, partition 2, and partition 3 are sequentially inserted into the linear table according to the above methods. The final result is shown in Fig. 4. As shown in Fig. 4, the binary bits of each node in partition 0 end with binary 1, the binary bits of each node in partition 1 end with binary 10, and the binary bits of each node in partition 2 end with binary 100. Nevertheless, the binary bits of each node in partition 3 is mixed. It is a minority. It is also one of the reasons for dynamic grouping in this paper.

Header table

item	Sup	Count	First_node_link / Last_node_link
b	6	1	1 / 1
d	5	2	2 / 13
j	5	2	3 / 15
g	4	4	7 / 17
h	4	3	4 / 20
m	4	4	5 / 22
i	3	3	9 / 21
k	3	3	6 / 19

Linear table

Index	(name, child, sibling, father, point, frequency, bit)	Partition
0	Ø	
1	(b, 2, 0, 0, 0, 6, 00000001)	partition 0
2	(d, 3, 7, 1, 0, 4, 00000011)	
3	(j, 4, 0, 2, 0, 3, 00000111)	
4	(h, 0, 5, 3, 0, 1, 00010111)	
5	(m, 6, 11, 3, 0, 1, 00100111)	
6	(k, 0, 0, 5, 0, 1, 10100111)	
7	(g, 8, 0, 1, 0, 1, 00001001)	
8	(m, 9, 0, 7, 5, 1, 00101001)	
9	(i, 10, 0, 8, 0, 1, 01101001)	
10	(k, 0, 0, 9, 6, 1, 11101001)	
11	(g, 12, 0, 3, 7, 1, 00001111)	
12	(i, 0, 0, 11, 9, 1, 01001111)	
13	(d, 14, 0, 0, 2, 1, 00000010)	partition 1
14	(g, 0, 0, 13, 11,1, 00001010)	
15	(j, 16, 0, 0, 3, 2, 00000100)	partition 2
16	(m, 0, 17, 15, 8, 1, 00100100)	
17	(g, 18, 0, 15, 14, 1, 00001100)	
18	(h, 0, 0, 17, 4, 1, 00011100)	
19	(k, 0, 0, 18, 10, 1, 10011100)	
20	(h, 21, 0, 0, 18, 2, 00010000)	partition 3
21	(i, 0, 0, 20, 12, 1, 01010000)	
22	(m, 0, 0, 0, 16, 1, 00100000)	

Fig. 4. Final header table and linear table.

Afterward, this paper builds the MPL. First, it is processed sequentially from the 0th item of the header table. The index of the last occurrence of the current item on the linear table is located according to the *Last_node_link* of the item in the header table. The node *N* information at this index position is copied to the last node of the MPL where the item is located. And only copy the support count and binary bits of the node. Similarly, the node information pointed to by *N.point* also needs to be copied to the MPL where the item is located in reverse. Repeatedly execute until the index number pointed to by *N.point* is less than *First_node_link*, and the processing of this item is terminated.

Continue the above processing for the following item in the header table until all items have been processed.

The linear table and header table shown in Fig. 4 is taken as an example. It is assumed that *HT* represents the header table, and *HT[m]* represents the *m*th item in the header table. And *MPL[m]* represents the MPL where the *m*th item is located, where $m \geq 0$. *LT* represents the linear table, and *LT[i]* represents the node with index *i*.

The 0th item *b* of the header table is processed first. It can be seen that *HT[0].Last_node_link* = 1, so it is necessary to copy the *LT[1].frequency* and *LT[1].bit* of the node at index 1 in the linear table to *MPL[0]*. Then *LT[1].point* at index 1 in the linear table is judged to be 0, which is less than *HT[0].First_node_link* = 1. So the processing of item *b* in the header table is ended.

Next, the first item *d* of the header table is processed. The same can be obtained for *HT[1].Last_node_link* = 13, so copy the *LT[13].frequency* and *LT[13].bit* of the node at index 13 to *MPL[1]*. Then, it is judged that *LT[13].point* = 2 at index 13, which is not less than *HT[1].First_node_link* = 2. It is also necessary to copy *LT[2].frequency* and *LT[2].bit* at index 2 to MPL[1]. Continue to judge *LT[2].point* = 0 at index 2 in the linear table, which is less than *HT[1].First_node_link* = 2. So the processing of item *d* in the header table is also terminated. The remaining items in the header table are processed sequentially in the same way, and the final MPL diagram is shown in Fig. 5. The node of the generated MPL is sorted according to the partition order of the linear table in Fig. 4.

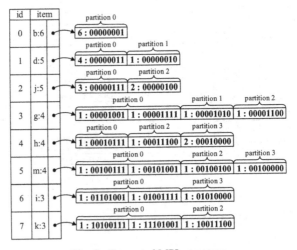

Fig. 5. Generated MPL structure.

3.4 Mining Based on MPL

Before beginning this chapter, a new bit splicing operation method proposed in this paper is introduced to obtain the candidate itemset. If the frequent itemsets that have been mined are $F = \{f_0, f_1, \ldots, f_{num}\}$, and $MPL[N]$ is currently being mined, where ($N \geq 0$). The algorithm uses binary to represent the itemset so that a new candidate itemset can be generated by the bitwise OR operation $f_t \mid 2^N$. Here 2^N is called the operator of the Nth mining stage, that is, $MPL[N]$, and $0 \leq t \leq num$.

First, the bit splicing operation generates candidate itemsets and then performs mining for each MPL sequentially from top to bottom. The length of the current MPL to be traversed is determined by the least significant bit of the candidate itemset. The bitwise AND operation determines whether the candidate itemset matches the node on the current MPL. If so, the support count on the node is accumulated. The total support count is then compared with the preset support count. Finally, the frequent itemsets are stored in the result table in turn.

Next, mining is described in detail with examples. This paper will continue to use the information in Table 2 and perform mining on the MPL in Fig. 5 in sequence. The result table before starting mining is null, as shown in Table 3. The *Index* column represents the sequence number of the frequent itemset in the result table. The *Itemset* column stores the actual frequent itemset, and the *Bit* column represents the binary form. The *Frequency* column records the corresponding support count.

Table 3. Initial result table

Index	ItemSet	Bit	Frequency
0	null	null	0

First, the 0th MPL in Fig. 5 is mined, and the operator in this stage is $2^0 = 00000001$. Because there is no frequent itemset in the result table at present, it is only necessary to save the information of $MPL[0].item$ to the result table. Therefore, the mining of the 0th MPL is over. This $MPL[0].item$ is item b, its corresponding binary is 00000001, and the support count is 6. Which is represented by a tuple (b, 00000001, 6) and inserted into the result table. At this time, the updated result table is shown in Table 4.

Table 4. Result table after mining MPL[0]

Index	ItemSet	Bit	Frequency
0	b	00000001	6

Next, the $MPL[1]$ in Fig. 5 is mined, and the operator in this stage is $2^1 = 00000010$. In the beginning, the information of $MPL[1].item$ is stored in the result table. That is, (d, 00000010, 5) is inserted into the result table. Table 4 is the result table after the last update, so only a bit splicing operation is required for frequent itemset in Table 4 to generate subsequent candidate itemset. It can be seen that there is only one record in Table 4, so it is only required to perform a bitwise OR operation on the binary bit at index 0 and the operator at this time to obtain a new candidate itemset. That is, 00000001

| 00000010 = 00000011, and the actual itemset it represents is *bd*. The binary of this candidate ends with 1. It only needs to be compared with nodes in the range of partition 0 in *MPL*[1]. There is only one node in partition 0 here, and a bitwise AND operation is performed on the candidate itemset generated above and the binary of this node. That is 00000011 & 00000011 = 00000011. It can be known that the current node matches the candidate itemset, and the support count on the node is accumulated. Consequently, the judgment of the candidate itemset 00000011 ends, it meets the minSup, and this itemset is inserted into the result table. The processing of the bit splicing operation of all the itemsets in Table 4 has been completed, and the updated result table is shown in Table 5.

Table 5. Result table after mining MPL[1]

Index	ItemSet	Bit	Frequency
0	b	00000001	6
1	d	00000010	5
2	bd	00000011	4

Next, the MPL[2] is mined, and the operator in this stage is $2^2 = 00000100$. The candidate itemsets of this stage are generated by performing a bit splicing operation on the itemsets in the latest result table (Table 5). First, the information of *MPL*[2].*item* (j, 00000100, 5) is stored in the result table.

The bit splicing is performed on the itemset at index 0 to obtain candidate itemset 00000101, whose actual itemset is *bj*. The candidate itemset ends with binary 1, so only the nodes in the range of partition 0 in *MPL*[2] need to be calculated. Then it is judged that the support count of the candidate itemset is 3, which is frequent.

The bit splicing is performed on the itemset at index 1 to obtain 00000110, and the corresponding itemset is *dj*. The candidate itemset ends with binary 10, then only the nodes in the range of partition 0 and partition 1 in *MPL*[2] need to be judged. Therefore, the support count for this candidate itemset is calculated as 3. It is frequent.

Similarly, bit splicing is performed on the itemset at index 2 to obtain 00000111. It is *bdj*. The candidate itemset also ends with 1, so it is only necessary to judge the nodes within the range of partition 0 in *MPL*[2]. The candidate itemsets are frequently by computation. The processing of Table 5 ends, and the updated result table is shown in Table 6. Continue to process the remaining MPL in Fig. 5 in the same way to obtain all frequent itemsets.

Table 6. Result table after mining MPL[2]

Index	ItemSet	Bit	Frequency
0	b	00000001	6
1	d	00000010	5

(continued)

Table 6. (*continued*)

Index	ItemSet	Bit	Frequency
2	bd	00000011	4
3	j	00000100	5
4	bj	00000101	3
5	dj	00000110	3
6	bdj	00000111	3

4 Pruning Strategy

4.1 Consecutive Supersets Pruning

After each mining phase executed by the BSFIM-DG algorithm, the number of consecutive supersets of frequent itemsets in this stage will be recorded. According to the Apriori principle, if the candidate itemset of a frequent itemset after the bit splicing operation is infrequent, its supersets must also be infrequent. Then the consecutive supersets of this item will not need to perform a bit splicing operation. That is, pruning is done.

For example, perform a bitwise OR operation of binary bit 0000 1000 with the operator 0100 0000 to obtain 0100 1000. If it is judged that 0100 1000 is infrequent, then items consecutively containing 0000 1000 will not need to be calculated. And the pruning is completed.

4.2 Grouping Pruning

In addition, grouping processing is performed on the result table. The binary bits stored in the nth group all end with binary 2^n. The candidate itemset generated by bit splicing of the first binary in the group with the operator of the current stage is judged to be infrequent. The remaining items in the current group are skipped to complete the pruning. For example, perform a bitwise OR operation on the binary bits at index 1(it is also the first item in group 1) in the result table with the operator 0100 0000 to get 1000 0010, which is judged to be infrequent. Then the candidates generated by the bit splicing operation on the remaining binary in group 1 will also be infrequent. Therefore, they will be directly skipped.

The pruning strategy is used to improve the execution efficiency of the algorithm considerably when mining is performed. The pseudocode of the algorithm after the pruning strategy is shown in Table 7.

Table 7. Pseudocode of BSFIM-DG algorithm

Pseudocode of BSFIM-DG algorithm
input: Dataset(D), Support(minSup)
output: result table(RT)
1) Dataset D is preprocessed
2) Constructing multi-partition list MPL by linear table
3) for i ← 0 to MPL_LT.size()
4)   insert the information of MPL[i] into RT'
5)   for r ← 0 to RT.size()
6)   tempItem ← Perform bit splicing operation on RT[r]
7)   MPL[i] is mined according to the binary end of tempItem
8)   if tempItem ⩾ minSup
9)   insert tempItem into RT'
10)   recording the number of consecutive supersets and grouping logically of tempItem
10)   else
11)   pruning operation of consecutive supersets and grouping
12) RT += RT'

5 Experiment

5.1 Test Environment and Dataset

The BSFIM-DG algorithm proposed is compared with the FP-growth, the Bit-combination, and the BCLT-O. During the experiment, the public datasets are used to evaluate each algorithm's performance. The soybean is a biological gene expression sequence. The accidents dataset is a series of real traffic accidents. Both chess and mushroom are derived from the UCI dataset and PUMSB. The retail has consisted of product information purchased. The T1 and T4 (that is, T10I4D100K, T40I10D100K) are generated using the generator from IBM Almaden Quest Labs.

Some datasets are dense, and some are sparse. Using fixed grouping will result in an imbalance in speed and memory usage. Therefore, this paper will conduct experiments and analyses on the above datasets to determine the coverage in a dynamic grouping and further calculate a reasonable number of groups for each dataset. Under the average support of each dataset, only the impact of coverage on algorithm running time is explored, but the impact on memory is not studied, as shown in Fig. 6. And it is because there is a linear relationship between coverage and the memory usage of the algorithm. The greater the coverage, the more sharing among items and less memory consumption.

Therefore, this paper balances the memory consumption of the algorithm under the condition of good running performance. It can be seen that the coverage is approximately between 70% and 90% maintaining good performance.

Next, this paper will choose to evaluate the algorithm at a coverage of 80%. Figure 7 shows the number of grouping for various datasets at this coverage.

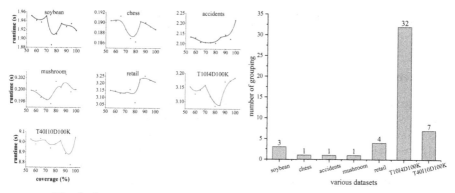

Fig. 6. Coverage experiment.   **Fig. 7.** Groups under 80% coverage.

5.2 Time Evaluation

Figure 8, 9, 10, 11, 12, 13 and 14 show the runtime results on each dataset. It can be seen that the BSFIM-DG algorithm proposed in this paper has the best running speed, followed by the BCLT-O and the FP-growth, and the worst is the Bit-combination algorithm.

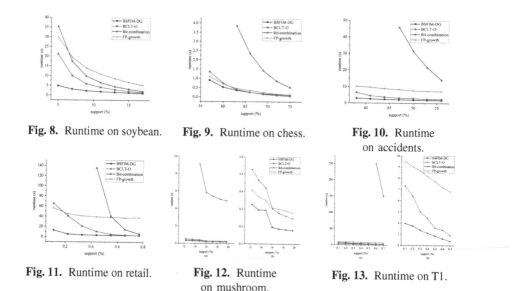

Fig. 8. Runtime on soybean.   **Fig. 9.** Runtime on chess.   **Fig. 10.** Runtime on accidents.

Fig. 11. Runtime on retail.   **Fig. 12.** Runtime on mushroom.   **Fig. 13.** Runtime on T1.

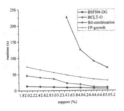
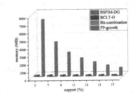
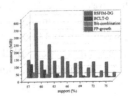

Fig. 14. Runtime on T4.   **Fig. 15.** Memory on soybean.   **Fig. 16.** Memory on chess.

5.3 Memory Evaluation

In this section, the memory usage of each algorithm is evaluated, as shown in Fig. 15, 16, 17, 18, 19, 20 and 21. It can be seen that the memory performance of the Bit-combination algorithm is better, followed by the BSFIM-DG algorithm, then the BCLT-O algorithm and the worst is the FP-growth algorithm. The running time of the Bit-combination algorithm increases exponentially, so taking into account the running efficiency and memory usage, the BSFIM-DG algorithm proposed in this paper has the best performance.

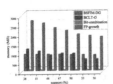
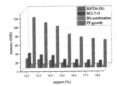
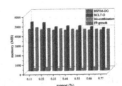

Fig. 17. Memory on accidents.   **Fig. 18.** Memory on mushroom.   **Fig. 19.** Memory on retail.

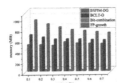
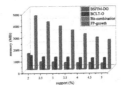

Fig. 20. Memory on T1.   **Fig. 21.** Memory on T4.

5.4 Result Analysis

The BSFIM-DG algorithm proposed in this paper has very outstanding runtime performance on all datasets, especially in the case of low support. This is because the BSFIM-DG algorithm uses MPL to replace the previous tree structure, which reduces the time cost of maintaining a large number of pointers. On the other hand, the algorithm uses bit splicing to obtain candidate itemsets. And cooperates with the pruning strategy, a large number of candidate itemsets are filtered out. The running time of the BSFIM-DG algorithm is close to that of the other three algorithms when the support is relatively large. The reason is that the support is set relatively large, the items to be mined are few, and the execution time is short. However, the time to construct the MPL has a more significant impact. Bit-combination algorithm and BCLT-O algorithm use bit traversal to generate candidate itemsets and repeatedly scan the entire original dataset to determine the frequency, which incurs a lot of time cost.

On each dataset, no matter how the support size is set, the memory occupied by the BSFIM-DG algorithm is almost uniform. The BSFIM-DG algorithm uses dynamic grouping to enhance the degree of sharing between nodes, thereby reducing memory consumption. Overall, the BSFIM-DG is superior to the BCLT-O and the FP-growth in the memory usage of various datasets. Nevertheless, there is still a particular gap with the Bit-combination algorithm. The reason is that the Bit-combination algorithm merely iterates over the original dataset without building a complex data structure. The FP-growth algorithm requires repeatedly constructing the FP-tree when performing mining, and a large number of pointers need to be maintained, which increases the memory overhead. In summary, the BSFIM-DG algorithm has better performance.

6 Conclusion

This paper proposes a new BSFIM-DG algorithm with an MPL structure for the frequent itemset mining problem. The MPL structure consists of arrays and stores the minimum valid information required for mining. Each node in MPL stores the corresponding combination of items instead of a single item, which saves the time cost of the traversal process. The BSFIM-DG algorithm uses coverage to calculate the number of groups dynamically. The length of the MPL to be traversed is calculated through the binary bit information of the candidate itemset. Experimental results on the different datasets demonstrate that the BSFIM-DG algorithm has outstanding performance in terms of running speed and memory usage compared with state-of-the-art algorithms. In particular, the algorithm utilizes bit splicing to generate candidate itemsets, and consecutive superset pruning and group pruning are used for additional filtering, which reduces the number of candidate itemsets. In addition, the algorithm proposed in this paper can also be practiced on topics such as frequent closed itemsets, maximum frequent itemsets, and top-k frequent patterns.

Acknowledgments. This research was supported by the Graduate Innovation Research Project of Heilongjiang University [grant number YJSCX2022-233HLJU].

References

1. Borgelt, C.: Frequent item set mining. Wiley Interdisc. Rev. Data-Min. Knowl. Discov. **2**(6), 437–456 (2012)
2. Agrawal, R., Srikant, R.: Fast algorithms for mining association rules. In: Proceedings of the 20th International Conference on Very Large Databases, VLDB. Citeseer (1994)
3. Han, J., et al.: Mining frequent patterns without candidate generation: a frequent-pattern tree approach. Data Min. Knowl. Discov. **8**(1), 53–87 (2004)
4. Aryabarzan, N., Minaei-Bidgoli, B., Teshnehlab, M.: NegFIN: an efficient algorithm for fast mining frequent itemsets. Expert Syst. Appl. **105**, 129–143 (2018)

5. Qu, J.-F., et al.: Efficient mining of frequent itemsets using only one dynamic prefix tree. IEEE Access **8**, 183722–183735 (2020)

6. Deng, Z.-H., Lv, S.-L.: PrePost+: an efficient N-lists-based algorithm for mining frequent itemsets via Children-Parent Equivalence pruning. Expert Syst. Appl. **42**(13), 5424–5432 (2015)

7. Deng, Z., Wang, Z., Jiang, J.: A new algorithm for fast mining frequent itemsets using N-lists. Sci. China Inf. Sci. **55**(9), 2008–2030 (2012)

8. Tseng, F.-C.: Mining frequent itemsets in large databases: the hierarchical partitioning approach. Expert Syst. Appl. **40**(5), 1654–1661 (2013)

9. Ahmed, S.A., Nath, B.: ISSP-tree: an improved fast algorithm for constructing a complete prefix tree using single database scan. Expert Syst. Appl. **185**, 115603 (2021)

10. Sun, J., et al.: Incremental frequent itemsets mining with FCFP tree. IEEE Access **7**, 136511–136524 (2019)

11. Lessanibahri, S., Gastaldi, L., Fernández, C.G.: A novel pruning algorithm for mining long and maximum length frequent itemsets. Expert Syst. Appl. **142**, 113004 (2020)

12. Lu, J., Zhao, R., Zhou, K.: Frequent item set mining algorithm based on bit combination. In: 2019 IEEE 4th International Conference on Cloud Computing and Big Data Analysis (ICCCBDA). IEEE (2019)

13. Zhou, K., Lu, J.: Research and implementation of bit combination-based frequent itemset mining algorithm for linear tables. Heilongjiang University (2020). https://doi.org/10.27123/d.cnki.ghlju.2020.000404

Entity Resolution Based on Pre-trained Language Models with Two Attentions

Liang Zhu, Hao Liu, Xin Song, Yonggang Wei, and Yu Wang[✉]

Hebei University, Baoding 071002, Hebei, China
wy@hbu.edu.cn

Abstract. Entity Resolution (ER) is one of the most important issues for improving data quality, which aims to identify the records from one and more datasets that refer to the same real-world entity. For the textual datasets with the attribute values of long word sequences, the traditional methods of ER may fail to capture accurately the semantic information of records, leading to poor effectiveness. To address this challenging problem, in this paper, by using pre-trained language model RoBERTa and by fine-tuning it in the training process, we propose a novel entity resolution model IGaBERT, in which interactive attention is applied to capture token-level differences between records and to break the restriction that the schema required identically, and then global attention is utilized to determine the importance of these differences. Extensive experiments without injecting domain knowledge are conducted to measure the effectiveness of the IGaBERT model over both structured datasets and textual datasets. The results indicate that IGaBERT significantly outperforms several state-of-the-art approaches over textual datasets, especially with small size of training data, and it is highly competitive with those approaches over structured datasets.

Keywords: Entity Resolution · Pre-trained Language Model · Interactive Attention · Global Attention

1 Introduction

Entity resolution (ER) is the process of identifying the records from one or more datasets that refer to the same real-world entity [1]. For example, r_1 and s_1 in Table 1 are two records referring to the products from Amazon and Google, respectively. The goal of ER is to identify whether the pair (r_1, s_1) refer to the same product. ER has become an important issue in big data research and the database community due to its applications in a wide variety of commercial, scientific and security domains [6, 7, 9]. Extensive research on ER has received considerable attention since 1946 [3], including rule-based [19, 23], machine learning [18], and crowd-sourcing [22] methods. For a structured dataset with explicit definition, format, and meaning of schema, its attribute values are generally "short strings" or "numeric fields", such as "*Mac*", "*199.95*" as shown in Table 1. Traditional techniques of ER can be used to handle structured datasets well based on their characteristics [14]. However, the attribute values of textual dataset correspond

X. Song et al. (Eds.): APWeb-WAIM 2023, LNCS 14333, pp. 433–448, 2024.
https://doi.org/10.1007/978-981-97-2387-4_29

to the original text and usually contain long sequences of words as shown in Table 2, and then it is a challenge for traditional methods of ER to process textual datasets effectively. To address this problem, several models for ER have been proposed in recent years, by using the mechanism of the family of Transformer-based pre-trained language models (TPLMs) including BERT [2], RoBERTa [12], etc.

Table 1. Record pair over structured datasets for ER [14].

	Title	Manufacturer	Price
r_1	Quickbooks pro 2007 for Mac (Mac)	Intuit	199.95
s_1	Intuit Quickbooks pro 2007 software for Mac finance software	NULL	189.95

Table 2. Record pair over textual datasets for ER [16].

	Description
r_2	zebra card printer zxp7 cod z71 rm0c0000em00 zebra zxp7 card printer color dye sublimation or thermal transfer printing usb ethernet interface single sided printing
s_2	zebra stampante card zxp7 magnetizzatore smart mifare contactless contact encoder cod z71 am0c0000em00

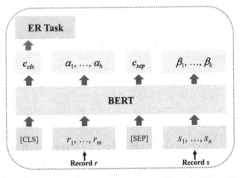

Fig. 1. Fine-tuning procedures for ER task.

TPLMs are based on the Transformer [21], which is an architecture using self-attention mechanism. TPLMs are firstly trained unsupervised on a large corpus, e.g., BERT is trained on BooksCorpus (800 M words) and English Wikipedia (2,500 M words). Then for the fine-tuning step, the TPLMs are first initialized with the pre-trained parameters, and all of the parameters are fine-tuned using small labeled data and small computing costs in the downstream tasks. Figure 1 shows a fine-tuning step for ER task, and the output of BERT is a set of 768-dimensional vectors, in which c_{cls} is considered to include all the contextual information of a record pair in the classification task, c_{sep} is a separator that has no useful information, and $(\alpha_1, \cdots, \alpha_h, \beta_1, \cdots, \beta_k)$ are the contextual vectors corresponding to the respective token, i.e., these vectors are encoded with contextual information. Moreover, most TPLMs can be obtained freely from Hugging Face library [24], and there are methods by applying TPLMs, such as Schema-agnostic EM [20] and DITTO [11], which only utilize the c_{cls} provided by BERT as input to the downstream model. However, the contextual information of a record pair may be lost in compressing a large amount of information into c_{cls} (i.e., a 768-dimensional vector), especially for long strings of record pairs, which leads to poor effectiveness.

To address the problems above, in this paper, we discuss the TPLM-based models of ER over structured and textual datasets. Our contributions are summarized below: (1) We propose a novel TPLM-based model with interactive attention and global attention. (2) To obtain relevant information from record pairs, we use interactive attention to capture token-level differences between record pairs, and global attention to determine the importance of these differences. (3) We conduct extensive experiments without injecting domain knowledge to measure the effectiveness of the proposed model and compare it with existing state-of-the-art methods. (4) Based on the LIME algorithm, we analyze both a matching and a non-matching decision of the proposed model, respectively.

The rest of this paper is organized below. Section 2 briefly reviews some related work. In Sect. 3, the problem definition is introduced, and the framework of IGaBERT is proposed. Section 4 presents the experimental results. Finally, we conclude the paper in Sect. 5.

2 Related Work

Numerous research works on ER have been proposed from different aspects such as declarative rules, crowd-sourcing, machine learning, and deep learning (DL). Here, we only review some related studies involving the ideas in DL.

In 2018, the model DeepER [4] first applied the ideas of DL to ER by employing GloVe [15] for word embedding and long short-term memory (LSTM) [8] for records representation. The DeepMatcher model [14] additionally used attention mechanism in LSTM. Most methods require that the records in the dataset(s) have the same schema. Unfortunately, records in real-world datasets usually come from different sources, which often have different schemas. In 2021, HierMatcher [5] introduced a hierarchical network that successfully breaks the limitations of the methods above.

In recent years, TPLMs have been studied extensively in natural language processing, and the main ideas of TPLMs are employed to extract the deep semantics of record pairs for ER. For instance, Schema-agnostic EM [20] treated ER as a simple binary classification task by feeding the c_{cls} provided by BERT into a binary classifier, which outperforms the model DeepMatcher [14]. Furthermore, DITTO [11] improved the performance of Schema-agnostic EM by manually injecting domain knowledge into the Schema-agnostic EM; however, this domain knowledge needs to be customized for different datasets and is not always available and effective.

3 Pre-trained Model with Two Attentions

The basic idea of the model IGaBERT (i.e., *Interactive Global attention BERT*) proposed in this paper is that ER can be processed accurately based on more information available from relevant data. As shown in Fig. 2, the architecture of IGaBERT is as follows: Firstly, the model gets contextual vectors of record pairs by using the RoBERTa model. Secondly, interactive attention is used to align all tokens of two records with extracting more detailed information and to calculate token-level comparison information between the two records. Thirdly, global attention is utilized to determine the weights of these comparisons for the final decision. Finally, a gate is designed to combine the above results with c_{cls}.

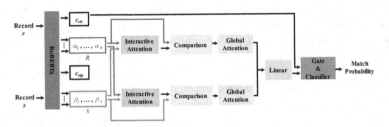

Fig. 2. Architecture of model IGaBERT.

3.1 Problem Definition

Given two records/tuples $r = (r_1, \cdots, r_m)$ and $s = (s_1, \cdots, s_n)$ from two data sources D_1 with schema $D_1(A_1, \cdots, A_m)$ and D_2 with schema $D_2(B_1, \cdots, B_n)$, where m may or may not be equal to n; A_i is the ith attribute of D_1; r_i is the attribute value of tuple r under the attribute A_i, i.e., r_i is a token sequence, and so are B_j and s_j ($1 \leq i \leq m$ and $1 \leq j \leq n$). Our model aims to decide whether r and s refer to the same real-world entity.

3.2 Framework of IGaBERT Model

Figure 2 illustrates the architecture of our model IGaBERT. We concatenate the attribute values of r and s to break the boundaries of attributes:

$$ConcatenateValues(r, s) = [CLS](r_1, \cdots, r_m)[SEP](s_1, \cdots, s_n) \qquad (1)$$

We take *ConcatenateValues*(r, s) as input of the RoBERTa model, and then we obtain c_{cls}, c_{sep} and the contextual vectors of each token α_i and β_j ($1 \leq i \leq h$, $1 \leq j \leq k$), in which RoBETRa is initialized with the parameters provided by Hugging Face library. Let $\Re^{d \times h}$ be the set of all $d \times h$ matrices over the field $\Re$ of real scalars. Suppose that $R = (\alpha_1, \cdots, \alpha_h) \in \Re^{d \times h}$ is the contextual matrix of r, h is the token number of r, and the column vector $\alpha_i = (a_{1i}, \cdots, a_{di})^T \in \Re^{d \times 1}$ is the ith contextual vector in R, $i = 1, 2, \cdots$, h. Similarly, $S = (\beta_1, \cdots, \beta_k) \in \Re^{d \times k}$ is the contextual matrix of s, k is the token number of s, and the column vector $\boldsymbol{\beta}_j = (b_{1j}, \cdots, b_{dj})^T \in \mathfrak{R}^{d \times 1}$ is the jth contextual vector in S, $j = 1, 2, \cdots, k$. Moreover, $\boldsymbol{c_{cls}} = (c^{(cls)}_1, \cdots, c^{(cls)}_d)^T \in \mathfrak{R}^{d \times 1}$, $\boldsymbol{c_{sep}} = (c^{(sep)}_1, \cdots, c^{(sep)}_d)^T \in \mathfrak{R}^{d \times 1}$, and $d = 768$ that is set by Liu et al. [12].

Unlike previous studies that only used $\boldsymbol{c_{cls}}$, we use all these contextual vectors output by RoBERTa in Fig. 2. First, we use $\boldsymbol{R}$ and $\boldsymbol{S}$ to obtain all token-level differences between r and s through the Interactive Attention component. Next, we use the Global Attention component to calculate the weights of these differences and to integrate these differences into a single vector. Then, we design a Gate component to combine the above results with $\boldsymbol{c_{cls}}$. Finally, we feed the fused vector into a binary classifier to generate the matching probability.

The IGaBERT breaks the attribute boundaries by $ConcatenateValues(r, s)$, which can process record pairs from the datasets of different schemas $D_1(A_1, \cdots, A_m)$ and $D_2(B_1, \cdots, B_n)$ with $m \neq n$. Focusing on token-level differences and assigning different weights to these differences, the Interactive Attention component and Global Attention component are effective against the interference of dirty data.

3.3 Interactive Attention

We illustrate the Interactive Attention component for each $\alpha_i \in \{\alpha_1, \cdots, \alpha_h\}$, $1 \leq i \leq h$. Suppose that $\boldsymbol{q}^{(t)} = (q^{(t)}_1, \cdots, q^{(t)}_d)^T \in \mathfrak{R}^{d \times 1}$ is a token-level query, $\boldsymbol{K} \in \mathfrak{R}^{d \times k}$ is a key matrix, and $\boldsymbol{V} \in \mathfrak{R}^{d \times k}$ is a value matrix. An interactive attention vector $\boldsymbol{e} = (e_1, \cdots, e_k)^T \in \mathfrak{R}^{k \times 1}$ is obtained using α_i, $\boldsymbol{q}^{(t)}$, and $\boldsymbol{K}$ as follows:

$$\boldsymbol{q}^{(t)} = \boldsymbol{W}^{(q)} \boldsymbol{a}_i \tag{2}$$

$$\boldsymbol{K} = \boldsymbol{W}^{(k)} \boldsymbol{S} \tag{3}$$

$$\boldsymbol{V} = \boldsymbol{W}^{(v)} \boldsymbol{S} \tag{4}$$

$$\boldsymbol{e} = \boldsymbol{K}^T \boldsymbol{q}^{(t)} \tag{5}$$

where matrices $\boldsymbol{W}^{(q)}, \boldsymbol{W}^{(k)}, \boldsymbol{W}^{(v)} \in \mathfrak{R}^{(d \times d)}$ are randomly initialized and jointly learned during training. Then we normalize $\boldsymbol{e}$ using the function $softmax(\cdot)$ to obtain $\boldsymbol{e}' = (e'_1, \cdots, e'_k)^T \in \mathfrak{R}^{k \times 1}$:

$$\boldsymbol{e}' = (e'_1, \cdots, e'_k)^T = softmax(\boldsymbol{e}) = (softmax(e_1), \cdots, softmax(e_k) \tag{6}$$

where $e'_i = softmax(e_i) = exp(e_i) / \left(\sum_{i=1}^{k} exp(e_i) \right)$. Then, $\boldsymbol{e}'$ and $\boldsymbol{V}$ are used to extract the information in $\boldsymbol{S}$ about α_i to obtain $\alpha'_i = \boldsymbol{V} \boldsymbol{e}' = (a'_{1i}, \cdots, a'_{di})^T \in \mathfrak{R}^{d \times 1}$, i.e., α'_i is the soft-aligned token for α_i in r with all tokens in s. In practice, matrix operations are used to obtain $\boldsymbol{R}' = (\alpha'_1, \cdots, \alpha'_h) \in \mathfrak{R}^{d \times h}$ directly by performing the above operations on each column in $\boldsymbol{R}$. Similarly, for $\boldsymbol{S} = (\boldsymbol{\beta}_1, \cdots, \boldsymbol{\beta}_k)$, we get $\boldsymbol{S}' = (\boldsymbol{\beta}'_1, \cdots, \boldsymbol{\beta}'_k) \in \mathfrak{R}^{d \times k}$.

3.4 Comparison

By using the above soft-aligned token matrix R' that contains the more detailed information, in the following process, we need to compare the differences between R and R'. Consider $R = (a_{ij})_{dh}$, $R' = (a'_{ij})_{dh}$, and $abs(R, R') = (|a_{ij} - a'_{ij}|)_{dh}$. We obtain their concatenation $[R, R', abs(R, R')]$, where the $[\cdot, \cdot]$ denotes the concatenation of vectors and $[\cdot, \cdot, \cdot] = [[\cdot, \cdot], \cdot]$. The concatenated vectors are fed into a fully-connected RELU network $FC(\cdot)$ as follows:

$$U^{(sr)} = FC([R, R', abs(R, R')]) \tag{7}$$

where $U^{(sr)} \in \Re^{d \times h}$ contains the differences from S to each column vector in R. Similarly, we can obtain $U^{(rs)} \in \Re^{d \times k}$ that contains the differences from R to each column vector in S.

3.5 Global Attention

The token-level differences obtained via the above process contribute differently to the final decision. The Global Attention component of IGaBERT model as shown in Fig. 2 can effectively identify which differences are more worthy of concern.

Let $q^{(g)} = (q^{(g)}_1, \cdots, q^{(g)}_d)^T \in \Re^{d \times 1}$ be a global-level query that is randomly initialized and jointly learned during training. Thus, $q^{(g)}$ can be seen as a representation of the query "which differences are more important". For $U^{(sr)}$, as an example, we utilize $q^{(g)}$ and $U^{(sr)}$ to obtain a global attention vector $g = U^{(sr)T} q^{(g)} = (g_1, \cdots, g_h)^T \in \Re^{h \times 1}$, and normalize g using the function $softmax(\cdot)$ to get $g' = softmax(g) = (g'_1, \cdots, g'_h)^T \in \Re^{h \times 1}$. Then, we use all the column vectors in $U^{(sr)}$ to generate a single vector $v^{(sr)} = U^{(sr)} g' = (v^{(sr)}_1, \cdots, v^{(sr)}_d)^T \in \Re^{d \times 1}$, where $v^{(sr)}$ represents the difference from s to r. Similarly, we can obtain $v^{(rs)} = (v^{(rs)}_1, \cdots, v^{(rs)}_d)^T \in \Re^{d \times 1}$, which represents the difference from r to s. Finally, $[v^{(sr)}, v^{(rs)}]$ can be generated by concatenating $v^{(sr)}$ and $v^{(rs)}$, and is fed into a fully-connected RELU network $FC(\cdot)$ to get $v = (v_1, \cdots, v_d)^T \in \Re^{d \times 1}$:

$$v = FC([v^{(sr)}, v^{(rs)}]) \tag{8}$$

3.6 Gate and Binary Classifier

In order to utilize the information of c_{cls}, inspired by [8], a Gate component is designed to fuse v and c_{cls} as follows:

$$sigmod(x) = 1/(1 + exp(-x)) \tag{9}$$

$$z = sigmod\left(w^{(1)} c_{cls} + w^{(2)} v + b\right) \tag{10}$$

$$v' = z c_{cls} + (1 - z)v \tag{11}$$

where row vectors $w^{(1)}$, $w^{(2)} \in \Re^{1 \times d}$, and real number $b \in \Re$ are randomly initialized and jointly learned during training, while $z \in (0, 1)$ is used for information flow control.

With the increase of z, c_{cls} will be more important in $v' \in \Re^{d \times 1}$ and vice versa. Then, we send v' to a FC($\cdot$) network followed by a SOFTMAX layer to get matching probabilities. Binary cross-entropy loss function below is used during the process of training.

$$loss = -(1/|T|) \sum\nolimits_{i=1}^{|T|} (y_i log(p) + (1 - y_i)log(1 - p)) \qquad (12)$$

where $|T|$ is the size of the training set T, $y_i \in \{0,1\}$ is the golden standard, and p is the probability output by our model.

4 Experimental Results

All the experiments are carried out on a server with Intel(R) Xeon(R) Gold 6330 CPU @ 2.00 GHz, 60 GB memory, and NVIDIA GeForce GTX 3090. We use PyTorch, an open-source machine learning framework, to train, validate, and test our model.

4.1 Parameters and Settings

In our model IGaBERT, RoBERTa, a pre-trained Transformer-based language model provided in [24], is used for the word embedding and can be fine-tuned during training. Moreover, the Adam algorithm is employed to optimize IGaBERT. The total number of parameters for IGaBERT is 140 M, and the training time of IGaBERT ranges from 0.02 h (for the dataset iA$_1$ as shown in Table 3) to 6.5 h (for the dataset Computers with Xlarge size in Table 4) depending on the size of training dataset. We report results based on a default setting: (1) The parameters of batch size, learning rate and dropout are 32, 1e-5 and 0.1, respectively. (2) Depending on the size of the datasets, the epoch number will be 15 or 30. (3) The F1 score is used as the metric. We present the average F1score for the results of three independent experiments.

4.2 Data Augmentation

This paper utilizes two approaches to enhance input data for our model IGaBERT. (1) TPLMs are sensitive to the order of input record pairs. Consequently, we double our training data by reversing the record pairs, e.g., (s, r) is the reverse of (r, s), meanwhile (s, r) retains the corresponding label of (r, s). (2) When input values consist of extremely lengthy strings, it is a challenge for a model to determine which components require attention during the matching process. TextRank [13] is a graph-based text summarization algorithm without requiring predefined keywords or phrases. Moreover, TextRank automatically extracts key information based on the content of the text, i.e., it is an unsupervised algorithm. In order to help our model IGaBERT make correct and effective decisions in the ER process, we employ the TextRank algorithm to filter out extraneous tokens and present the most significant ones to the model.

Compared to the model DITTO, the optimization approach used for our IGaBERT is an unsupervised manner, and it does not necessitate any customization based on the dataset, nor does it require additional domain knowledge to be injected.

4.3 Datasets

Five structured datasets are used in our experiments, as illustrated in Table 3, which are publicly available real datasets iTunes-Amazon$_1$ (i.e., iA$_1$), Walmart-Amazon$_1$ (WA$_1$), DBLP-ACM (DA$_1$), DBLP-Scholar$_1$ (DS$_1$) and Amazon-Google (AG) provided in [14], and all attributes of each dataset will be used in the experiments. Moreover, four dirty datasets in Table 3 are used to measure the robustness of our model, which are iTunes-Amazon$_2$ (iA$_2$), Walmart-Amazon$_2$ (WA$_2$), DBLP-ACM$_2$ (DA$_2$), and DBLP-Scholar$_2$ (DS$_2$) provided in [14]. These dirty datasets are generated from their respective original ones (e.g., iA$_2$ is generated from iA$_1$) by randomly moving the value of each attribute to the attribute *title* in the same tuple with 50% probability. #Positive is the number of the matching record pairs in the dataset. #Attr means the number of attributes in the dataset. Each dataset is split into training, validation, and test sets by the ratio of 3:1:1.

For the textual data, the same datasets used in [11] are used in this paper. The four datasets are from WDC Product Data Corpus [16], as shown in Table 4. The Corpus provides 4400 manually created labels of record pairs for four datasets: Computers, Watches, Cameras, and Shoes. Each dataset has 300 positive and 800 negative record pairs as a test set. There are four subsets of each dataset depending on the size of the data labeled Small, Medium, Large, and Xlarge. Each subset is split into a training set and a validation set with a ratio of 4:1. Our experiments use four attributes *title*, *brand*, *description*, and *specTableContent* during model training, where the latter two attributes contain long word sequences.

Table 3. Statistics of structured datasets.

Type	Dataset	Domain	Size	#Positive	#Attr
Structured	iA$_1$	music	539	132	8
	WA$_1$	electronics	10242	962	5
	AG	software	11460	1167	3
	DA$_1$	citation	12363	2220	4
	DS$_1$	citation	28707	5347	4
Structured (dirty)	iA$_2$	music	539	132	8
	WA$_2$	electronics	10242	962	5
	DA$_2$	citation	12363	2220	4
	DS$_2$	citation	28707	5347	4

4.4 Comparison with Existing Models

We will compare directly the experimental results of our model with those of three existing models reported in [11] and/or [14]. (1) *Magellan* [10]: A non-deep learning ER model. The experimental results of Magellan for structured datasets were reported in

Table 4. Statistics of textual datasets.

Datasets	Small	Medium	Large	Xlarge	Test
Computers	2834	8094	33359	68461	1100
Watches	2255	6413	27027	61569	1100
Cameras	1886	5255	20036	42277	1100
Shoes	2063	5805	22989	42429	1100

[14]. Notice that no results of Magellan over textual datasets were reported in [11, 14]. (2) *DeepMatcher* [14]: A DL-based ER framework which provides a categorization of DL solutions and defines a design space for these solutions. It requires that the records to be matched have the same pattern. The experimental results of DeepMatcher over structured datasets and textual datasets were reported in [11, 14]. (3) *DITTO* [11]: A DL-based ER solution using Transformer-based pre-trained language models. DITTO proposes three optimization techniques, injecting domain knowledge, summarizing long strings, and augmenting training data. The experimental results of DITTO were reported in [11].

Results Over Structured Datasets. Using the results in Table 5, we compare the performance of IGaBERT with those of the above three methods over nine structured datasets. (1) Compared with Magellan and DeepMatcher, our IGaBERT (significantly) outperforms them over all the nine datasets. For the three datasets iA_1, DA_1, and DS_1, all methods in Table 5 have excellent performances, and then the F1 score of IGaBERT is a little higher than that of the two methods. For datasets WA_1 and AG, IGaBERT significantly outperforms Magellan and DeepMatcher with the F1 score improving from 5.41 to 25.61. The main reason is that the data in WA_1 and AG are descriptions of products with few shared terms, in which case it becomes critical to understand the semantics of the record pairs. IGaBERT can extract more contextual information from record pairs by using attention mechanisms that can discriminate which information is vital for the final decision. (2) IGaBERT slightly under-performs DITTO over the three datasets iA_1, AG and DA_1, but slightly outperforms DITTO for the six datasets WA_1, DS_1, iA_2, DS_2, WA_2 and DA_2. The characteristics of the training sets play an important role in the performance of a model, e.g., the size of the training set for iA_1 is small and the attribute values of their records are short strings, which cause the information of the training sets is not enough, then DITTO can obtain better performance by injecting domain knowledge during training. Consequently, our IGaBERT without injecting additional domain knowledge achieves competitive performance compared to DITTO over these structured datasets. (3) IGaBERT is quite robust for addressing the challenge of dirty data. Because the four dirty datasets are generated by randomly moving the value of each attribute to the attribute *title* in the same tuple with 50% probability, IGaBERT can capture token-level difference information across attributes by using Interactive Attention component.

Results for Textual Dataset. We compare the performance of IGaBERT with those of two methods DeepMatcher and DITTO over four textual datasets. From Table 6, the following results are obtained. (1) Compared with DeepMatcher and DITTO, our

Table 5. Results of F_1 score over nine structured datasets.

Type	Datasets	Model F_1 score			
		Magellan	DeepMatcher	DITTO	IGaBERT
Structured	iA$_1$	91.2	88.5	**97.06**	95.48
	WA$_1$	71.9	67.6	86.76	**87.31**
	AG	49.1	69.3	**75.58**	74.71
	DA$_1$	98.4	98.4	**98.99**	98.97
	DS$_1$	92.3	94.7	95.6	**95.94**
Structured	iA$_2$	46.8	79.4	95.65	**96.42**
(dirty)	WA$_2$	37.4	53.8	85.69	**87.02**
	DA$_2$	91.9	98.1	99.03	**99.10**
	DS$_2$	82.5	93.8	95.75	**95.87**

IGaBERT (significantly) outperforms both of them over all the four datasets; moreover, DITTO outperforms DeepMatcher over the four datasets except for the dataset Shoes with "Large Size" (88.07 vs 90.39) and "Xlarge Size" (90.11 vs 92.61). The main reasons are as follows: (i) The data in all the four datasets are mainly about the description of commodities. The word sequences are longer strings and contain more noise than that of the structured datasets. (ii) The two attention components employed in IGaBERT can automatically distinguish between noise and information. (2) When the size of the training sets of the four datasets is "Small", our IGaBERT is significantly better than DITTO, which means that IGaBERT can perform well even when training data is insufficient. Generally, the cost of manual annotation grows rapidly as the amount of labeled data increasing, i.e., a bigger size of training set will lead to a larger cost of processes including computation and manual work. Thus, our model IGaBERT will be applied in a wider range of scenarios.

Note: It is a manual process to inject domain knowledge in DITTO; however, the technique of data augmentation used for IGaBERT is automatic, which can be implemented by a simple program without human intervention. As a result, IGaBERT exhibits more flexibility than DITTO.

4.5 Ablation Study

To gain more insights regarding the behavior of the part(s)/component(s) used in IGaBERT, including Data Augmentation (denoted by DA), Interactive Attention (IA), and Global Attention (GA), we conduct ablation study by comparing IGaBERT with its variants without some part(s)/component(s).

The experiments include: (1) IGaBERT-DA. No Data Augmentation is performed on the input data. (2) IGaBERT-IA. We directly perform the operations of Global Attention on the contextual vectors of tokens except c_{cls} and c_{sep}. (3) IGaBERT-GA. We assign

Table 6. Results of F1 score over four textual datasets with various sizes of the training set.

Datasets	Size	Model F_1 score		
		DeepMatcher	DITTO	IGaBERT
Computers	Small	70.55	80.76	**89.81**
	Medium	77.82	88.62	**93.56**
	Large	89.55	91.70	**95.85**
	Xlarge	90.80	95.45	**97.01**
Cameras	Small	68.59	80.89	**87.56**
	Medium	76.53	88.09	**92.53**
	Large	87.19	91.23	**94.33**
	Xlarge	89.21	93.78	**94.82**
Watches	Small	66.32	85.12	**90.56**
	Medium	79.31	91.12	**92.91**
	Large	91.28	95.69	**96.37**
	Xlarge	93.45	96.53	**96.89**
Shoes	Small	73.86	75.89	**81.03**
	Medium	79.48	82.66	**86.23**
	Large	90.39	88.07	**91.48**
	Xlarge	92.61	90.11	**92.85**

the same weight to all token-level differences. As an example, we only report the results over textual datasets, in which the same settings in Sect. 4.1 are used.

From Table 7, the following results are obtained. (1) IGaBERT achieves the highest F1 scores over all datasets with all sizes, and removing the part DA and the components IA, GA from IGaBERT results in an average decrease of 1.90, 1.45, and 1.93 in F1 scores, respectively. Therefore, each of the part(s)/component(s) plays an important role in IGaBERT and contributes to the model's performance. (2) When size of the dataset is "Small", compared to other dataset sizes, removing the part DA results in the most significant decrease in model performance over the four datasets Computers, Cameras, Watches, and Shoes, with a decrease in F1 scores of 3.29, 2.58, 3.14, and 4.63, respectively. However, the components GA and IA are not size sensitive, i.e., their contributions to the model IGaBERT are independent of the size of a training set.

4.6 Case Study

LIME (Local Interpretable Model-agnostic Explanations) [17] is an explainable AI algorithm used to interpret deep neural network, by approximating it locally with an interpretable model. Based on the LIME algorithm, we create explanations for both a matching decision and a non-matching decision, respectively.

Table 7. F1 score of IGaBERT and its variants over four textual datasets.

Datasets	Size	Model F_1 score			
		IGaBERT-DA	IGaBERT-IA	IGaBERT-GA	IGaBERT
Computers	Small	86.52	88.65	87.09	**89.81**
	Medium	92.55	90.38	89.64	**93.56**
	Large	95.04	93.97	94.45	**95.85**
	Xlarge	96.24	96.02	95.91	**97.01**
Cameras	Small	84.98	85.95	85.73	**87.56**
	Medium	90.20	92.14	92.51	**92.53**
	Large	92.74	94.04	93.13	**94.33**
	Xlarge	93.04	93.53	93.85	**94.82**
Watches	Small	87.42	89.73	90.06	**90.56**
	Medium	91.65	91.75	90.21	**92.91**
	Large	95.97	95.69	93.18	**96.37**
	Xlarge	96.01	95.53	95.09	**96.89**
Shoes	Small	76.40	79.45	79.43	**81.03**
	Medium	84.61	84.16	84.79	**86.23**
	Large	89.42	89.12	87.65	**91.48**
	Xlarge	90.61	90.47	90.21	**92.85**

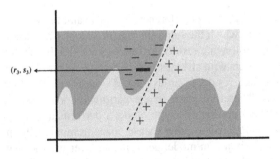

Fig. 3. Primary intuition of the explanation for IGaBERT with a pair (r, s).

Given a pair (r, s), Fig. 3 depicts the primary intuition for the explanation of our model IGaBERT(r, s), here, which is denoted by $\varphi(r, s)$ for simplicity. In Fig. 3, the green zone indicates that the result of $\varphi(r, s)$ is matching for the record pair (r, s), i.e., $\varphi(r, s) = $ " $+$"; while the orange zone (with two parts) means that the result of $\varphi(r, s)$ is not matching, i.e., $\varphi(r, s) = $ " $-$". For instance, the long bold negative sign "$-$" means that the record pair (r_3, s_3) is non-matching, and the other positive signs {" $+$"} or negative signs {" $-$"} around "$-$" indicate the other pairs generated from (r_3, s_3) may or may not be matching. The dash line is a simple decision function $\psi(r, s)$ of logistic regression model, which is assumed to approximate $\varphi(r, s)$ locally.

The above non-matching record pair (r_3, s_3) is selected randomly from textual dataset Computers with r_3 = "*corsair 256 gb flash voyager slider x2 usb 3 0 drive cmfsl3x2 c ocuk*" and s_3 = "*corsair 128 gb flash voyager slider x2 usb 3 0 drive cmfsl3x2 c ocuk*", where $r_3 = s_3$ except for the non-matching substrings "*256 gb*" of r_3 and "*128 gb*" of s_3. Firstly, a set of several perturbations $\{(r^{(i)}_3, s^{(i)}_3): i = 1,..., K\}$ is generated by eliminating some word(s) from r_3 and s_3 respectively, where $r^{(i)}_3$ is a substring of r_3 and $s^{(i)}_3$ is a substring of s_3. Secondly, inputting $\{(r^{(i)}_3, s^{(i)}_3)\}$ into $\varphi(r, s)$, we obtain their corresponding labels. Thirdly, by using the strategy of Bag of Words, we vectorize the perturbations, and use the set of these vectors to train a simple logistic regression model $\psi(r, s)$, which approximates the original $\varphi(r, s)$ model locally. Finally, the coefficients of the logistic regression model $\psi(r, s)$ are extracted to represent the weights of words when $\varphi(r, s)$ makes decisions.

Similarly, we process the pair (r_2, s_2), which is a matching pair as shown in Table 2. Figure 4 illustrates the weights of words for (r_2, s_2) and (r_3, s_3), which are the descriptions of products. For the matching pair (r_2, s_2), our model IGaBERT makes a "matching" decision based on the brand ("*zebra*") and the model ("*z71*", "*zxp7*", "*rm0c0000em00*", and "*am0c0000em00*"). For the non-matching pair (r_3, s_3), however, IGaBERT makes a "non-matching" decision based on the difference in storage capacity ("*128 gb*" and "*256 gb*"). For both (r_2, s_2) and (r_3, s_3), therefore, the results and rules of $\varphi(r, s)$ are consistent with those of human cognition.

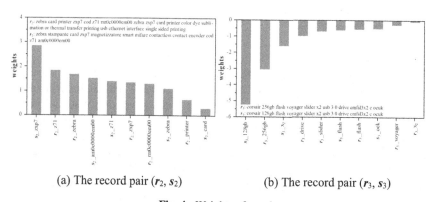

(a) The record pair (r_2, s_2) (b) The record pair (r_3, s_3)

Fig. 4. Weights of words.

To illustrate the effect of the Global Attention and Interactive Attention components used in our IGaBERT, we take (r_2, s_2) as an example to visualize attention scores of each word (i.e., e' and g' mentioned in Sects. 3.3 and 3.5). Using the Subword Tokenization Algorithms [24], the tokenizer can decompose "*annoyingly*" into "*annoying*" and "*ly*". For a split-up word, we sum the attention scores over its tokens to display word-level attention scores. From Table 8, "*zxp7*" in s_2 receives the highest global attention score of 0.5406, "*zebra*" and "*am0c0000em00*" receives scores of 0.2262 and 0.1248, respectively. The global attention scores of the other words are quite close to zero.

These results demonstrate that IGaBERT can focus on important information. Then, for "*zxp7*" with the highest global attention score in s_2, we display the interactive attention scores of all words in r_2. From Table 9, for "*zxp7*" in s_2, our Interactive Attention component mainly selects "*zxp7*" in r_2 as its alignment. Thus, IGaBERT is capable of comparing the difference between r_2 and s_2 in token-level.

Table 8. Global attention scores for the record s_2 in (r_2, s_2).

s_2	**zebra** stampante card **zxp7** magnetizzatore smart mifare contactless contact encoder cod z71 **am0c0000em00**
global attention score	**0.2262** 0.0168 0.012 **0.5406** 0.021 0.0003 0.0042 0.0024 0.0014 0.0035 0.0166 0.047 **0.1248**

Table 9. Interactive attention scores of all words in r_2 to "*zxp7*" in s_2.

r_2	zebra card printer **zxp7** cod z71 rm0c0000em00 zebra **zxp7** card printer color dye sublimation or thermal transfer printing usb ethernet interface single sided printing
interactive attention score	0 0 0 **0.33** 0 0.08 0 0 0 **0.59** 0 0 0 0 0 0 0 0 0 0 0 0 0 0 0

5   Conclusion

For entity resolution (ER), based on pre-trained language models with interactive attention and global attention, we proposed a new model IGaBERT without injecting domain knowledge. Interactive attention is used to capture token-level comparison information between record pairs, and global attention enables the model to focus on the critical information. Extensive experiments are conducted on nine structured datasets and four textual datasets to measure the effectiveness of IGaBERT, and to compare with existing state-of-the-art methods Magellan, DeepMatcher, and DITTO. The experimental results show that our IGaBERT significantly outperforms Magellan and DeepMatcher for all the datasets. For the nine structured datasets, the results indicate that IGaBERT is highly competitive with DITTO, meanwhile IGaBERT outperforms DITTO over the four textual datasets, especially when the training data is insufficient, i.e., the size of training dataset is "Small". Moreover, our IGaBERT is quite robust for addressing the challenge of dirty data, and can be applied in a wider range of scenarios. In future work, we plan to explore Blocking algorithms and integrate IGaBERT into an ER system.

References

1. Christophides, V., Efthymiou, V., Palpanas, T., Papadakis, G., Stefanidis, K.: An overview of end-to-end entity resolution for big data. ACM Comput. Surv. **53**(6), 1–42 (2020)
2. Devlin, J., Chang, M.W., Lee, K., Toutanova, K.: BERT: pre-training of deep bidirectional transformers for language understanding. arXiv:1810.04805 (2018)
3. Dunn, H.L.: Record linkage. Am. J. Public Health Nations Health **36**(12), 1412–1416 (1946)
4. Ebraheem, M., Thirumuruganathan, S., Joty, S., Ouzzani, M., Tang, N.: Distributed representations of tuples for entity resolution. Proc. VLDB Endowment **11**(11), 1454–1467 (2018)
5. Fu, C., Han, X., He, J., Sun, L.: Hierarchical matching network for heterogeneous entity resolution. In: Proceedings of the Twenty-Ninth International Conference on International Joint Conferences on Artificial Intelligence, pp. 3665–3671 (2021)
6. Gal, A.: Uncertain entity resolution: re-evaluating entity resolution in the big data era: tutorial. Proc. VLDB Endowment **7**(13), 1711–1712 (2014)
7. Getoor, L., Machanavajjhala, A.: Entity resolution for big data. In: Proceedings of the 19th ACM SIGKDD International Conference on Knowledge Discovery and Data Mining, p. 1527 (2013)
8. Hochreiter, S., Schmidhuber, J.: Long short-term memory. Neural Comput. **9**(8), 1735–1780 (1997)
9. Kejriwal, M.: Entity resolution in a big data framework. In: Proceedings of the Twenty-Ninth AAAI Conference on Artificial Intelligence, pp. 4243–4244 (2015)
10. Konda, P., et al.: Magellan: toward building entity matching management systems over data science stacks. Proc. VLDB Endowment **9**(13), 1581–1584 (2016). https://doi.org/10.14778/3007263.3007314
11. Li, Y., Li, J., Suhara, Y., Doan, A., Tan, W.C.: Deep entity matching with pre-trained language models. Proc. VLDB Endowment **14**(1), 50–60 (2020)
12. Liu, Y., et al.: Roberta: a robustly optimized BERT pretraining approach. arXiv:1907.11692 (2019)
13. Mihalcea, R., Tarau, P.: TextRank: bringing order into text. In: Proceedings of the 2004 Conference on Empirical Methods in Natural Language Processing, pp. 404–411 (2004)
14. Mudgal, S., et al.: Deep learning for entity matching: a design space exploration. In: Proceedings of the 2018 International Conference on Management of Data, pp. 19–34 (2018)
15. Pennington, J., Socher, R., Manning, C.D.: Glove: global vectors for word representation. In: Proceedings of the 2014 Conference on Empirical Methods in Natural Language Processing, pp. 1532–1543 (2014)
16. Primpeli, A., Peeters, R., Bizer, C.: The WDC training dataset and gold standard for large-scale product matching. In: Companion Proceedings of the 2019 World Wide Web Conference, pp. 381–386 (2019)
17. Ribeiro, M.T., Singh, S., Guestrin, C.: " Why should I trust you?" Explaining the predictions of any classifier. In: Proceedings of the 22nd ACM SIGKDD International Conference on Knowledge Discovery and Data Mining, pp. 1135–1144 (2016)
18. Sarawagi, S., Bhamidipaty, A.: Interactive deduplication using active learning. In: Proceedings of the eighth ACM SIGKDD International Conference on Knowledge Discovery and Data Mining, pp. 269–278 (2002)
19. Singh, R., et al.: Synthesizing entity matching rules by examples. Proc. VLDB Endowment **11**(2), 189–202 (2017). https://doi.org/10.14778/3149193.3149199
20. Teong, K.S., Soon, L.K., Su, T.T.: Schema-agnostic entity matching using pre-trained language models. In: Proceedings of the 29th ACM International Conference on Information & Knowledge Management, pp. 2241–2244 (2020)

21. Vaswani, A., et al.: Attention is all you need. In: Proceedings of the 31st International Conference on Neural Information Processing Systems, pp. 6000–6010 (2017)
22. Wang, J., Kraska, T., Franklin, M.J., Feng, J.: CrowdER: crowdsourcing entity resolution. Proc. VLDB Endowment 5(11), 1483–1494 (2012)
23. Wang, J., Li, G., Yu, J.X., Feng, J.: Entity matching: how similar is similar. Proc. VLDB Endowment 4(10), 622–633 (2011)
24. Wolf, T., et al.: Huggingface's transformers: State-of-the-art natural language processing. arXiv:1910.03771 (2019)

A High-Performance Hybrid Index Framework Supporting Inserts for Static Learned Indexes

Yuquan Ding and Xujian Zhao$^{(\boxtimes)}$

Southwest University of Science and Technology, Mianyang, China
yuquanding@mails.swust.edu.cn, jasonzhaoxj@gmail.com

Abstract. The learned index is a new index structure that uses a trained model to directly predict the position of a key and thus has high query performance. However, static learned indexes cannot handle insert operations. Although static PGM-index uses a dynamic data structure to support inserts, it faces a serious read amplification problem under read-write workloads, as the inefficient lookup process of the buffers diminishes the learned indexes. Besides, this structure also leads to periodic retraining of the internal PGM-indexes because the buffers and the learned indexes are strongly coupled, which is unacceptable for those static learned indexes that need tuning. Obviously, this structure is not an ideal general framework. In this paper, we propose a two-layer Hybrid Index Framework (HIF) to address such issues. Specifically, the dynamic layer is used as a buffer for inserts, and the static layer consisting of static learned indexes is used for lookups only. HIF effectively alleviates read amplification by searching the static layer directly. And with this hierarchical structure, HIF isolates learned indexes from insert operations. Thus HIF can completely avoid the retraining of the learned indexes by transformation strategy from the dynamic layer to the static layer. Moreover, we provide a self-tuning algorithm for the learned indexes that cannot be built in a single pass over the data, allowing them to be applied to dynamic workloads with low training overhead. We have conducted experiments using multiple datasets and workloads and the results show that on average, three HIF-based static learned indexes, HLI, PGM, and RMI, achieve up to $1.8\times$, $1.7\times$, and $1.5\times$ higher throughput than the original dynamic PGM-index for insert ratio below 70%.

Keywords: Static Learned Index · Index Framework · Insert Operation

1 Introduction

The learned index is a new type of index that incorporates machine learning techniques, which consider an index as a model that takes a key and outputs the position of that key in an array. As the first work in this area, Recursive Model Index (RMI) [1] improves the lookup performance by 1.5–3× compared to the

B+Tree. However, this index does not support insert operations. This is because newly inserted data destroy existing data distribution, which will invalidate the saved errors of the models and make it impossible to find the correct position. Like RMI, several subsequent works, such as FITing-tree [2] and PGM-index [3], are still static indexes even though they have improved space occupation or prediction accuracy, which means they face the same problem. Therefore, it is necessary to re-learn the data distribution to create a new learned index for each insert operation to guarantee the validity of the learned index, which is obviously an unacceptable cost.

To solve this problem, PGM-index uses a series of arrays as a buffer to receive inserted data. When the buffer is full, these arrays are merged into a larger array on which a PGM-index is trained. Although this method avoids frequent retraining under workloads containing inserts, it has two problems. First, the inefficient search method leads to a serious read amplification problem, since a single lookup is likely to require multiple binary searches on multiple arrays. Second, due to the high coupling between the learned indexes and the buffers, the insert operations of this structure inevitably lead to periodic retraining of the learned indexes which blocks all operations until it has finished. It can be seen that an inappropriate data structure completely fails to exploit the potential of static learned indexes. So, it is necessary to design a dedicated dynamic index structure for static learned indexes.

Aiming at these issues, we propose a Hybrid Index Framework (HIF) for all static learned indexes that make them workable for read-write workloads. This framework is a two-layer structure consisting of a dynamic layer and a static layer, where the dynamic layer uses a B+Tree for inserts and the static layer consists of multiple static learned indexes for lookups only. To avoid traditional search methods weakening the performance of learned indexes, HIF first searches the static layer. This greatly alleviates read amplification because a large amount of data will be transferred to the static layer under read-write workloads. Obviously, it is not wise to check the buffer once for all lookups. When the amount of buffer data reaches a threshold, the B+Tree is directly transformed into a learned index in the static layer. The process is easy because the storage of the B+Tree naturally satisfies the requirement for sorted data for learned indexes. Therefore, subsequent inserts will never have an impact on the existing learned indexes. Furthermore, to improve the generality of the framework, we provide a general self-tuning algorithm for the learned indexes that need to improve their performance by changing the number of models. With this algorithm, this class of indexes can get a suitable result quickly, thus avoiding long blocking times. Overall, the main contribution of this paper can be summarized as follows.

- We propose a general Hybrid Index Framework, HIF, to solve the problem that static learned indexes do not support insert operations. To protect the existing learned indexes, this two-layer structure uses B+Trees as dynamic layers to receive data. Further, the HIF avoids the traditional search method to weaken the lookup performance of static learned indexes by searching the static layer consisting of learned indexes first, which effectively alleviates the read amplification problem of the original dynamic PGM-index.

- We propose an efficient transformation strategy based on the HIF. Since the data organization of the B+Tree naturally meets the requirement for learned indexes that the underlying data must be sorted, the B+Tree can be conveniently transformed into a static learned index, avoiding affecting other learned index structures. Therefore, there is no retraining in HIF.
- We propose a self-tuning algorithm for the static learned indexes that need to be tuned under read-write workloads. With this algorithm, such static learned indexes can automatically and quickly adjust its own structure to obtain better performance, which greatly alleviates the blocking problem caused by adjusting the learned index structure.
- We conduct extensive experiments with both real and synthetic datasets under varying read-write workloads, and compare three HIF-based static learned indexes with the dynamic PGM-indx, XIndex, Masstree, and B+Tree. Specifically, HLI and PGM, on average, achieve up to 1.8× and 1.7× higher throughput than dynamic PGM-index for insert ratio below 70% and is far better than B+Trees in all cases. And RMI also shows good performance in all but two datasets with extremely complex data distribution.

2   Motivation

2.1   Dynamic PGM-Index

The PGM-index adopts a dynamic data structure to support inserts (see Fig. 1). This structure consists of several ordered arrays and the higher-ordered arrays have a larger capacity. To share the cost of retraining as much as possible, this structure does not build a learned index on each array, but only selects those arrays with sufficient capacity. Because the arrays on the front will be filled up quickly with new data due to their small size, and therefore need to copy the data stored in it to a larger array behind them to leave new space. Obviously, the learned indexes built on these small arrays will be destroyed immediately, which brings a huge retraining overhead.

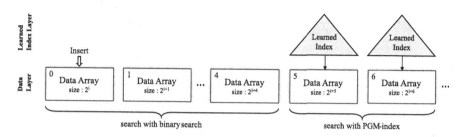

Fig. 1. The dynamic PGM-index.

Specifically, insert operations may trigger a merge of data in multiple arrays (Fig. 2). When the first array has no space left, its internal data and the newly inserted data form a sorted dataset, and then this array is cleared. The dynamic structure then carries this sorted dataset to detect the array behind it sequentially. In this process, the dataset is merged and sorted with each array along the way to form a new sorted dataset, and then empty this array. At the same time, it determines whether the current array has enough space for this new dataset. If the array is large enough, it puts the dataset into the current array and ends the merge process. Instead, the process continues until it finds a larger array that can hold all the data involved in the process.

Fig. 2. Periodic retraining of the dynamic PGM-index.

2.2   The Problem

Read Amplification. As can be seen in Fig. 1, lookups need to traverse multiple arrays until the target key is found in a certain array, which causes the problem of read amplification. In addition, the front of the array set can only be searched by binary search, which does not even make use of any index structure. Obviously, this method is undoubtedly inefficient.

Periodic Retraining of the Learned Index. The learned index is an auxiliary structure that depends on a large array. Therefore, the array to which the learned index is attached also needs to receive all the data from all arrays in front of it many times (Fig. 2). The data distribution of each array will be changed after receiving a new dataset, which causes the existing learned index to be invalid, so a newly learned index must be learned on the new dataset.

2.3   The Reason

The reasons for these problems can be summarized as follows. First, the too-inefficient lookup method weakens the excellent lookup performance of the static learned index. Because, under read-write workloads, a large amount of data is stored in large-size arrays with a learned index at the end of the array set. Obviously, it is not wise to check many small buffers using a binary search for all lookups. Second, the buffers and static learned indexes are strongly coupled. Specifically, each array is both a buffer for the next array and the underlying storage structure of the learned index. To summarize, the dynamic data structure of PGM-index cannot take full advantage of static learned indexes.

3   Hybrid Index Framework

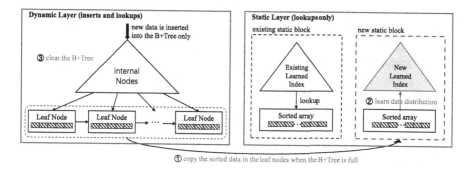

Fig. 3. The HIF's transformation process from the dynamic layer to the static layer.

3.1   Overview

In summary, the goal of this paper is to alleviate read amplification and decouple the insert operation from the retraining. This requires us to redesign a dedicated dynamic index structure for static learned indexes, so we propose a two-layer hybrid index framework (Fig. 3).

To support inserts without affecting the learned indexes, this paper adopts the same idea as PGM-index, i.e., using a buffer as the dynamic layer to receive new data while protecting the learned index structure. In this case, the traditional index structure naturally fits the requirements of learned indexes. Because the traditional index structure does not consider the distribution of data, which means that they do not need to be retrained. So under workloads containing inserts, traditional indexes protect the learned indexes while ensuring a lower bound on search performance.

Besides, the static layer consists of multiple static learned indexes that are used for lookups only. To maximize the performance of static learned indexes, the lookup operation starts from the static layer first, which is to avoid the read amplification problem caused by traditional search methods. Further, to prevent the framework from degenerating into a B+Tree, the framework can transform the B+Tree into a learned index without affecting other learned indexes, thus avoiding the retraining of the learned indexes.

3.2   The Dynamic Layer for Inserts

There are two advantages to using a single B+tree as a buffer. On the one hand, the B+Tree has good lookup performance in that it locates leaf nodes quickly by traversing internal nodes, which is more efficient than performing a binary search within a large array. On the other hand, the use of B+Tree reduces the cost of converting buffers to a learned index. This is because the data in the leaf

---

Algorithm 1: Insert Algorithm

Input: *key*: key to be inserted, *value*: the payload, *buffer_max_size*: threshold
      of B+Tree size, *uesed_level*: the number of existing indexes

1   *btree.insert(key, value)*   // insert into static layer
2   **if** *btree.count() < buffer_max_size* **then**
3 | *return*

4   *slot_required ← buffer_max_size + 1*
5   **for** *i in used_levels* **do**   // find an empty static block
6 | **if** *data_arrays[i] == 0* **then**
7 | | *empty_level ← i*
8 | | *break*

9   **if** *empty_level == used_levels* **then**
10 | *used_levels ← used_levels + 1*

11 *create a static learned index of size slots_required in empty_level*

---

nodes of the B+Tree are globally ordered and the leaf nodes are connected by pointers, which naturally meets the requirement that the underlying data must be sorted for the learned index. Therefore, the data can be quickly copied to the static layer by sequentially traversing the leaf nodes when building the learned index, which greatly reduces the cost of data preprocessing.

In addition, to avoid the framework degenerating into a B+Tree, we will not let the size of the B+Tree grow indefinitely. So, after inserting a certain amount of data, the B+Tree will be transformed into a learned index (see Fig. 3). Therefore, the size of the B+Tree needs to be chosen carefully. A too large threshold can lead to two problems. First, the height of the B+Tree will increase, which causes a degradation of the B+Tree query performance; second, a too-large B+Tree means a larger dataset and a more complex data distribution. It will increase the difficulty of fitting and leads to a large prediction error in the learned index. On the contrary, a small size leads to too many learned indexes in the static layer. Obviously, it brings the problem of read amplification. So it is important to limit the amount of data input to the static layer reasonably. Empirically, we set the size at the 10 million level, which ensures good performance of the learned index with any dataset.

The insert algorithm of this hybrid indexing framework is shown in Algorithm 1. First, the data is inserted into the B+Tree regardless of whether the amount of data in the B+Tree reaches the threshold. However, when the amount of data in the B+Tree reaches the threshold, all the data inside the B+Tree is copied to a new static block in the static layer, and then the current B+Tree is cleared. Subsequently, a static learned index is trained based on the data in that static block. Eventually, the static layer generates a new static block containing all the data of the B+Tree and a learned index that indexes these data.

Algorithm 2: Lookup Algorithm

Input: *key*: key, *min_index_level*: start position of the learned indexes,
    uesed_level: the number of existing indexes
Output: *value*

```
1  for i ← min_index_level to used_levels do
2  │   it ← LI.search(key)              // search using static learned index
3  │   if it.first == key then
4  │   └   return it.second
5  it ← btree.search(key)                // final search buffer
6  if it.first == key then
7  └   return it.second
8  return NULL
```

3.3 The Static Layer for Lookups

As shown in Fig. 3, the static layer consists of multiple static blocks that contain sorted data and a learned index for indexing this data. In this layer, the size of the static block is almost the same as the size of the B+Tree, i.e., the maximum capacity of the B+Tree plus one newly inserted data. In the beginning, there is only one static block containing initialized data in the static layer. As described in the previous section, the data from the B+Tree of the dynamic layer will enter the static layer after reaching a threshold. In order to avoid the huge overhead caused by retraining, the input data of the dynamic layer will not be merged with the data of any of the existing static blocks but will be put into a new static block, and a new learned index will be learned in it afterward. Thus the number of static blocks in the static layer gradually increases under a workload containing insert operations. This process can be considered as a direct transformation of the B+Tree into a newly learned index without additional overhead. It can be seen that the existing learned indexes in the static layer will not be destroyed. In other words, these indexes will not be affected by subsequent insert operations.

It can be seen that as the amount of inserted data increases, a large amount of data will move into the static layer with the transformation strategy. In dynamic PGM-index, for most of the data located in the static layer, each lookup requires traversing the buffers first. Obviously, this step is meaningless. To alleviate the read amplification, we must avoid the impact of an inefficient search process. Therefore, the lookup operation of HIF starts with the first static learned index in the static layer, and only searches the dynamic layer after determining that the target key does not exist in the static layer (see Algorithm 2).

With a two-layer design and transformation strategy, learned indexes do not need to consider the problems caused by insertion, so they can maximize their advantages in lookup. However, storing data in chunks causes the data to be globally unordered. This leads to multiple indexes that may need to be traversed for a lookup of a key, which undoubtedly slows down the query speed. In this case, learned indexes are more suitable for this chunked data organization than

---

Algorithm 3: Self-Tuning Algorithm

Input: *init_n*: Initial number of models, *model_n*: the number of underlying
models used in the next iteration, *step_size*: the number of models
added at each time

Output: *the learned index*

1 *train_LI(init_n)*
2 *prev_mean_error* ← *cal_avg_err(sampled_data[])*
3 *min_model_n* ← *init_n*
4 **while** *true* **do**
5   *train_LI_buttom_level(model_n)*   `// only retrain the buttom level`
6   *mean_error* ← *cal_avg_err(sampled_data[])*
7   **if** *mean_error* < *prev_mean_error* **then**
8   *min_error* ← *mean_err*
9   *prev_mean_error* ← *mean_err*
10   *min_model_n* ← *model_n*
11   **else**
12   *break*   `// end immediately when the error cannot be reduced`
13   *model_n* ← *model_n* + *step_size*   `// tuning the number of models`
14 *train_LI_buttom_level(min_model_n)*

---

traditional indexes due to their extremely fast query speed. Obviously, the HIF effectively alleviates the problem of read amplification compared to the dynamic PGM-index that requires multiple binary searches in multiple arrays. In addition, due to the transformation strategy, we can limit the number of indexes in the static layer by adjusting the size of the B+Tree, which further reduces the read amplification.

3.4 Self-tuning Algorithm

Some existing static learned indexes, such as RMI, need to be tuned by the number of underlying models to obtain better query performance. This is because the number of models determines the granularity of the fit to the data distribution, and too many models do not mean that better performance can be obtained. Therefore, for a given dataset, a static learned index needs to be iterated to determine the final structure. Specifically, building a learned index requires multiple retraining processes. This requires that the hybrid index framework must provide an auto-tuning strategy for this class of indexes to allow them to be applied to workloads containing insert operations.

In order to obtain the best results possible with the shortest training time possible. The goal of the self-tuning algorithm under dynamic workloads is not to obtain an optimal result, but a suboptimal result. In this requirement, the self-tuning strategy for the learned index is shown in Algorithm 3. In the beginning, the algorithm trains a learned index using the preset number of the underlying models. In the calculation of the error of the index, the algorithm takes a sample of data from the training set at equal intervals and then uses the trained index to predict all the sampled data and finally calculates an average error. This method greatly reduces the overhead of testing the learned indexes and the entire training time. In addition, the algorithm approximates the desired result quickly by substantially increasing the number of models used for the next iteration. The above steps are executed continuously and a minimum error and the corresponding result are maintained until the average error of the newly generated index is greater than the minimum error, at which point the tuning process ends and the final learned index is trained based on the saved number of models.

4   Performance Evaluation

4.1   Experimental Settings

Baselines. To test the performance of the hybrid index framework, we put three different types of static learned indexes into the HIF: (1) HLI [4] is a static learned index that combines the respective advantages of PGM and RMI. (2) PGM [3], a data-aware learned index that can capture linear patterns in data distributions. (3) RMI [1], a two-level Recursive Model Index consists entirely of linear models. In our experiments, we directly use the names of these learned indexes to represent their hybrid structures when combined with HIF. Besides, we compared HIF against four baselines: (1) originally dynamic PGM-Index (denoted by O-PGM) [3], which uses a series of arrays to receive inserted data and contains some PGM-indexes inside. (2)XIndex uses a simplified masstree as a buffer to receive new data, and a background thread to perform retraining of the learned index. (3) Masstree is a combination of B+Tree and Trie. (4) Standard B+Tree, as implemented in the STX B+Tree [5]. All indexes are implemented in C++.

Empirically, we set the error threshold for HLI to 32. For the error threshold of PGM, we follow the guidance in their paper. For HIF-based HLI and RMI, the number of models can be automatically adjusted by the self-tuning algorithm. Instead, PGM can be built in a single pass and thus does not require any tuning.

Moreover, different results can be obtained for the same index structure by changing the tuning granularity of the self-tuning algorithm, so we evaluate variants of the same structure as well. These variants are named directly using the parameters of the tuning algorithm, e.g., HLI-100 means that the initial number of models is 100 and the step size is 100.

Datasets. We evaluate HIF on 8 datasets from SOSD [6] and GRE [7]: (1) books: Each key represents the popularity of a particular book. (2) osm: cell IDs from Open Street Map. Each key represents an embedded location. (3) libio: repository ID from libraries.io. (4) fb: Each key represents a Facebook User ID. (5) lognormal: a dataset is generated artificially according to the log-normal distribution. (6) normal: a dataset is generated artificially according to the log-normal distribution. (7) uniform dense: uniformly distributed dense integers. (8) uniform sparse: uniformly distributed sparse integers. The first four are real-world datasets and the last four are synthetic datasets. Each dataset consists of 200 million unique unsigned 64-bit integer keys, and we generate 8-byte payloads for each key. For each lookup, we return a value corresponding to that key. Each dataset has a different data distribution, which means that the difficulty of learning is different. In general, real-world datasets are more complex than synthetic datasets. The figures of data distribution for these datasets can be viewed in [6]. Note that fb contains key with the value 2^{64-1}, which is not supported by PGM. Therefore we subtract all keys of fb by one when using PGM.

Workloads. The main metric is the average throughput of the indexes. To demonstrate the performance of the hybrid index framework, we evaluate the throughput on different workloads. Specifically, we set 10 different read-write ratios, with the percentage of insert operations increasing from 10% to 100% in 10% intervals. In addition, the lookup keys are randomly selected from the existing keys of the index according to the Zipfian distribution, and the insertion keys are selected from the remaining keys. For a given dataset, we use 100 million keys for initialization and then run the workload for 60 s to insert the remaining keys. We execute each workload 5 times and eventually obtain an average throughput. Note that we shuffled each dataset once randomly in each experiment to simulate real-world scenarios.

Environments. We implement HIF in C++. Experiments are conducted on an Ubuntu Linux machine with a 2.1 GHz Intel Xeon and 64 GiB memory, only using a single thread.

4.2   Performance Evaluation Under Read-Write Workloads

As shown in Fig. 4, for read-write workloads, the HIF-based static learned indexes achieve the best performance in most cases. Specifically, HLI-5k and PGM, on average, achieve up to 1.8× and 1.7× higher throughput than O-PGM for insert ratio below 70%. Also, on average, these two HIF-based indexes achieve up to 1.4×, 2.1×, and 2.7× higher throughput than the B+Tree, XIndex, and Masstree, respectively. This is because the HIF's lookup operations start at the static layer, thus avoiding inefficient search approaches to weaken the performance of the learned index. In addition, the framework provides good insert performance while ensuring efficient lookups in the buffer by using a traditional index, thus further alleviating the read amplification that exists in O-PGM. Further, the HIF avoids retraining learned indexes and new learned indexes can be built quickly by the self-tuning algorithm.

In addition, on all datasets except osm and fb, RMI-10k achieves up to 1.5× higher throughput compared to O-PGM for insert ratios below 70%. However, RMI performs poorly on osm and fb. Because the data distribution of the two

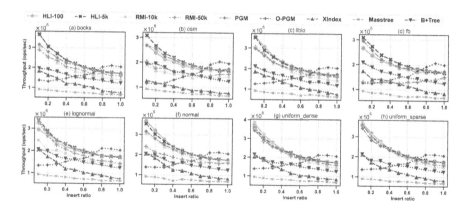

Fig. 4. Comparison of throughput under read-write workloads. Hybrid Index Framework (HLI, RMI, and PGM) vs. dynamic data structure used by PGM (O-PGM).

datasets is highly non-uniform. Unlike HLI and PGM, RMI cannot extract linear patterns from this distribution, so the underlying model of RMI does not fit well.

It can be seen that the performance of O-PGM is even much worse than B+Tree when the insert ratio is lower. This is because the lookup operation of O-PGM face the problem of serve read amplification. In addition, the insert operation causes frequent retraining of the internal learned index, resulting in blocking. On the contrary, the performance of the PGM-index has a huge improvement when combined with the HIF.

However, HIF cannot maintain its advantage after the insert ratio exceeds 70%, and the O-PGM shows the best performance. This is because the performance of the index is gradually dominated by insert performance as the insert ratio increases. Thus the O-PGM with the best insert performance outperforms other indexes. However, when the percentage of inserts approaches 100%, the performance of O-PGM degrades. This is because after all the data is inserted, the index will start to perform lookup operations, and at this point the performance of the index is once again dominated by the lookup performance.

4.3 Lookup Performance Under Read-Write Workloads

Figure 5 shows that on average, HIF-5k, PGM, and RMI-10k achieve up to 2.74×, 2.28×, and 2.29× higher throughput than O-PGM. This is because the HIF significantly alleviates the read amplification problem that exists in PGMs. This framework significantly improves the query performance of the buffer by using a single index structure. In contrast, the dynamic PGM-index need to lookup keys in multiple arrays of varying sizes, and in most arrays can only be searched using inefficient binary search. In addition, the HIF's transformation strategy limits the number of indexes, thus further reducing read amplification.

The lookup performance of all HIF-based learned indexes decreases significantly as the insert ratio increases. Because the number of static learned indexes

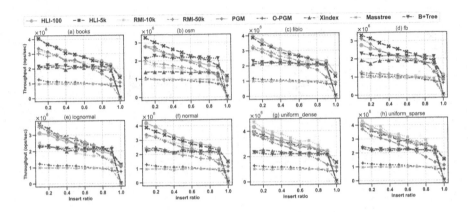

Fig. 5. Comparison of lookup performance under read-write workloads.

within the HIF increases with a large amount of data inserted. Therefore, lookups need to traverse more static learned indexes, which leads to the problem of read amplification. On the contrary, the B+Tree and Masstree is a general index structure that does not consider the data distribution so its lookup performance is more stable. And for XIndex, it uses a background thread to constantly rebuild the existing learned indexes, so its lookup performance is more stable than HIF. In addition, on the osm, fb, and lognormal datasets, the lookup performance of HIF is lower than that of the B+Tree after the insert ratio exceeds 80%. Besides the factor of write amplification, another reason is that the data distribution of these three datasets is too complex and thus difficult to fit for learned indexes.

4.4   Insert Performance Under Read-Write Workloads

Unsurprisingly, Fig. 6 shows the O-PGM has the best insert performance. This is because O-PGM sacrifices lookup performance for extremely high insert performance. The dynamic data structure of O-PGM is just some arrays, so the overhead of insertion comes almost entirely from reordering and copying the data. In practice, however, real-world workloads tend to be read-heavy [7], so we believe that the data structure used by PGM is unacceptable.

Except for O-PGM, the HIF-based PGM-index has the best insert performance. This is because the streaming algorithm [8] can build the underlying models of the PGM-index by traversing the dataset once. Therefore, PGM incurs less blocking time compared to other learned indexes that require tuning. It can be seen that HLI has almost the same insert performance as PGM. This is because HLI uses the same method as PGM to build the underlying models, thus the complexity of tuning is greatly reduced. Because only the upper level of the HLI needs to be tuned based on the keys extracted from these models.

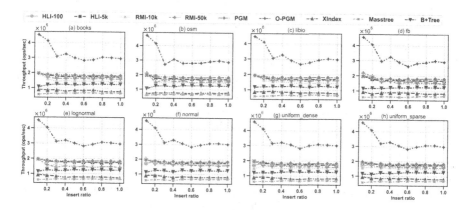

Fig. 6. Comparison of insert performance under read-write workloads.

4.5 Learning Overhead of Hybrid Index Framework

In this subsection, we use three metrics to evaluate the learning overhead of the hybrid index framework, namely, the number of times to create a learned index, the total time of copying data, and the total time of training. After initializing the HLI with 100M key, we insert the remaining 100M keys at a 100% insert ratio until all keys are inserted. As shown in Table 1, on average, compared to the original dynamic data structure, HIF-based HLI has an 9.4× reduction in the number of creating a learned index, a 6.6× reduction in copying time, and a 9.8× reduction in the training time. As described in Sect. 2, the dynamic data structure leads to frequent copying and merging of data between arrays. Thus causing the change of the data distribution, which inevitably leads to the retraining of the learned index. On the contrary, HIF's transformation strategy based on the two-layer structure does not change the data distribution, thus avoiding the retraining of the learned index. Therefore the build of the learned index will only be triggered when the capacity of the dynamic layer reaches a threshold. In addition, the self-tuning algorithm can quickly obtain a high-performance index structure, thus further reducing the training time.

Table 1. Learning overhead of HLI-5k under two dynamic data structures. Hybrid index framework vs. original dynamic data structure adopted by PGM.

dataset	HIF-based HLI			dynamic HLI		
	# of learned index creations	copying (s)	training (s)	# of learned index creations	copying (s)	training (s)
books	5	1.7	3.2	47	11.4	32.5
osm	5	1.7	4.9	47	11.2	47.6
libio	5	1.7	3.8	47	11.2	36.5
fb	5	1.7	4.8	47	11.2	48.1

5   Related Work

The Learned Index [1] is a new type of index structure that uses machine learning methods to build a specialized index structure for a specific dataset. While retaining the core idea, subsequent works have improved it differently. To reduce the model error in Learned Index, many works use greedy algorithms to extract linear patterns in the data distribution to improve the accuracy of linear models. ASLM [9] uses an algorithm that divides the data based on the similarity between the data. FINEdex [10] proposes a learning probe algorithm that divides data matching the same linear distribution into the same sub-datasets. FITing-tree [2] generates models directly using an algorithm called Shrinking Cone, which guarantees that the maximum prediction error of each model does not exceed a predefined value. PGM-Index [3] also limits the prediction error of the models, with the difference that it generates a smaller number of models.

In addition, the subsequent works use approaches to solving the problem that the Learned Index fails to handle insert operations. The first way is to use a buffer to receive new data. The learned indexes described above all use this method. For example, FINEdex uses a variant of the B+Tree as a buffer. Notably, PGM-index uses multiple buffers to share the insertion overhead. The second method is to leave gaps in the nodes of the tree. ALEX [11] uses a gapped array layout so that the inserted data can be put directly into the empty slot based on the model's predictions. But when the predicted position is already occupied, ALEX needs to shift the data to make a gap. Instead, LIPP [12] creates a new node directly at the conflicting position, so the model-based prediction is accurate.

Further, some learned indexes focus on other aspects. Xindex [13] and FINEdex [10] are dedicated to making learned indexes that can concurrently process requests. APEX [14] and PLIN [15] explore the use of learned indexes in non-volatile memory scenarios. CARMI [16] uses a cache-aware design to reduce the number of memory access. RadixSpline [17] can quickly build an index structure by traversing the data once. In addition, Marcus et al. [6] proposed a benchmark to evaluate static learned indexes in detail, and Wongkham et al. [7] performed a comprehensive evaluation on updateable learned indexes with multiple real-world datasets.

6   Conclusion

In this paper, we propose a high-performance hybrid index framework that effectively solves the problem that static learned indexes do not support insert operations without destroying the existing learned index. For the severe read amplification problem in the dynamic data structure, the HIF greatly alleviates this problem by searching the static learned index first. In addition, HIF completely avoids the periodic retraining of static learned indexes by the transformation strategy from the traditional index to the learned index. Further, we provide a self-tuning algorithm that enables the learned indexes that need tuning to be efficiently applied to read-write workloads. The experiments show that HIF exhibits good performance when combined with multiple static learned indexes.

Acknowledgements. This paper is supported by the Humanities and Social Sciences Foundation of the Ministry of Education (17YJCZH260), and the Sichuan Science and Technology Program (2020YFS0057).

References

1. Kraska, T., Beutel, A., Chi, E.H., Dean, J., Polyzotis, N.: The case for learned index structures. In: SIGMOD, pp. 489—504 (2018)
2. Galakatos, A., Markovitch, M., Binnig, C., Fonseca, R., Kraska, T.: Fiting-tree: A data-aware index structure. In: SIGMOD, pp. 1189—1206 (2019)
3. Ferragina, P., Vinciguerra, G.: The PGM-index: a fully-dynamic compressed learned index with provable worst-case bounds. Proc. VLDB Endow. **13**, 1162–1175 (2020)
4. Ding, Y., Zhao, X., Jin, P.: An error-bounded space-efficient hybrid learned index with high lookup performance. In: DEXA, pp. 216–228. Springer (2022)
5. Bingmann, T.: STX B+ Tree (2013). https://panthema.net/2007/stx-btree
6. Marcus, R., et al.: Benchmarking learned indexes. Proc. VLDB Endow. **14**, 1–13 (2020)
7. Wongkham, C., Lu, B., Liu, C., Zhong, Z., Lo, E., Wang, T.: Are updatable learned indexes ready? Proc. VLDB Endow. **15**, 3004–3017 (2022)
8. Xie, Q., Pang, C., Zhou, X., Zhang, X., Deng, K.: Maximum error-bounded piecewise linear representation for online stream approximation. VLDB J. **23**, 915–937 (2014)
9. Li, X., Li, J., Wang, X.: Aslm: Adaptive single layer model for learned index. In: DASFAA Workshops, pp. 80–95 (2019)
10. Li, P., Hua, Y., Jia, J., Zuo, P.: Finedex: a fine-grained learned index scheme for scalable and concurrent memory systems. Proc. VLDB Endow. **15**, 321–334 (2021)
11. Ding, J., et al.: ALEX: an updatable adaptive learned index. In: SIGMOD, pp. 969–984 (2020)
12. Wu, J., Zhang, Y., Chen, S., Wang, J., Chen, Y., Xing, C.: Updatable learned index with precise positions. Proc. VLDB Endow. **14**, 1276–1288 (2021)
13. Tang, C., et al.: Xindex: a scalable learned index for multicore data storage. In: PPoPP, pp. 308—320 (2020)
14. Lu, B., Ding, J., Lo, E., Minhas, U.F., Wang, T.: Apex: a high-performance learned index on persistent memory. Proc. VLDB Endow. **15**, 597–610 (2021)
15. Zhang, Z., et al.: Plin: a persistent learned index for non-volatile memory with high performance and instant recovery. Proc. VLDB Endow. **16**, 243–255 (2022)
16. Zhang, J., Gao, Y.: Carmi: a cache-aware learned index with a cost-based construction algorithm. Proc. VLDB Endow. **15**, 2679–2691 (2021)
17. Kipf, A., Marcus, R., van Renen, A., Stoian, M., Kemper, A., Kraska, T., Neumann, T.: RadixSpline: a single-pass learned index. In: aiDM@SIGMOD, pp. 1–5 (2020)

A Study on Historical Behaviour Enabled Insider Threat Prediction

Fan Xiao[1] [ID], Wei Hong[2] [ID], Jiao Yin[3(✉)] [ID], Hua Wang[3] [ID], Jinli Cao[4] [ID],
and Yanchun Zhang[3] [ID]

[1] College of Computer Science and Technology, Harbin Engineering University,
Harbin 150001, China
[2] School of Artificial Intelligence, Chongqing University of Arts and Sciences,
Chongqing 402160, China
[3] Institute for Sustainable Industries and Liveable Cities, Victoria University,
Melbourne, VIC 3011, Australia
{jiao.yin,hua.wang,yanchun.zhang}@vu.edu.au
[4] Department of Computer Science and Information Technology,
La Trobe University, Melbourne, VIC 3086, Australia
j.cao@latrobe.edu.au

Abstract. Insider threats have been the major challenges in cybersecurity in recent years since they come from authorized individuals and usually cause significant losses once succeeded. Researchers have been trying to solve this problem by discovering the malicious activities that have already happened, which offers not much help for the prevention of those threats. In this paper, we propose a novel problem setting that focuses on predicting whether an individual would be a malicious insider in a future day based on their daily behavioral records of the previous several days, which could assist cybersecurity specialists in better allocating managerial resources. We investigate seven traditional machine learning methods and two deep learning methods, evaluating their performance on the CERT-r4.2 dataset for this specific task. Results show that the random forest algorithm tops the ranking list with f1 = 0.8447 in the best case, and deep learning models are not necessarily better than machine learning models for this specific problem setting. Further study shows that the historical records from the previous four days around can offer the most predicting power compared with other length settings. We publish our codes on GitHub: https://github.com/mybingxf/insider-threat-prediction.

Keywords: Insider threat · Predictive model · Deep learning

1 Introduction

For a long time, cybersecurity concerns have overwhelmingly focused on defending against attacks from outside while omitting that threats from inside have the potential to cause more detrimental harm [17,26,30]. According to a recent report from PwC[1], the majority of surveyed companies consider that threats from insiders pose far more risk to cybersecurity, even though they only account for a small part of the total.

An insider refers to an individual who has or had authorized access to an organization's critical assets [2]. Due to the particularity of these individuals, their behaviors are often considered safe and harmless by traditional security systems. Therefore, when there are malicious behaviors from authorized individuals, the impact could be more lasting and profound than external intrusions [23,24]. To mitigate this kind of threat effectively, many enterprises and research institutions have increasingly invested in insider threat detection techniques in recent years [4,32].

Academically, the majority of researchers have focused their efforts on detecting threats that have already occurred. Typically, logged user activities are analyzed, and features are extracted in the first step, and then a detection model is introduced to distinguish between malicious and normal activities [22]. However, this "remedy-based" strategy would inevitably breed a lag in corrections, which could pose huge risks for large or critical organizations. Therefore, it is more beneficial to adopt a "prevention-based" strategy in the insider threat research domain, which means setting the research goal as a better prediction rather than detection.

Considering that individuals within organizations typically have workloads that follow the calendar day, we follow the practice in paper [14] and propose a daily activity-based insider threat prediction study in this paper. This is also because, for a cybersecurity analyst, monitoring the risk level on a daily basis often aligns with the typical operation pattern of an organization [29]. Specifically, by analyzing the characteristics of a user's behavior a few days before a particular day, we aim to predict whether there will be any malicious behavior on that day. This may facilitate IT specialists in allocating monitoring resources in advance and help reduce the loss caused by insider threats, if not prevent them from occurring [27].

We summarize the main contributions of this paper as follows:

- Compared with the widely-discussed "detection" problem setting, we propose a more practical problem setting, which is to predict whether a user will have abnormal behavior on a particular day by analyzing the user behavior a few days prior.
- We conduct a comprehensive experimental study by implementing various machine learning models on an open-source insider threat dataset, CERT4.2. The results show that different models perform differently on the same problem settings, and better performance relies on better fine-tuning specific to the dataset.
- We investigate the influence of different time windows on the performance of prediction and find that the 4-6 day time window setting yields the best results. This possibly could be echoing the fact that there are five working days per week, and people's behaviour has a latent weekly pattern.

---

[1] https://www.pwc.dk/da/publikationer/2021/cybercrime-survey-2020-en.html.

The rest of the paper is organized as follows. Section 2 introduces related works on the history, datasets, and methods of insider threat research. Section 3 presents the problem setting of malicious behavior prediction proposed by this paper, followed by the prediction framework and predictive models used in this paper. The dataset, data processing, experimental results and analysis are described in Sect. 4. Finally, Sect. 5 concludes the paper and points out future works.

2   Related Works

Insider and Insider Threat. As one of the most challenging problems in cybersecurity research, insider threat research has attracted attention as earlier as 2001 in the academic world when Schultz and Shumway [21] considered "insider attack" as "the intentional misuse of computer systems by users who are authorized to access those systems and networks." At the same time, Garinkel et al. [8] define an insider as "a subject of the database who has personal knowledge of information in the confidential field."

Since then, the scope of definition for insiders and insider threats has been expanding along the time. Greitzer and Frincke [11] emphasized that historical access privilege should also be included in the definition of "insider" in 2010. Pleeger et al. expanded the victim of insider threat from just system and network to "an organization's data, processes, or resources". Currently, the most widely-used terms for "insider" and "insider threat" are maintained by the CERT research team of the Software Engineering Institute at Carnegie Mellon University [2], which refers to an "insider" as "an individual who has or had authorized access to an organization's critical assets" and defines a "insider threat" as "the potential for an individual who has or had authorized access to an organization's critical assets to use their access, either maliciously or unintentionally, to act in a way that could negatively affect the organization".

Dataset-Based Insider Threat Research. Other than theoretical taxonomy, a large number of previous works have conducted their research taking an operational approach, which emphasizes practical solutions related to certain datasets [13].

As earlier as 2001, Schonlau, M. et al. [20] introduced a command history-based dataset SEA and adopted a statistical approach to detect insider threats. Garg, A. et al. [9] then introduced a new dataset based on mouse operation history, extending insider detection context from a single command line to a graphical user interface (GUI). The SVM method was also employed in this work, achieving an accuracy rate up to 96% high with few false positives. Salem and Stolfo [19] later introduced the RUU dataset that oriented to masquerader detection and also used SVM modeling techniques to reveal user intent.

Richer datasets in later times started to boost the academic research related to insider threats. Particularly, CERT from Carnegie Mellon University has worked with other partners to generate a synthetic dataset for insider threat research. This dataset contains a large amount of logon data, browsing history, access logs, emails, device usage and so on [10]. Previous works that relate most to our paper are mainly based on this comprehensive dataset, though under different problem settings.

Among them, Anagi Gamachchi and Serdar Boztas [7] constructed a hierarchy graph for an organization and then used attributed graph clustering techniques to rank outliers. They consider logged activities across a long time as a whole, setting the research goal as to distinguish insiders from all users.

Chattopadhyay et al. [3] constructed a feature vector based on statistics of single-day features over a period of time and tried to discover insider threats that aligned with different threat scenarios. Yuan, F. et al. [31] aim to find out insiders by examining users' activities across a time window more than one year long, and designed a framework to extract fixed-size feature representations using the LSTM model. Another later work [18] applied LSTM auto-encoder as a classifier based on the reconstructed loss to do detection. However, it still takes all activities across a long time as a whole. In paper [15], Jiang, J. et al. constructed a weighted graph based on a specially-designed similarity function and introduced a GCN model to detect malicious insiders. Liu et al. [16] define the insider detection problem as finding malicious activities in their work. Hong et al. [14] proposed a novel problem setting to detect malicious user-day rather than some person or some activities.

We could observe that the problem settings in the majority of those works are either to discover malicious insiders or to discover malicious activities, both of which aim at detection rather than prediction. However, research work that focuses more on how to predict and prevent insider threats would benefit industrial applications more.

3   Methodology

In this section, we describe in detail the problem we define, the overall framework, and the machine learning and deep learning models we use.

3.1   Problem Setting

Traditionally, the problem setting in insider threat research mainly focused on detecting threats from past behaviors, which is done by analyzing the characteristics of activities from a certain user [14]. For example, if a user's login time is considerably longer than in previous days, the detection system may categorize this behavior as malicious. Another example is if a user visits a website that is largely different from the ones visited by his or her colleagues, the detection system may also consider it a threat. Since these detection tasks are based on actions that have already taken place, it is relatively easy to implement but less significant for practical application.

Based on our preliminary research, we argue that whether a malicious behavior would occur on a certain day could be inseparable from that user's historical behaviors in the previous days. We offer two possible explanations for this:

(1) Some previous activities are the causes of malicious behavior in the present day. For example, if employees are mistreated by their bosses by demanding to work overtime consistently, it could make them perform some malicious activities for retaliation. Moreover, if abnormal access in previous days already exposed the system to a dangerous environment, it would be more possible for the system to be misused in the current day.

(2) Some previous activities are the preparations for malicious behavior in the present day. For example, some attacks can only be launched after certain access is granted to employees or after some necessary actions are performed by employees, such as clicking on certain websites first.

Therefore, what we have designed is to predict whether there might be malicious behaviors today based on the behavioral characteristics of a certain user in the past few days (such as login time, device usage times, whether there were abnormalities, etc.).

Considering that an individual inside an organization generally has a working load that follows the calendar day and one could act maliciously one day and benignly the rest days, we follow the practice in paper [14] to formulate our problem based on the user-day setting. Specifically, our goal is to predict whether a certain user would conduct malicious behavior on a certain day based on daily activity logs in previous days.

3.2 Framework

The proposed user-day based insider threat prediction framework is shown in Fig. 1. Originally, behavioral logs were organized according to different types of activities. However, our method aims to predict whether a user will conduct malicious behavior on a certain day. Therefore, as shown in Fig. 1, in the first step of the proposed method, we pre-process the whole dataset by reorganizing all activity logs into daily activity logs corresponding to each user in the system. For user $i(1 \leq i \leq I)$, $A_i = \{D_1, D_2, \cdots, D_N\}$ denotes all the activities logged by the system across N days, where $D_n(1 \leq n \leq N)$ denotes the set of activities logged during a certain calendar day n and I denotes the maximum user number.

In order to train the prediction model for each user, we need to "cut" the activity log $A = \{D_1, D_2, \cdots, D_N\}$ into many "pieces" and select appropriate features for model training and testing, and we denote the length of each "piece" as t, which is also the window length for historical behavior. After this reconstruction and feature engineering process, for a user i, we will have a set of daily activity sequence $S_i = \{s_{i1}, s_{i2}, \cdots, s_{ij}, \cdots\}$, where $s_{ij}(1 \leq i \leq I, 1 \leq j \leq N-t)$ stands for the sequence data used to predict whether user i would conduct

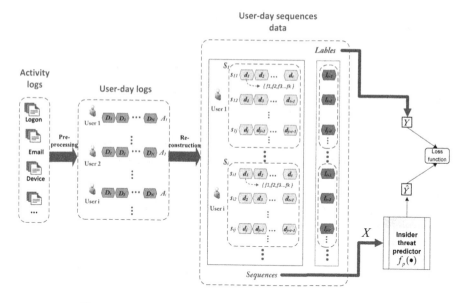

Fig. 1. User-day based insider threat prediction framework

malicious behavior on the day $j + t$. Furthermore, for each sequence data s, $d = \{f_1, f_2, \cdots, f_K\}$ represents the feature vector of one day, and $f_k (1 \leq k \leq K)$ denotes one feature element in the vector. Particularly, l denotes the label for the user-day we want to predict.

After obtaining the reconstructed sequence data, all the sequences will serve as model input X during training and testing, which is to say, $X = \{S_1, S_2, \cdots, S_I\}$ $(1 \leq i \leq I)$. In the final stage, we choose different models as the insider threat predictor, and the procedure can be described by Eq. (1).

$$\hat{Y} = f_p(X, \Theta_p), \tag{1}$$

where $\hat{Y}$ is the predicted results on whether a user-day is malicious or not, f_p is the mapping function of the chosen model, and Θ_p is the trainable parameters. The parameters Θ_p can be optimised on the training set by comparing the predicted results $\hat{Y}$ with the true labels Y and minimising the loss function.

3.3   ML and DL Models

In the proposed framework, the prediction task could be performed by any chosen model. Here we lay out a brief description of the models that will be adopted in the experimental part.

- K-Nearest Neighbor (K-NN): K-NN is a type of classification according to the category of the nearest K sample points.
- Multi-layer Perceptron (MLP): MLP is a fully connected class of feedforward artificial neural network. It can distinguish data that are not linearly separable [12, 28].

- Logistic Regression (LR): LR estimates the probability of an event occurring, such as voting or not voting, based on a given dataset of independent variables.
- Decision Tree (DT): DT is a decision-making analysis method for judging the feasibility of various situations on the basis of known occurrence probabilities.
- Random Forest (RF): RF is an ensemble learning method for classification, regression and other tasks that operates by constructing a multitude of decision trees at training time [1].
- AdaBoost Classifier (ABC): ABC is a meta-estimator that begins by fitting a classifier on the original dataset and then fits additional copies of the classifier on the same dataset [6].
- Gaussian Naive Bayes (GNB): GNB is a probabilistic classification algorithm based on applying Bayes' theorem with strong independence assumptions.
- Convolutional Neural Networks (CNN): CNN is a class of artificial neural networks, most commonly applied to analyze visual imagery [25].
- Long Short-Term Memory (LSTM): LSTM is a type of recurrent neural network suitable for processing and predicting important events with very long intervals and delays in time series [5].

4 Experiments

4.1 Dataset and Data Processing

Our proposed methods and experiment are based on the widely-recognized CERT 4.2 dataset, which contains multiple behavioral types of 1,000 users around a one-year time frame, including email, device login time, network access status, etc. The CERT Division, in partnership with ExactData, LLC, and under sponsorship from DARPA I2O, has generated a collection of synthetic insider threat test datasets. These datasets provide both synthetic background data and data from synthetic malicious actors. We follow the practice of [14] to extract the process the data and perform feature engineering. Among them, the device.csv and logon.csv files are used, and six behavioral features are selected as shown in Table 1.

Following the framework described in Sect. 3.2, we reconstructed the original dataset into a new one containing 356,000 pieces of sequence along with their corresponding label, removing all weekends. Because the ratio of abnormal behavior in the original dataset is extremely low, and our goal is to test the feasibility of the proposed problem setting, we further down-sampled a balanced dataset by matching every malicious user-day with a benign one, which has a size of 1908. In the end, we use 70% of them for training and the other 30% for testing.

4.2 ML and DL Performance Comparison

At the first stage, we set the historical window length t to be 5, considering the fact that one week has 5 working days, and the pattern may reveal itself

Table 1. Details of selected features

Feature name	Description	Process method	Value type
F1	First log on time	Map the timestamp of the first login activity to the range of [0, 1] , where 0 stands for 00:00, 1 stands for 24:00	Float: 0-1
F2	Last log off time		
F3	First device activity time		
F4	Last device activity time		
F5	Number of off-work device activities	Count the number of the device activities (connect or disconnect) during off-work time (18:00 pm - 8:00 am)	Int: times
F6	Malicious user-day tag	If malicious activities happened during this day tagged as 1, otherwise 0	Bool: 1 or 0

following this time circle. We implemented all seven machine learning models through scikit-learn package[2], which include K-Nearest Neighbor (KNN), Multi-layer Perceptron (MLP), Logistic Regression (LR), Decision Tree (DT), Random Forest (RF), AdaBoost Classifier (ABC), Gaussian Naive Bayes (GNB). For deep learning models, PyTorch[3] is chosen to implement a basic CNN and an LSTM model separately. The results for all deep learning and machine learning models are shown in Table 2. We adopted default parameter settings for all models unless otherwise specified.

Table 2. Performance of different models(window length=5)

	KNN	MLP	LR	DT	RF	ABC	GNB	CNN	LSTM
accuracy	0.8133	0.8325	**0.8377**	0.7801	0.8202	0.8325	0.8010	0.8168	0.8255
precision	0.8139	0.8325	0.8392	0.7803	0.8209	0.8333	0.8013	0.7891	**0.8811**
recall	0.8132	0.8324	0.8376	0.7801	0.8203	0.8324	0.8010	**0.8636**	0.7517
f1-score	0.8132	0.8324	**0.8375**	0.7801	0.8202	0.8323	0.8010	0.8247	0.8113

As we can see, for the machine learning group, almost all models achieved satisfying results across all metrics. Particularly, logistic regression from the machine learning group tops the list both in accuracy and f1 score, while Decision Tree ranks the worst unanimously. However, Random Forest enjoyed much better performance compared with Decision Tree. As we all know, Random Forest works in a way that averages multiple deep decision trees, trained on different parts of

---

[2] https://scikit-learn.org/stable/.
[3] https://pytorch.org/docs/stable/index.html.

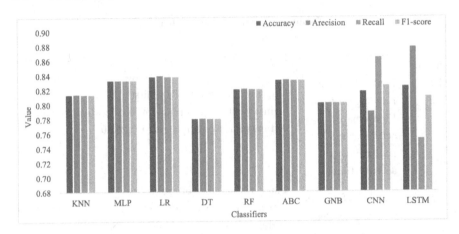

Fig. 2. Performance Bar-chart for models (window length = 5)

the same training set, with the goal of reducing the variance. Random forests solve the problem of over-fitting which happens in the Decision Tree, which is possibly why it achieved better performance here for this task.

For the deep learning group, in general, both of the two models demonstrated inferior performance compared with most of the machine learning models, which failed our expectations. We think this situation is caused by the small amount of the sample dataset. Another thing that countered our expectation is, for f1 scores, even a simple CNN model achieved better results than the LSTM model. Originally, we expected there to be some sequential pattern that could be discovered by the LSTM model. But results seem to indicate there is no strong factor related to time order. However, it could also be because the window length is very short. Nevertheless, as shown by the bar chart from Fig. 2, LSTM achieved the highest precision score for this task, while CNN performs well in recall score. In practical applications, different models can be selected according to specific requirements.

4.3   The Impact of Historical Window Length

In previous experiments, we set the historical window length as 5, which means to predict whether a user would conduct malicious behavior on the present day, we use daily activity features from the previous 5 days to perform the prediction. In order to find out the possible impact brought by historical window length, we further conducted an ablation study across all models over different window lengths.

In the beginning, the window length equaling 5 was selected intuitively because there are 5 working days in a week and we removed weekend data in pre-processing. To avoid the possible circling effect, we limited the maximum window length in our ablation study to 9 (10 is two times 5). Thus, we tested

Table 3. F1 scores for all models across different window lengths

No	Machine Learning								Deep Learning		
	KNN	MLP	LR	DT	RF	ABC	GNB	ML_avg	CNN	LSTM	DL_avg
1	0.8143	0.8244	0.8244	0.6997	0.8040	0.8176	0.8244	0.8013	0.8098	0.8076	0.8087
2	**0.8272**	0.8097	0.8197	0.7450	0.8167	0.8200	0.8132	0.8074	0.8124	0.8038	0.8081
3	0.8150	0.8307	0.8358	0.7906	0.8272	0.8166	0.8185	0.8192	0.8153	0.8083	0.8118
4	0.7958	0.8220	<u>0.8392</u>	**0.7958**	**0.8447**	0.8272	**0.8272**	**0.8217**	**0.8283**	<u>0.8113</u>	**0.8198**
5	0.8132	**0.8377**	0.8375	0.7871	0.8271	**0.8324**	0.8010	0.8194	0.8247	0.8098	0.8173
6	0.8097	0.8307	0.8340	0.7574	0.8270	0.8271	0.8011	0.8124	0.8289	0.8084	0.8186
7	0.8219	0.8324	0.8375	0.7661	0.8254	0.8201	0.7958	0.8142	0.8222	0.8106	0.8164
8	0.8062	0.8237	**0.8410**	0.7818	0.8235	0.8061	0.7906	0.8104	0.8257	0.8076	0.8166
9	0.8097	0.8255	0.8340	0.7799	0.8303	0.8115	0.7853	0.8109	0.8249	**0.8144**	0.8197

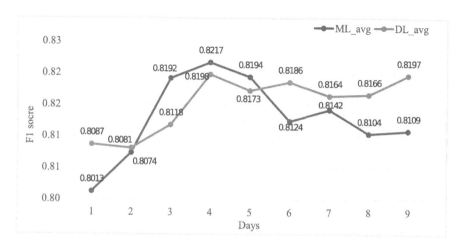

Fig. 3. Performance trend of different model groups across window lengths

all models with window length increasing from 1 to 9 while keeping all the other settings the same. We present the f1 score in Table 3.

From Table 3, we could observe that for most of the models setting window length to 4 or 5 would yield the best f1 performance. If we average the performance within each model group, both would give the consistent result that features from the previous 4 days have the strongest predicting power.

If we plot the average performance of each model group corresponding to a different length, we will have a clearer view of how the performance change along with the length setting. As Fig. 3 shows, both lines take the convex curve shape obviously. When the window length is smaller than 4, the f1 scores is continuously rising along with increased length, while after passing the peak both lines start to take a downtrend.

This result fulfills our expectations and interestingly echoes our intuition in the first part of the experiment. Our explanation is, since in the pre-processing stage we omitted all weekends, if a user tends to have a similar work schedule on the same day of each week, then the gap should be around 5.

5   Conclusion

In this paper, we proposed a novel problem setting in insider threat research that aims to "predict" threat rather than "detect" threat, which is potentially more of practical significance. Based on the open-sourced CERT 4.2 dataset, we implemented our framework and tested various machine learning and deep learning models on this task. Results show that most of the tested models achieved satisfying performance, and deep learning models are not necessarily better than machine learning models for this specific problem setting. We also studied the impact of window length on the predicting performance and found that for most models the best performance is achieved around length 4 or 5. This finding shows that richer historical data do not guarantee stronger predicting power.

However, there are still some limitations that could be addressed in the future. First, we only selected features from part of the activity log types, and it could be more promising if extra activity types are considered in the feature. Second, in practice, insider threat research is a highly imbalanced research problem, and we need to face this huge challenge in real-world applications. Third, an online learning strategy would be more suitable for insider threat scenarios, and this could be our prior working direction in the future.

References

1. Breiman, L.: Classification and regression trees. Routledge (2017)
2. Center, C.N.I.T.: Common sense guide to mitigating insider threats. Carnegie Mellon University, 7th edn. (2022)
3. Chattopadhyay, P., Wang, L., Tan, Y.P.: Scenario-based insider threat detection from cyber activities. IEEE Trans. Comput. Soc. Syst. 5(3), 660–675 (2018)
4. Cui, D., Piao, Y.: A study on the privacy threat analysis of PHI-code. In: Gao, Y., Liu, A., Tao, X., Chen, J. (eds.) APWeb-WAIM 2021. CCIS, vol. 1505, pp. 93–104. Springer, Singapore (2021). https://doi.org/10.1007/978-981-16-8143-1_9
5. Duan, J., Zhang, P.F., Qiu, R., Huang, Z.: Long short-term enhanced memory for sequential recommendation. World Wide Web 26(2), 561–583 (2023)
6. Freund, Y., Schapire, R.E.: A decision-theoretic generalization of on-line learning and an application to boosting. J. Comput. Syst. Sci. 55(1), 119–139 (1997)
7. Gamachchi, A., Boztas, S.: Insider threat detection through attributed graph clustering. In: 2017 IEEE Trustcom/BigDataSE/ICESS, pp. 112–119. IEEE (2017)
8. Garfinkel, R., Gopal, R., Goes, P.: Privacy protection of binary confidential data against deterministic, stochastic, and insider threat. Manage. Sci. 48(6), 749–764 (2002)
9. Garg, A., Rahalkar, R., Upadhyaya, S., Kwiat, K.: Profiling users in gui based systems for masquerade detection. In: Proceedings of the 2006 IEEE Workshop on Information Assurance, vol. 2006, pp. 48–54 (2006)

10. Glasser, J., Lindauer, B.: Bridging the gap: a pragmatic approach to generating insider threat data. In: 2013 IEEE Security and Privacy Workshops, pp. 98–104. IEEE (2013)

11. Greitzer, F.L., Frincke, D.A.: Combining traditional cyber security audit data with psychosocial data: towards predictive modeling for insider threat mitigation. In: Insider threats in cyber security, pp. 85–113. Springer (2010)

12. Hastie, T., Tibshirani, R., Friedman, J.H., Friedman, J.H.: The elements of statistical learning: data mining, inference, and prediction, vol. 2. Springer (2009)

13. Homoliak, I., Toffalini, F., Guarnizo, J., Elovici, Y., Ochoa, M.: Insight into insiders and it: a survey of insider threat taxonomies, analysis, modeling, and countermeasures. ACM Comput. Surv. (CSUR) **52**(2), 1–40 (2019)

14. Hong, W., et al.: Graph intelligence enhanced bi-channel insider threat detection. In: Network and System Security: 16th International Conference, NSS 2022, Denarau Island, Fiji, December 9–12, 2022, Proceedings, pp. 86–102. Springer (2022)

15. Jiang, J., et al.: Anomaly detection with graph convolutional networks for insider threat and fraud detection. In: MILCOM 2019-2019 IEEE Military Communications Conference (MILCOM), pp. 109–114. IEEE (2019)

16. Liu, F., Wen, Y., Zhang, D., Jiang, X., Xing, X., Meng, D.: Log2vec: A heterogeneous graph embedding based approach for detecting cyber threats within enterprise. In: Proceedings of the 2019 ACM SIGSAC Conference on Computer and Communications Security, pp. 1777–1794 (2019)

17. Miller, S.: 2017 u.s. state of cybercrime highlights. Carnegie Mellon University's Software Engineering Institute Blog (Jan 17, 2018 [Online]). http://insights.sei.cmu.edu/blog/2017-us-state-of-cybercrime-highlights/. Accessed 23 Aug 2022

18. Paul, S., Mishra, S.: Lac: Lstm autoencoder with community for insider threat detection. In: 2020 the 4th International Conference on Big Data Research (ICBDR'20), pp. 71–77 (2020)

19. Salem, M.B., Stolfo, S.J.: Masquerade attack detection using a search-behavior modeling approach. Columbia University, Computer Science Department, Technical Report CUCS-027-09 (2009)

20. Schonlau, M., DuMouchel, W., Ju, W.H., Karr, A.F., Theus, M., Vardi, Y.: Computer intrusion: Detecting masquerades. Statistical science, pp. 58–74 (2001)

21. Schultz, E., Shumway, R.: Incident response: a strategic guide to handling system and network security breaches. Sams (2001)

22. Shi, Y., Wang, S., Zhao, Q., Li, J.: A hybrid approach of http anomaly detection. In: Web and Big Data: APWeb-WAIM 2017 International Workshops: MWDA, HotSpatial, GDMA, DDC, SDMA, MASS, Beijing, China, July 7-9, 2017, Revised Selected Papers 1, pp. 128–137. Springer (2017). https://doi.org/10.1007/978-3-319-69781-9_13

23. Sun, X., Wang, H., Li, J.: Injecting purpose and trust into data anonymisation. In: Proceedings of the 18th ACM Conference on Information and Knowledge Management, pp. 1541–1544 (2009)

24. Wang, H., Sun, L.: Trust-involved access control in collaborative open social networks. In: 2010 Fourth International Conference on Network and System Security, pp. 239–246. IEEE (2010)

25. Wang, W., Wang, W., Yin, J.: A bilateral filtering based ringing elimination approach for motion-blurred restoration image. Current Optics Photonics **4**(3), 200–209 (2020)

26. Yin, J., Tang, M., Cao, J., You, M., Wang, H.: Cybersecurity applications in software: Data-driven software vulnerability assessment and management. In: Emerging Trends in Cybersecurity Applications, pp. 371–389. Springer (2022)

27. Yin, J., Tang, M., Cao, J., You, M., Wang, H., Alazab, M.: Knowledge-driven cybersecurity intelligence: software vulnerability co-exploitation behaviour discovery. IEEE Trans. Ind. Inform. (2022)
28. Yin, J., You, M., Cao, J., Wang, H., Tang, M., Ge, Y.F.: Data-driven hierarchical neural network modeling for high-pressure feedwater heater group. In: Databases Theory and Applications: 31st Australasian Database Conference, ADC 2020, Melbourne, VIC, Australia, February 3–7, 2020, Proceedings 31, pp. 225–233. Springer (2020)
29. You, M., Yin, J., Wang, H., Cao, J., Miao, Y.: A minority class boosted framework for adaptive access control decision-making. In: Web Information Systems Engineering–WISE 2021: 22nd International Conference on Web Information Systems Engineering, WISE 2021, Melbourne, VIC, Australia, October 26–29, 2021, Proceedings, Part I 22. pp. 143–157. Springer (2021)
30. You, M., et al.: A knowledge graph empowered online learning framework for access control decision-making. World Wide Web, pp. 1–22 (2022)
31. Yuan, F., Cao, Y., Shang, Y., Liu, Y., Tan, J., Fang, B.: Insider threat detection with deep neural network. In: International Conference on Computational Science, pp. 43–54. Springer (2018)
32. Yuan, S., Wu, X.: Deep learning for insider threat detection: review, challenges and opportunities. Comput. Secur. **104**, 102221 (2021)

PV-PATE: An Improved PATE for Deep Learning with Differential Privacy in Trusted Industrial Data Matrix

Hongyu Hu[1], Qilong Han[1], Zhiqiang Ma[1(✉)], Yukun Yan[1], Zuobin Xiong[2], Linyu Jiang[1], and Yuemin Zhang[1]

[1] Harbin Engineering University, Harbin, China
{hhyu,hanqilong,mazhiqiang,yanyukun,linyujiang,zhangyuemin}@hrbeu.edu.cn
[2] Georgia State University, Georgia, USA
zxiong2@student.gsu.edu

Abstract. Differential privacy (DP) has been widely used in many domains of statistics and deep learning (DL), such as protecting the parameters of DL models. The framework Private Aggregation of Teacher Ensembles (PATE) is a popular solution for privacy protection that effectively avoids membership inference attacks in model training. However, in Trusted Industrial Data Matrix (TDM) where privacy budgets are constrained and information sharing between models is required, existing works using PATE have two issues. First, the data utility is reduced due to the overfitting problem resulting from insufficient knowledge transfer from teachers to students. Second, teachers cannot share information, thus creating an information silo problem. In this paper, we first proposed the Personalized Voting-based PATE framework (PV-PATE) in TDM to solve the above-mentioned issues. It includes Teacher Credibility that reduces sensitivity by changing voting weights and an Adaptive Voting mechanism based on teachers voting. In addition, we propose a Model Sharing mechanism to achieve model cloning and elimination. We conduct extensive experiments on MNIST dataset and SVHN dataset to demonstrate that our approach achieves not only outstanding learning performance but also provides strong privacy guarantees.

Keywords: Differential privacy · Deep learning · Trusted industrial data matrix · Private Aggregation of Teacher Ensembles · Personalized voting

1 Introduction

Recently, the current circulation of industrial data sharing faces difficulties in quality, sovereignty, and transaction. In order to achieve the maximal utility of data, the data must be able to be applied across data providers and industries. As a result, the International Data Spaces Association (IDSA) has partnered with government, industry, and research to create a reference architecture and formal standards for secure and trusted Data Spaces [9]. One of the classic data spaces is the Trusted Industrial Data Matrix (TDM) system architecture, which

X. Song et al. (Eds.): APWeb-WAIM 2023, LNCS 14333, pp. 477–491, 2024.
https://doi.org/10.1007/978-981-97-2387-4_32

is divided into four main parts from the perspective of business, as shown in Fig. 1. Moreover, the widespread development of DL models has raised increasing privacy concerns. For example, DL models can subconsciously "memorize" sensitive data. That is, once the model is released, the privacy of sensitive data is vulnerable to membership inference attacks [12]. Since the adversary may get access to the parameters of the DL models, the privacy of user and data may be exposed for some specific datasets, such as job interviews [6] and healthcare data [16].

DP has become the gold standard of privacy protection, which is widely used in data and model privacy protection. At present, Differentially Private Stochastic Gradient Descent (DP-SGD) algorithm [13] and the Private Aggregation of Teacher Ensembles (PATE) framework [10] successfully apply DP to DL models. The first approach adds noise to the gradient during model training, and the privacy budget ε becomes larger as the number of model iterations increases. The second approach adds noise to the teachers' aggregated results by combining knowledge distillation with semi-supervised learning to finally publish the student models. For PATE, because different teachers can use different DL algorithms to participate in the training of student models, the method can be applied not only to centralized learning with non-convex models but also to distributed learning scenarios. Moreover, when we train more complex DL models since the privacy bounds depend on the model parameters, the data utility is reduced in DP-SGD while it is not the case for PATE [10]. However, in TDM, for PATE, there will be the following problems. First, due to expensive and constrained privacy budgets in TDM and the addition of fixed noise per vote, teachers can only answer limited student queries with poor data utility [14], which results in overfitting problems when we train student models with only the semi-supervised learning [8,17]. Second, information cannot be shared among teacher models, which is not in line with the concept of data circulation among data providers in TDM scenario.

In this work, in order to solve the above issues, we first propose a Personalized Voting-based PATE framework (PV-PATE) in TDM which is able to achieve personalized privacy budget allocation through an Adaptive Voting mechanism, reduce sensitivity Δf by changing teachers' voting weights, and enable information transfer between data providers through Model Sharing mechanism. Then, we evaluate the proposed PV-PATE framework on two commonly used datasets, MNIST and SVHN. The experimental results show that our work provides the first practical PATE solution for DL with DP guarantee in TDM. The contributions of this work are summarized as follows.

- We introduce the Personalized Voting Mechanism, including the Adaptive Voting Mechanism and the aggregation mechanism with Teacher Credibility.
- We introduce the Model Sharing Mechanism and realize information sharing between data providers, which can break the information silos problem in TDM.
- We propose the novel framework PV-PATE that is specified for data utility and privacy protection in TDM.

Fig. 1. Fusion mode of Trusted Industrial Data Matrix. Image from Trusted Industrial Data Space System Architecture 1.0.

- Extensive experimental evaluations are conducted on DL tasks on utility and privacy for MNIST and SVHN dataset.

2 Related Work

2.1 DP Mechanisms for DL

DP ensures that data is used for research or analysis without causing data leakage. We can incorporate DP mechanisms during the training of DL models by model service providers in TDM, which can effectively avoid various attacks [12], and one of the most classical approaches is the Differentially Private Stochastic Gradient Descent (DP-SGD) [13]. In DP-SGD, the noise is added to clipped gradients when training DL models. Privacy costs of DP-SGD are accounted for through the moments accountant [1]. Another popular method is PATE, The first PATE framework [10] adds Laplace noise to teachers voting, passes the noisy labels to students, and finally, we use semi-supervised learning to train the ultimate student models. The scalable PATE [11] replaced Laplace noise with Gaussian noise while using Rényi differential privacy advanced composition theorem for privacy tracking to the original PATE is improved, while Jordan et al. uses the idea of under-sampling to divide the samples at a finer granularity and introduces a personalized moments accountant to achieve separate DP bounds for each data point [4].

2.2 DP-SGD vs PATE

In realistic scenarios, such as TDM, we will train highly sophisticated models, and the privacy bounds of the DL models applying DP-SGD are limited with the complexity of the models [10], which affects the data utility. Furthermore, Bagdasaryan et al. demonstrate that the accuracy of the DL models drops dramatically when the neural network is trained using DP-SGD for the less represented groups. If the original DL model is unfair, then it is even more unfair after applying DP [2]. However, Uniyal et al. empirically show that PATE is much less severe than DP-SGD in these cases [15]. For PATE, since the DL algorithms of teachers can be different, it lays the foundation for us to apply PATE to distributed scenarios that meet future trends. Therefore, we choose PATE instead of DP-SGD and make improvements to PATE to address the model privacy leakage problem that exists in TDM.

3 Background and Notation

This section provides the theoretical basis for the current work and describes the notation used in this paper.

3.1 Differential Privacy

Differential privacy, introduced by Dwork et al. [3], has been embraced by multiple research communities as a commonly accepted notion of privacy for algorithms on statistical databases [7]. As applications of DP begin to emerge, DP will be increasingly applied to DL model privacy calculations and tracking.

Definition 1 $((\varepsilon, \delta)\text{-}DP)$. *A randomized mechanism* $H: \mathcal{D} \mapsto \mathcal{R}$ *offers* (ε, δ)-*DP if for any adjacent dataset* $D, D' \in \mathcal{D}$ *and* $\mathcal{T} \in \mathcal{R}$

$$Pr[H(D) \in \mathcal{T}] \leq e^{\varepsilon} Pr[H(D') \in \mathcal{T}] + \delta \tag{1}$$

We call $\mathcal{D}$ and $\mathcal{R}$ the set of all possible data points. The symbol ε is called the privacy budget. δ is a slack term and indicates that it is acceptable for DP to be unsatisfied to some extent. $Pr[\cdot]$ represents the probability of an event according to an appropriate probability measure.

3.2 Rényi Differential Privacy

Rényi Differential Privacy (RDP) is a natural slack of DP and is a strictly stronger definition of privacy. Most importantly, RDP can be intuitively combined with the application of advanced composition theorems.

Definition 2 $((\alpha, \varepsilon)\text{-}Rényi Differential Privacy)$. *A randomized mechanism* H *is said to have* ε-*RDP of order* α, *if for any adjacent datasets* D *and* D',

$$D_{\alpha}(H(D) \parallel H(D')) \leq \varepsilon \tag{2}$$

In DL, we often process data multiple times in order to train a model. This process can be thought of as a composition of mechanisms that each with a privacy cost [18]. The composition theorem of RDP is as follows.

Proposition 1 *(Composition). If a mechanism* H *consists of mechanisms* H_1 *and* H_2, *and* H_i *guarantees* (α, ε_i)-*RDP, then* H *guarantees* $(\alpha, \varepsilon_1 + \varepsilon_2)$-*RDP.*

Proposition 2 *(From RDP to* (ε, δ)-*DP). If* f *is an* (α, ε)-*RDP mechanism, it also satisfies* $(\varepsilon + \frac{log1/\delta}{\alpha - 1}, \delta)$-*differential privacy for any* $0 < \delta < 1$.

3.3   PATE and PV-PATE

The PATE consists of three main components, Teacher models, Aggregation mechanism and the Student model. Each teacher completes the model training on their own subsets which are disjoint and have the same distribution. We then feed incomplete public data to teachers, and each teacher raises a vote for a concrete class. The aggregation mechanism adds noise to teachers voting and then selects the class with the most votes to students. To reduce the total privacy budgets, we need to use semi-supervised learning to reduce the number of student queries to teachers and train unlabeled data with labeled data.

In this paper, we still use the Gaussian mechanism that satisfies the RDP to add noise to the voting results, i.e., the Gaussian NoisyMax (GNMax) aggregation method.

Definition 3 *(cf. [11], Sec. 4.1). For a sample x and classes 1 to k, let $t_j(x) \in [k]$ denote the j-th teacher model's prediction on x and $c_j(x)$ be the vote count for the j-th class. The GNMax aggregation mechanism is as follows.*

$$GNMax_\sigma(x) \triangleq \underset{j}{argmax} \left\{ c_j(x) + \mathcal{N}(0, \sigma^2) \right\} \tag{3}$$

The Gaussian noise is sampled from a normal distribution $\mathcal{N}(\mu, \sigma^2)$ with mean $\mu = 0$ and variance σ^2.

Corollary 1 *If real-valued function f has sensitivity Δf, then the Gaussian mechanism $G_\sigma f(D) = f(D) + \mathcal{N}(0, \sigma^2)$ satisfies $(\alpha, (\Delta f \cdot \alpha)/(2\sigma^2))$-RDP.*

For PV-PATE, we use the Improved Confident-GNMax Aggregator described in Algorithm 2 to filter queries that do not reach a sufficiently strong consensus of the teachers. In order to perform privacy calculations and analysis in Sect. 4, we first give the definitions of the following methods.

Definition 4 *(Teacher Credibility). Teacher Credibility is defined as*

$$credibility_i(cre_i) = acc_i \cdot k \tag{4}$$

Let acc_i be the accuracy of the i-th teacher, k is the moderating factor to adjust the weight of teacher votes.

Definition 5 *(Teachers Voting Gap). Let $g_{i\text{-}th}$ be the gap of teachers voting results for the i-th query of size $n \in \mathbb{N}$. Teachers Voting Gap is defined as*

$$g_{i\text{-}th} = v_1(i) - v_2(i) \tag{5}$$

Let $v_1(i)$ be the maximum value of voting results for the i-th query and $v_2(i)$ is the second largest value.

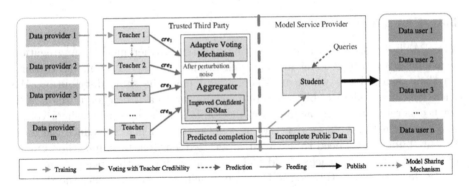

Fig. 2. privacy protection framework PV-PATE in TDM. Using this framework, we can effectively protect the parameters and sensitive data in Model Service Provider.

4   Personalized Voting for PATE in TDM

In order to solve the privacy leakage problem of model service providers in TDM, we propose a privacy protection framework PV-PATE in TDM, as shown in Fig. 2.

We first introduce Teacher Credibility from the perspective of sensitivity reduction in Sect. 4.1 to differentiate teacher models, describe the Adaptive Voting mechanism in Sect. 4.2, which consumes fewer privacy budgets and provides more stringent privacy guarantees, and describe the Model Sharing mechanism in Sect. 4.3 to achieve model clone and elimination.

4.1   Teacher Credibility

Although each teacher acquires an equally distributed data set, there is a gap between teacher models. The accuracy of the teachers on MNIST and SVHN is shown in Fig. 3. At present, the teacher models in PATE all have a voting weight of 1. In order to reduce the sensitivity Δf, we replace the original voting weight with Teacher Credibility as defined in Definition 4, and the weight of each teacher's vote will be decided by itself to realize personalized voting. For Confident-GNMax aggregator, its Δf changes from 2 to $2cre_{max}(< 2)$. According to the Gaussian noise mechanism as defined in Corollary 1, with the same noise, less privacy budgets will be consumed due to the reduction of sensitivity.

In Definition 4, if only accuracy is used to define Teacher Credibility, due to the reduction of the voting weight for each teacher, the total number of votes for each classification for a particular sample is decreased. This causes the gap between the total number of votes for each classification to be smaller than before. If we add noise to the voting process, according to Definition 3, we are more likely to obtain incorrect classification results. Therefore, we introduce a moderating factor k to compensate for this drawback. To ensure that the global sensitivity Δf is less than 2, we need to ensure that $2cre_{max} < 2$, which means that $k < 1/acc_{max}$. However, in order to reduce the number of incorrect

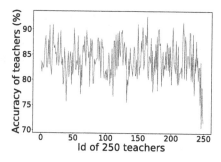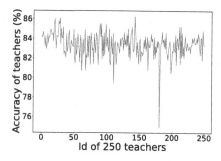

Fig. 3. Accuracy of 250 teacher models on MNIST(left) and SVHN(right). The maximum value of teacher model accuracy on MNIST (left) is 92.56% and the minimum value is 70.51%, with a large variation. The accuracy on SVHN (right) have a maximum of 86.27% and a minimum of 74.94%, which is smoother relative to the MNIST dataset.

classification results, we want k to be greater than or equal to 1. Therefore, $k \in [1, 1/acc_{max})$.

4.2   Adaptive Voting Mechanism

In TDM, because some data providers have restricted privacy budgets, the student cannot make all the queries and teachers pass relatively little knowledge to the student, leading to the overfitting problem that student is prone to due to insufficient data with labels. PATE is identified as having no upper boundary for privacy budget, only to achieve higher accuracy with the smallest possible privacy budget, and if the privacy budget is limited, i.e., the number of queries is limited, the accuracy of PATE is bounded to decrease.

In order to solve this problem, we first need to adaptively allocate privacy budgets for teachers' votes results based on Teachers Voting Gap in Definition 5. When the gap is larger, the randomness is smaller, the label utility is better, we need to add less noise and allocate a larger privacy budget ε, on the contrary, increase the noise and allocate a smaller privacy budget as shown in Algorithm 1. For all queries, the algorithm first needs to keep all Teachers Voting Gap in Definition 5 in list G, and B is the list of Gaussian noise with the standard deviation σ_2 perturbed by standard normal distribution in this paper, which is the same size as list G. Specifically, we set σ_2 to be 40, consistent with Papernot et al [11]. paper, which means that list B is a collection of perturbed samples from the standard normal distribution with $\sigma_2 = 40$, and the element in the list is denoted as $\hat{\sigma}$.

Algorithm 1: Adaptive Voting

Input: input the number of queries n, lists G and B

1 **for** $i = 1, 2, .., n$ **do**
2   positive sort for $g_{i\text{-}th} \in G$;
3   reverse sort for $\hat{\sigma}_{i-1} \in B$.
4 **for** $i = 1, 2, .., n$ **do**
5   get index l of $g_{i\text{-}th}$;
6    $\hat{\sigma}_l^2 \in B$ as the Gaussian noise variance of the i-th query, denoted as $\hat{\sigma}_{i\text{-}th}^2$.

Output: $\hat{\sigma}_{i\text{-}th}$ for the i-th query

For a certain dataset, after Algorithm 1 we can obtain the noise added by the student model to each query of the teacher model, and for the i-th query, Gaussian noise with 0 as the mean and $\hat{\sigma}_{i\text{-}th}^2$ as the variance is added, and by Corollary 1 and Proposition 2 we are able to compute the privacy budget consumed by each query. However, in the paper by Papernot et al [11]. it is shown that when there is a lack of consensus among teachers, i.e., the Gap in Definition 5 is small, the aggregator is likely to output the wrong labels and answering such queries consumes too much privacy budget for training from a privacy perspective. In contrast, queries with strong consensus can achieve tight privacy bounds. Therefore, we need a algorithm to filter out queries that do not have strong consensus among teacher models.

We then make improvements to the Confident-GNMax algorithm proposed by Papernot et al [11]. The improved algorithm is shown in Algorithm 2. To select queries with overwhelming consensus, the algorithm checks whether the $GNMax_\sigma(x)$ for a sample x in Definition 3 exceeds a threshold T. The comparison is performed after adding Gaussian noise with variance σ_1^2. If the i-th query passes the threshold check, the aggregator with variance $\hat{\sigma}_{i\text{-}th}^2$ generated by Algorithm 1 implements the usual GNMax mechanism in Definition 3. For queries that do not pass the noise threshold check, the aggregator will simply return $\perp$, and the student model will discard this example in its training [11].

We typically choose σ_1^2 to be much larger than any $\hat{\sigma}_{i\text{-}th}^2$ because for each query we need to check whether the maximum number of classification votes after adding Gaussian noise exceeds the threshold T. Therefore, we want to ensure that the privacy budget consumed by each check is not too large, i.e., σ_1^2 needs to be large. Based on the privacy utility tradeoff, we generally choose T to be between 60% and 80% of the number of teachers.

Algorithm 2: Improved Confident-GNMax Aggregator

Input: input x, threshold T, noise parameters σ_1 and $\hat{\sigma}_{i\text{-}th}$ of i-th query

1 **if** $\max_i \{c_j(x)\} + \mathcal{N}(0, \sigma_1^2) \geq T$ **then**
2   return $argmax_j \{c_j(x) + \mathcal{N}(0, \hat{\sigma}_{i\text{-}th}^2)\}$;
3 **else**
4   return $\perp$;

From the perspective of teacher voting, the two methods above lend to allowing for personalised differential privacy [5], and we refer to them collectively as Personalized Voting-based PATE framework (PV-PATE). We illustrate the advantages of our proposed framework based on experimental results in Sect. 5 and provide a rigorous privacy analysis and proof in Sect. 4.4.

4.3   Model Sharing Mechanism

In TDM scenario, as shown in Fig. 2, each data provider will send data to Trusted Third Party, and the Third Party will choose to share the teacher model or not according to the data autonomy of each data provider. In the case of sharing, after training each teacher model on independently and identically distributed datasets, we can obtain its accuracy and loss values, and Trusted Third Party can use these metrics to identify the best-performing teacher model as well as several inferior ones. By copying the optimal teacher model, including its structure and parameters, and replacing the inferior models, the Model Sharing mechanism can be implemented. Finally, we obtain better utility for the aggregated teacher models. The specific implementation of this mechanism and the experimental results under this mechanism are presented in Sect. 5.4.

4.4   Privacy Analysis

In this section we perform privacy calculations and analysis of PV-PATE. We assume that N is the number of all queries from students to teachers and $\widetilde{N}$ is the number of queries answered by teachers. For the i-th query, the Gaussian noise with variance σ_1^2 added by the aggregator to determine whether it is worth answering, and for queries that are definitely answered, we add a Gaussian noise variance with $\widehat{\sigma}_{i\text{-}th}^2$.

Theorem 1 *Suppose that on adjacent data sets D and D', the label counts c_j are differ from each other by only one sample. For the i-th query, let $\mathcal{F}$ be the mechanism that generates $argmax_j \left\{ c_j + \mathcal{N}(0, \widehat{\sigma}_{i\text{-}th}^2) \right\}$. Then $\mathcal{F}$ satisfies $(\alpha, (cre_{max} \cdot \alpha)/\widehat{\sigma}_{i\text{-}th}^2)$-RDP. Therefore, for any $\alpha \geq 1$, the GNMax aggregator in PV-PATE guarantees*

$$D_\alpha \left(\mathcal{F}(D) \| \mathcal{F}(D') \right) = \frac{1}{\alpha - 1} \log \mathbb{E}_{x \sim \mathcal{F}(D)} \left[\left(\frac{\Pr[\mathcal{F}(D) = x]}{\Pr[\mathcal{F}(D') = x]} \right)^{\alpha - 1} \right] \leq \frac{cre_{max} \cdot \alpha}{\widehat{\sigma}_{i\text{-}th}^2}$$

Thus over N queries, according to the combination theorem of RDP as proposed in Proposition 1, we get $(\alpha, \frac{cre_{max} \cdot \alpha \cdot N}{\sigma_1^2} + \sum_{i=1}^{\widetilde{N}} \frac{cre_{max} \cdot \alpha}{\widehat{\sigma}_{i\text{-}th}^2})$-RDP, $\widehat{\sigma}_{i\text{-}th}^2$ can be calculated by Algorithm 1.

Theorem 2 *Comparison of the privacy bounds of PATE and PV-PATE with Personalized Voting Mechanism. For PATE, the privacy bounds are $\frac{\alpha \cdot N}{\sigma_1^2} + \frac{\alpha \cdot \widetilde{N}}{\sigma_2^2}$, Moreover, the privacy bounds of PV-PATE can be obtained from Theorem 1, which is smaller than PATE. The proof is as follows.*

Proof We first subtract privacy bounds for PATE and PV-PATE.

$$(1 - cre_{max}) \cdot \frac{\alpha \cdot N}{\sigma_1^2} + \alpha \sum_{i=1}^{\tilde{N}} \left(\frac{1}{\sigma_2^2} - \frac{cre_{max}}{\hat{\sigma}_{i-th}^2} \right)$$

For the term $(1 - cre_{max}) \cdot \frac{\alpha \cdot N}{\sigma_1^2}$, it must be > 0, we only need to prove that the term $\alpha \sum_{i=1}^{\tilde{N}} \left(\frac{1}{\sigma_2^2} - \frac{cre_{max}}{\hat{\sigma}_{i-th}^2} \right) > 0$.

$$\alpha \sum_{i=1}^{\tilde{N}} \left(\frac{1}{\sigma_2^2} - \frac{cre_{max}}{\hat{\sigma}_{i-th}^2} \right) = \alpha \sum_{i=1}^{\tilde{N}} \left(\frac{1}{\sigma_2} + \frac{\sqrt{cre_{max}}}{\hat{\sigma}_{i-th}} \right) \left(\frac{1}{\sigma_2} - \frac{\sqrt{cre_{max}}}{\hat{\sigma}_{i-th}} \right) >$$

$$\alpha \sum_{i=1}^{\tilde{N}} \left(\frac{\sqrt{cre_{max}}}{\sigma_2} + \frac{\sqrt{cre_{max}}}{\hat{\sigma}_{i-th}} \right) \left(\frac{1}{\sigma_2} - \frac{1}{\hat{\sigma}_{i-th}} \right) > \alpha\sqrt{cre_{max}} \sum_{i=1}^{\tilde{N}} \left(\frac{1}{\sigma_2} + \frac{1}{\hat{\sigma}_{i-th}} \right) \left(\frac{1}{\sigma_2} - \frac{1}{\hat{\sigma}_{i-th}} \right)$$

$$> \dots > \alpha (\sqrt{cre_{max}})^{+\infty} \sum_{i=1}^{\tilde{N}} \left(\frac{1}{\sigma_2} + \frac{1}{\hat{\sigma}_{i-th}} \right) \left(\frac{1}{\sigma_2} - \frac{1}{\hat{\sigma}_{i-th}} \right) = 0$$

5   Experiments

This section presents the implementation details of our proposed approach. We experiment with MNIST and SVHN datasets described in Sect. 5.1 with the same test and training sets as in previous DP work. In Sect. 5.2, we evaluate the improvements of Teacher Credibility on Confident-GNMax, and demonstrate the importance of reduced sensitivity to privacy budgets. In Sect. 5.3, we then prove the effects of PATE with Adaptive Voting mechanism added on these two datasets. Afterwards, we experimentally demonstrate that PV-PATE based on Personalized Voting mechanism proposed in Sect. 4 can be applied to TDM scenarios, which can achieve high accuracy within strict privacy bounds. In Sect. 5.4, we evaluate the Model Sharing mechanism to achieve information transfer of teacher models through model cloning and elimination.

5.1   Datasets and Models

Because a more detailed description of the experimental setup can be found in Papernot et al. [10,11], we only provide a brief overview here. For both datasets, we first trained 250 teacher models and then trained student models. For MNIST, the student model is given 9,000 samples, and we label the noise-added subsets of either 80, 100, 160, 200, 250, and 320 samples through the teacher aggregator, evaluating the performance of the student model on the remaining 1000 samples. For SVHN, the student model is given 10,000 training input samples, and we use the aggregator to label 500, 1,000, 1,500, 2,000 and 3,000 samples, and test the performance of the student model on the remaining 16,032 samples.

For MNIST, we stack two convolutional layers with max-pooling and one fully connected layer with ReLUs to train teacher and student models. For SVHN, we add two hidden layers. The structure of these neural networks remains the same as Papernot(2018) et al [11]. paper.

Our experiments are compared to the model accuracy and privacy budgets consumed in the Papernot et al [11]. paper. However, the dataset of teacher voting results used in that paper was lost, so we have to train our own teacher models, which also resulted in the final accuracy and privacy budgets consumed by the student model being inconsistent with the original paper. We will replicate that paper's experiments with our own trained teacher models and ensure that the experimental results are as close to the original paper as possible.

5.2 Aggregation Mechanism with Teacher Credibility

We introduced the concept of Teacher Credibility in Sect. 4.1. By changing the teacher voting weights, not only can we reduce sensitivity, but teachers can also personalize their votes based on their accuracy. The value range of the moderating factor k in Definition 4 is determined by the maximum accuracy of teacher models. We should let $acc_{max} \leq k \cdot acc_{max} < 1$, so that we can ensure that the sensitivity is less than 2. On each dataset, we can obtain the accuracy and privacy budgets of PATE according to the range of k. Based on the maximum teacher model accuracy for each dataset shown in Fig. 3, we found that for MNIST dataset, $1 \leq k < 1.08$, and for SVHN dataset, $1 \leq k < 1.16$. We randomly selected some values of k and found the value of k that resulted in the highest accuracy for the student model, which will be used in Sect. 5.3. As shown in Fig. 4, when k takes the value of 1.02, the highest accuracy is obtained on MNIST, but for SVHN, k is taken as 1.10 to achieve the best results, while the privacy cost is relatively small as shown in Fig. 5.

5.3 Student Training with Adaptive Voting Mechanism

In our experiments, keeping T and σ_1 constant with Papernot et al [11]. paper, we perturb σ_2 with standard normal distribution when σ_2 equals 40, resulting in

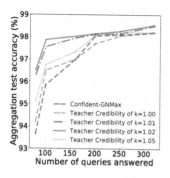
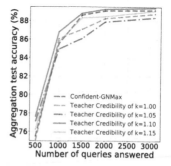

Fig. 4. Accuracy corresponding to different moderators k in Teacher Credibility for 80, 100, 160, 200, 250, and 320 answers on MNIST(left) and for 500, 1000, 1500, 2000, 3100 answers on SVHN(right).

Table 1. Accuracy and privacy bounds of two datasets with four different mechanisms.

Dataset	Queries answered	Confident-GNMax (Papernot et al., 2018)		Teacher Credibility		Adaptive Voting		Personalized Voting	
		Accuracy	Privacy bound	Accuracy	Privacy bound	Accuracy	Privacy bound	Accuracy	Privacy bound
MNIST	80	93.65%	0.99	96.52%	0.954	96.09%	0.973	**96.82%**	**0.942**
	100	95.85%	1.10	97.87%	1.075	97.35%	1.090	**97.94%**	**1.063**
	160	97.00%	1.394	98.03%	1.342	97.86%	1.365	**98.21%**	**1.323**
	200	98.05%	1.575	98.15%	1.524	98.12%	1.550	**98.23%**	**1.507**
	250	98.10%	1.758	98.26%	1.709	98.18%	1.735	**98.30%**	**1.685**
	320	98.20%	2.04	98.54%	1.977	98.28%	2.012	**98.58%**	**1.952**
SVHN	500	75.52%	1.826	77.65%	1.777	77.01%	1.802	**78.15%**	**1.754**
	1000	85.51%	2.638	86.79%	2.566	87.07%	2.604	**88.67%**	**2.534**
	1500	88.61%	3.283	88.81%	3.192	89.11%	3.240	**89.91%**	**3.151**
	2000	89.08%	3.843	89.26%	3.735	89.19%	3.793	**90.53%**	**3.688**
	3100	89.27%	4.932	89.41%	4.791	90.02%	4.868	**91.13%**	**4.729**

a range of values for σ_2 between 35 and 45. We calculate the Teachers Voting Gap in Definition 5 and add smaller σ_2 if the Gap is larger as shown in Algorithm 1. For MNIST, the student model will get 8000 samples with perturbed noise. For SVHN, the student model obtains 10,000 samples with noise addition. The student model accuracy and its privacy cost are shown in Table 1 and Fig. 5.

The Personalized Voting mechanism, mentioned in Sect. 4.2, is a combination of Teacher Credibility and Adaptive Voting mechanism. The results of comparing each mechanism such as Confident-GNMax with the Personalized Voting mechanism are given in our experiments, as shown in Table 1. For MNIST, the student model has the highest accuracy when k=1.02, as shown in Fig. 4, so we added the Adaptive Voting mechanism to this. For SVHN, on the basis of k=1.10, we composed the Personalized Voting mechanism, and the privacy cost is shown in Fig. 5.

5.4 Student Training with Model Sharing Mechanism

As outlined in Sect. 4.3, we eliminate some poor teacher models and replace them with better models. We simply use the accuracy and loss values as metrics to evaluate the models and set a high priority on model accuracy, i.e., remove the models with lower accuracy if given the same loss values.

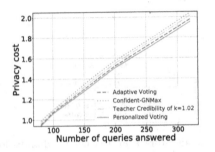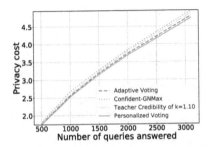

Fig. 5. Privacy cost of different mechanisms on MNIST(left) and SVHN(right).

For MNIST and SVHN, we use the teacher model with the highest accuracy to replace the worst 5, 10, and 25 teacher models. Detailed information such as the accuracy of this mechanism at different number of responses are shown in Table 2. We use this method to simulate the process of information sharing among data providers which is based on their own data autonomy.

We can observe that as the number of excellent teacher models increases, the accuracy of the student model does not always increase, but it is still higher than that of the Confident-GNMax mechanism. The reason for this phenomenon is that the excellent teacher models reach a consensus on the misclassification of certain samples, which leads to a greater number of votes for misclassification than for correct classification in the aggregation mechanism. This results in feeding the misclassified samples to the semi-supervised learning model, leading to a decrease in the accuracy of the student model.

Table 2. Accuracy of two datasets with Confident-GNMax Mechanism (Papernot et al., 2018) and Model Sharing Mechanism.

Dataset	Queries answered	Accuracy			
		Confident-GNMax	Model Sharing Mechanism		
			Out of 5	Out of 10	Out of 25
MNIST	80	93.65%	**97.76%**	97.50%	96.75%
	100	95.85%	**97.91%**	97.51%	97.40%
	160	97.00%	**98.12%**	97.90%	97.92%
	200	98.05%	98.18%	**98.20%**	98.12%
	250	98.10%	98.27%	98.40%	**98.43%**
	320	98.20%	98.36%	**98.54%**	98.48%
SVHN	500	75.52%	76.24%	76.88%	**77.50%**
	1000	85.51%	86.03%	**87.02%**	86.50%
	1500	88.61%	88.70%	**89.66%**	88.91%
	2000	89.08%	89.18%	**89.84%**	89.16%
	3100	89.27%	**90.12%**	89.91%	89.86%

6   Conclusion

PATE framework as a popular DP-based learning method has attracted a lot of research attention, but it is restricted in the application of TDM due to some identified issues. In this work, we try to solve these problems with a novel improvement on the original PATE framework and achieve both high data utility and privacy guarantee in TDM. Based on this, we propose a flexible PV-PATE framework that combines teacher credit, adaptive voting, and model sharing mechanisms to effectively protect the parameters of DL models, thus preventing adversaries from obtaining raw and sensitive data. Specifically, we analyze the privacy bounds of TDM theoretically and evaluate its utility experimentally on two datasets. Our results show that PV-PATE has higher accuracy and consumes less privacy budget. Thus, the PV-PATE framework can effectively protect individual data providers' model parameters and sensitive information in TDM, achieving data autonomy for individual data providers, breaking information silos, and meeting the IDSA criteria.

Acknowledgments. This work was supported by the National Key R&D Program of China under Grant No. 2020YFB1710200, and the National Natural Science Foundation of China under Grant No. 62072136.

References

1. Abadi, M., et alL.: Deep learning with differential privacy. In: Proceedings of the 2016 ACM SIGSAC Conference on Computer and Communications Security, pp. 308–318 (2016)
2. Bagdasaryan, E., Poursaeed, O., Shmatikov, V.: Differential privacy has disparate impact on model accuracy. Adv. Neural Inform. Process. Syst. **32** (2019)
3. Dwork, C., McSherry, F., Nissim, K., Smith, A.: Calibrating noise to sensitivity in private data analysis. In: Halevi, S., Rabin, T. (eds.) TCC 2006. LNCS, vol. 3876, pp. 265–284. Springer, Heidelberg (2006). https://doi.org/10.1007/11681878_14
4. Jordon, J., Yoon, J., van der Schaar, M.: Differentially private bagging: Improved utility and cheaper privacy than subsample-and-aggregate. Adv. Neural Inform. Process. Syst. **32** (2019)
5. Jorgensen, Z., Yu, T., Cormode, G.: Conservative or liberal? personalized differential privacy. In: 2015 IEEE 31St International Conference on Data Engineering, pp. 1023–1034. IEEE (2015)
6. Mahmoud, A.A., Shawabkeh, T.A., Salameh, W.A., Al Amro, I.: Performance predicting in hiring process and performance appraisals using machine learning. In: 2019 10th International Conference on Information and Communication Systems (ICICS), pp. 110–115. IEEE (2019)
7. Mironov, I.: Rényi differential privacy. In: 2017 IEEE 30th Computer Security Foundations Symposium (CSF), pp. 263–275. IEEE (2017)
8. Miyato, T., Maeda, S.i., Koyama, M., Ishii, S.: Virtual adversarial training: a regularization method for supervised and semi-supervised learning. IEEE Trans. Pattern Analy. Mach. Intell. **41**(8), 1979–1993 (2018)
9. Otto, B., et al.: Reference architecture model for the industrial data space (2017)
10. Papernot, N., Abadi, M., Erlingsson, U., Goodfellow, I., Talwar, K.: Semi-supervised knowledge transfer for deep learning from private training data. arXiv preprint arXiv:1610.05755 (2016)
11. Papernot, N., Song, S., Mironov, I., Raghunathan, A., Talwar, K., Erlingsson, Ú.: Scalable private learning with pate. arXiv preprint arXiv:1802.08908 (2018)
12. Shokri, R., Stronati, M., Song, C., Shmatikov, V.: Membership inference attacks against machine learning models. In: 2017 IEEE Symposium on Security and Privacy (SP), pp. 3–18. IEEE (2017)
13. Song, S., Chaudhuri, K., Sarwate, A.D.: Stochastic gradient descent with differentially private updates. In: 2013 IEEE Global Conference on Signal and Information Processing, pp. 245–248. IEEE (2013)
14. Tramer, F., Boneh, D.: Differentially private learning needs better features (or much more data). arXiv preprint arXiv:2011.11660 (2020)
15. Uniyal, A., et al.: Dp-sgd vs pate: which has less disparate impact on model accuracy? arXiv preprint arXiv:2106.12576 (2021)

16. Wiens, J., Shenoy, E.S.: Machine learning for healthcare: on the verge of a major shift in healthcare epidemiology. Clin. Infect. Dis. **66**(1), 149–153 (2018)
17. Wu, B., et al.: P3sgd: patient privacy preserving sgd for regularizing deep cnns in pathological image classification. In: Proceedings of the IEEE/CVF Conference on Computer Vision and Pattern Recognition, pp. 2099–2108 (2019)
18. Yu, D., Kamath, G., Kulkarni, J., Yin, J., Liu, T.Y., Zhang, H.: Per-instance privacy accounting for differentially private stochastic gradient descent. arXiv preprint arXiv:2206.02617 (2022)

LayerBF: A Space Allocation Policy for Bloom Filter in LSM-Tree

Jiaoyang Li[1,2], Zhixin Fan[1,2], Yinliang Yue[3(✉)], Zekun Yao[1,2], Jinzhou Liu[1,2], and Jiang Zhou[1,2]

[1] School of Cyberspace Security, University of Chinese Academy of Sciences, Beijing, China
lijy62@chinatelecom.cn
[2] Institute of Information Engineering, Chinese Academy of Sciences, Beijing, China
{fanzhixin,yaozekun,liujinzhou,zhoujiang}@iie.ac.cn
[3] Zhongguancun Laboratory, Beijing 100080, People's Republic of China
yueyl@zgclab.edu.cn

Abstract. LSM-Tree based key-value stores commonly suffer from the issue of read amplification, as the retrieval of a particular key typically requires examination of multiple layers of SSTables. To enhance query performance, a bloom filter is commonly employed, although it is susceptible to the problem of false positives, which leads to additional I/Os. To mitigate the issue of false positives, the bloom filter size can be increased, but this in turn results in higher memory consumption. In response, we have developed LayerBF, a space allocation strategy for layered bloom filters. By leveraging access frequency, LayerBF dynamically allocates bits-per-key of bloom filters in each layer. Hotter layers are allocated a larger space, while colder layers are allocated a smaller space. This approach reduces the average false positive rate, improves storage read performance, and simultaneously minimizes memory consumption. We have implemented LayerBF in the widely used RocksDB key-value store and evaluated its performance with and without LayerBF on both hard disk drives (HDDs) and solid-state drives (SSDs). The evaluation results demonstrate that LayerBF improves read performance by 5% to 14% and reduces the false positive rate by 8% to 10%.

Keywords: LSM-Tree · Bloom Filter · Key-Value Database

1 Introduction

Persistent key-value stores are used widely across various applications, such as online shopping [7], search indexing [9,10], advertising [8], and cloud systems [13]. However, as data volume increases, writing to large-scale data becomes challenging for disks. To address this challenge and improve write performance, especially random write performance, most modern key-value stores use the Log-Structured Merge-Tree (LSM-Tree) data structure. For instance, key-value stores like BigTable [2], RocksDB [8], HBase [9], LittleTable [20], and Cassandra [14]

X. Song et al. (Eds.): APWeb-WAIM 2023, LNCS 14333, pp. 492–506, 2024.
https://doi.org/10.1007/978-981-97-2387-4_33

all use LSM-Tree. LSM-Tree [19] is a well-known index structure for key-value stores that can convert random writes into sequential writes.

To reduce the read amplification in a key-value store that uses an LSM-Tree, many storage systems like Leveldb [10] and RocksDB [8] incorporate a bloom filter. The purpose of the bloom filter is to quickly determine whether a key is present in an SSTable or not. However, the bloom filter is prone to false positive problem [1], meaning that it may report a key as present even if it is not actually stored in the SSTable. This can result in unnecessary I/O.

A Bloom filter consists of a large bit array, where each key-value pair's key corresponds to several bits in the array. Generally, the larger the bit array, the lower the probability of false positives. To reduce the negative impact of the bloom filter's false positive, dayan [6] proposed a method to increase the number of bits of the underlying array of bloom filters. This method can indeed reduce the false positive, but it increases the overall memory consumption of bloom filter. For example, under the setting of 10 bits per key, a 10TB key-value storage with 100B KV pairs requires 128GB memory space to store bloom filters. If bloom filters are stored on disk, they will add additional I/Os and increase the read amplification of the system. Therefore, it is a good idea to increase the space of the bloom filter for some frequently accessed SSTables, while decreasing the space for SSTables with low access frequency. In the actual access process, we find that the access frequency of SSTable decreases with the decrease of layer, which means the access frequency of L_0 to L_n is gradually decreased. So the lower layer needs more bloom filters than the upper layer.

Therefore, we propose LayerBF. LayerBF assigns varying bits_per_key to the bloom filter of SSTables based on access frequency, allocating more bits to filters with higher access frequency and fewer bits to filters with lower access frequency. We implemented LayerBF on RocksDB and compared the performance of RocksDB with LayerBF and without LayerBF. The experimental results show that LayerBF improves the read performance by 2%–14% and reduces read amplification by 7%. We also evaluate the performance of LayerBF using YCSB [4] in default workloads. Similarly, LayerBF also performs better in most scenario.

2   Background and Motivation

In this section, we use RocksDB [8] as an illustrative example to explain the structure of LSM-Tree and its read amplification issue. We also examine the limitations of the current strategy for allocating space to bloom filters and the regularity of file access frequency.

2.1   LSM-Tree

LSM-Tree is a storage structure that optimizes random writes by batching them in memory, converting them into sequential operations to leverage the high sequential write bandwidth of storage devices. RocksDB is a key-value database

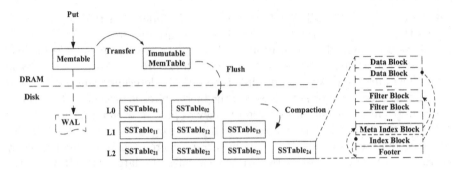

Fig. 1. The data organization structure of LSM-Tree.

that uses LSM-tree structure. It has two components, one in memory (MemTable and Immutable MemTable) and the other on disk (SSTables). SSTables are partitioned into different levels, with each subsequent level having ten times more capacity than the previous level.

To insert or update KV pairs in RocksDB, the storage first writes them to the Memtable, which is a fixed-size memory buffer that maintains the KV pair order using a skip list. Once the Memtable reaches its maximum capacity, it becomes an Immutable Memtable and is flushed to the L_0 layer as a Sorted String Table (SSTable).

RocksDB sorts KV pairs only within each SSTable in the L_0 layer, but both within and across SSTables in L_i $(i \geqslant 0)$.

When a layer reaches its maximum capacity of sstables, the database compacts and merges sstables into the next layer. The compaction process involves the following steps:

- Initially, the database selects an SSTable from L_i and another SSTable with overlapping key ranges from L_{i+1}. If additional SSTables in L_i fall within the key range of the compaction, they are also included in the selection.
- Subsequently, the key-value (KV) pairs of these SSTables are read into memory and sorted based on the order of keys. The sorted output is reorganized into new SSTables, discarding all invalid KV pairs (i.e., KV pairs that have been deleted or replaced by new updates).
- Finally, the newly generated SSTable is written to L_{i+1}.

LSM-Tree based KV stores like RocksDB use multi-layer storage. In L_0, SSTables may have overlapping key ranges, so compaction involves multiple SSTables. But in higher layers (L_i where $i > 0$), SSTables have non-overlapping key ranges, so compaction involves only one SSTable from the current layer and multiple SSTables from the next layer.

RocksDB stores multiple versions of a key using out-place updates. Searching for KV pairs involves traversing the layers from L_0 to L_{k-1} and returning the latest version found. The process involves searching Memtable, Immutable MemTable, and SSTables.

In L_0, SSTables may have overlapping keys, requiring scanning multiple files. RocksDB identifies a candidate SSTable, examines its bloom filter, and searches for the KV pair. If not found, it searches the lower layers directly.

2.2 Bloom Filter

A Bloom filter is frequently utilized to ascertain the presence of a query element in a collection. This filter is implemented using a bit array and multiple hash functions. The length of the array is equivalent to the number of bits in the Bloom filter. Increasing the number of hash functions results in greater accuracy in determining membership, but also increases the storage overhead.

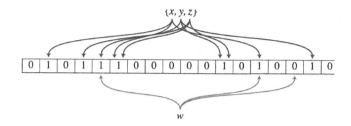

Fig. 2. The structure of bloom-filter.

Figure 2 illustrates the structure of a Bloom filter that employs three hash functions to map each element. Take the element w in the figure as example. During the insertion of an element, the Bloom filter initializes a bit array by setting each bit to 0. Next, the element w is mapped to three hash values, which are then used as indices in the array. The corresponding positions are subsequently marked with a value of 1. To determine if an element exists in the set, the Bloom filter maps the element to three positions in the bit array using the three hash functions. If the values of all three positions are 1, the element w is likely to exist in the set. Conversely, if any of these positions are 0, then the element is definitely not in the set. In RocksDB, the Bloom filter determines whether an SSTable needs to be accessed from disk, and false positives result in additional I/Os for key comparison.

The false positive rate of a Bloom filter can be calculated using the formula $FPR = (1 - e^{-\frac{N_{hash}}{m_i}})^{N_{hash}}$, where m_i denotes the number of bits assigned to each SSTable and N_{hash} denotes the number of hash functions employed. As evidenced by the formula, assigning a greater number of bits to each element can help reduce the false positive rate.

2.3 Motivation

LSM-Tree involves reads of SSTables from disk to memory, which results in read amplification. To mitigate this issue, RocksDB incorporates bloom filters.

However, increasing the number of bits assigned to each key-value pair in the bloom filter array (i.e., bits-per-key) to minimize the false positive rate and reduce extra I/Os leads to a significant increase in memory usage. For instance, in a 4TB KV storage containing 100B key-value pairs, elevating the number of bits from 5 to 10 reduces the false positive rate from 10% to 1%, but results in a memory overhead of 25 GB, which has now increased by 100% to 50 GB.

In this study, we conducted experiments with RocksDB to investigate the access frequency pattern of KV stores. Specifically, we loaded a 100GB storage with 1 KB KV pairs using db_bench [8]. The size of each SSTable was set to 64 MB, and the size of each successive level was set to ten times that of the previous level ($i \geqslant 0$). In order to analyze file access more accurately, we disabled the bloom filters. We then generated benchmarks with both Uniform and Zipfian distributions and issued 1M(one million) *Get* requests to the storage. The results in Table 1 indicate that the file access frequency of upper-level files is significantly higher than that of lower-level files.

These findings shed light on the access pattern of LSM-tree-based KV stores and have implications for optimizing their performance.

Table 1. Average access frequencies of files at each level

level	L_0	L_1	L_2	L_3	L_4	L_5
zipform	89.97%	9.51%	0.45%	0.05%	0.01%	0.01%
uniform	86.06%	13.51%	0.35%	0.04%	0.02%	0.02%

Allocating the same bits-per-key to the bloom filter of each layer in LSM-Tree neglects the difference in data access frequency among different layers. To achieve a more efficient use of space and improve the accuracy of the bloom filters, it is crucial to design a reasonable bloom filter space allocation strategy based on the access frequency of each layer.

3   Design

LayerBF is a technique for allocating Bloom filters based on how frequently files are accessed. This technique helps to reduce the average false positive rate of Bloom filters, improving read performance, without requiring additional space.

3.1   Main Idea

Figure 3 depicts the architectural framework of RocksDB with LayerBF, which comprises a storage module and an allocation module. The storage module is constructed using the Log-Structured Merge-Tree (LSM-Tree) and is character-ized by the presence of a bloom filter in each SSTable. The primary function of the bloom filter is to preclude false positives by determining the existence of a key in the SSTable. The allocation module is underpinned by the LayerBF strategy, which governs the allocation of space to each bloom filter.

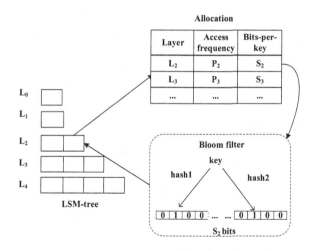

Fig. 3. The architecture of RocksDB with LayerBF.

3.2 Minimize Average False Positive Rate

In order to enhance the read performance of Key-Value (KV) stores, it is crucial to mitigate the I/O counts caused by the false positive rate of bloom filters. To this end, it is necessary to minimize the average false positive rate. This can be achieved by reducing the bits-per-key, as indicated by the equation $FPR = (1 - e^{-\frac{N_{hash}}{m_i}})^{N_{hash}}$. In this regard, it is pertinent to examine the relationship between the access frequency and the allocated bits-per-key, which will be analyzed in the following section by means of a model-based approach.

Consider a Key-Value (KV) store comprising N SSTables, containing a total of N_{kv} key-value pairs. Let the total space allocated to the bloom filter be denoted by M, where m_i bits-per-key are allocated to the bloom filter of $SSTable_i$, with an access frequency of p_i. The average false positive rate of all bloom filters in the KV store can be expressed as follows, by means of Eq. (1):

$$\sum_{i=1}^{N}(1 - e^{-\frac{N_{hash}}{m_i}})^{N_{hash}}p_i \tag{1}$$

The average false positive rate of all bloom filters can be computed as the summation of the product of the error rate of each bloom filter and its corresponding access frequency, where N_{hash} denotes the number of hash functions used in the bloom filter.

To minimize the average false positive rate of all bloom filters, it is essential to minimize the expression $e^{-m_i}p_i$. This can be achieved by transforming the problem into a combinatorial optimization problem.

$$\arg\min_{m_i, i \in N} \sum_{i=1}^{N} e^{-m_i} p_i,$$

$$s.t. \quad \sum_{i=1}^{N} (m_i * N_{kv}) = M, \quad \sum_{i=1}^{N} p_i = 1. \tag{2}$$

As the exponential function e^x is a convex function, the optimal solution can be obtained when $e^{-m_i} p_i = e^{-m_j} p_j$. By solving the combinatorial optimization problem (represented by Eq. (2)), the following conclusions can be drawn:

Theorem 1. *For the KV storage based on LSM-Tree, the bloom filter of SSTable$_i$ allocates m_i bits-per-key, and when the m_i meets:*

$$m_i = \log p_i - b, \qquad i = \{1, \cdots, N\} \tag{3}$$

We would get the minimum average false positive rate.

Moreover, the false positive rate is always less when we set the same bits-per-key for each bloom filter ($m_i = m/n$).

$$\sum_{i=1}^{N} e^{-m_i} p_i = N e^{-b} \leq \sum_{i=1}^{N} e^{-\frac{M}{N * N_{kv}}} p_i.$$

Note that $b = \frac{M}{N} - \frac{1}{N} \sum_{j=1}^{N} \log p_j$ is a constant related to access frequency.

3.3   LayerBF Allocation Strategy

Based on the modeling and analysis presented above, it can be concluded that when the number of bits allocated to each key by the bloom filter is logarithmically proportional to the file access frequency, the system achieves the lowest average false positive rate, which is consistently better than the minimum value of the average false positive rate obtained by the UniBF allocation strategy. In order to investigate the pattern of file access frequency in the LSM-Tree, this section provides further analysis.

According to the locality principle, data access in a system generally follows the *2–8 theorem*, whereby 20% of the content accounts for 80% of the page views, and data access follows a power-law distribution (Zipfian distribution). However, in practical applications, a large-scale system often employs a cache (such as Redis [21]) to store frequently accessed data. In this scenario, the data access of the backend database presents a uniform distribution. This section conducts experiments under both scenarios and derives the following observations: for LSM-Tree, the access frequency of L_0-L_1 decreases by 6–10 times, that of L_1-L_2 decreases by 20–40 times, that of L_2-L_3 decreases by 8–10 times, and that of L_3-L_4 decreases by 2–5 times. There is no significant difference between the access frequency of L_4-L_5.

The LSM-Tree storage capacity gradually decreases, with a tenfold decrease from one layer to the next. Allocating additional space to the upper layer has a

Table 2. Relationship of bit_per_key with false positive

bit_per_key	2	3	5	10	15
false positive	38.3%	23.7%	10.1%	1.3%	0.1%

minimal impact on the overall storage cost. For instance, in a key-value storage system with 100 million key-value pairs, the L_0 layer only contains 100 key-value pairs. If each key in the L_0 layer's bloom filter is allocated space, increasing the bits-per-key from 2 to 10, as shown in Table 2, reduces the false positive rate for this layer from 38% to 1% with only 0.8 KB of additional storage overhead.

Based on the experimental observations above, we define the access frequency as $P = [20000k, 2000k, 50k, 5k, k, k]$ by taking the maximum access frequency multiple of the two adjacent layers, where k is an arbitrary constant. Once the access frequency is determined, the number of bits allocated by the bloom filter for each key can be calculated by finding the value of the constant b using the formula $m_i = \log p_i - b$.

In complex and dynamic industrial production environments, engineers often need to adjust the size of bloom filters according to the specific requirements of the system. To improve the read performance of the system, the size of the bloom filter can be increased for read-intensive tasks. However, in scenarios with frequent add, delete, and modify operations, a larger bloom filter may increase the time required to generate new files. To address this, the LayerBF allocation policy considers the number of bits allocated to each key by the L_0 layer bloom filter (m_0) as a system parameter, and provides an external interface for users to configure as required. Based on the user-defined value of m_0, the system automatically calculates the value of the constant b, as well as the number of bits allocated for each key in the other layers, and completes the space allocation of the bloom filters accordingly.

In summary, the algorithm of LayerBF as Algorithm 1 shows.

4   Evalution

4.1   Experimental Setup

All experiments were conducted on a system equipped with a 24-core Intel(R) Xeon(R) Platinum 8260M CPU @ 2.40 GHz processor, 64 GB RAM, and running CentOS 8.3 LTS with the Linux 4.18.0 kernel and the ext4 file system. The system includes a 1TB HDD and a 128GB SSD, and experiments were performed on the SSD by default.

We implemented the LayerBF allocation policy in the key-value store RocksDB and conducted experiments to evaluate its performance. Specifically, we compared the performance of RocksDB with and without LayerBF to assess the impact of LayerBF on system performance.

Algorithm 1: LayerBF algorithm

Data: Initial $L_0(bits - per - key) = Option(bits - per - key)$. Initial
$P = [20000k, 2000k, 50k, 5k, k, k]$, where k is an arbitrary constant.

1 $b = \log(p_0) - m_0$;
2 **if** *Flush key-value pairs from memory to L_0* **then**
3 | $m_0 = L_0(bits - per - key)$;
4 | $Space(FilterBlock_0) = m_0 * Num_{key}(SSTable)$;
5 | save SSTable in L_0;
6 **end**
7 **if** *Compaction between L_{i-1} and L_i* **then**
8 | Sort keys in SSTables from L_{i-1} and L_i;
9 | $m_i = \log p_i - b$;
10 | $Space(FilterBlock_i) = m_i * Num_{key}(SSTable)$;
11 | save SSTable in L_i;
12 **end**

To provide a comprehensive analysis of LayerBF's behavior, we employ the YCSB benchmark [4]. All experiments utilize the default configuration, including 64 MB Memtables/SSTables, 256 MB L_0 size. The size of sstables at L_{i+1} is 10 times larger than the size of sstables at L_i. Additionally, we set the size of each key-value pair to 1 KB and load data with randomly generated distinct keys. Moreover, we allow the key-value stores to utilize all available capacity on our disk space, resulting in significant overheads from the read and write amplification in LSM-tree management. Specifically, we set the average bloom filter space for each key (i.e., bits-per-key for RocksDB) to 8 bits, and RocksDB with LayerBF allocates 15 bits-per-key for the L_0 bloom filter.

4.2   Average False Positive Rate

The number of I/O requests is a critical factor that affects read performance, and bloom filters can effectively reduce this number. However, the false positive rate limits the benefits of bloom filters. To evaluate the effectiveness of the LayerBF strategy in reducing the average false positive rate of bloom filters, we conducted experiments on a 100GB storage system.

We conducted read-only benchmarks to perform get requests and calculated the average false positive rate when using LayerBF allocation strategies. The experiments recorded the total number of I/O requests, denoted by *hit_count*, made by the storage, and the number of counts that the bloom filter correctly judged as positive, denoted by. We calculated the average false positive rate using the following equation:

$$FPR = 1 - \frac{true\_positive}{hit\_count}$$

Figure 4 demonstrates that when the average bloom filter space per key (referred to as *bitsperkey* in the context of RocksDB) is set to 8 bits, the average false positive rate of the key-value store with LayerBF implementation remains below 3%. In contrast, the key-value store without LayerBF exhibits a significantly higher false positive rate of 13%.

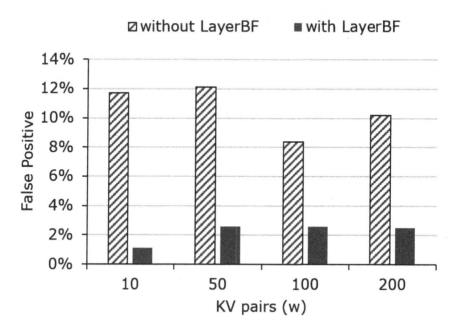

Fig. 4. The average false positive of the bloom filters.

4.3   Overall Performance Evaluation

This section presents an evaluation of the LayerBF strategy's performance through micro-benchmarks conducted on both HDD and SSD platforms. The evaluation encompasses assessments of both read throughput and read amplification to validate the effectiveness of the LayerBF strategy. Additionally, We measured the number of IO requests initiated during the read process.

Read Performance. As the primary motivation behind the LayerBF strategy is to optimize read performance, we conducted experiments to measure the same. Specifically, we randomly loaded 100GB of key-value pairs into RocksDB using two distinct allocation strategies. We then executed a read benchmark with a specific number of point query requests (i.e., 100K, 500K, 1M, 2M), and analyzed the results to evaluate read throughput. The outcomes of these experiments, as depicted in Fig. 4.3, indicate that the LayerBF strategy improves read performance in RocksDB. The read throughput increased by 3%–15% when compared to scenarios where the LayerBF strategy was not utilized.

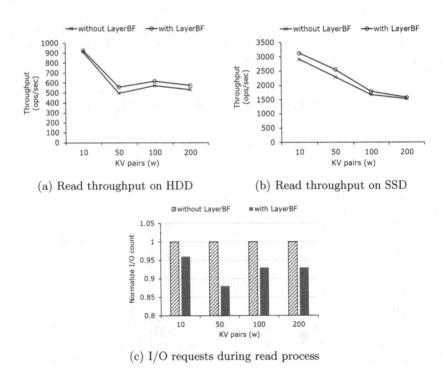

(a) Read throughput on HDD   (b) Read throughput on SSD

(c) I/O requests during read process

Fig. 5. BenchMarks

To assess the potential influence of storage medium on the efficacy of the LayerBF strategy, we conducted further experiments on SSD. Following a similar methodology, we randomly loaded 50 GB of key-value pairs into each KV storage system and evaluated read throughput with a certain number of point query requests (i.e., 100K, 500K, 1M, 2M). As depicted in Fig. 4.3, the read throughput of RocksDB with LayerBF outperformed that without the strategy, with a performance improvement of 2%–14%. This improvement is consistent with the findings from the HDD experiments. The experimental results indicate that LayerBF helps enhance read performance and is not affected by the underlying storage medium, demonstrating good cross-media performance.

To further evaluate the effectiveness of the two allocation strategies, we measured the I/O requests generated by workloads that randomly read some key-value pairs on HDD. Specifically, we used duplicate test keys to issue the same number of *Get* requests to the storage system with and without LayerBF, and then computed the I/O requests of SSTables during the reading process. As demonstrated in Fig. 4.3, the storage system utilizing LayerBF required only about 93% of the I/O requests of the system not using it, effectively reducing I/O requests by 7%. LayerBF reduces the average false positive rate of the bloom filter, as demonstrated in Sect. 4.2, which helps decrease I/O requests resulting from misjudgment. As a result, LayerBF improves read performance.

4.4   YCSB Benchmarks

The Yahoo Cloud Serving Benchmark(YCSB) [4] is a widely used micro-benchmark tool and is the industry standard in evaluating KV stores. YCSB consists of six workloads that capture different real-world scenarios, and the detail of these workloads is described in Table 3.

Table 3. YCSB core workloads description

Workloads	Description
Load	100% writes
YCSB A	50% reads and 50% updates
YCSB B	95% reads and 5% updates
YCSB C	100% reads
YCSB D	95% reads(latest writes) and 5% updates
YCSB E	95% range queries and 5% updates
YCSB F	50% reads and 50% reads-modify-writes

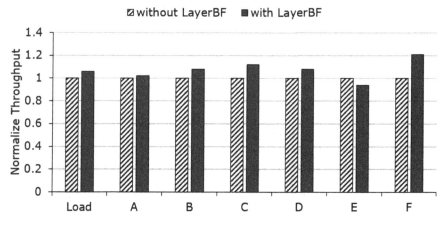

Fig. 6. YCSB performance

We conducted a YCSB benchmark with Zipfian key distribution, using a memtable size of 128 MB and an SSTable size of 64 MB. We limited the compaction process to a maximum of 4 threads. A 100GB dataset comprising 1 KB key-value pairs was randomly written into the storage system. Workloads A to F were evaluated with one million key-value pairs. Our results, as presented in Fig. 6, demonstrate that LayerBF significantly enhances the system's performance for all workloads except for workload E.

Our study reveals that RocksDB with LayerBF performs notably better for read-intensive workloads, namely workload B, C, D, and F. Specifically, LayerBF improves the throughput by 8% for workload B and D. For the read-only workload C, the optimized bloom filter allocation strategy leads to a 12% improvement. The improvement in workload B is inferior to that in workload C due to the Zipfian key distribution, which causes most *Get* requests to fall on the upper layer of the LSM-Tree after updates, thereby requiring fewer SSTables to be scanned during reads. As for workload F, which involves a *Get* operation before a *Put* operation, LayerBF achieves 21% better throughput by optimizing both reading and writing processes.

5   Related Work

In recent years, many studies have proposed new designs based on LSM-Tree. This algorithm uses the FPGA [22] to speed up every compaction process of LSM-Tree. It also proposes key-value separation and index-data block separation strategies to take advantage of the pipeline mechanism of the FPGA. Wisckey [17] proposes a key-value separation technique to manage keys and values, and it stores keys in LSM-Tree and values in a log, reducing the size of the LSM-Tree and disk I/O. Bourbon [5] uses machine learning methods to speed up the LSM-Tree search, uses greedy stepwise linear regression to learn the key distribution, and achieves a fast search with minimal computation. Ackey [23] modifies RocksDB's cache algorithm to implement an adaptive cache that adjusts the size of individual cache components based on the workload. SFM [15] proposes space fragmentation LSM-Tree, which delays the merge operation of non-read intensive key space, and identifies the read intensity of each key space by dynamically estimating the read/write heat of each key space, thus improving IO throughput and reducing read latency.

Other studies focus on using emerging storage media to improve the performance of storage systems. NoveLSM [11] takes advantage of NVM's byte addressing capability to reduce read and write latency. Implementing an in-place updatable memtable in NVM reduces the amount of data written and fail-over time. It introduces optimistic parallel reads that can simultaneously access multiple levels of LSM in NVM or SSD, thereby decreasing read request latency. MatrixKV [24] utilizes NVM to manage L_0 in matrix containers, designs fine-grained column compression, decreases the amount of compressed data, increases the width of each level, and reduces the depth of the LSM-Tree, thereby reducing write magnification. SLM-DB [12] uses byte-addressable persistent memory to store B+ Tree index and LSM-Tree to insert KV pair in PM cache to improve throughput. OurRocks [3] uses NVM and GPU to offload scan operations to the GPU, eliminating the data transfer bottleneck between direct memory access (DMA) and devices, effectively utilizing NVMe SSDs and GPUs, and significantly improving query speed.

Of course, there are also studies focusing on optimizing Bloom filters. Succinct Range Filters [25] reduce I/O by filtering requests for point queries and range queries based on a Succinct data structure. Rosetta [18] reduces the false positive of the filter by sacrificing the detection time of the filter, which improves the range query performance without losing the point query performance Monkey [6] uses a coarse-grained scheme to assign the same number of filters to the same level of SSTables. ElasticBF [16] goes a step further and adjusts dynamically based on data heat, taking advantage of fine-grained access positioning and reducing additional access I/O.

6   Conclusion

This paper introduces LayerBF, a dynamically adjusted the bloom filter space allocation strategy. LayerBF can adjust the space usage of the bloom filter of each layer according to the access frequency of different layers, thus improving the read performance of the entire storage system without increasing memory usage. Finally, we did several experiments to prove LayerBF's effectiveness, showing a 2% to 14% improvement in read performance and reduces the false positive rate by 8% to 10%.

References

1. Bloom, B.H.: Space time trade-offs in hash coding with allowable errors. Commun. ACM **13**(7), 422–426 (1970)
2. Chang, F., et al.: Bigtable: a distributed storage system for structured data. TOCS
3. Choi, W.G., Kim, D., Roh, H., Park, S.: Ourrocks: Offloading disk scan directly to GPU in write-optimized database system. IEEE Trans. Computers **70**(11), 1831–1844 (2021). https://doi.org/10.1109/TC.2020.3027671, https://doi.org/10.1109/TC.2020.3027671
4. Cooper, B.F., Silberstein, A., Tam, E., Ramakrishnan, R., Sears, R.: Benchmarking cloud serving systems with YCSB. In: Proceedings of the 1st ACM Symposium on Cloud Computing. pp. 143–154
5. Dai, Y., Xu, Y., Arpaci-Dusseau, R.H.: From wisckey to bourbon: A learned index for log-structured merge trees. In: OSDI (2020)
6. Dayan, N., Athanassoulis, M., Idreos, S.: Monkey: Optimal navigable key-value store. In: Proceedings of the 2017 ACM International Conference on Management of Data (2017)
7. DeCandia, G., Hastorun, D., Pilchin, A., Vogels, W.: Dynamo: amazon's highly available key-value store. In: Bressoud, T.C., Kaashoek, M.F. (eds.) SOSP. ACM (2007)
8. FaceBook: Rocksdb documentation (2012). http://rocksdb.org/
9. George, L.: HBase: the definitive guide: random access to your planet-size data. O'Reilly Media, Inc. (2011)
10. Google: Leveldb documentation (2021). https://github.com/google/leveldb
11. Huang, H., Ghandeharizadeh, S.: Nova-LSM: a distributed, component-based LSM-tree key-value store. In: Proceedings of the 2021 International Conference on Management of Data, pp. 749–763 (2021)

12. Kaiyrakhmet, O., Lee, S., Nam, B., Noh, S.H., ri Choi, Y.: SLM-DB: single-level key-value store with persistent memory. In: 2019 USENIX Conference on File and Storage Technologies, pp. 191–205. Boston, MA (2019)

13. Lai, C., Jiang, S., Yang, L., Hou, Z., Cui, C., Cong, J.: Atlas: Baidu's key-value storage system for cloud data. In: MSST

14. Lakshman, A., Malik, P.: Cassandra: a decentralized structured storage system. ACM SIGOPS Oper. Syst. Rev. **44**(2), 35–40 (2010)

15. Lee, H., Lee, M., Eom, Y.I.: Sfm: Mitigating read/write amplification problem of LSM-tree-based key-value stores. IEEE Access **PP**(99), 1–1 (2021)

16. Li, Y., Tian, C., Guo, F., Li, C., Xu, Y.: Elasticbf: elastic bloom filter with hotness awareness for boosting read performance in large key-value stores. In: 2019 USENIX Annual Technical Conference, pp. 739–752 (2019)

17. Lu, L., Pillai, T.S., Gopalakrishnan, H., Arpaci-Dusseau, A.C., Arpaci-Dusseau, R.H.: Wisckey: Separating keys from values in SSD-conscious storage. ACM Trans. Storage **13**(1), 1–28 (2017)

18. Luo, S., Dayan, N., Qin, W., Idreos, S.: Rosetta: A robust space-time optimized range filter for key-value stores. In: SIGMOD. ACM (2020)

19. O'Neil, P., Cheng, E., Gawlick, D., O'Neil, E.: The log-structured merge-tree (LSM-tree). Acta Informatica **33**(4), 351–385 (1996)

20. Rhea, S., Wang, E., Wong, E., Atkins, E., Storer, N.: Littletable: A time-series database and its uses. In: Salihoglu, S., Zhou, W., Chirkova, R., Yang, J., Suciu, D. (eds.) SIGMOD (2017)

21. Sanfilippo, S.: Redis documentation (2021). https://redis.io/

22. Sun, X., Yu, J., Zhou, Z., Xue, C.J.: Fpga-based compaction engine for accelerating LSM-tree key-value stores. In: ICDE, pp. 1261–1272 (2020)

23. Wu, F., Yang, M., Zhang, B., Du, D.H.C.: Ac-key: Adaptive caching for LSM-based key-value stores. In: Gavrilovska, A., Zadok, E. (eds.) 2020 USENIX Annual Technical Conference, USENIX ATC 2020, July 15-17, 2020, pp. 603–615. USENIX Association (2020)

24. Yao, T., et al.: Matrixkv: Reducing write stalls and write amplification in LSM-tree based KV stores with matrix container in NVM. In: 2020 USENIX Annual Technical Conference, pp. 17–31 (2020)

25. Zhang, H., Lim, H., Pavlo, A.: Succinct Range Filters. ACM Trans, Database Syst (2020)

HTStore: A High-Performance Mixed Index Based Key-Value Store for Update-Intensive Workloads

Jinzhou Liu[1,2], Yinliang Yue[3(✉)], Jiang Zhou[1,2], Zhixin Fan[1,2], and Zekun Yao[1,2]

[1] School of Cyberspace Security, University of Chinese Academy of Sciences, Beijing, China
{liujinzhou,zhoujiang,fanzhixin,yaozekun}@iie.ac.cn
[2] Institute of Information Engineering, Chinese Academy of Sciences, Beijing, China
[3] Zhongguancun Laboratory, Beijing 100080, People's Republic of China
yueyl@zgclab.edu.cn

Abstract. In this paper, we propose a high-performance Mixed Index based key-value store named HTStore to improve the write and read performance in update-intensive workloads of LSM-tree based key-value stores. The key idea of HTStore is to build a global index, called Mixed Index, in the DRAM and NVM hybrid storage, which saves keys and their latest positions. HTStore judges the key's version participating in flush or compaction is the latest by accessing the Mixed Index. If the key-value pair is expired redundant old version data, HTStore filters it away, which helps reduce the flush and compaction overhead caused by such expired redundant data, leading to improved write throughput. Additionally, the Mixed Index helps quickly locate the level of the key, avoiding layer-by-layer search. The Mixed Index comprises a HashTable in DRAM and a Trie-tree in NVM. We implemented HTStore on RocksDB and compared it with the original RocksDB. Our performance evaluation showed that write throughput increased by up to 73–105% in update-intensive workloads and read throughput by up to 80%. Compared with MatrixKV and PebblesDB, HTStore also demonstrated specific performance improvements.

Keywords: Key-Value Store · LSM-tree · NVM · Remove Redundancy

1 Introduction

The amount of data being generated by the interaction between users and Internet applications is increasing rapidly. With higher performance standards required for these applications, traditional relational storage systems are no longer capable of handling the vast amount of unstructured data being generated. Unstructured data lacks a predefined data model, making it difficult to be stored in a two-dimensional logical table in a relational database. Additionally, expanding these databases by adding more nodes is challenging due to the relationship between database tables.

© The Author(s), under exclusive license to Springer Nature Singapore Pte Ltd. 2024
X. Song et al. (Eds.): APWeb-WAIM 2023, LNCS 14333, pp. 507–521, 2024.
https://doi.org/10.1007/978-981-97-2387-4_34

To address this issue, persistent key-value store systems such as BigTable, HBase, Cassandra, and DynamoDB have emerged as a solution for ample data storage. These systems are designed to handle unstructured data efficiently and have the ability to scale easily. The write-optimized key-value store, which is based on the log-structured merge-tree (LSM-tree) [6] implementation, is a popular solution and includes LevelDB, RocksDB [2], and HBase.

The widely used key-value store is currently being run on a DRAM-Disk setup, but the system is unable to utilize more DRAM to improve its performance. The maximum capacity of a single DRAM memory stick is 128 GB, while a single Non-volatile Memory (NVM) [1] stick can hold up to 512 GB. Furthermore, enterprise-level DRAM memory modules with 128 GB capacity cost around $4,500, while NVM memory modules with 512 GB capacity cost around $6,500, making NVM much more cost-effective. Therefore, it is imperative to explore the use of NVM to enhance system performance.

In this paper, we present a novel approach to optimize the performance of RocksDB by utilizing NVM. RocksDB employs a delayed update mechanism, known as Out-of-Place Update, which results in the accumulation of a significant amount of redundant expired data in the system. This redundant data negatively impacts operations such as compaction, get, and scan, leading to increased write amplification and reduced read and write performance. Additionally, the multi-level design of RocksDB causes layer-by-layer search problems, further decreasing read performance. To address these issues, we propose an improved version of RocksDB called HTStore. HTStore leverages a global index called the Mixed Index, which is constructed using HashTable in DRAM and Trie-tree in NVM. HTStore uses the Mixed Index to record the latest version locations of some keys in the system. By accessing the Mixed Index, HTStore significantly reduces the amount of redundant data in the system, thereby reducing write amplification caused by redundant data and improving the system's write performance in update-intensive workloads. Additionally, HTStore's Mixed Index enables it to quickly locate the key location, avoiding layer-by-layer search and improving read performance.

2   Background and Motivation

2.1   Log-Structured Merge-Tree (LSM-Tree)

The InnoDB storage engine of MySQL is widely used in the industry, but it is not suitable for write-intensive workloads due to its In-Place update approach, which writes data randomly to disk. This results in low throughput that falls short of write-intensive workload requirements. To address this issue, O'Neil proposed the LSM-tree index data structure in 1996, which takes advantage of the fact that "the sequential write performance of disks far exceeds the random write performance." LSM-tree design has been applied to various products such as LevelDB, RocksDB, and HBase.

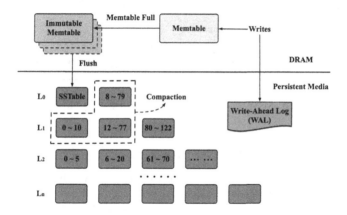

Fig. 1. LSM-tree architecture

LSM-Tree Components. Figure 1 illustrates the main components of LSM-tree, which includes Memtable, Immutable Memtable, SSTable, Write Ahead Log (WAL). Memtable and Immutable Memtable are kept in the DRAM memory, while LSM-tree persists the remaining components to disk. To insert key-value pair data, it is first appended to the WAL and then inserted into the Memtable. LSM-tree sorts the data in Memtable according to the key order, which is implemented using Skiplist. When the Memtable becomes full, it is converted to a readonly Immutable Memtable and written sequentially to disk (referred to as flush), forming an SSTable file.

LSM-Tree Compaction. As data accumulates, more SSTables are created on the disk, and their key ranges overlap, leading to deteriorating read performance. To address this issue, LSM-trees employ a compaction strategy that merges and sorts unordered SSTables on the disk at appropriate intervals, creating newly ordered SSTables to maintain the data's ordering in the system. However, compaction incurs system overhead since it involves three steps: (1) reading the SSTables on the disk, (2) merging and sorting them to create new SSTables, and (3) writing the new SSTables back to the disk.

Why Multi-level Design. In Fig. 2, when a large number of SSTables are involved in compaction, significant system resources such as CPU, disk I/O, and memory are consumed, competing with the main thread for resources and reducing system write throughput. To address this issue, LSM-tree implements a multi-level design that distributes SSTables across several levels. The number of SSTables increases as the level gets deeper. Figure 1 shows that only a few SSTables from two adjacent levels are needed for compaction, reducing the workload and system overhead. Fine-grained compaction is a technique that reduces the system overhead of a single compaction, thereby preventing performance resource competition with the main thread.

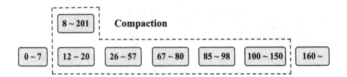

Fig. 2. Motivation for multi-level design of LSM-tree

2.2 Non-volatile Memory (NVM)

In recent years, the demand for data storage and processing in internet technology has continued to grow, creating a pressing need for Non-volatile Memory (NVM). To address this, Intel has introduced the Intel Optane DC Persistent Memory, which boasts byte-addressability, low latency, high bandwidth, large capacity, and non-volatility. The read performance of NVM is similar to that of DRAM. While write performance maybe 2–3 times slower than DRAM, the capacity is greater and the power is non-volatile. The Persistent Memory Development Kit (PMDK) can be used to read and write to NVM. PMDK uses memory-mapped files to read and write NVM, which has good read-and-write performance. Table 1 presents a performance comparison of NVM with other storage media.

Table 1. Comparison of NVM with other storage media

Performance	DRAM	NVM	SSD	HDD
Volatile	Yes	No	No	No
Read Latency	60 ns	≈60 ns	25 us	10 ms
Write Latency	60 ns	150 ns	300 us	10 ms
Addressing Mode	Byte	Byte	Block	Block
Capacity	128 GB	512 GB	TB	TB

2.3 Motivation

Remove Redundant Data. To achieve efficient write performance by converting random disk writes to sequential writes, LSM-tree updates key-value pairs using Out-of-Place update. When updating the value of an existing target key in the LSM-tree system, the new version of the target key-value pair is stored in a new physical disk space instead of overwriting the original physical location. This avoids the disk random access overhead caused by querying the target key and writing back the new version of the key-value pair. However, this design can result in multiple versions of the value for the same key in the system, and only the latest version is valid while the others are redundant and invalid.

As shown in Fig. 3a, the latest version data of K_1, K_2, and K_3 are in the L_0 and L_1 layers of the LSM-tree, and their redundant old version data is in the L_2 and L_3 layers. At this point, these old version data have become invalid, but they still participate in the compaction system operation, increasing the amount of data participating in compaction and exacerbating the write amplification problem, thereby reducing write performance. The system cannot determine if the key-value pairs participating in compaction have a newer version in the upper layers, so it cannot judge if the data is valid. If this redundant data is not cleaned up in time, it will be saved in the new generated SSTable after compaction and continue to participate in subsequent compaction operations.

These redundant data not only participate in compaction operations but also participate in point queries and range queries. When queried, redundant data will extend the search path, exacerbating the read amplification problem and reducing read performance. The problem of redundant data participating in system operations is more serious under update-intensive workloads.

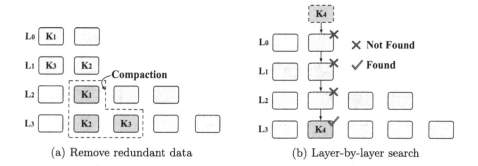

(a) Remove redundant data   (b) Layer-by-layer search

Fig. 3. HTStore motivation

Layer-by-Layer Search. The LSM-tree adopts a multi-level design. As more and more data is written into the system, the number of levels gradually increases. By default, the maximum number of levels in the disk is 7.

This multi-level design leads to the problem of layer-by-layer search. If the target key does not exist in the Memtable, it will be searched layer by layer in the SSTable on the disk. Since the SSTable in L_0 is unordered, it is necessary to search in each SSTable. If it does not exist, continue to search in the SSTable in L_1. Since L_1 to L_6 are ordered, the target SSTable that may contain the target key can be quickly located by using binary search, and then detailed search is performed in this SSTable. If it still does not exist, the same method is used to search the next level until the target key is found or the last level is reached.

As shown in Fig. 3b, it took 4 levels of search to find the target key K_4. Each level of search on the disk will cause a random read I/O, exacerbating the problem of read amplification and reducing read performance. The layer-by-layer search problem caused by the multi-level design will be more severe in read-intensive workloads.

3   HTStore Design

The HTStore key-value store system is based on the extension of RocksDB, breaking the original memory-disk architecture of RocksDB and introducing new storage hardware, Intel Optane NVM, between memory and disk devices. The architecture of HTStore is shown in Fig. 4, which introduces a new component, a Mixed Index, on the basis of the architecture of RocksDB. The Mixed Index consists of a HashTable in DRAM and a Trie-tree in NVM. The Mixed Index records part of the keys in the HTStore and their latest version's level position in the format of <Key, Value>. HTStore achieves the elimination of redundant data and acceleration of point query performance by accessing the Mixed Index. At the same time, HTStore migrates the write-ahead log (WAL) from disk to NVM, leveraging the low write latency and non-volatile nature of NVM to accelerate WAL writes.

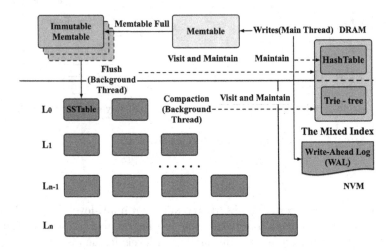

Fig. 4. HTStore architecture

3.1   Mixed Index Design

The Mixed Index is composed of a HashTable in DRAM and a Trie-tree in NVM, and is used to record the latest level position of some keys in the system. The data stored in the Mixed Index is in the format of <Key, Value>. As shown in Table 2, it is the enumeration corresponding to all positions. If the latest version of the target key is in the Memtable of HTStore, the Mixed Index will be recorded as <Key, −2>; if the latest version of the target key is in the Immutable Memtable, the Mixed Index will be recorded as <Key, −1>; if the latest version of the target key is in an SSTable located in L_0, the Mixed Index will be recorded as <Key, 0>, and so on.

Table 2. Enumeration of key-value pair positions

Memtable	Immu Mem	L_0	L_1	L_2	L_3	L_4
−2	−1	0	1	2	3	4

HashTable is located in DRAM and records a small number of keys and their latest positions in HTStore. It is only responsible for recording the location information of keys in the Memtable and Immutable Memtable in memory. Data updates in the HashTable are jointly maintained by the main thread and the background flush thread. The HashTable can be accessed by the background flush thread and the background compaction thread. As shown in Fig. 5a, the HashTable has a logical structure composed of arrays and linked lists. The elements in the array are called hash buckets, which record the head node pointer of the linked list. The linked list is composed of several hash nodes linked together, and the hash node records the key and its latest level position.

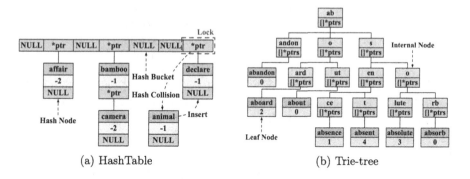

(a) HashTable   (b) Trie-tree

Fig. 5. Components of the Mixed Index.

Trie-tree is located in NVM and is used to record the keys and their latest positions in the first five levels of the SSTables in HTStore, ranging from L_0 to L_4. Data updates in the Trie-tree are jointly maintained by the background flush thread and the background compaction thread. The Trie-tree can only be accessed by the background compaction thread. As shown in Fig. 5b, the Trie-tree has a logical structure consisting of internal nodes and leaf nodes. Internal nodes record the common prefix of keys and an array of pointers to all child nodes. Leaf nodes record the key and its latest level position. The Trie-tree is a data structure with stable performance, and the time complexity of operations such as insertion or update can be kept constant at $O(N)$, where N is the length of the key, in the worst case scenario.

3.2 Eliminating Redundant Data

HTStore can eliminate redundant data by accessing Mixed Index, thereby reducing write amplification and improving write performance. The implementation of data deduplication involves three key processes: the foreground main thread performs write operations, the background thread performs flush operations, and the background thread performs compaction operations. Figure 6 illustrates these three steps. The HashTable located above the horizontal line in the Mixed Index is in DRAM, while the Trie-tree located below the horizontal line is in NVM.

The Foreground Main Thread Performs Write Operations: As shown in Fig. 6c and d, the latest version of K_1 is in L_1 and is recorded as $<K_1, 1>$ in the Trie-tree of the Mixed Index. The main thread inserts new version data of K_1 into the Memtable and inserts a $<K_1, -2>$ record into the HashTable. At this point, the old version of K_1 in L_1 is redundant and invalid, and the $<K_1, 1>$ record in the Trie-tree is also invalid. When a transition from Memtable to Immutable Memtable occurs, HTStore modifies the HashTable position enumeration information of the participating keys from -2 to -1.

The Background Thread Performs Flush Operations: As shown in Fig. 6a and b, K_0 already has the latest version in the Memtable and is recorded as $<K_0, -2>$ in the HashTable. During the flush operation of the Immutable Memtable performed by the background thread, accessing the Mixed Index reveals that there is a newer version of K_0 in the Memtable. At this point, the version of K_0 participating in the flush is redundant and invalid, so its old version is filtered out and prevented from being written to disk. K_2 in the Immutable Memtable also participates in the flush. By accessing the HashTable, it is found that there is no newer version of K_2, so the K_2 in the HashTable is deleted. After K_2 falls into the SSTable in L_0, a $<K_2, 0>$ record is inserted into the Trie-tree of the Mixed Index.

The Background Thread Performs Compaction Operations: As shown in Fig. 6d and e, there are three SSTables between L_1 and L_2 participating in the compaction operation. During the compaction performed by the background thread, accessing the HashTable reveals that K_1 already has the latest version in the Memtable. By accessing the Trie-tree, it is found that K_2 already has the latest version in the SSTable located in L_0. Therefore, the redundant data of K_1 and K_2 participating in the compaction is filtered out and prevented from being written to the new SSTable, and the position information of the invalid K_1 record in the Trie-tree is deleted.

As shown in Fig. 6e and f, there are also three SSTables between L_2 and L_3 participating in the compaction. During the compaction performed by the

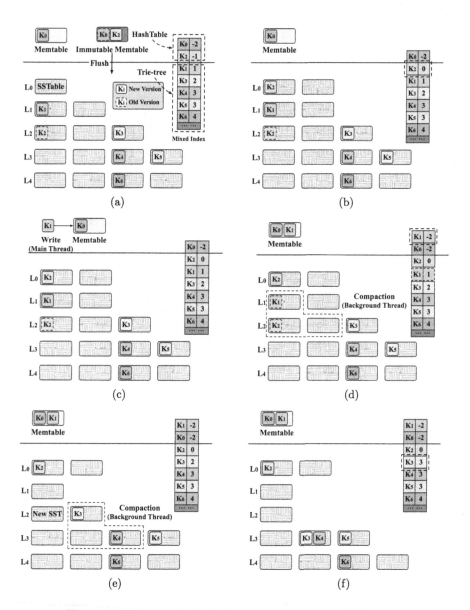

Fig. 6. The process of eliminating redundant data in HTStore

background thread, accessing the HashTable and Trie-tree in the Mixed Index reveals that K_3 and K_4 participating in the current compaction are already the latest version of the data. Therefore, they are allowed to be written to the new SSTable, and the position information of K_3 in the Trie-tree is changed from 2 to 3.

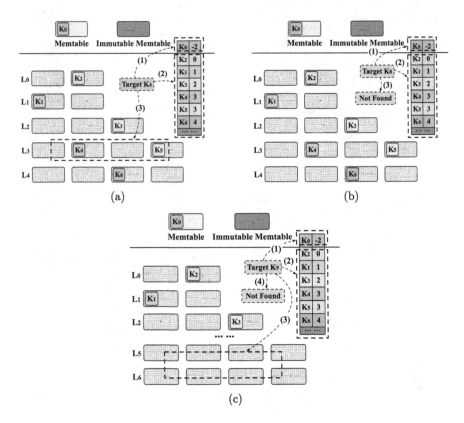

Fig. 7. The process of querying acceleration points in HTStore

3.3   Accelerating Point Query

HTStore can quickly locate the level where the target key is located by accessing the Mixed Index, thus avoiding searching layer by layer, reducing random access to disks, and improving read performance. HTStore introduces a global variable, **SST_Level_Num**, to record the number of levels of SSTables in the system, which is maintained by the background compaction thread. Currently, RocksDB's maximum number of levels is set to 7, which can cover most industrial application scenarios, so the maximum number of levels in HTStore is also 7. The Trie-tree in HTStore only records the position information of keys in the first 5 levels of SSTables. There are three main scenarios where HTStore uses Mixed Index to accelerate read performance: when the target key exists in the system, when the target key does not exist in the system and SST_Level_Num is less than or equal to 5, and when the target key does not exist in the system and SST_Level_Num is greater than 5.

When the Target Key Exists in the System: As shown in Fig. 7a, the main thread's target for the query is the step of K_4: (1) Firstly, access the HashTable in the Mixed Index and find that K_4 does not exist. (2) Then, access the Trie-tree and find the location record information $<K_4, 3>$ for the latest version of K_4. (3) Finally, the main thread directly performs a binary search in L_3 and returns the key-value pair data of K_4. By accessing the Mixed Index, the main thread avoids searching for Memtable, SSTables located in L_0–L_2, reducing read amplification and speeding up point queries.

When the Target Key Does Not Exist and SST_Level_Num is Less Than or Equal to 5: As shown in Fig. 7b, the main thread's target for the query is K_8, which does not exist in the HTStore. The number of levels of SSTables in the HTStore is 5, which belongs to the case where SST_Level_Num is less than or equal to 5. The Trie-tree in the mixed index records the location information of the keys in the first 5 levels of SSTables. Therefore, the current Mixed Index records the location information of all keys in the system. The main thread's steps for querying K_8 are as follows: (1) Firstly, access the HashTable and find that K_8 does not exist. (2) Then, access the Trie-tree and find that K_8 still does not exist. (3) Finally, determine that the current SST_Level_Num is less than or equal to 5, so K_8 must not exist in the HTStore, and return a null value directly. By accessing the Mixed Index, the main thread avoids searching for all components, greatly reducing read amplification and speeding up point queries.

When the Target Key Does Not Exist and SST_Level_Num is Greater Than 5: As shown in Fig. 7c, the main thread's target for the query is K_9, which does not exist in the HTStore. The number of levels of SSTables in the HTStore is 7, which belongs to the case where SST_Level_Num is greater than 5. The Trie-tree only records the location information of the keys in the first 5 levels of SSTables. Therefore, the current Mixed Index records the location information of non-full keys in the system. The main thread's steps for querying K_9 are as follows: (1) Firstly, access the HashTable and find that K_9 does not exist. (2) Then, access the Trie-tree and find that K_9 still does not exist. (3) Immediately afterwards, determine that the current SST_Level_Num is greater than 5, so it is impossible to directly determine whether K_9 exists in the HTStore through the Mixed Index. (4) Finally, follow the original point query search process of RocksDB, and search for K_9 in the SSTables located in L_5 and L_6. The search results indicate that K_9 does not exist in either of the two levels, and a null value is returned directly. By accessing the mixed index, the main thread avoids searching for Memtable, SSTables located in L_0 to L_4, greatly reducing read amplification and speeding up point queries.

4 Performance Evaluation

HTStore is implemented based on RocksDB v6.24.2. The development and stress testing of HTStore were conducted on the Linux, which is equipped with a 256 GB Optane non-volatile memory stick and a 3.6 TB mechanical hard disk.

Update-Intensive Experiment: The Fig. 8 shows update-intensive experiments with different value sizes, including 500 B, 1 KB, and 5 KB. The horizontal axis represents the number of update operations executed, and the vertical axis represents throughput. Firstly, in experiments with a Value size of 500 B, as shown in Fig. 8a, HTStore's write throughput improved by 45%–63% compared to RocksDB, 38% compared to PebblesDB, and 27% compared to MatrixKV. Additionally, experiments were conducted on write amplification of HTStore and RocksDB, as shown in Fig. 8d, and the write amplification of HTStore was found to reduce by an average of 35% compared to RocksDB. Secondly, in experiments with a Value size of 1 KB, as shown in Fig. 8b, HTStore's write throughput improved by 73%–105% compared to RocksDB, 45% compared to PebblesDB, and 34% compared to MatrixKV. Finally, in experiments with a Value size of 5 KB, as shown in Fig. 8c, HTStore's write throughput improved by 66%–71% compared to RocksDB, 37%–59% compared to PebblesDB, and 28%–37% compared to MatrixKV.

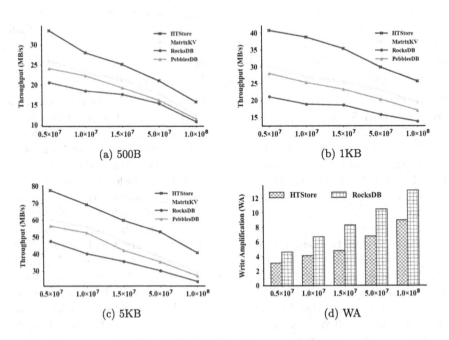

(a) 500B

(b) 1KB

(c) 5KB

(d) WA

Fig. 8. Update-intensive experiment

Read-Intensive Experiment: Figure 9a shows the point query experiment, where the horizontal axis represents the number of times the Get operation is executed, and the vertical axis represents throughput. The point query experiment is conducted on top of random update operations with value sizes of 1 KB, ranging from 0.5×10^7 to 5.0×10^7. HTStore's throughput is 80% higher than that of RocksDB and 125% higher than that of PebblesDB. Moreover, as the dataset size increases, HTStore's performance advantage becomes increasingly apparent.

The range query experiment performs Scan operations on top of random update operations with value sizes of 1 KB, totaling 10.0×10^7 executions. The experiment measures latency as a quantified performance metric, with the number of Scan operations ranging from 100 to 2000. As shown in Fig. 9b, HTStore's read latency is up to 16% lower than that of RocksDB and up to 24% lower than that of PebblesDB in the range query experiment. HTStore reduces redundant data, reduces the number of blocks, and shortens the scan search path, resulting in improved performance for scan.

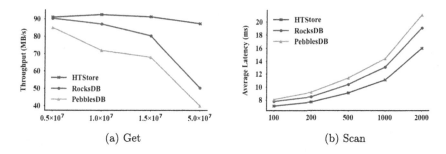

(a) Get   (b) Scan

Fig. 9. Read-intensive experiment

YCSB Experiment: Figure 10 shows that HTStore performs better than RocksDB and PebblesDB in most YCSB workloads. Specifically, in workload A, HTStore achieves a 102% performance improvement over RocksDB and a 26% improvement over PebblesDB. In B, it outperforms RocksDB by 50% and PebblesDB by 25%. In C, it outperforms RocksDB by 37% and PebblesDB by 5%. In D, HTStore achieves an 11% improvement over RocksDB and a 32% improvement over PebblesDB. In E, it only outperforms PebblesDB by 37%. Finally, in F, it performs 48% better than RocksDB and 18% better than PebblesDB.

5 Related Work

In order to reduce the frequency of I/O operations during compaction, dCompaction [7] utilizes a delayed merge method where actual compaction only occurs

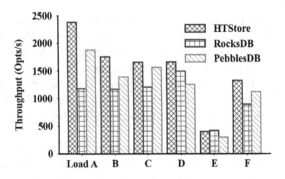

Fig. 10. YCSB experiment

when the number of SSTables exceeds a certain threshold. This means that virtual compaction is performed instead in cases where the threshold is not met. PebblesDB [8] takes inspiration from the idea of a SkipList and introduces a Guard to divide each level into several areas. The SSTables of different guards on the same level do not overlap in terms of their key ranges. SSTables within the same Guard can overlap in terms of their key ranges, but the SSTables in the next level do not need to participate in compaction. While this approach helps reduce write amplification, it has a negative impact on key globality and scan performance.

Key-value separation is a technique employed by several database systems to improve their performance. Wisckey [5] separates the key and value components of a key-value pair, storing the value in vLog alone. During compaction, only the Key is required to participate, not the value, which reduces write amplification overhead. DiffKV [4] proposes a novel key-value separation strategy that manages different value sizes differently. Large-sized values are stored in vLog, medium-sized Values in vTree, and small-sized values in the LSM-tree. However, this approach can compromise both write performance and range query performance.

NoveLSM [3] attempts to reduce the number of compactions in the short term by building a larger Memtable in NVM. However, as the amount of written data increases, it exacerbates problems of later write amplification and write stalls. MatrixKV [9] proposes optimizing the L_0–L_1 compaction algorithm by storing L_0 in NVM, reconstructing the storage structure, and disorganizing the SSTable of the L_0 of LSM-tree. This reduces the amount of data involved in L_0–L_1 compaction and improves write throughput while reducing write amplification. ChameleonDB [10] uses NVM as a block device to store key-value pairs on a hierarchical LSM-tree-like structure through log appending. It also adopts a new compaction strategy that merges multiple levels, reducing the overhead caused by compaction and write amplification. However, this approach weakens scan performance.

6   Conclusions and Future Work

In recent years, key-value store systems proposed by the academic community often sacrifice other aspects of performance or consume more system resources in order to improve a certain aspect of performance. This paper proposes a key-value store system called HTStore based on a Mixed Index. HTStore improves system performance by eliminating redundancy and avoiding layered searching through access to the Mixed Index. In the future, we will further optimize the scan performance of HTStore by leveraging the data orderliness of the Trie-tree in the Mixed Index.

References

1. Bez, R., Pirovano, A.: Non-volatile memory technologies: emerging concepts and new materials. Mater. Sci. Semicond. Process. **7**(4–6), 349–355 (2004)
2. Cao, Z., Dong, S.: Characterizing, modeling, and benchmarking RocksDB key-value workloads at Facebook. In: 18th USENIX Conference on File and Storage Technologies (FAST 2020), pp. 209–223 (2020)
3. Kannan, S., et al.: Redesigning LSMs for nonvolatile memory with NoveLSM. In: 2018 USENIX Annual Technical Conference (USENIX ATC 2018), pp. 993–1005 (2018)
4. Li, Y., et al.: Differentiated key-value storage management for balanced I/O performance. In: USENIX Annual Technical Conference, pp. 673–687 (2021)
5. Lu, L., et al.: WiscKey: separating keys from values in SSD-conscious storage. ACM Trans. Storage (TOS) **13**(1), 1–28 (2017)
6. O'Neil, P., et al.: The log-structured merge-tree (LSM-tree). Acta Inform. **33**, 351–385 (1996)
7. Pan, F., Yue, Y., Xiong, J.: dCompaction: delayed compaction for the LSM-tree. Int. J. Parallel Prog. **45**, 1310–1325 (2017)
8. Raju, P., et al.: PebblesDB: building key-value stores using fragmented log-structured merge trees. In: Proceedings of the 26th Symposium on Operating Systems Principles, pp. 497–514 (2017)
9. Yao, T., et al.: MatrixKV: reducing write stalls and write amplification in LSM-tree based KV stores with a matrix container in NVM. In: Proceedings of the 2020 USENIX Conference on USENIX Annual Technical Conference, pp. 17–31 (2020)
10. Zhang, W., et al.: ChameleonDB: a key-value store for Optane persistent memory. In: Proceedings of the Sixteenth European Conference on Computer Systems, pp. 194–209 (2021)

Author Index

Printed in the United States
by Baker & Taylor Publisher Services